Field Guid
Micro
moths
of Great Britain and Ireland

General Editor: Phil Sterling

Main Contributor: Mark Parsons

Illustrated by Richard Lewington

BLOOMSBURY

Contents

British Wildlife Publishing
An imprint of Bloomsbury Publishing Plc

50 Bedford Square
London
WC1B 3DP
UK

www.bloomsbury.com

British Wildlife Publishing, BLOOMSBURY and
the Diana logo are trademarks of Bloomsbury
Publishing Plc
First published 2012
Reprinted 2012, 2015

Phil Sterling and Mark Parsons have asserted their
right under the Copyright, Designs and Patents Act,
1988, to be identified as Authors of this work.

British Library Cataloguing-in-Publication Data
A catalogue record for this book is available from
the British Library.

ISBN (hardback) 978-0-9564902-2-3
ISBN (paperback) 978-0-9564902-1-6

4 6 8 10 9 7 5 3

Printed in China by C&C Offset Printing Co., Ltd.

MIX
Paper from
responsible sources
FSC® C008047

Dedications

To my late father, Col. Dougie Sterling, and to my brother Mark, for the fun we have had together studying micro-moths since we started in 1977, and to my mother, Audrey, for putting up with three entomologists in the family; and to my family, Carol, Hannah and Alistair, for their support and encouragement throughout the project. *PS*

To Helen and the children, Charlotte and Daisy, with thanks for their help and support over the years. *MP*

General Editor: Phil Sterling, lead ecological advisor for Dorset County Council, and a lifelong lepidopterist
Main Contributor: Mark Parsons heads up Butterfly Conservation's work on moths
Illustrated by Richard Lewington, one of Europe's leading natural history illustrators

Acknowledgements

This field guide has been written and illustrated in response to growing frustration among many people who see micro-moths and want to be able to identify them without having to purchase a library to do so. We are therefore indebted to Andrew and Anne Branson of British Wildlife Publishing for initiating the project, their guidance throughout, and for co-ordinating design, layout and proof-reading. The writing was a joint effort of the authors and a number of contributors, and special thanks are owed to Rich Andrews, Mike Bailey and Dr Barry Henwood for their contributions to the species accounts, and particularly to Mike Bailey who has been the workhorse behind the production of all the thumbnail maps and genitalia photomontages. We are indebted to Martin Corley for his original work on the key to the families, and to Ben Smart, who has collated the hundreds of photographs from many people in Britain and across Europe. Steve Palmer and Dr Mark Young were originally appointed as regional reviewers to provide perspectives on the micro-moth fauna of, respectively, northern England and Scotland, but both did much more than that; by providing independent, and sometimes frank commentary, the final work is much the better for their contribution.

We would also like to thank all those entomologists and photographers who have assisted with enthusiasm and generosity by lending specimens, giving access to vast collections of photographs, and offering comments and their unpublished knowledge for use in this project. In particular, we thank Martin Albertini and Peter Chandler (BENHS, Dinton Pastures); Dr Peter Buchner; Matthew Deans; Rob Edmunds; Dave Foot; Dave Green; Bob Heckford; Dr Barry Henwood; Jeff Higgott; Ian Kimber; Dr John Langmaid; Chris Manley; Darren Mann, James Hogan and Zoë Simmons (Oxford University Museum of Natural History); Guy Meredith; Trevor and Dilys Pendleton; Tony Prichard; Tony Rayner; Stuart Read; Neil Sherman; and George Tordoff.

The Checklist of British and Irish micro-moths has been a particular challenge to compile and we owe our thanks to a number of entomologists who have assisted us, including Rev. Dr David Agassiz; Stella Beavan; Martin Corley; Bob Heckford; Ole Karsholt (Zoological Museum, Natural History Museum of Denmark, Copenhagen); Martin Honey (Natural History Museum, London); Dr Matthias Nuss (Senckenberg Naturhistorische Sammlungen, Dresden); and Dr Erik van Nieukerken (Netherlands Centre for Biodiversity Naturalis, Leiden).

Richard Lewington would also like to express his gratitude to the Leverhulme Trust, without whose support this book would be much less extensive in its detail, and to Sir David Attenborough, Prof. Jeremy Thomas and Dr Martin Warren (Butterfly Conservation) who have given their full support and encouragement.

Phil Sterling, Mark Parsons and Richard Lewington

Introduction

This field guide caters for similar audiences to its companion volumes, *Field Guide to the Moths of Great Britain and Ireland* (Waring *et al*. 2009 2nd edition) and *Concise Guide to the Moths of Great Britain and Ireland* (Townsend *et al*. 2007), and is written for both amateur naturalists and professional ecologists. It is partly aimed at those who have been 'bitten by the bug', who have perhaps mastered Britain's macro-moths and have now turned their passion towards the micro-moths, but who may have hit a somewhat impenetrable wall of rather expensive requirements. Until now, to gain a reasonably comprehensive coverage of micro-moths one needed to acquire a long list of reference books, monographs and papers published in journals, not all written in English, as well as a hand lens and microscope.

This guide is for those who simply do not know where to start to identify the apparently bewildering array of smaller moths, as well as those who have already started out on the journey of discovery. It covers 1,033 of the 1,627 species recorded in Great Britain and Ireland, bringing together all the families into one volume, and covering most of the more regularly encountered and easily identifiable species. It provides not only a starting point for beginners, but an *aide-mémoire* for the more experienced. Perhaps this guide might be kept in the glove compartment of the car, or in a field bag, for those all too frequent moments when the name of what one is looking at is temporarily lost from recall. We have also included references for more detailed information, and directions to websites for comparative material.

Micro-moths exhibit a fascinating diversity of form and habit, much more so than their 'big brothers', since their generally smaller size allows them to exploit a much wider range of habitat niches. They are found everywhere, from shoreline to mountaintop, underwater and on dry land. Several species are associated with humans and manmade habitats. Whilst most are associated with higher plants, some are found on algae and fungi, mosses and ferns, and on decaying plant or animal materials.

As adults, many are very beautiful indeed and, within a species, may be so variable that the unsuspecting would not believe them to belong as one. They also exhibit a wide variety of resting postures, which are often helpfully diagnostic. Of course, there are a good number of species that fit a sceptic's view of them as 'small brown jobs', but a little care taken to examine their detail usually enables a more informed conclusion to be drawn.

We have illustrated the micro-moths at rest, rather than with wings spread as in a 'set' specimen, following the successful change in approach adopted by the macro-moth guides. Many recorders today do not want to kill and set moths, but are keen to identify what they have found. This is possible for many micro-moth species, providing they are in good condition. However, for a substantial minority of species there is little option other than to retain a voucher specimen as the best way to secure identification, and to allow verification, if required, by county recorders and experts at a later date.

In this field guide, illustrations are backed up by a concise description of key features of each species, a comment on abundance, flight season, habitat and foodplant, together with a section on similar species where appropriate, and all but a few are supported with a distribution map.

We have also included photographs of the early stages of a sample of species from almost all families to give an insight into the variety of larvae, and the artefacts they make as they feed, such as leaf mines, leaf folds and spinnings. A number of these are so characteristic that the species of micro-moth can be identified with certainty from the artefact. We also hope this guide will encourage you to search for the early stages.

What defines a micro-moth?

What makes a species a micro-moth rather than a macro-moth? This is custom-and-practice, rather than a matter rooted in taxonomy. Families of moths which tend to feature small species go under micro-moths, and those which contain mostly larger species are called macro-moths. Current taxonomic thinking considers most micro-moth families to be more primitive than macro-moths, a few exceptions being Hepialidae (swifts), Cossidae (the leopards and Goat Moth), Zygaenidae (burnets and foresters), Limacodidae (Festoon and Triangle) and Sesiidae (clearwings) which are 'honorary' macro-moths. These families are thus covered in the macro-moth books, the *Field Guide* (Waring *et al.* 2009) and *Concise Guide* (Townsend *et al.* 2007), and other well-known books such as Skinner (2009).

Using this field guide

How to use this guide to identify a micro-moth

Before the main species text is an 'at-a-glance' guide to families of micro-moths, showing one or a few species from each family. It should be possible to compare a living micro-moth with these illustrations and quickly get to one or a few candidate families.

It is then difficult to resist the temptation simply to dive into the illustration plates in the middle of the book to find the best match available, and for some distinctive species this may suffice. However, a more systematic key is also provided, and recommended, as there may be considerable variation within a species, and species within different groups may look remarkably similar. To begin identifying micro-moths, choose only examples in good condition.

Those moths which hold their wings in a roof-like posture over the abdomen are best compared from a side view, while those holding their wings flat to the ground should be examined from above. In most families, resting posture is reasonably consistent between species and the illustrations are straightforward to compare. However, where there is variation, for example in Oecophoridae and Elachistidae, we have chosen to illustrate each species in the posture we consider most helpful for identification, even if this has meant that neighbouring species are shown in a different orientation.

Carefully check the posture and wing pattern of the species you have with the illustration; micro-moths often also show characteristic wing folds and scale-tufts, and are rather more 'three-dimensional' in appearance than macro-moths. Turn to the appropriate species text to find more detail, whether there are similar species, and how to tell them apart. Use all the information provided, including flight period, foodplant and distribution, before settling on a diagnosis, since this additional information may help you rule in or out species on your tentative list.

Identification may require much more than simple comparison with illustrations and a quick read of several lines of text, and this field guide cannot hope to encompass all the variation there is within and between species. For the 'difficult-to-do' examples, and for the more enquiring, there are two additional sections to this field guide.

The first, following this introduction, is a new style of key separating families (p.30), based mostly on characters other than wing markings, such as resting posture, wing shape, length and shape of antenna, arrangement of scales on the head, and mouthparts. These features can be seen relatively easily with a x10 hand lens or low-power binocular microscope. Using this key, it should be possible for almost any micro-moth found in Britain and Ireland to be identified at least to a family.

The second, after the main species texts, introduces the subject of genitalia examination (p.387), together with pictures of a few regularly encountered pairs or groups of species almost identical superficially, but which require internal examination to separate .

How many species are included?

A total of 1,033 species are covered, of which 927 species are illustrated, from the 1,627 micro-moths on the British list. Most species in most families are described, but coverage in five major families (Nepticulidae, Gracillariidae, Elachistidae, Coleophoridae and Gelechiidae) is only partial. In Nepticulidae, *Parornix* and *Phyllonorycter* (Gracillariidae), and *Elachista* (Elachistidae), the adults are very small and easily confused, so most have been omitted. Many Coleophoridae are superficially similar and most adults require very careful study, often involving genitalia examination, unless they have been reared from larvae. Adults of many *Depressaria* (Elachistidae) and Gelechiidae are brown, with variable darker and paler patterns, and differences between similar species are very subtle, so only the more widespread and/or recognisable species have been included.

Species which are very rare, confined to a single site, or which have not been seen in recent decades have usually been excluded, as have the rarer immigrant species and most adventives or accidental imports. However, in a few cases, rarities and adventives are described for the sake of completeness, or because they are very similar to a commoner species, and we consider it helpful to distinguish them e.g. *Falseuncaria ruficiliana* and *F. degreyana* (Tortricidae).

It would be unfortunate and an unintended consequence if this guide led to misidentification through readers trying to shoehorn every micro-moth found into species illustrated here. Where we consider there is a significant risk of this happening, we have chosen either to omit detailed coverage of those groups, such as *Parornix* (Gracillariidae), or have chosen to include sufficient information to allow identification to a group of similar species, such as within *Coleophora* (Coleophoridae) and *Cnephasia* (Tortricidae). In many cases, we have recommended that genitalia dissection is required to determine the species; records are unlikely to be accepted without this level of confirmation.

In the end, the decision to include or exclude a species or group has been a compromise between several factors.

Format of the species accounts

Taxonomic order

The taxonomic list of British and Irish micro-moths we have used follows the hierarchy adopted by *Fauna Europaea* (FE). Within a genus, however, the order of the species largely follows Bradley (2000), with subsequent changes included where appropriate. Note that although each species is given a number, the Bradley list is not in numerical sequence, reflecting taxonomic changes which were adopted at the time of publication. The list in FE has the hierarchy of species within families sometimes reversed from that which we are used to in Britain, and in other cases species are simply listed in alphabetical order where no taxonomic distinction has been made. We have decided against adopting the FE species order, in consultation with a number of expert micro-lepidopterists in Britain. Thus, the order of species within a genus will be largely familiar to experienced micro-moth recorders.

A full taxonomic Checklist is provided at the end of the guide. Species listed in bold in the Checklist are covered in detail within the guide. Species may additionally be annotated as presumed extinct, immigrant, adventive/accidental imports, Channel Islands or Ireland only, or doubtfully British.

Introduction to the family

An introduction to every family is provided, giving an account of the characteristics of family members, a summary of life histories and other notable habits of the species which are found in the British Isles. The family description usually starts with the typical resting posture. This may be a key feature of a living micro-moth, and four basic wing positions can be distinguished: **roof-like**, where the wings are arranged like two sides of a pitched roof or tent (also known as tectiform); **wrapped or rolled around the abdomen**, sometimes giving the moth a cylindrical appearance, and often more so towards the back end of the moth (also known as involute); laid **flat** over the body, the forewings usually, but not always, partly overlapping, with the hindwings completely covered, although the wings are rarely completely flat but tend to be slightly angled downwards; or less often are **extended**, revealing the hindwings. In addition to wing position, the moth may adopt an **inclining posture**, or head-up position, with the front end raised at an angle from the substrate, or a **declining posture**, or head-down position, with the rear end raised at an angle. Occasionally, the wings are held flat but above the substrate. Unusually, one pair of legs may be held out perpendicular to the body (Stathmopodidae), or the wings are extended and rolled on themselves (*Agdistis*, Pterophoridae).

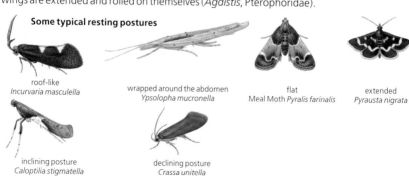

Some typical resting postures

roof-like
Incurvaria masculella

wrapped around the abdomen
Ypsolopha mucronella

flat
Meal Moth *Pyralis farinalis*

extended
Pyrausta nigrata

inclining posture
Caloptilia stigmatella

declining posture
Crassa unitella

The family description continues with more detailed features of the structure of the moth. The head can appear rough or smooth-scaled, or somewhere between, according to the direction of alignment of the scales on the top of the head (crown), and on the face. Four categories are distinguishable here: **spiky**, with long, erect, bristle-like scales, giving the effect of a bottle-brush; **erect**, with a dense mat of erect scales, resembling a carpet; **raised**, with scales that rise from the head but are directed forwards and downwards; or **smooth**, with appressed or flattened scales. Frequently, the face is smooth, but the top of the head is not. Species with a smooth face and crown may have erect scales on other parts of the head, particularly around the compound eyes and on the neck.

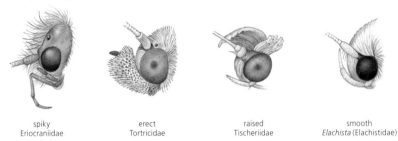

spiky	erect	raised	smooth
Eriocraniidae	Tortricidae	Tischeriidae	*Elachista* (Elachistidae)

The **compound eyes** do not provide useful characters for separation of families. However, there is a pair of **ocelli** in some groups, situated above the compound eyes and just behind the antenna base. In a few cases these are quite evident, but often they are very small or more or less obscured by scales (e.g. Tortricidae). Although sometimes difficult to see, the presence of ocelli can be an important and defining character.

The mouthparts consist of two pairs of palps. The outer pair of **labial palps** is usually larger and composed of three segments, with the first segment very short. The inner pair are the **maxillary palps**, usually smaller and sometimes much reduced, but when fully developed they are 5- or 6-segmented, and sometimes folded. In most families, there is a **tongue** which is coiled when fully developed. A few of the most primitive families have mandibles, which are functional but very small in Micropterigidae.

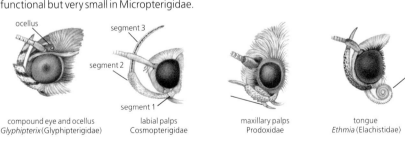

compound eye and ocellus	labial palps	maxillary palps	tongue
Glyphipterix (Glyphipterigidae)	Cosmopterigidae	Prodoxidae	*Ethmia* (Elachistidae)

The **antennae** have an enlarged basal segment, the **scape**, and a thread-like, many-segmented **flagellum**. The scape may be clothed with a broad cloak of scales forming an **eye-cap**, or it may have one or a few downward-projecting bristle-like scales, forming a **pecten**. This pecten has nothing to do with **pectinate antennae**, in which the segments of the flagellum have lateral branches that together form a comb, or **bipectinate antennae** with

Types of antennae

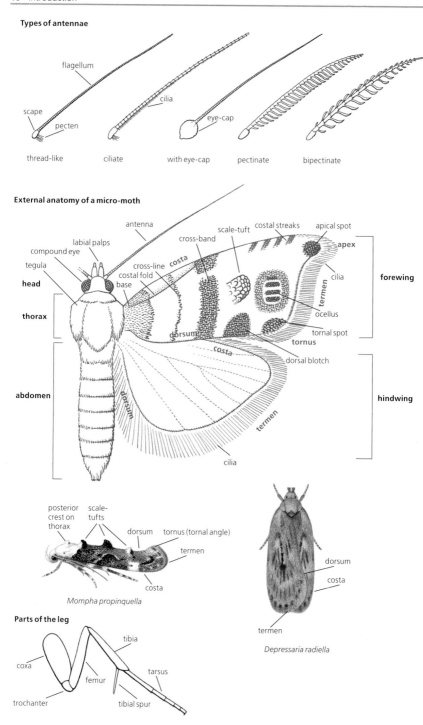

thread-like   ciliate   with eye-cap   pectinate   bipectinate

External anatomy of a micro-moth

Mompha propinquella

Depressaria radiella

Parts of the leg

a double comb. **Ciliate antennae** have hairs on each segment.

There are three pairs of **legs**: fore-, mid- and hindlegs, and each has a number of segments. There is a **coxa** joining the leg to the thorax, a short **trochanter**, a long **femur**, a long **tibia** and a **tarsus** with five segments, the last with a pair of claws. Frequently, the legs have bristles, and most species have one pair of spurs at the end of the mid-tibia and two pairs of spurs on the hind-tibia, one pair in the middle and one at the distal end.

The **forewing** is essentially triangular, the three sides being the **costa** (the leading edge), the **dorsum** (the side lying on the back of the abdomen, sometimes called the trailing edge) and the **termen** (the outer edge). The **apex** is the angle between the costa and the termen, the **tornus** is the angle between the termen and the dorsum. The area of the wing closest to the thorax is known as the **base**. In narrow-winged species the termen and dorsum may merge into one, so there is no discernible tornus. The outer margin of the termen has a fringe of **cilia** (hair-like scales), which may extend along the dorsum. The **hindwing**, which may be broad or narrow, rarely has a defined tornus; the cilia extend along the dorsum, and in narrow-winged species the dorsal cilia are usually longer than the width of the wing.

There is a variety of markings on the forewing. The most common descriptive words or phrases we have used are:

- **streak** – a short dash, line or bar
- **cross-band/cross-line** – often complete or sometimes an interrupted band or line between the costa and the dorsum
- **tornal spot** – a marking at the tornus
- **ocellus** – especially in Tortricidae, an oval, metallic or shining mark, often containing dark spots or short streaks
- **scale-tufts** – patches of raised scales
- **costal fold** – especially in Tortricidae, a narrow flap folded over the wing along the costa from the base
- **one-third/one-half** – approximate distance from the base to the feature being described

The forewings are joined to the **thorax**, with connections protected by the **tegulae**. In some species, the thorax has a significant **posterior crest**. The thorax is joined to the **abdomen**, which sometimes supports diagnostic features on its surface, and the **genitalia** are within the posterior segments of the abdomen.

At the end of the each family introduction is a short list of references to direct the reader to further information on British and Irish species and European species in that family, and to any specific websites.

Scientific names

Micro-moths in this guide are listed by scientific name. Names are in binomial form, that is, they start with genus (= generic name), which groups the moth with its close relatives, followed by the species (= specific name). By convention, both are in italics, the genus starting with a capital letter. Although most specific names can be used alone to classify a micro-moth, it is the binomial that describes it uniquely. For example, there are two micro-moths with the specific name *betulae*: one is within the genus *Parornix* (Gracillariidae), the other within *Ortholepis*

(Pyralidae); and similarly there are *Acleris rufana* and *Celypha rufana* (both Tortricidae).

The binomial should usually be followed by the author who described the species, with the date of publication. For convenience, the author and date have been omitted in the main body of text, but are given in full in the Checklist at the back of the guide.

In the species accounts the scientific name is followed by two numbers. The first is the Bradley number; many existing recorders have memorised Bradley numbers, identify species by them, and use them for entering records into databases such as MapMate and Recorder. The second number is the European checklist number after Karsholt & Razowski (1996). Vernacular or English names given in Bradley (2000) and Hart (2011) have also been included.

Scientific names do change from time to time as taxonomic revisions provide more insight into relationships between species, and literature reviews reveal earlier combinations which take precedence. While this can be frustrating for field naturalists, changes should be seen as progress towards an understanding of the true relationships between organisms. Changes since Bradley (2000) have been incorporated where appropriate, and in the Checklist, where there is a new name, the old Bradley name is given underneath in square brackets.

Distribution maps

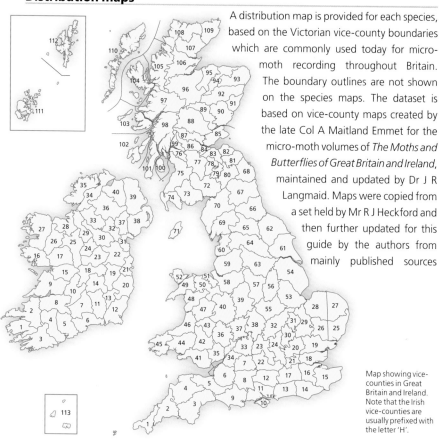

A distribution map is provided for each species, based on the Victorian vice-county boundaries which are commonly used today for micro-moth recording throughout Britain. The boundary outlines are not shown on the species maps. The dataset is based on vice-county maps created by the late Col A Maitland Emmet for the micro-moth volumes of *The Moths and Butterflies of Great Britain and Ireland*, maintained and updated by Dr J R Langmaid. Maps were copied from a set held by Mr R J Heckford and then further updated for this guide by the authors from mainly published sources

Map showing vice-counties in Great Britain and Ireland. Note that the Irish vice-counties are usually prefixed with the letter 'H'.

since 2006, but we have also consulted a variety of websites which hold maps, including Butterfly Conservation East Scotland, and Moths of Ireland. We have aimed to make the maps up to date to at least the end of 2010, and have included some data from 2011.

The maps are intended to provide a quick reference to distribution. The data presented have not been verified with vice-county recorders, so presence should not be taken as a definitive statement that a species has been confirmed from a particular vice-county.

Abundance

General comments are provided on the relative abundance of each species in four categories: common, local, scarce and rare, and the comment should be read in conjunction with the distribution map. For example, the map may indicate a species is widely distributed, yet it is rarely seen; another species might have a narrow geographic range on the map, but is found commonly in that area.

A species that is described as 'common' is very likely to be encountered in suitable habitat within its known geographic range. 'Local' refers to a species which is found rather less frequently in its known habitats; it is not a comment on its overall geographic distribution. 'Scarce' refers to a species which is very infrequently seen, even in its known haunts, and 'rare' is reserved for those species which are known or believed to be confined to very few sites. Sometimes further qualifying comments have been added. However, distribution and abundance of most micro-moths are imperfectly known, so comments are based on the personal experience of recorders and are, inevitably, somewhat subjective. None of the comments on scarcity is given according to the definition of rare and scarce in *British Red Data Books 2. Insects* (Shirt 1987).

Forewing length (FL)

Forewing length, that is the distance between wing apex and wing base, is the measurement used in this guide.

General description

A brief description of each species is given, using a few anatomical terms (shown on p.10) widely used in the study of Lepidoptera (Scoble 1995). The use of this terminology is preferred here since there is no English equivalent for structures of the head and mouthparts, thorax, some parts of the wings, legs and genitalia, all of which may be used in identifying a species, and are more important in identifying micro-moths than macro-moths.

Where there is a pair of very similar species, the description of the first is extended and emboldened where it provides the detail on how to separate them. In cases where there are three or more look-alikes, these are described in the similar species section. Occasionally, the text describing similar species is located in the description of the species most likely to be encountered. In a few instances, the similar species section describes a little more about those rare species which are not illustrated and do not have a separate account.

Descriptions of other species in similar pairs or groups are often brief and refer the reader back to the species where the diagnostic features can be found. We hope this will encourage the reader to concentrate on working out the differences between closely-related species.

Flight season (FS)

The number of life cycles, or generations, each species undergoes in a year is stated, together with the approximate time of year when the adults may be expected to appear. Note that this brief description cannot comprehensively cover appearance across a geographic range which spans the Channel Islands and Shetland. Species which are double-brooded (two generations per year) in southern England may well be single-brooded (one per year) in Scotland. In the past two decades, there has also been an increasing tendency for species to appear earlier in the year, or to have a second or third generation, which were not previously recorded. It is also difficult to know whether some species are genuinely having more broods, or whether their time of appearance has become more spread throughout the year. There is also a comment on adult behaviour, whether diurnal or nocturnal, if attracted to light, and any other notable habit.

Habitat (Hab)

The main habitats that the species is known from are listed, along with any preference for soil type (acidic or calcareous), inland or coastal, lowland or montane, or other habitat preferences that may help determine the likelihood of finding a species in a location. If more than one habitat is listed, the sequence does not indicate preference. Despite their small size, many micro-moth species appear to be just as capable of long-distance flight as macro-moths. It is therefore possible for a species to turn up in an apparently unsuitable location. However, the observer should be particularly cautious about recording a species well away from its preferred haunts.

Foodplant (Fp)

The foodplant of the larva is an important feature which aids identification, perhaps more so for micro-moths than macro-moths, and the part of the plant eaten is listed if it is not the leaf, along with brief details of larval habits. Micro-moths are more often confined to one or a few foodplants than are macro-moths, so a working knowledge of the flora of a location is a distinct advantage. Our knowledge of the foodplants of micro-moths in Britain is surprisingly good; the reason for this is that it is actually far easier to become an expert in micro-moths by rearing adults from larvae. However, in a few cases the foodplant in Great Britain and Ireland is not known, and this information has been drawn from Continental literature. The nomenclature of the larval and nectar plants in the guide follows Stace (2010).

Many species rarely come to light, and if they do, they may not be in good condition having been battered among a trap full of macro-moths. Searching for larvae and then rearing them is a good way of seeing the adult in fine condition, and provides a means of improving one's botanical skills. Not all micro-moth larvae eat plants, of course, so the term foodplant covers other foodstuffs, such as fungi and animal tissues.

Illustrations

For most species, we have relied on one illustration being sufficient to guide identification. Richard Lewington has painted each artwork by viewing a combination of live examples (where possible), alongside photographic images and specimens in collections. He has thus tried to reflect natural variation in his artwork. Additional illustrations are provided where there is marked variation, although for some particularly variable species (e.g. *Acleris cristana*,

Tortricidae, where over 130 forms are known in Britain) it has not been practicable to represent the range adequately. In some cases, variation within a species is shown by one or more fore-wings, rather than the whole moth, which are shown as if the forewing were part of a set specimen. The flattened wing can appear a slightly different shape to that of the forewing wrapped around the body (e.g. *Ypsolopha ustella*, Ypsolophidae). Forewing-only illustrations have also been used within the Coleophoridae and in *Cnephasia* (Tortricidae). However, identification of these species based on forewing characters alone is very difficult or impossible, and without examination of genitalia it will be possible only to get to a group of similar species.

Most micro-moths, by virtue of their small size, need to be illustrated at a scale larger than life size. The magnification is given at the top of each section or page. Some recent specialist works have shown the micro-moths so that they all appear precisely the same size. This approach has not been adopted here, since relative size is an important determinant used in the guide. In groups where there is much size variation between species, and this variation is helpful to show, we have necessarily had to strike a balance between not illustrating some species too small, whilst not allowing large species to take too much space.

Illustrations are laid out mainly in two forms: those moths holding their wings in a roof-like posture, or rolled around the abdomen, are best compared from a side view. Those holding their wings flat to the ground, or spread out, are shown from above.

Identifying worn examples

Examples of micro-moths which have lost many wing scales, or characteristic tufts of scales, will usually be much more difficult to determine to species level. However, resting posture is not normally affected by loss of scales and with practice it should at least be possible to put a moth into its higher grouping. There may be particular features that remain in good condition, such as colour and markings on the palps, or at the base of the forewing, which help rule in or out particular species. The worst examples should either be subject to genitalia examination, or disregarded.

▼ Examining the genitalia of a micro-moth through a low-power binocular microscope.

Genitalia examination

A separate section devoted to this subject follows the main text and illustrations (p. 387). There is a growing band of enthusiastic recorders who spend much of the winter months with their heads over binocular microscopes, examining the internal mating structures of moths, especially micro-moths. It is a most rewarding pastime, and more recorders should take it up. The late Col Dougie Sterling, Phil Sterling's father, was an avid collector of small, scale-less moths found in the bottom of a moth trap after a warm night in the summer, which he would store up for later examination and dissection. He used to identify many unusual species which were wandering a long way from their normal habitats. Species groups which have always proved difficult to sort on wing pattern are often surprisingly easy to identify on genitalia. The structures, at least in males, are reasonably robust, and, as a beginner, it is best to start by examining a male.

There are a number of groups of micro-moth which generally require genitalia examination to determine field-caught examples of adults. For example, most of the Nepticulidae are quite easy to determine; their genitalia are large compared with the minute size of the adult. Most of the genera *Parornix* (Gracillariidae), *Coleophora* (Coleophoridae), *Scrobipalpa* (Gelechiidae) and *Cnephasia* (Tortricidae) should usually be dissected to determine which species has been found.

Further reading, systematic references and websites

In this field guide, the most relevant identification sources are given in the introduction to each family. For convenience, a list of systematic references covering micro-moths across Europe is given below, together with useful works and general websites.

Selected identification references for British and Irish micro-moths

Beirne, B P 1952 *British pyralid and plume moths.* Warne & Co, London

Bradley, J D, Tremewan, W G, & Smith, A 1973 *British Tortricoid Moths. Cochylidae and Tortricidae: Tortricinae.* The Ray Society, London. Also available as a CD-ROM.

Bradley, J D, Tremewan, W G, & Smith, A 1979 *British Tortricoid Moths. Tortricidae: Olethreutinae.* The Ray Society, London. Also available as a CD-ROM.

Clifton, J, & Wheeler, J 2011 *Bird-dropping Tortrix moths of the British Isles. A Field Guide to the Bird-dropping Mimics.* Clifton & Wheeler, Norfolk

Emmet, A M (ed) 1996 *The Moths and Butterflies of Great Britain and Ireland. Volume 3: Yponomeutidae – Elachistidae.* Harley Books, Essex

Emmet, A M, & Langmaid, J R (eds) 2002a *The Moths and Butterflies of Great Britain and Ireland. Volume 4(1): Oecophoridae – Scythrididae (excluding Gelechiidae).* Harley Books, Essex

Emmet, A M, & Langmaid, J R (eds) 2002b *The Moths and Butterflies of Great Britain and Ireland. Volume 4(2): Gelechiidae.* Harley Books, Essex

Goater, B 1986 *British Pyralid Moths.* Harley Books

Hart, C 2011 *British Plume Moths. A guide to their identification and biology.* British Entomological & Natural History Society, London

Heath, J (ed) 1976 *The Moths and Butterflies of Great Britain and Ireland. Volume 1: Micropterigidae – Heliozelidae.* Blackwell Scientific Publications & Curwen Press, Oxford

Heath, J, & Emmet, A M (eds) 1985 *The Moths and Butterflies of Great Britain and Ireland. Volume 2: Cossidae to Heliodinidae.* Harley Books, Essex

Manley, C 2008 *British Moths and Butterflies: A photographic guide.* A&C Black, London. Includes many micro-moths and available as an iPhone app.

Meyrick, E [1928] *A Revised Handbook of British Lepidoptera.* Watkins & Doncaster, London. Very outdated nomenclature, and no pictures, but still a useful reference.

A selection of European references including Britain and Ireland

Bengtsson, B Å 1997 *Scythrididae*. In: Huemer, P, &
Karsholt, O (eds) *Microlepidoptera of Europe 2.*
Apollo Books, Stenstrup
Bengtsson, B Å, & Johansson, R 2011
Nationalnyckeln till Sveriges flora och fauna.
Fjärilar: Bronsmalar-rullvingemalar. Lepidoptera:
Roeslerstammiidae-Lyonetiidae. ArtDatabanken,
Uppsala
Bengtsson, B Å, Johansson, R, & Palmqvist, G 2008
Nationalnyckeln till Sveriges flora och fauna.
Fjärilar: Käkmalar-säckspinnare. Lepidoptera:
Micropterigidae-Psychidae. ArtDatabanken,
Uppsala
Elsner, G, Huemer, P, & Tokár, Z 1999 *Die
Palpenmotten (Lepidoptera, Gelechiidae)
Mitteleuropas.* F Slamka, Bratislava
Gielis, C 1996 *Pterophoridae.* In: Huemer, P,
Karsholt, O, & Lyneborg, L (eds) *Microlepidoptera
of Europe 1.* Apollo Books, Stenstrup
Goater, B, Nuss, M, & Speidel, W 2005 *Pyraloidea
I (Crambidae: Acentropinae, Evergestinae,
Heliothelinae, Schoenobiinae, Scopariinae).* In:
Huemer, P, & Karsholt, O (eds) *Microlepidoptera
of Europe 4.* Apollo Books, Stenstrup
Huemer, P, & Karsholt, O 1999 *Gelechiidae I.* In:
Huemer, P, Karsholt, O, & Lyneborg, L (eds)
Microlepidoptera of Europe 1. Apollo Books,
Stenstrup
Huemer, P, & Karsholt, O 2010 *Gelechiidae II.* In:
Huemer, P, Karsholt, O, & Lyneborg, L (eds)
Microlepidoptera of Europe 6. Apollo Books,
Stenstrup
Johansson, R, Nielsen, E S, van Nieukerken, E J,
& Gustafsson, B 1990 The Nepticulidae and
Opostegidae (Lepidoptera) of North West Europe.
Fauna Entomologica Scandinavica 23 (parts 1 &

2). Scandinavian Science Press, Leiden
Karsholt, O, & Razowski, J 1996 *The Lepidoptera of
Europe – A distributional checklist.* Apollo Books,
Stenstrup
Koster, S, & Sinev, S 2003 *Momphidae s.l.* In:
Huemer, P, & Karsholt, O (eds) *Microlepidoptera
of Europe 5.* Apollo Books, Stenstrup
Kuchlein, J H, & Bot, L E J 2010 *Identification keys to
the Microlepidoptera of the Netherlands.* TINEA
Foundation & KNNV Publishing, Zeist
Palm, E 1986 *Nordeuropas Pyralider (Lepidoptera:
Pyralidae).* Fauna Bøger, Copenhagen
Palm, E 1989 *Nordeuropas Prydvinger (Lepidoptera:
Oecophoridae).* Fauna Bøger, Copenhagen
Razowski, J 2002 *Tortricidae of Europe. Volume
1. Tortricinae and Chiladanotinae.* F Slamka,
Bratislava
Razowski, J 2003 *Tortricidae of Europe. Volume 2.
Olethreutinae.* F Slamka, Bratislava
Slamka, F 1997 Die Zünslerartigen (Pyraloidea)
Mitteleuropas. F Slamka, Bratislava
Slamka, F 2006 *Pyraloidea of Europe 1/Pyralinae,
Galleriinae, Epipaschiinae, Cathariinae &
Odontiinae.* F Slamka, Bratislava
Slamka, F 2008 *Pyraloidea of Europe 2/Crambina &
Schoenobiinae.* F Slamka, Bratislava
Slamka, F 2010 *Pyraloidea of Central Europe.*
F Slamka, Bratislava
Tokár, Z, Lvovsky, A, & Huemer, P 2005 *Die
Oecophoridae s.l.* (Lepidoptera) *Mitteleuropas.*
F Slamka, Bratislava
Traugott-Olsen, E, & Schmidt Nielsen, E 1977 The
Elachistidae (Lepidoptera) of Fennoscandia and
Denmark. *Fauna Entomologica Scandinavica* 6.
Scandinavian Science Press, Klampenborg

Useful guides to micro-moths and macro-moths

Emmet, A M 1988 (2nd ed) *A Field Guide to Smaller
British Lepidoptera.* British Entomological &
Natural History Society, London
Leverton, R 2001 *Enjoying Moths.* T & AD Poyser,
London
Majerus, M E N 2002 *Moths.* New Naturalist Series,
HarperCollins, London
Parenti, U 2000 *A Guide to the Microlepidoptera
of Europe.* Museo Regionale di Scienze Naturali,
Torino
Scoble, M J 1995 *The Lepidoptera: Form, Function
and Diversity.* Oxford University Press, Oxford

Skinner, B 2009 (3rd ed, revised) *The Colour
Identification Guide to Moths of the British Isles.*
Apollo Books, Strenstrup
Townsend, M, Waring, P, & Lewington, R 2007
*Concise Guide to the Moths of Great Britain and
Ireland.* British Wildlife Publishing, Dorset
Waring, P, Townsend, M, & Lewington, R 2009 *Field
Guide to the Moths of Great Britain and Ireland*
(2nd ed). British Wildlife Publishing, Dorset
Young, M R 1997 *The Natural History of Moths.*
T & AD Poyser, London

Other references

Bradley, J D 2000 *Checklist of Lepidoptera recorded from the British Isles* (2nd ed). D J & M J Bradley, Fordingbridge

Davis, A M 2012 *A Review of the Status of Microlepidoptera in Britain.* (Butterfly Conservation Report No. S12-02). Butterfly Conservation, Wareham

Fry, R, & Waring, P 2001 A guide to moth traps and their use. *Amateur Entomologist 24* (2nd ed). Amateur Entomologists' Society, London

Parsons, M S, & Clancy, S P 2002 *Dioryctria sylvestrella* (Ratz.) – New to Britain and Ireland, and the identification of the British *Dioryctria. Atropos* 15: 16-19

Shirt, D B (ed) 1987 *British Red Data Books 2. Inseccts.* Nature Conservancy Council, Peterborough

Sokoloff, P 1980 *Practical Hints for Collecting and Studying the Microlepidoptera.* Amateur Entomologists' Society, London

Stace, C 2010 *New Flora of the British Isles* (3rd ed). Cambridge University Press, Cambridge

Townsend, M C, Clifton, J, & Goodey, B 2010 *British and Irish moths: an illustrated guide to selected difficult species.* Butterfly Conservation, Wareham

Selected websites covering micro-moths

British Leafminers: www.leafmines.co.uk Photographs of leaf mines, some larvae and pupae; covers some non-lepidopterous mines; free newsletter available by e-mail.

Butterfly Conservation East Scotland: www. eastscotland-butterflies.org.uk Includes maps of the moths and butterflies of the whole of Scotland.

Fauna Europaea: www.faunaeur.org The scientific names of all European land and freshwater animals brought together in one authoritative database.

Finnish moths: www.ftp.funet.fi/index Photographs of many Lepidoptera; links to other sites, including UKMoths.

Finnish moths: www.kimmos.freeshell.org Photographs of mainly pinned adult micro-moths.

Lepidoptera Dissection Group: www. dissectiongroup.co.uk/page44.html Photographs of male and female genitalia of many species of macro- and micro-moth, including some species not on the British list.

Lepiforum (Germany): www.lepiforum.de Very large website, in German, but Google will translate into English automatically; coverage of Central European families.

microlepidoptera.nl (Holland): www. microlepidoptera.nl/index.php In Dutch, but Google will translate into English automatically; coverage of all families; useful feature is the Quick Start section on common species; has photos of set specimens, larvae and leaf mines.

Moths of Ireland: www.mothsireland.com Includes maps of the moths and butterflies of the whole of Ireland.

Swedish Museum of Natural History: www2. nrm.se/en/svenska_fjarilar Photographs of set specimens with links to UKMoths; some genitalia illustrated.

UK Lepidoptera: www.ukleps.co.uk Images complementing UKMoths; covers all families; section on feeding signs.

ukmicromoths (Yahoo group): www.pets. groups.yahoo.com/group/ukmicromoths Mailing list containing experienced and knowledgeable members willing to try to identify from photos; free to join.

UKMoths: ukmoths.org.uk Images, short descriptions and key features; all families covered; available as an iPhone app.

▲ Tapping vegetation onto a beating tray can dislodge both adults and larvae of micro-moths.

Field study of micro-moths

Searching for adults by day

Searching by day for micro-moths can be very productive and remarkably little equipment is required, although there is a tendency to want to acquire more paraphernalia over time. At its simplest, all that is needed is a small kit bag containing a few transparent plastic boxes for larvae, glass or plastic tubes for adult moths, a roll of small plastic bags for foodplants and leaf mines, a small collecting net, a x10 hand lens, and a stout stick for tapping or brushing through vegetation.

For daytime recording the addition of a beating tray or a sweep net can be an advantage, but beware of sweeping through wet vegetation, as this leaves small moths inevitably stuck with moisture to the material and impossible to identify from wing markings. Use of a bee-smoker may sound antiquated, but it can be remarkably effective in certain situations; on sand dunes, if the smoke is wafted into grassy overhangs, it can cause dozens of tiny moths, mainly in the Gelechiidae, to stream forth from their hideouts. Another neat trick for some species in the genus *Elachista* (Elachistidae) is to place a large sheet of black cloth over the habitat and to leave it for a few minutes. This simulates dusk, which encourages the moths to move up grass stems to sit on the cloth.

Warm, sultry weather is usually good in that it encourages flight, and wandering through good quality semi-natural habitats, such as chalk grassland and heathland, can be very rewarding. Most micro-moths are not fast flying and do not fly far if disturbed.

However, cold conditions can also be helpful for recording some groups. Tapping branches of bushes onto a beating tray on cold, windless days in April and May allows adult moths to fall onto the tray, where they remain motionless and can be examined easily. This is a good way of finding species in the Eriocraniidae, or rarely seen species such as *Eucosma pauperana*, *Pammene agnotana* and others in the genus *Pammene* (Tortricidae).

Any time between dawn and dusk may be fruitful for micro-moths. There are groups which are predominantly day-flying, e.g. *Pyrausta* (Crambidae). Some species prefer flying at dawn, e.g. *Semioscopis steinkellneriana* (Elachistidae), or in morning sunshine, e.g. *Bankesia conspurcatella* (Psychidae), but late afternoon on a warm summer day is generally good for many, particularly for Tortricidae and Crambidae. Dusk is also excellent for some species, e.g. subfamily Cochylinae (Tortricidae).

Searching particular plants for adult moths at rest can be instructive, as it is likely that the plant is used by those species in their life cycle. For example, Tansy Plume *Gillmeria ochrodactyla* (Pterophoridae) nectars at dusk on the flowers of Tansy, the larval foodplant, while Yarrow Plume *Gillmeria pallidactyla* is usually among its foodplants, Yarrow and Sneezewort. Adults of these species from light traps are hard to tell apart without careful examination of the hindleg.

Pheromones

While moths may use a range of chemicals to communicate with other individuals within a species, it is currently only pheromones, or sex-chemicals, which are of value to naturalists. A number of artificial pheromone lures have been developed for monitoring and control of pest species, and in micro-moths, mainly for species in the Tortricidae and Pyralidae. Although each artificial pheromone is produced for a target species, other species are sometimes attracted. Occasionally, *Neosphaloptera nubilana* (Tortricidae) is attracted to the pheromone for Clearwing moths (Sesiidae), and *Pammene albuginana* to that designed for Plum Fruit Moth *Grapholita funebrana* (both Tortricidae).

Searching for adults at night

Adult micro-moths nectar on very similar flower species to macro-moths. Throughout the season, there is a succession of flowers which should be examined by torchlight. In spring, male catkins of Grey Willow and Goat Willow are good for species emerging from hibernation, such as *Agonopterix ocellana* (Elachistidae). During summer, a variety of plants may produce results, including Common Ragwort, willowherbs and Wild Teasel. Buddleia can be particularly good; in south-east Dorset, a single flowering bush may support dozens of *Udea fulvalis* (Crambidae) after dark. By autumn, Ice-plant and Ivy blossom are good sources of nectar for plume moths, such as Twenty-plume Moth *Alucita hexadactyla* (Alucitidae), and Beautiful Plume *Amblyptilia acanthadactyla* and Common Plume *Emmelina monodactyla* (both Pterophoridae).

Light-trapping

Much has already been written on this subject (e.g. Fry & Waring 2001; Leverton 2001), with new designs and new bulbs regularly coming on the market to suit different situations. Whatever means is chosen to attract moths overnight, there is always an element of hopeful expectation when approaching the trap the following morning. This pursuit offers potentially more exciting prospects for finding micro-moths than macro-moths as there is a higher chance of finding something interesting, since there is a greater number of species about which generally less is known.

However, there are several micro-moth families which are much less well represented at light than the rest (e.g. Micropterigidae, Nepticulidae, Incurvariidae, Prodoxidae, Psychidae and Scythrididae). Some individual species just do not seem to come to light much; given

▲ A sociable evening trapping moths around a lamp on a sheet.

the billions of Horse-chestnut Leaf Miner *Cameraria ohridella* (Gracillariidae) that plague the Horse-chestnut trees throughout much of mainland Britain at the moment, this moth is surprisingly poorly recorded at light. Occasionally one species is regularly found at light, even though congeners are not, e.g. *Nematopogon metaxella* (Adelidae) and *Dichrorampha acuminatana* (Tortricidae), and this can aid identification.

Searching for early stages

It is the specialisation of the larval stage of micro-moths as well as the niches they are able to occupy as herbivores that allows them to stand head and shoulders above macro-moths in terms of diversity of structure and behaviour. While this guide concentrates on the adult form, those moth enthusiasts wishing to extend their identification skills into the search for larvae, and rearing them on, are in for a treat that should last a lifetime. There is an added bonus that some life histories have yet to be unravelled, offering the prospect of publishing one's find-

▼ Searching for larvae of micro-moths among coastal vegetation on Alderney, Channel Islands.

▲ The larval gall of *Mompha divisella* in the stem of a willowherb.

▲ The free-living larva of Dusky Plume *Oidaematophorus lithodactyla* on Ploughman's Spikenard.

ings in the entomological journals and magazines.

Almost every part of a plant has a micro-moth which eats it. The skill in searching for larvae of micro-moths is to detect the clues that they leave behind, and to separate these clues from damage or disease. Some larvae cause their hostplant to change growth form. Several species in the Momphidae cause the stem of willowherb plants within which they feed to form a small swelling or gall, to branch more readily, and even to change colour from green to reddish.

The typical free-living, leaf-eating larva commonly seen in macro-moths is a rarity in micro-moths, seen frequently only in the plume moths (Pterophoridae). Most micro-moth larvae produce silk which is used to spin leaves together in which to hide, sometimes in a rough spinning like that produced by Large Fruit-tree Tortrix *Archips podana* (Tortricidae) on various shrubs, or in a neat conical roll, as in *Caloptilia semifascia* (Gracillariidae) on Field Maple. Larvae may live within a dense tangle of silk, such as the gregarious Bird-cherry Ermine *Yponomeuta evonymella* (Yponomeutidae).

The habit of leaf-mining, that is, burrowing within the lamina of a leaf and leaving behind a characteristic trace of feeding and frass, has been adopted within many families of micro-moth. The form of the mine, together with the plant species, can often be diagnostic of a particular species; even weeks after the larva has vacated, its creator can sometimes be confirmed from the empty mine.

For the smallest of species, larval growth can be completed as a leaf miner, e.g. Eriocraniidae, Nepticulidae, Incurvariidae, Lyonetiidae and some Elachistidae. The larva of many species in the Gracillariidae changes feeding habit as it grows. For the first two or three instars it is a sap feeder, with semicir-

◄ Webs of Bird-cherry Ermine *Yponomeuta evonymella* on Bird Cherry.

▲ The leaf mine of *Stigmella aurella* on bramble.

▲ The transportable larval case of *Coleophora discordella* on Bird's-foot-trefoil.

cular mouthparts which point forward, scything into the plant cells as the head moves from side to side. Later, the head shape changes to a more familiar form, with chewing mouthparts, and the larva chews the parenchyma tissues within the leaf. Many of the Coleophoridae are also leaf miners, but in this family larvae live within a transportable case constructed of silk, leaves or leaf fragments. Usually, the case is moved by the larva from one leaf to the next, fixed in position with silk, from where the larva begins mining the leaf. The larval case can often be characteristic and can be a useful identification aid.

Flowers, fruits and seeds all provide a source of food, e.g. Cocksfoot Moth *Glyphipterix simpliciella* (Glyphipterigidae) feeds on the seeds of Cock's-foot grass, while Pea Moth *Cydia nigricana* (Tortricidae) feeds in developing pods of several pea species. The flower-feeding habit may be transitory, e.g. *Nemophora cupriacella* (Adelidae) feeds on the flowers of Devil's-bit Scabious in its first instar before constructing a portable case, dropping to the ground and continuing to feed on fresh or wilted leaves. A few species feed on different parts of the plant, depending on what is available, e.g. *Selania leplastriana* (Tortricidae) can be found in the main stem, side shoots, flowers and seeds of Wild Cabbage.

There are species which specialise in buds, e.g. Ash Bud Moth *Prays fraxinella* (Yponomeutidae) on Ash, and stems, e.g. *Spuleria flavicaput* (Elachistidae) in the terminal twigs of hawthorns. A range of species are associated with leaf litter, dead wood and fungi, especially bracket fungi. The family illustrating the greatest variety in this respect are the Tineidae, and some species in this family have further specialised to live on animal material, such as the Case-bearing Clothes-moth *Tinea pellionella* on animal fibres.

Quite a range of species feed on lower plants. Within the Tineidae, *Infurcitinea argentimaculella* makes silk tubes among lichens in the genus *Lepraria*, and in the Crambidae, *Eudonia lineola* feeds among the lichen *Xanthoria parietina* on rocks and bark. Mosses are widely eaten, with representatives in Micropterigidae, Psychidae, Oecophoridae, Gelechiidae, Tortricidae and Crambidae, but only one species in Britain, *Celypha aurofasciana* (Tortricidae), is recorded living on liverworts.

Rearing the early stages to adults

Probably the most efficient way to become competent in identifying micro-moths is to spend the first few years collecting larvae only, rearing them on to adults. Freshly-emerged adults which have not flown show all the diagnostic features that should be present on trap-caught examples, but often have been lost.

For larvae that make a spinning, try to keep the hostplant fresh, minimising disturbance while exchanging the plant for fresh material. It is usually best to collect only the largest larvae, as it may be only a matter of days before pupation occurs. Growing or buying potted plants of appropriate species, releasing larvae onto them, and netting the pot to prevent escape, is a very effective way of rearing many species, especially those which feed late into the autumn before pupating, or feed in autumn and spring.

It is important to understand the best medium for successful pupation. Whilst larvae of many species which feed in the spring and summer, and hatch the same year, are happy to pupate in folds in paper tissue placed at the bottom of the rearing container, the overwintering of fully fed larvae is often tricky. For example, many species of *Cydia* (Tortricidae) seem to prefer soft dead wood in which to make their cocoon. However, dead wood often contains numerous small invertebrates, including predatory ones, which live and hide among the fibres of the wood and may feed on micro-moth larvae introduced for pupation. It is best to bake the wood beforehand in an oven for a while. Once the micro-moth larvae have disappeared into the wood, the wood should be placed in a flowerpot, covered top and bottom with fine mesh netting to prevent unwelcome predatory visitors during the winter, and placed outdoors in dappled shade and half buried in soil, until the following spring.

Leaf miners need careful treatment over winter. Many Gracillariidae, especially *Phyllonorycter*, hibernate within the leaf mine. Intact mines are best placed in a plastic flowerpot covered with a netting or stocking sleeve and suspended above ground in a sheltered, shady position. There are two methods for looking after Nepticulidae, almost all larvae of which vacate the mine to pupate. For most *Stigmella* it is sufficient to allow the larvae to pupate in a jam jar half filled with loose, slightly damp, moss, removing the leaves once the larvae have vacated, storing the sealed jar in a cold shed until spring. However, for *Ectoedemia* the moss should be laid on compost in a flowerpot, covered with netting or stocking as above, and then placed outdoors, in shade.

Collecting and preparing material as vouchers

Identifying with certainty every species of micro-moth one encounters requires experience and patience, and the need from time to time to take voucher specimens. Vouchers enable the county recorder or others to confirm the identity subsequently, especially by dissection.

Freshly killed specimens which require subsequent identification should at least be pinned through the thorax and appropriately labelled, and should preferably have the wings spread. Specimens rolling around dead in the bottom of a plastic tube tend to lose their markings and are not an exciting prospect for an expert willing to help a new recorder. Similarly, unpinned micro-moths laid on cotton wool tend to adopt a 'head-down' position within the fibres and are difficult to extract without losing scales. A specimen should have a small label pinned to it giving details of location (ideally with six-figure grid reference), date, how it was found and any life-history details, and the recorder's name. Further information on preparing micro-moths can be found in Sokoloff (1980).

Photography

The advent of affordable digital cameras with in-built high-quality macro lenses has allowed recorders to snap away at a trap full of moths, and many species can be identified in this way. Digital images are also easy to share and compare, and numerous websites, chat groups and recorder blogs are now available. However, such images have limitations, especially when trying to identify micro-moths. Without a scale it can be difficult to work out the true size of the moth, and wing markings may be obscured by annoying reflections and shadows. A moderate file size is also important; tiny, thumbnail images are often too small to distinguish detail, while large images do not necessarily make it any easier, as identification is rarely dependent on the presence or absence of one or two scales on a wing. Remember, a photograph or digital image is no substitute for a voucher. However good the picture may be, it cannot be dissected!

Changes in distribution and conservation of micro-moths

The patterns of change which have been charted in the population and distribution of British butterflies and macro-moths are also reflected in micro-moths, but the data supporting these changes are nowhere near as detailed. Similarly, the effort nationally that has been invested in the conservation of butterflies outstrips that for macro-moths, but by comparison for micro-moths the subject is very much in its infancy. This is not to say that the case for the conservation of micro-moths is any weaker. Indeed, with the large range of species in Britain and Ireland, there are many that appear threatened and some that are vulnerable to extinction.

There are changes in distribution of micro-moth species happening every bit as fast as with macro-moths, and the reasons appear similar. There are plenty of species showing huge extension of range. For example, the Horse-chestnut Leaf Miner *Cameraria ohridella* (Gracillariidae) has spread throughout much of north-western Europe in the past decade, illustrating phenomenal powers of dispersal. Other species expanding their range appear to be doing so a little more modestly, at observed rates of between 2km and 10km per year.

Reliable data on reductions in the range of micro-moths are hard to come by. For a number of years, the Joint Nature Conservation Committee (JNCC) commissioned reviews of scarce and threatened micro-moth species in its UK Nature Conservation series. Sadly, insufficient resources were allocated to these reviews and they ceased publication in 2000. A few micro-moths are also included in Shirt (1987), and, more recently, there has been a review by Davis (2012) which proposes a conservation status for each micro-moth species in Britain.

Thus, just at a time when there is great effort being placed on conservation of species and habitats, there is a relatively poor story to tell on micro-moths. Fortunately, in the review of the UK Biodiversity Action Plan (BAP) in 2007, and through the efforts of the conservation charity, Butterfly Conservation, there are now several micro-moths recognised as UK BAP Priority species, which are highlighted as priorities for conservation action. These are:

- *Agonopterix atomella* (Elachistidae)
- *Agonopterix capreolella* (Elachistidae)
- *Agrotera nemoralis* (Crambidae)
- *Anania funebris* (Crambidae)
- *Aplota palpella* (Oecophoridae)
- *Celypha woodiana* (Tortricidae)
- *Coleophora hydrolapathella* (Coleophoridae)
- *Coleophora tricolor* (Coleophoridae)*
- *Coleophora vibicella* (Coleophoridae)
- *Coleophora wockeella* (Coleophoridae)
- *Epermenia insecurella* (Epermeniidae)
- *Eudarcia richardsoni* (Tineidae)
- *Grapholita pallifrontana* (Tortricidae)
- *Lampronia capitella* (Prodoxidae)
- *Nemapogon picarella* (Tineidae)
- *Nematopogon magna* (Adelidae)
- *Nemophora fasciella* (Adelidae)
- *Phyllonorycter sagitella* (Gracillariidae)
- *Phyllonorycter scabiosella* (Gracillariidae)
- *Pyrausta sanguinalis* (Crambidae)
- *Sciota hostilis* (Pyralidae)
- *Scythris siccella* (Scythrididae)
- *Stigmella zelleriella* (Nepticulidae)
- *Syncopacma albipalpella* (Gelechiidae)
- *Syncopacma suecicella* (Gelechiidae)

*UK BAP species from 1999 onwards

Reasons behind the decline of micro-moths are probably no different to declines in macro-moths or butterflies. Rapid changes to the countryside in the past century, the disappearance of widespread tracts of semi-natural grassland, woodlands and hedgerows, and the change to plantation forestry and intensive agriculture all have their part to play. In essence, there has been a massive decline in areas of semi-natural habitats, and that which remains has not been adequately managed owing to lack of incentives to landowners to look after land with marginal economic returns.

There are some reasons for optimism for the future. Environmental land management schemes, for agricultural and forestry systems, do now recognise and value biodiversity and the need for management to maintain structural and floristic diversity of habitats for invertebrates. There have been some notable success stories. Lowland heathland in Dorset, which has been largely protected from further destruction and fragmentation since the 1980s, appeared to be in inexorable decline as a result of the spread of scrub and secondary woodland across the open heaths. Management effort across almost all the remaining sites in the past 20 or more years has now arrested this decline, and is starting to put in place long-term management solutions, such as light grazing. It is a delight to walk on these heaths in summer, teeming as they are with invertebrate life, including micro-moths.

We hope this field guide will play its part in helping further to raise the profile and highlight the importance of micro-moths to conservation organisations and landowners, to remove the relative but unwarranted obscurity of this fascinating and diverse group of invertebrates and to give them wider recognition in wildlife conservation decision-making.

Societies and local recording groups

Amateur Entomologists' Society A society for all ages and experience. Publishes *The Bulletin*, books and leaflets, holds an annual exhibition and trade fair. Contact: AES, PO Box 8774, London SW7 5ZG; www.amentsoc.org.

British Entomological & Natural History Society The national society for field entomologists, including many active micro-moth recorders. Publishes the *British Journal of Entomology and Natural History*, organises an annual exhibition and series of lectures and field meetings. Maintains its own library, reference collection and lecture space in Reading, Berkshire. Contact: The Secretary, c/o The Pelham-Clinton Building, Dinton Pastures Country Park, Davis Street, Hurst, Reading, Berks RG10 0TH; www.benhs.org.uk.

Butterfly Conservation The national charity for the conservation of moths and butterflies. Lead partner for the majority of moths in the UK Biodiversity Action Plan, and runs the National Moth Recording Scheme, part of the Moths Count project. Publishes a members' magazine *Butterfly* and organises national recording meetings, members' days and international symposia. It has 31 branches across the UK run by volunteers, who organise over 700 events per year, many on moths. Contact: Butterfly Conservation, Manor Yard, East Lulworth, Wareham, Dorset BH20 5QP; www.butterfly-conservation.org.

Royal Entomological Society The society for professional entomologists and amateurs interested in research. Publishes various journals, organises international symposia and maintains a comprehensive library. Contact: RES, The Mansion House, Chiswell Green Lane, St Albans, Herts AL2 3NS; www.royensoc.co.uk.

Local Records Centres The focal point in several counties for the management and dissemination of wildlife records. A few centres work as a conduit between local moth-recording groups, County Moth Recorders and Butterfly Conservation to co-ordinate records management. Find your local records centre via the Association of Local Records Centres at www.alerc.org.uk.

Local moth-recording groups, County Moth Recorders and county lists

There are many local moth-recording groups which are self-organised by dedicated volunteers, and some of these maintain websites offering information on the status of species within a county or area, and report on latest sightings of macro- and micro-moths. Almost all counties have a County Moth Recorder, in some cases having separate individuals for micros and macros. They collate and verify records at the county (or vice-county) level, some also producing newsletters, organising field and indoor meetings etc. Often these individuals are key co-ordinators of the local recording group. Records should be forwarded to the relevant County Moth Recorder at least at the end of each year. You can find the contact details for your County Moth Recorder via the Moths Counts website at www.mothscount.org. Many individual county moth lists have been published, and these provide a good guide to species that may occur at particular localities. While some have not been updated since the *Victoria History of the Counties of England* (the *Victoria County History*) series which began publication in 1899, there are now many modern authoritative works covering a large part of England.

Journals

There are three main journals, well known to most UK and Irish lepidopterists, which regularly publish articles, and notes and observations, on micro-moths: *Atropos*, the *Entomologist's Gazette*, and the *Entomologist's Record and Journal of Variation*. The *Entomologist's Record* publishes an annual review of new vice-county records and interesting finds. *British Wildlife* has reports on mainly macro-moth sightings in each issue, and articles on moths.

Macro-moths and other insects that look like micro-moths

There are a few macro-moths and species in other insect Orders which are commonly mistaken for micro-moths. In particular, this includes smaller moth species in the Nolidae and Erebidae (note that all these species were in Noctuidae). Illustrations of the most frequent culprits are shown below.

Oak Nycteoline		
 Oak Nycteoline *Nycteola reveyana*	 *Acleris hastiana*	Oak Nycteoline *Nycteola reveyana* (Nolidae) resembles several species in the Tortricidae, especially the genera *Acleris* (p.271) and *Archips* (p.286). Oak Nycteoline has long, straight palps.

Cream-bordered Green Pea		
 Cream-bordered Green Pea *Earias clorana*	 Green Oak Tortrix *Tortrix viridana*	Cream-bordered Green Pea *Earias clorana* (Erebidae) is very similar to Green Oak Tortrix *Tortrix viridana* (p.271). Cream-bordered Green Pea has a distinct white line along the costa and yellowish, sometimes reddish, cilia.

Erebidae				
 Small Marbled *Eublemma parva*	 Straw Dot *Rivula sericealis*	 Pinion-streaked Snout *Schrankia costaestrigalis*	 Marsh Oblique-barred *Hypenodes humidalis*	Several species in the Erebidae, including Small Marbled *Eublemma parva*, Straw Dot *Rivula sericealis*, Pinion-streaked Snout *Schrankia costaestrigalis* and Marsh Oblique-barred *Hypenodes humidalis* resemble species in both Crambidae and Pyralidae, but the macro-moths lack scale tufts at the base of the proboscis.

Owl-midges		
 x4 Owl-midge *Pyscoda* spp. (Diptera)	 *Whittleia retiella* (Psychidae)	Owl-midges or moth-flies (Diptera: Psychodidae) which, as the name suggests, have a moth-like appearance and flight, are probably most similar to micro-moths in the Psychidae (p.66). However, owl-midges, like all true flies, have only one pair of wings. The long hairs on the wings, abdomen and thorax give these flies a furry appearance.

Whiteflies

Cabbage Whitefly
Aleyrodes proletella
(Hemiptera)

Elachista argentella
(Elachistidae)

Whiteflies (Hemiptera: Aleyrodidae) are somewhat moth-like in flight, and have four wings, perhaps resembling white species of micro-moths in the Elachistidae, but the wings are waxy, with no scales. These bugs are also tiny, usually less than 2-3mm, and very few moths are this small.

Caddisflies

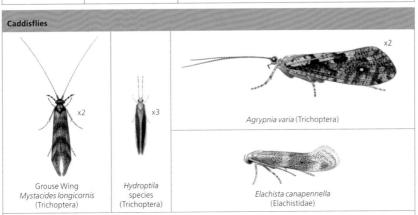

Agrypnia varia (Trichoptera)

Grouse Wing
Mystacides longicornis
(Trichoptera)

Hydroptila
species
(Trichoptera)

Elachista canapennella
(Elachistidae)

Caddisflies (Trichoptera) are frequently attracted to moth-trap lights and so are commonly found among moths. A number of the larger species have a passing resemblance to Pyralidae, and those with long antennae could be confused with Adelidae. However, the wings, thorax and body of all caddisflies have a sparse covering of hairs, not plate-like scales. The micro-caddisflies (Hydroptilidae) are frequently confused with dark species of micro-moths in the Elachistidae, as they are of similar size, but can be separated easily under a hand lens. These micro-caddisflies run faster than micro-moths.

Lacewings and Alderflies

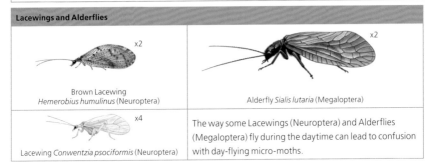

Brown Lacewing
Hemerobius humulinus (Neuroptera)

Alderfly *Sialis lutaria* (Megaloptera)

Lacewing *Conwentzia psociformis* (Neuroptera)

The way some Lacewings (Neuroptera) and Alderflies (Megaloptera) fly during the daytime can lead to confusion with day-flying micro-moths.

Key to families

This section of the field guide provides a key to separate micro-moths at a higher taxonomic level. We hope it will encourage readers to look at the whole moth, not just the wing pattern, and to investigate differences in the external structure of moths in different families. When identifying moths we tend to rely rather too heavily on wing markings to separate species, and we can struggle to put a name to a micro-moth with plain grey or brown forewings. A quick examination of structures and features other than forewing markings may rule out most of the look-alikes.

There are several keys to families already available, including Meyrick ([1928]) and Heath & Emmet (1985), covering most British and Irish Lepidoptera, and more recently by Kuchlein & Bot (2010), covering micro-moth families in The Netherlands. This key, based on original work by Martin Corley, is not presented in the form of a traditional dichotomous key. It is intended to help the reader assign a micro-moth to the most likely family, subfamily or genus.

Micro-moth families are defined by a range of characters. Several of these are not readily seen externally, or require dissection, such as ocelli, wing venation, tympanal organs, genitalia and other abdominal features, but they can be among the most important characters for identification. Examination of these characters usually requires the moth to be dead, and can involve invasive and time-consuming techniques which are rarely practised by amateurs, except for genitalia dissection. This key has been written to assist those who would rather not kill or damage moths, and avoids invasive examination. However, some practice will be needed to find tiny characters such as the ocelli, and for some characters, such as examination of the shape and width of the hindwing, it is often helpful to anaesthetise the moth temporarily by placing it in a carbon dioxide- or nitrous oxide-rich atmosphere in a stoppered tube for several seconds. The CO_2 and N_2O dispensers and cartridges used in the food and drinks trades are not expensive and serve this purpose well.

In this synopsis, features such as resting position of the antennae, wings and body of a living example are preferred. The position of the antennae in a moth at rest may be different from that in a moth that is restless but not moving, so antennae are best examined after a moth has been cooled for an hour or so in a fridge. Other characters should be visible with a hand lens or a low-power binocular microscope. There are two pairs of families which cannot be separated using this key: Crambidae and Pyralidae, and Argyresthiidae and Yponomeutidae.

The key covers the 45 micro-moth families described in the main text of this field guide. These families embrace the 1,627 species on the British list, of which just over 1,300 (about 80%) belong to just 12 families. In theory, it should be straightforward to familiarise oneself with the general 'jizz' of these 12 families, and by inference over three-quarters of our micro-moth species, but some of these families are very diverse. The remaining 33 families, each containing small numbers of species, include some of the most abundant British moths, such as Diamond-back Moth *Plutella xylostella* (Plutellidae), Cocksfoot Moth *Glyphipterix simpliciella* (Glyphipterigidae) and Nettle-tap *Anthophila fabriciana* (Choreutidae). The family Heliodinidae has been omitted entirely from this field guide; the one species ever recorded in

Britain is presumed extinct. The key applies only to British and Irish micro-moths. Non-British species of some families may have characters that differ from those used here. Furthermore, there are many families worldwide that are not represented in the British fauna. Indeed, the Continental European fauna includes more than a dozen additional families.

The number of families, and which genera each contains, are matters that change every time a new classification of Lepidoptera is produced. The 45 families treated here were confined to just 32 families in Heath (1976). Since that time, there has been a tendency to split the more diverse families into smaller units, although there have also been some amalgamations. The proliferation of families means that each family is less diverse, which should make it easier to recognise a family, but it also means that the differences between families are smaller than before. Throughout this field guide, we have used the hierarchy adopted by *Fauna Europaea*.

Working through the key

There are 17 groups, each containing one or more families. We advise starting with Group A, moving on to B, and so on through the key. If the characters of the example being examined appear to match a group, work through all the characters in that group to check for a good match. If they do not match, move on to the next group. Once a plausible family has been found for the moth, check the illustrations adjacent and then turn to that family in the main section of the book.

Groups A to D use very obvious characters. Groups E and F encompass the large and diverse families, Tortricidae and Crambidae/Pyralidae, which include a high proportion of the most familiar micro-moths, and most of the larger species. Because these families are so diverse, it has not been possible to pick just one or two characters which define them. Groups G to Q use characters requiring close examination by a hand lens or low-power binocular microscope; the moths will normally need to be torpid, except Group I, where the activity of wings of the moths is characteristic. Following the key there is a miscellaneous group covering a range of conspicuous features that are worth learning as useful field characters. Within the key there are three genera/species which do not exhibit all the characters that have been used here to distinguish the families in which they are classified. These are listed as exceptions in the appropriate part of the key.

Technical terms are used alongside illustrations to guide the reader to the part of the moth that should be examined. A full explanation of the terms used is provided in the Introduction (p.9).

Group A

Females with wings absent or much reduced

i Wingless*

    a Terrestrial → **Psychidae** (part) (p.66)

    b Aquatic → *Acentria* (**Crambidae**) (p.364)

ii Wings reduced*

    a Labial palps well developed → **Chimabachidae** (p.133)

    b Labial palps shorter than diameter of the eye → *Exapate* (**Tortricidae**) (p.260)

    * Several macro-moths in the Geometridae also have females with wings absent or reduced

i wingless female
Acanthopsyche atra (Psychidae)

ii female of *Dasystoma salicella*
(Chimabachidae)

Group B

Forewings at rest usually held away from the body, with the costa perpendicular or obtusely angled to the body; wings either divided into 'fingers' or entire and rolled into a tube

i Wings divided into fingers

    a Wings divided into six fingers each; legs short; when in torpor wings may be held flat over the body

    → **Alucitidae** (p.186)

    b Forewings divided towards the termen into two fingers, sometimes with the dorsal finger tucked under the costal; hindwings with three fingers; legs very long

    → **Pterophoridae** (part) (p.187)

ii Wings rolled into a tube, obtusely angled to the body and raised at an angle; legs very long → *Agdistis* (**Pterophoridae**) (p.187)

ia *Alucita hexadactyla*
(Alucitidae)

ib *Amblyptilia acanthadactyla*
(Pterophoridae)

ii *Agdistis bennetii*
(Pterophoridae)

Group C

Antennae > 1.5x length of the forewing (exception: *Cauchas fibulella*); adult rests roof-like with the forewing apex rounded and termen oblique; mainly diurnal species

→ **Adelidae** (p.56)

Nemophora degeerella (Adelidae)

Cauchas fibulella (Adelidae)

Group D
Male antennae distinctly pectinate or bipectinate*

i Antennae pectinate → *Incurvaria* (**Incurvariidae**) (p.60)

ii Antennae bipectinate

    a pectinations long, sparsely covered with short individual hairs → **Psychidae** (part) (p.66)

    b pectinations short, each with pencil of long hairs → *Philedone, Philedonides* (**Tortricidae**) (p.260)

* Diurnea lipsiella, Dasystoma *(Chimabachidae) have ciliate pectinations not branches of the antennal segments*

i *Incurvaria pectinea* (Incurvariidae)   **iia** *Epichnopterix plumella* (Psychidae)   **iib** *Philedonides lunana* (Tortricidae)

Group E
Top of the head with dense erect (like a carpet pile) or raised scales; labial palps usually pointing straight forward, the second segment densely scaled, the third segment small; ocelli present, but small and hard to observe; hindwings broad, forewings more or less broad (note *Sparganothis* has long, forward-pointing palps)

→ **Tortricidae** (part) (p.260)

head of Tortricidae   *Archips podana*   *Sparganothis pilleriana*   *Agapeta zoegana*
                (Tortricidae)   (Tortricidae)   (Tortricidae)

Group F
Head with raised scales on the crown, face smooth or with raised or erect scales; labial palps variable, often directed forwards or upwards, second segment sometimes long or very long, clothed in scales in a slender tubular form, not triangular or ovoid; maxillary palps developed, sometimes concealed by labial palps; forewings broad or narrow, hindwings broad; wings more or less wrapped around the abdomen, or held flat and overlapping or extended; legs may be long

→ **Pyralidae** (p.348), **Crambidae** (part) (p.364)

head of Crambidae   *Pyrausta nigrata*   *Hypsopygia glaucinalis*   *Myelois circumvoluta*
              (Crambidae)   (Pyralidae)   (Pyralidae)

Group G
Moth rests with hindlegs in air

i Wings rolled round the abdomen; hindlegs held in front of mid-legs → **Stathmopodidae** (p.148)

ii Wings held flat, slightly diverging →
Schreckensteiniidae (p.198)

i *Stathmopoda pedella*   **ii** *Schreckensteinia festaliella*
(Stathmopodidae)   (Schreckensteiniidae)

Group H
Moth rests with antennae directed forwards, often parallel to one another

i Top of head smooth

     a Forewings with a cross-band or opposed costal and tornal spots → *Esperia* (**Oecophoridae**) (p.126)

     b Forewings, if marked, then lacking a cross-band or opposed spots → **Coleophoridae** (p.150)

ii Top of head with erect or raised scales

    → *Ypsolopha* **Ypsolophidae** (p.112), **Plutellidae** (p.106)

ia *Esperia sulphurella* (Oecophoridae)

ib *Coleophora albitarsella* (Coleophoridae)

i head of Oecophoridae

ib *Coleophora lineolea* (Coleophoridae)

ii *Ypsolopha parenthesella* (Ypsolophidae)

ii head of Ypsolophidae   ii head of Plutellidae

ii *Rhigognostis annulatella* (Plutellidae)

Group I
At rest, moth characteristically raises and lowers its wings; forewings dark; ocelli present; diurnal species

→ **Glyphipterigidae** (part) (p.108)

head of Glyphipterigidae

Glyphipterix haworthana (Glyphipterigidae)

Group J
Moth rests with wings held flat and a little apart; ocelli present (exception: in *Tebenna* forewings are held somewhat roof-like, the apices tucked down and pressed together)

→ **Choreutidae** (p.202)

head of Choreutidae

Anthophila fabriciana (Choreutidae)

Tebenna micalis (Choreutidae)

Group K

Antennae with more or less conspicuous eye-cap at base; typically, small or very small moths (FL 1.5-6mm)

i Antennae as long as or longer than the wings → *Lyonetia* (**Lyonetiidae**) (p.119)

ii Antennae half to three-quarters forewing length

    a Top of head with more or less spiky scales

        1 Face short, not projecting far below the eyes; eye-caps contrasting in colour with head; wing scales rather coarse → **Nepticulidae** (p.50)

        2 Face long, projecting well below the eyes; eye-caps similar in colour to head (except *Bucculatrix thoracella*); wing scales not coarse → **Bucculatricidae** (p.80)

    b Top of head with smooth scales

        1 Wings at rest roof-like, at shallow angle; eye-cap conspicuous when moth active; maxillary palps well developed → **Opostegidae** (p.54)

        2 Wings at rest roof-like, at steep angle; all palps rudimentary

            + Forewings with metallic spot at the tornus → *Leucoptera* (**Lyonetiidae**) (p.119)

            ++ Forewings with conspicuous black dot near the apex → *Phyllocnistis* (**Gracillariidae**) (p.82)

i head of Lyonetiidae   **i** *Lyonetia clerkella* (Lyonetiidae)   **iia1** head of Nepticulidae   **iia1** *Stigmella aurella* (Nepticulidae)

iia2 head of Bucculatricidae   **iia2** *Bucculatrix thoracella* (Bucculatricidae)   **iib1** head of Opostegidae

iib1 *Pseudopostega crepusculella* (Opostegidae)   **iib2+** *Leucoptera laburnella* (Lyonetiidae)   **iib2++** *Phyllocnistis xenia* (Gracillariidae)

Group L

Top of head appearing smooth, but with a ridge between the base of the antennae; typically small moths (FL 3.5-6mm)

i Scales on top of the head directed forwards forming a shelf over a smooth face; antennal scape with a pecten of a single stout-based bristle → **Tischeriidae** (p.64)

ii Scales between the antennae forming a ridge, but not a shelf over the face; antennal scape without a pecten → *Oinophila* (**Tineidae**) (p.72)

pecten

i head of Tischeriidae   **i** *Coptotriche marginea* (Tischeriidae)   **ii** head of *Oinophila* (Tineidae)   **ii** *Oinophila v-flava* (Tineidae)

Group M

Top of head with spiky or erect scales; no eye-cap at base of the antenna

i Hindwings broad, with dorsal cilia not longer than width of the wing

    a Head densely covered with long spiky scales

        1 Face spiky-scaled, shape and size of forewing and hindwing similar ➔ **Eriocraniidae** (p.48)

        2 Face smooth-scaled ➔ *Ochsenheimeria* (**Ypsolophidae**) (p.112)

    b Head with erect scales

        1 Face with erect scales

            + Maxillary palps long, folded; mandibles present;
              shape and size of the forewing and hindwing similar ➔ **Micropterigidae** (p.46)

            ++ Mandibles absent; labial palps well developed, with lateral bristles ➔ **Tineidae** (part) (p.72)

            +++ Mandibles absent; labial palps rather weak, lacking lateral bristles

                • Forewings opaque, covered with plate-like scales, the veins hardly visible when lit from
                  beneath ➔ *Phylloporia* (**Incurvariidae**) (p.60), **Prodoxidae** (p.62)

                •• Forewings slightly or moderately translucent, sparsely covered with plate-like or hair-like
                  scales, the veins clearly visible when lit from beneath ➔ **Psychidae** (part) (p.66)

        2 Face smooth

            + Antennae without cilia ➔ **Argyresthiidae** (p.100), **Yponomeutidae** (p.92)

            ++ Antennae dark with fine whitish cilia, scape with a
              pecten on the basal third ➔ *Amphisbatis* (**Lypusidae**) (p.132)

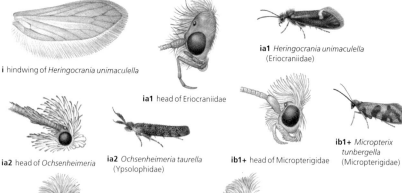

i hindwing of *Heringocrania unimaculella*

ia1 *Heringocrania unimaculella*
(Eriocraniidae)

ia1 head of Eriocraniidae

ia2 head of *Ochsenheimeria*

ia2 *Ochsenheimeria taurella*
(Ypsolophidae)

ib1+ head of Micropterigidae

ib1+ *Micropterix tunbergella*
(Micropterigidae)

bristle

ib1++ head of Tineidae

ib1++ *Tinea trinotella*
(Tineidae)

ib1+++• head of Prodoxidae

ib1+++• *Lampronia corticella*
(Prodoxidae)

ib1+++•• head of
Psychidae

ib1+++•• *Psyche casta*
(Psychidae)

ib2+ head of
Yponomeutidae

ib2+ *Yponomeuta rorrella*
(Yponomeutidae)

ib2++ antenna
of *Amphisbatis
incongruella*
(Lypusidae)

ii Hindwings narrow, with dorsal cilia at least twice as long as width of the wing

 a Forewings with distinct pattern → **Gracillariidae** (part) (p.82)

 b Forewings speckled but without a pattern → **Bedelliidae** (p.118)

ii hindwing of *Caloptilia rufipennella* (Gracillariidae)

iia *Cameraria ohridella* (Gracillariidae)

iib *Bedellia somnulentella* (Bedelliidae)

Groups N to Q have top of head smooth or with raised scales, and eye-caps lacking or not conspicuous; the position of the palps in live examples may be different from that on dead specimens

Group N

Labial palps directed forwards or downwards, or slightly curved and weakly ascending

i Hindwings narrower than forewings; small moths (FL 2.5-6.5mm)

 a Resting position nearly horizontal; FL 3-6.5mm; labial palps slender, weakly ascending → *Elachista* (**Elachistidae**) (p.136)

 b Resting position with front end raised; FL 2.5-4mm

  1 Forewings shining grey or with metallic markings consisting of a cross-band and two spots; labial palps short and drooping → **Heliozelidae** (p.55)

  2 Blackish-grey moths, with or without tornal white spot or fine white cross-line; labial palps short, forward pointing → **Douglasiidae** (p.122)

ii Hindwings as wide as or wider than forewings; small to medium-sized moths (FL 3-10mm)

 a Terminal segment of labial palps small, slender and pointed, ascending from a long second joint → **Oecophoridae** (part) (p.126)

 b Terminal joint of labial palps pointing forwards or downwards → **Chimabachidae** (males) (p.133)

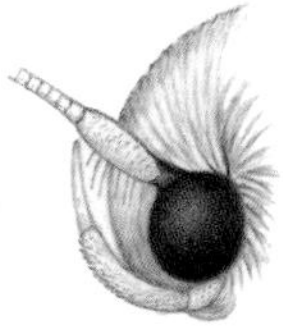

ia head of Elachistidae

ia *Elachista argentella* (Elachistidae)

ib1 head of Heliozelidae

ib1 *Antispila metallella* (Heliozelidae)

ib2 head of Douglasiidae

ib2 *Tinagma ocnerostomella* (Douglasiidae)

iia head of Oecophoridae

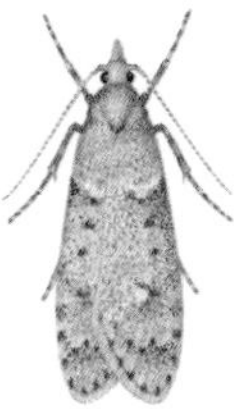

iib *Diurnea fagella* (Chimabachidae)

iia *Pleurota bicostella* (Oecophoridae)

iib head of Chimabachidae

Group O

Labial palps curving up around face

i Labial palps pointed or blunt, not or barely extending to top of the head; tongue not scaled

    a Antennae as long, or almost as long, as the forewings → **Gracillariidae** (part) (p.82)

    b Antennae about half to three-quarters forewing length

        1 Thorax with a posterior crest → *Digitivalva–Acrolepia* (**Glyphipterigidae**) (p.108)

        2 Thorax without a posterior crest

            + Dorsum of the forewing with one or more scale-teeth → **Epermeniidae** (part) (p.199)

            ++ Dorsum of the forewing without scale-teeth

                • Forewings plain metallic bronze; antennae blackish, usually with a white section before the apex → **Roeslerstammiidae** (p.79)

                •• Forewings straw-coloured with termen straight; large species (FL 9-14mm) → *Orthotelia* (**Glyphipterigidae**) (p.108)

                ••• Forewings coloured otherwise; small to medium species (FL 4-8mm) → **Praydidae** (p.116), **Epermeniidae** (part) (p.199)

ii Labial palps long, usually pointed and reaching nearly to top of the head or beyond; tongue scaled

    a Hindwing apex drawn into a finger-like extension; forewing apex rounded, termen more or less oblique → **Gelechiidae** (p.171)

    b Hindwings broad, length of dorsal cilia less than twice width of wing → **Group P**

    c Hindwings narrow, length of dorsal cilia at least twice width of wing → **Group Q**

ia head of Gracillariidae

ia *Caloptilia betulicola* (Gracillariidae)

ib1 *Acrolepia autumnitella* (Glyphipterigidae)

ib2+ *Epermenia chaerophyllella* (Epermeniidae)

i2++• *Roeslerstammia erxlebella* (Roeslerstammiidae)

i2++•• *Orthotelia sparganella* (Glyphipterigidae)

i2++••• *Phaulernis fulviguttella* (Epermeniidae)

iia head of Gelechiidae

iia hindwing of *Aproaerema* (Gelechiidae)

iib hindwing of *Acompsia* (Gelechiidae)

iic hindwing of *Carpatolechia* (Gelechiidae)

Group P

Labial palps curving up, pointed; hindwings broad

i Thorax with conspicuous black dots → *Ethmia* (**Elachistidae**) (p.136)

ii Forewings broad, at rest overlapping and nearly flat

    a Antennae as long as the forewing (often hidden at rest) → **Peleopodidae** (p.135)

    b Antennae shorter than the forewing → *Semioscopis–Hypercallia* (**Elachistidae**) (p.136), **Oecophoridae** (part) (p.126)

iii Forewings blackish with two or three pale yellow cross-bands or several spots and a pale thorax → **Autostichidae** (p.123)

iv Moth rests roof-like; thorax without conspicuous black dots → *Pseudatemelia* (**Lypusidae**) (p.132), **Oecophoridae** (part) (p.126)

i head of *Ethmia* (Elachistidae) **i** *Ethmia dodecea* (Elachistidae) **iia** *Carcina quercana* (Peleopodidae) **iib** *Semioscopis steinkellneriana* (Elachistidae)

iib *Endrosis sarcitrella* (Oecophoridae) **iii** *Oegoconia quadripuncta* (Autostichidae) **iv** *Pseudatemelia josephinae* (Lypusidae) **iv** *Crassa tinctella* (Oecophoridae)

Group Q
Labial palps curving up, pointed; hindwings very narrow to fairly narrow

i Forewing four to five times as long as wide

  a Dorsal surface of the abdomen with a narrow transverse band on every segment devoid of plate-like scales but sparsely covered with long setae; male antennae with weak eye-cap and notch on second segment → **Blastobasidae** (p.124)

  b Abdomen fully clothed in plate-like scales, without a visible band; antennae of both sexes with thickened first segment, without eye-cap and notch

    1 Forewings with metallic markings or raised scales → *Pancalia* (**Cosmopterigidae**) (p.167), **Momphidae** (part) (p.160), *Blastodacna–Chrysoclista* (**Elachistidae**) (p.136)

    2 Forewings without metallic markings or raised tufts of scales → **Momphidae** (part) (p.160), **Scythrididae** (p.165), **Oecophoridae** (part) (p.126)

      ii Forewing six to eight times long as wide

        a Forewings with metallic marks or raised tufts of scales → **Cosmopterigidae** (part) (p.167)

        b Forewings without metallic scales or obvious scale tufts → *Limnaecia* (**Cosmopterigidae**) (p.167), **Batrachedridae** (p.149)

ia body and base of antenna (male) of *Blastobasis* (Blastobasidae) **ia** *Blastobasis adustella* (Blastobasidae) **ib1** head of Cosmopterigidae **ib1** *Pancalia leuwenhoekella* (Cosmopterigidae)

ib1 head of Momphidae **ib1** *Mompha locupletella* (Momphidae) **ib1** head of *Blastodacna* (Elachistidae) **ib1** *Chrysoclista linneella* (Elachistidae)

Group Q continued

ib2 *Mompha subbistrigella* (Momphidae)

ib2 *Scythris grandipennis* (Scythrididae)

ib2 *Batia lunaris* (Oecophoridae)

iia *Cosmopterix pulchrimella* (Cosmopterigidae)

iib *Limnaecia phragmite* (Cosmopterigidae)

iib *Batrachedra praeangusta* (Batrachedridae)

Miscellaneous conspicuous features of some micro-moths

i Moth declines at rest with head down → Most Argyresthiidae; *Zelleria*, some *Swammerdamia* and related genera (Yponomeutidae); some *Ypsolopha* (Ypsolophidae); *Crassa unitella* (Oecophoridae); *Cosmopterix* (Cosmopterigidae) decline slightly

i *Zelleria hepariella* (Yponomeutidae)

i *Ypsolopha lucella* (Ypsolophidae)

ii Antennae unusually long (about as long as forewing or longer) → Adelidae; Bedelliidae; genera *Caloptilia* to *Leucospilapteryx* (Gracillariidae); *Lyonetia* (Lyonetiidae); Peleopodidae

ii *Adela croesella* (Adelidae)

ii *Gracillaria syringella* (Gracillariidae)

iii Forewings with raised non-metallic scales on their surface → *Stenoptinea* (Tineidae); some Ypsolophidae; Chimabachidae; *Teleiodes* and related genera, *Psoricoptera* (Gelechiidae); Momphidae; *Sorhagenia* (Cosmopterigidae); *Luquetia, Blastodacna, Spuleria* (Elachistidae); *Phtheochroa, Acleris* (Tortricidae); some Pyralidae.

iii *Mompha conturbatella* (Momphidae)

iii *Acleris cristana* (Tortricidae)

iv Forewings with metallic silver or golden scales on their surface → *Micropterix aureatella* (Micropterigidae); some *Stigmella* (Nepticulidae); *Antispila* (Heliozelidae); some *Phyllonorycter* (Gracillariidae); some Choreutidae; some Glyphipterigidae; *Leucoptera* (Lyonetiidae); a few Coleophoridae; a few Elachistidae; a few Oecophoridae; a few Gelechiidae; a few Momphidae; *Cosmopterix* (Cosmopterigidae); *Chrysoclista* (Elachistidae); some Tortricidae; a few Crambidae.

iv *Micropterix aureatella* (Micropterigidae)

iv *Cosmopterix orichalcea* (Cosmopterigidae)

At-a-glance guide
to micro-moths

This section provides a quick reference to all 45 of the micro-moth families in Britain and Ireland, illustrating one, few or several species from each family, to help beginners find their way through the diversity of micro-moths.

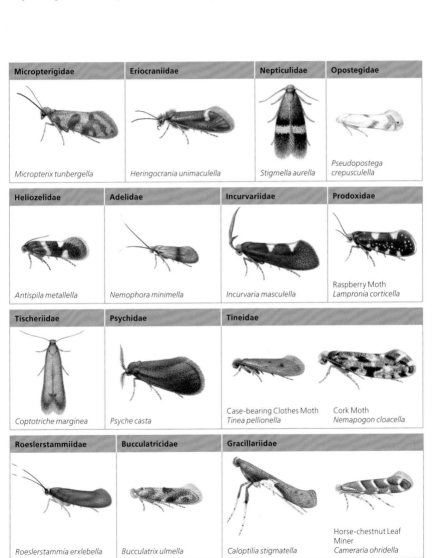

Micropterigidae	Eriocraniidae	Nepticulidae	Opostegidae
Micropterix tunbergella	*Heringocrania unimaculella*	*Stigmella aurella*	*Pseudopostega crepusculella*

Heliozelidae	Adelidae	Incurvariidae	Prodoxidae
Antispila metallella	*Nemophora minimella*	*Incurvaria masculella*	Raspberry Moth *Lampronia corticella*

Tischeriidae	Psychidae	Tineidae	
Coptotriche marginea	*Psyche casta*	Case-bearing Clothes Moth *Tinea pellionella*	Cork Moth *Nemapogon cloacella*

Roeslerstammiidae	Bucculatricidae	Gracillariidae	
Roeslerstammia erxlebella	*Bucculatrix ulmella*	*Caloptilia stigmatella*	Horse-chestnut Leaf Miner *Cameraria ohridella*

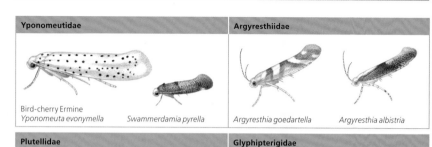

Yponomeutidae

Bird-cherry Ermine
Yponomeuta evonymella   *Swammerdamia pyrella*

Argyresthiidae

Argyresthia goedartella   *Argyresthia albistria*

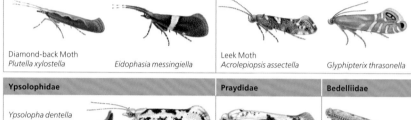

Plutellidae

Diamond-back Moth
Plutella xylostella   *Eidophasia messingiella*

Glyphipterigidae

Leek Moth
Acrolepiopsis assectella   *Glyphipterix thrasonella*

Ypsolophidae

Ypsolopha dentella

Ypsolopha sequella

Praydidae

Prays fraxinella

Bedelliidae

Bedellia somnulentella

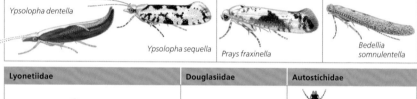

Lyonetiidae

Laburnum Leaf Miner
Leucoptera laburnella   Apple Leaf Miner
Lyonetia clerkella

Douglasiidae

Tinagma ocnerostomella

Autostichidae

Oegoconia quadripuncta

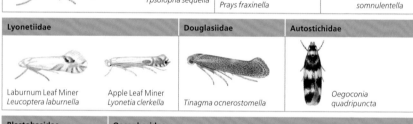

Blastobasidae

Blastobasis adustella

Oecophoridae

White-shouldered
House-moth
Endrosis sarcitrella   *Crassa unitella*

Esperia sulphurella

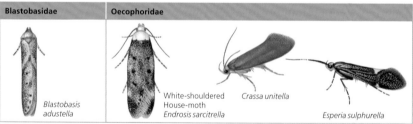

Lypusidae

Pseudatemelia josephinae

Chimabachidae

Diurnea fagella

Peleopodidae

Carcina quercana

Elachistidae

Elachista atricomella

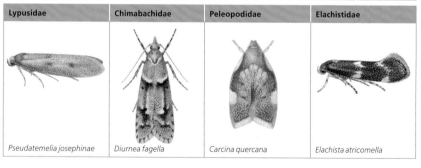

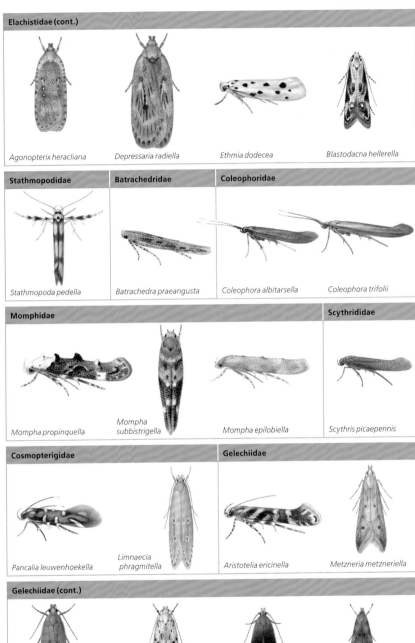

Elachistidae (cont.)

Agonopterix heracliana

Depressaria radiella

Ethmia dodecea

Blastodacna hellerella

Stathmopodidae

Stathmopoda pedella

Batrachedridae

Batrachedra praeangusta

Coleophoridae

Coleophora albitarsella

Coleophora trifolii

Momphidae

Mompha propinquella

Mompha subbistrigella

Mompha epilobiella

Scythrididae

Scythris picaepennis

Cosmopterigidae

Pancalia leuwenhoekella

Limnaecia phragmitella

Gelechiidae

Aristotelia ericinella

Metzneria metzneriella

Gelechiidae (cont.)

Bryotropha terrella

Carpatolechia proximella

Syncopacma taeniolella

Brachmia blandella

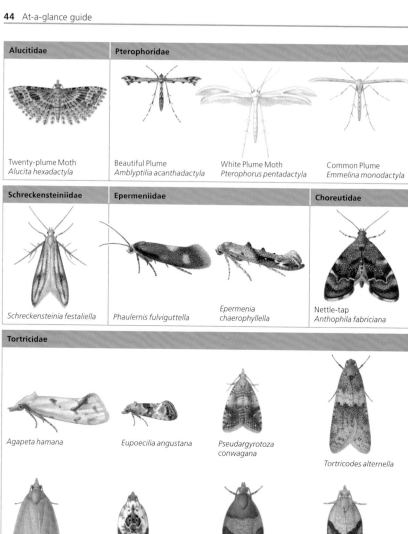

Alucitidae	Pterophoridae		
Twenty-plume Moth *Alucita hexadactyla*	Beautiful Plume *Amblyptilia acanthadactyla*	White Plume Moth *Pterophorus pentadactyla*	Common Plume *Emmelina monodactyla*

Schreckensteiniidae **Epermeniidae** **Choreutidae**

Schreckensteinia festaliella   *Phaulernis fulviguttella*

Epermenia chaerophyllella

Nettle-tap
Anthophila fabriciana

Tortricidae

Agapeta hamana

Eupoecilia angustana

Pseudargyrotoza conwagana

Tortricodes alternella

Green Oak Tortrix
Tortrix viridana

Garden Rose Tortrix
Acleris variegana

Barred Fruit-tree Tortrix
Pandemis cerasana

Light Brown Apple Moth
Epiphyas postvittana

Bactra lancealana

Apotomis turbidana

Celypha lacunana

Ancylis badiana

Tortricidae (cont.)

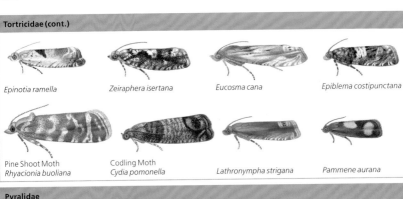

Epinotia ramella

Zeiraphera isertana

Eucosma cana

Epiblema costipunctana

Pine Shoot Moth
Rhyacionia buoliana

Codling Moth
Cydia pomonella

Lathronympha strigana

Pammene aurana

Pyralidae

Bee Moth
Aphomia
sociella
(female)

Gold Triangle
Hypsopygia costalis

Dioryctria abietella

Acrobasis advenella

Pyralidae (cont.)

Thistle Ermine
Myelois circumvoluta

Phycitodes
saxicola

Crambidae

Scoparia pyralella

Chrysoteuchia culmella

Crambidae (cont.)

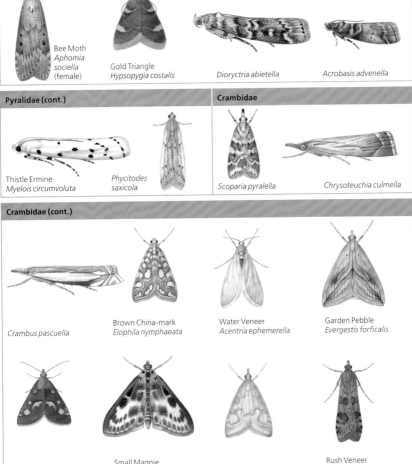

Crambus pascuella

Brown China-mark
Elophila nymphaeata

Water Veneer
Acentria ephemerella

Garden Pebble
Evergestis forficalis

Pyrausta purpuralis

Small Magpie
Anania hortulata

Udea lutealis

Rush Veneer
Nomophila noctuella

Micropterigidae

Micropterix tunbergella

There are five species in this family. At rest, the wings are held roof-like, at a steep angle. All species are very small, with forewing length 2.5-5.5mm. The forewings are glossy bronze or purplish, sometimes with markings. The hindwings are as broad as the forewings. Uniquely amongst moths, adults of this family have chewing mouthparts and eat the pollen of herbaceous plants and trees. They can occur in great abundance, swarming over the flowers of

Micropterix calthella

sedges, or sitting on buttercup flowers. Larvae, which have well-developed antennae, have been found in young shoots of plants, and in soil where they probably feed on leaf litter or fungal hyphae, but little is known about their habits.

Further reading
British and Irish species: Heath *et al.* (1976)
European species: Bengtsson *et al.* (2008)

▲ Adults of *Micropterix calthella* in a buttercup flower, feeding on pollen.

◄ Adult of *Micropterix aureatella*.

Micropterix tunbergella page 204 1 (38)

Common. Local or scarce further north. **FL** 4-5.5mm. Head yellowish. Forewing bronzy golden with variable markings; four more or less complete reddish-purple cross-bands from the costa at the base, at one-half, three-quarters and before the termen, the outer two often merging. **FS** Single-brooded, late April-June. Active by day, occasionally swarming around trees, and also flies in evening sunshine. Adult feeds on pollen of oaks, Sycamore, Cherry Laurel and hawthorns. **Hab** Woodland, grassland. **Fp** Not known.

Micropterix mansuetella page 204 2 (28)

Very local. Scarce in Scotland. **FL** 3.5-4.5mm. Head black. Forewing golden, bronzy purple on the costa near the base, in a broad cross-band at one-third, and in the apical third where often indistinct. **The black head distinguishes this species from other** *Micropterix*. **FS** Single-brooded, May-June. Active by day and has been recorded at light. Adult feeds on pollen of various herbaceous plants, especially sedges. **Hab** Damp and wet woodland, carr, fen. **Fp** Not known.

Micropterix aureatella page 204 3 (8)

Local. Common in Scotland. **FL** 4-5mm. Head orange or yellowish. Forewing purplish, markings variable, whitish tinged golden, often with two cross-bands and a large spot towards the apex. **FS** Single-brooded, May-June. Active by day, flying in sunshine. Adult feeds on pollen of various herbaceous plants, especially sedges. **Hab** Damp woodland, damp heathland, high moorland. **Fp** A single larva has been found in mixed oak, Beech and Bilberry leaf litter among fungal hyphae.

Micropterix aruncella page 204 4 (7)

Common. **FL** 2.5-4.5mm. Head golden. Forewing bronzy golden, costal half of the base purple; only the male has two silvery white cross-lines and sometimes a small spot or streak near the apex, the female is unmarked. **Similar species *M. calthella* male and female have the base purple along the costa and the dorsum**; male *M. aruncella* could also be confused with *Phylloporia bistrigella* (Incurvariidae). **FS** Single-brooded, May-August. Active by day and has been recorded at light. Adult feeds on pollen of many herbaceous plants and Hawthorn, and is sometimes abundant. **Hab** Downland, dry grassland, woodland. **Fp** Young shoots of herbaceous plants.

Micropterix calthella page 204 5 (10)

Common. **FL** 3.5-5mm. Head yellowish. Forewing bronzy golden, distinctly purple at the base, otherwise unmarked. **Similar species** *M. aruncella* female. **FS** Single-brooded, late April-early July. Active by day. Adult feeds on pollen of various herbaceous plants and is often found on the flowers of buttercups, Marsh-marigold, Dog's Mercury and sedges, and is sometimes abundant.
Hab Damper parts of woodland rides and clearings, meadows, waste ground.
Fp Young shoots of herbaceous plants.

► Adults of *Micropterix calthella* on a spike of sedge flowers, feeding on pollen.

Eriocraniidae

There are eight species in this family. Adults are diurnal, flying at dawn and in sunshine, from late March to May, often around host trees. They sometimes come to light but are most readily found on cold, sunny days by tapping branches of trees over a beating tray, when they fall onto the cloth and sit motionless. At rest, the wings are held roof-like, at a steep angle. The

Dyseriocrania subpurpurella

Heringocrania unimaculella

forewing length is 4-7mm. The forewings are golden or purple, often with mottled or reticulated markings. The hindwings are as broad as the forewings. The head has spiky scales on the top and on the face. The mouthparts are reduced to rounded lobes; labial palps are present and the tongue is weak. As adults, **six of the seven mainly purple-coloured species may be difficult to separate with certainty, and some require genitalia examination; these are not illustrated**. Only *Dyseriocrania subpurpurella* and *Heringocrania unimaculella* are covered below, as they can usually be identified in the field.

The larvae form large blotch mines containing intertwining strands of frass, on oaks, birches, Hornbeam or Hazel, depending on species, these appearing shortly after the leaves are fully expanded in mid spring. The British Leafminers website provides photographs and key characteristics of larvae of Eriocraniidae and their leaf mines, and identification of all British species can be made with reasonable certainty from the tenanted mines. Pupation takes place in a cocoon in soil. Adults can be difficult to rear from larvae in captivity.

Further reading
British and Irish species: Heath *et al.* (1976)
European species: Bengtsson *et al.* (2008)
British Leafminers: http://www.leafmines.co.uk/

Dyseriocrania subpurpurella page 204   6 (48)

Common. Local in Scotland. **FL** 5-7mm. Forewing golden with variable purplish and bluish speckling, and a small, sometimes obscure, pale tornal spot. **FS** Single-brooded, April-May. Flies in sunshine and at night; rests on branches and trunks of oak trees in dull weather. Sometimes numerous. **Hab** Woodland. **Fp** Oaks.

Heringocrania unimaculella page 204   8 (50)

Local. **FL** 4-5mm. Forewing purple with a distinct silvery whitish tornal spot, often elongated into a short line extending half-way across the wing. **The only one of the Eriocraniidae with purple forewings without distinct golden mottling, and with the tornal spot whitish**; in other species the tornal spot, if present, is golden. **FS** Single-brooded, March-April. **Hab** Heathland, woodland. **Fp** Birches.

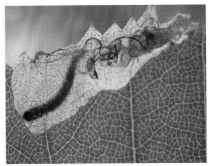

▲ Leaf mine of *Eriocrania sangii* on Silver Birch. Note the grey larva in the mine, the only one of the birch-feeding group with a dark larva.

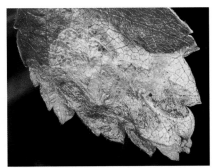

▲ Old leaf mine of an *Eriocrania* species on Silver Birch; vacated mines are difficult to determine to species.

▲ Leaf mine of *Paracrania chrysolepidella* on Hazel.

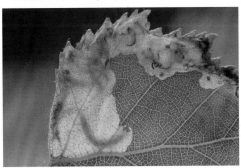

▲ Tenanted leaf mine of *Heringocrania unimaculella* on Silver Birch. Note the backward-pointing extensions of the head of the larva, which are characteristic of this species.

▼ Tenanted leaf mine of *Eriocrania cicatricella* on Silver Birch. Often a single mine of this species contains two or three larvae, whereas other species are solitary.

▲ Adult of *Eriocrania sparrmannella*.

▼ Old leaf mine of *Dyseriocrania subpurpurella* on oak.

Nepticulidae

Stigmella aurella   Ectoedemia
decentella

There are 100 species in this family. At rest, the wings are held roof-like, at a very shallow angle. All species are tiny, with forewing length 1.5-4.5mm, although the forewing scales are rather large, giving the wing a coarse appearance. This group includes *Enteucha acetosae*, Britain's smallest moth, with a forewing length of just 1.5mm. The forewings are variable in colour, usually dark, often with pale spots or a central cross-band, and some species have attractive metallic reflections. The hindwings are narrower than the forewings, with long dorsal cilia. The head has spiky scales on top; the face hardly projects below the eyes. The antennae are rather short, one-third to two-thirds the length of the forewing, with an **eye-cap at the base**, usually contrasting in colour with the head. The labial palps are reduced and the tongue is absent. The neck has spiky scales directed posteriorly over the front of the thorax, often different in colour from the head scales.

Owing to their small size and the similarity of many species, the majority are difficult to separate with certainty as adults, and may require examination of the genitalia. The most ubiquitous species is *Stigmella aurella*. Most species seem not to be attracted regularly to light, but occasionally on warm nights some may occur in large numbers, such as *Ectoedemia decentella* and *E. heringella*.

The larva forms a gallery or blotch mine (which often starts as a gallery) in leaves, although the mine may begin in the petiole. A few species mine the buds and winged fruits of Sycamore and its relatives, the outer (cortical) layer of stems of Broom, or the young bark of oaks and elms. The egg, which is surprisingly large for a small moth, is visible even to the naked eye at the start of the mine. In nearly all species, the larva vacates the mine to pupate in a small, pale-coloured silken cocoon.

With experience, identification of the majority of species is possible with reasonable certainty at the leaf-mine stage, often long after the larva has vacated the mine to pupate. A fair botanical knowledge is needed; larvae feed on many trees, shrubs and herbs, but not on grasses. Most species are restricted to one or a few foodplants, which is helpful. The larva usually leaves behind a distinctive track or pattern, and the distribution and colour of the frass, position of the egg on the under- or upperside of the leaf, colour of the larva, and time of year of appearance may be important. Rearing adults from larvae is surprisingly easy for some species, and may help to confirm identity. The British Leafminers website provides photographs and key characteristics of many species in this family.

Further reading
British and Irish species: Heath *et al.* (1976)
European species: Johansson *et al.* (1990); Bengtsson *et al.* (2008)
British Leafminers: http://www.leafmines.co.uk/

Stigmella aurella page 204 50 (152)

Common. Scarcer in the north. **FL** 2-3mm. Head orange. Forewing coppery purple with a shining golden cross-band beyond the middle, edged proximally by a deep blue-black cross-band. There are several similar-looking species, but *S. aurella* is by far the most regularly encountered at light or at rest on leaves. More readily found as a larval mine, although similar mines are made by other closely related species. Occasionally mines are abundant, and tenanted mines can be found throughout winter. **FS** Two or three broods per year, depending on latitude and severity of winter, April-September. **Hab** Wherever the foodplants occur. **Fp** Brambles, particularly evergreen varieties, Agrimony, avens, strawberries; in a linear mine.

Ectoedemia decentella page 204 20 (246)

Common. Scarcer in the north. **FL** 2.5-3mm. Head black. Forewing ground colour black with shining white markings, including a broad cross-band before the middle, often merged with a basal spot along the dorsum, sometimes obscuring the ground colour, and with white tornal and dorsal spots. **Similar species May be confused with *E. sericopeza* (photo p.52) and *E. louisella* (not illustrated), which have a reddish-brown, not black, head and a narrower cross-band before the middle, and in *E. louisella* the markings are yellowish white.** FS Two overlapping broods per year, June-August. Occasionally found in abundance at light on warm nights. **Hab** Potentially wherever the foodplant occurs. **Fp** Sycamore; in the fruiting key.

▲ Leaf mines of *Enteucha acetosae* on Common Sorrel.

▲ Vacated leaf mine of *Stigmella aurella* on Bramble. Often there is only one mine per leaf.

▲ Tenanted and vacated leaf mines of *Stigmella auromarginella* on Bramble. Several mines per leaf and reddish or purplish suffusion at the edge of the mines may indicate this species, but otherwise the mine is very similar to that of *S. aurella*.

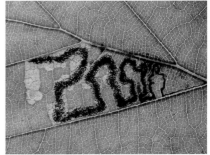

▲ Vacated leaf mine of *Stigmella tityrella* on Beech.

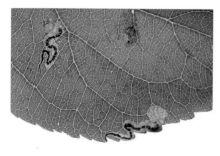

▲ Vacated leaf mines of *Stigmella incognitella* on apple.

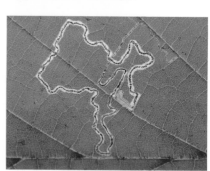

▲ Tenanted and vacated leaf mines of *Stigmella catharticella* on Purging Buckthorn.

◄ Tenanted leaf mine of *Stigmella oxyacanthella* on Hawthorn.

▲ Tenanted leaf mine of *Stigmella hybnerella* on Hawthorn.

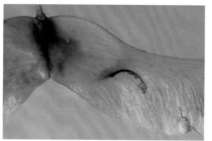

▲ Tenanted and vacated leaf mines of *Stigmella betulicola* on birch.

▲ Tenanted leaf mine of *Stigmella microtheriella* on Hazel.

▲ Mine of *Ectoedemia sericopeza* on key of Norway Maple. Note the orange oval pupal cocoon which is sometimes found on the key near the mine.

▲ Adult of *Ectoedemia sericopeza* on key of Norway Maple.

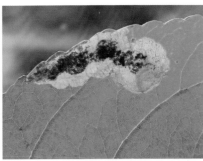

▲ Tenanted leaf mine of *Bohemannia pulverosella* on apple.

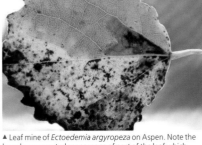

▲ Leaf mine of *Ectoedemia argyropeza* on Aspen. Note the larva has prevented senescence of part of the leaf which remains as a green island, allowing the larva to continue feeding after the leaf has fallen from the tree.

▶ Leaf mine of *Ectoedemia turbidella* on Grey Poplar. Note the mine starts in the petiole and progresses to the green island in the leaf once fallen from the tree.

▲ Tenanted leaf mine of *Ectoedemia atricollis* on Hawthorn.

▶ Leaf mines of *Ectoedemia occultella* on Silver Birch. The larva usually concentrates its frass into a circular mass in the middle of the mine, under which the larva hides when not feeding. Several mines may join in a leaf.

▲ Tenanted leaf mine of *Ectoedemia intimella* on willow.

▲ Mine of either *Ectoedemia atrifrontella* or *E. longicaudella* in thin bark of oak. These two species cannot be easily separated from the form of the mine.

▲ Tenanted leaf mine of *Ectoedemia septembrella* on St John's-wort.

Opostegidae

Opostega salaciella

There are four species in this family, only two of which are likely to be encountered in the field, either at light or flying weakly on a warm evening. At rest, the wings are held roof-like, at a shallow angle, and the moth appears short and squat as the forewings are relatively

Pseudopostega crepusculella

broad. The forewing length is 3-6mm. The forewings are white or pale, with few, if any, markings. The hindwings are narrower than the forewings, but have long dorsal cilia. The head has flat scales, apart from a tuft of erect scales between the antennae. The antennae are thread-like, about two-thirds the length of the forewing, **with characteristically large scapes, forming broad white or pale eye-caps**. The labial palps are short and straight, and the tongue is rudimentary. The presence of large eye-caps makes Opostegidae appear similar to Nepticulidae, but the latter are generally smaller and darker, with eye-caps that contrast in colour with the thorax. The early stages are little known.

Further reading
British and Irish species: Heath *et al.* (1976)
European species: Johansson *et al.* (1990); Bengtsson *et al.* (2008)

Opostega salaciella page 204   119 (315)

Local. **FL** 4.5-6mm. Head, eye-caps and forewing white, otherwise unmarked. **Similar species *Elachista argentella* (Elachistidae), which rests with wings roof-like, at a steep angle, does not have distinct eye-caps, and in the south flies about a month earlier. FS** Single-brooded, June-July. Most often found on warm evenings, flying weakly in grassland containing Sheep's Sorrel, although also where this plant is absent. Comes to light occasionally. **Hab** Open habitats, especially grasslands. **Fp** Sheep's Sorrel.

Pseudopostega crepusculella page 204   121 (319)

Very local. Scarce in Scotland. **FL** 3.5-5mm. Head, eye-caps and forewing white, with brown oblique dorsal and costal marks at about one-half, sometimes joined to form an angulated cross-band, and along the termen, with a black dot in the terminal cilia. **FS** Single-brooded, June-mid August. Comes to light and is occasionally found flying around Water Mint in the evening. **Hab** Damp habitats such as fens, marshes, wet woodland. **Fp** Unknown, but probably Water Mint.

▼ Adult of *Pseudopostega crepusculella*. Note the large scape is difficult to see when the adult is resting.

Heliozelidae

Antispila metallella. Note this species is very similar to *Antispila treitschkiella.*

There are five species in this family. Adults can be found in numbers on sunny days in spring and early summer, flying around trees and shrubs; they also settle on flowers. Adults rarely fly far from their host tree, and this can be a guide to identification. The forewing length is 2.5-4mm. Adults rest with the front end raised at least slightly, with the wings held roof-like, at a steep angle. The three species in the genus *Heliozela* have forewings shining grey or bronzy, with white dorsal dots; the two species in the genus *Antispila* have forewings dark copper-coloured with violet reflections, with silver metallic spots or cross-bands. The hindwings are slightly narrower than the forewings, with long dorsal cilia. The head has smooth scales. The antennae are thread-like and two-thirds the length of the forewing. The labial palps are short and drooping, and the tongue is well developed. Owing to their small size and predominantly day-flying habit, these species are often overlooked, and adults are difficult to tell apart. All *Heliozela* species are widespread, but *Antispila* species have a more southerly distribution.

The larvae of the three *Heliozela* species mine, respectively, the petiole of oaks, birches and Alder, and it is from the vacated mines that most records are made of the species. The mine is very difficult to detect until just before the larva vacates, when it makes an oval mine in the leaf near the midrib. The oval is cut out to form a pupal case which falls to the ground, and the vacated mine with the excision is then relatively easy to observe. The larvae of the two *Antispila* species make obvious blotch mines in the leaves of Dogwood and cut out similar pupal cases to *Heliozela*.

Further reading
British and Irish species: Heath *et al.* (1976)
European species: Bengtsson *et al.* (2008)

▲ (top right) Vacated leaf mine of *Heliozela hammoniella* on birch. Note the oval-shaped portion of the leaf which has been cut out to form the pupal case.

▶ Vacated leaf mine of *Antispila treitschkiella* on Dogwood. The oval gap left in the leaf after excision of the pupal case is obvious.

▶▶ Tenanted leaf mine of *Antispila treitschkiella* on Dogwood.

Adelidae Longhorns

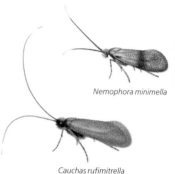

There are 15 species in this family. At rest, the wings are held roof-like, at a steep angle. The forewing length is very variable, 3.5-11mm; the female is often slightly smaller than the male. The forewings are broad, mostly single coloured, sometimes with a cross-band or other markings, and several species have beautiful green, golden green or purple reflections. The hindwings are similarly broad but smaller than the forewings, with narrow and broad

Nemophora minimella

Cauchas rufimitrella

scales, depending on species and sex, sometimes aiding identification. The head has a smooth face with upward-pointing hair-scales. **Characteristically, in all but one species, the antennae are long or very long, up to four times the length of the forewing, giving rise to the colloquial name for this group, the Longhorns.** The labial palps in *Nemophora* are curved upwards, with long, forward-pointing hairs, while in *Adela* they are forward-pointing and somewhat hairy. The tongue is short in all genera. All species are active by day, and fly in sunshine, sometimes in large aggregations around bushes and tree-tops. A few species are attracted to light, but rarely in large numbers.

Most species are relatively easy to identify, although care is required with *Nematopogon*, especially examples from moorland habitats.

The eggs are laid on vegetation, usually flowers, around which the adults fly. The young larvae begin feeding in flowers and developing seeds. They then construct a portable case in which they drop to the ground, where feeding continues on green

or wilted leaves, or amongst leaf litter. As the larva grows, leaf fragments are added to the case. Cases, which are usually in the shape of the body of a violin, and up to 2cm long, can be found by searching among surface layers of leaf litter or around the base of the plants around which the adults fly.

Further reading
British and Irish species: Heath *et al.* (1976)
European species: Bengtsson *et al.* (2008)

◀ Adults of *Adela reaumurella* aggregating in sunshine.

▲ Old larval case/pupal exuvium of *Nemophora degeerella*.

▲ Adult of *Nemophora degeerella*.

◄ Old larval case/pupal exuvium of *Adela reaumurella*.

► Larva and case of *Nemophora minimella*.

Nemophora fasciella   page 205   144 (352)

Very local. Recently recorded only from the east. **FL** 6-7.5mm. Antenna twice length of forewing in the male, about the same length as forewing in the female. Forewing metallic, golden reddish with purplish reflections, golden green near the base, with a wedge-shaped dark purplish-black cross-band at one-half. **FS** Single-brooded, June-July. Flies in sunshine over the foodplant and other flowers; in dull weather, rests on the foodplant or nearby vegetation. **Hab** Wasteland, roadside verges, embankments, marshes, downland. **Fp** Black Horehound, possibly White Dead-nettle; at first on seeds, later on lower leaves.

Nemophora minimella   page 205   145 (356)

Very local. **FL** 5-7mm. Antenna less than twice length of forewing in the male and a little longer than forewing in the female. Forewing metallic, golden green in the basal two-thirds with a short black basal streak, an indistinct dark cross-band beyond, and the outer part of the wing golden reddish. Hindwing of the male white, more especially in southern England populations, or dark purplish, rarely intermediate in colour. **Similar species** *N. cupriacella* is bigger, on average, without a basal streak and forewing cross-band. **FS** Single-brooded, July-August. Flies in sunshine and has been found at rest on scabious and other flowers. **Hab** Chalk downland, woodland clearings, marshy grassland, lowland raised bogs. **Fp** Devil's-bit Scabious, Small Scabious; at first on seeds, later on lower leaves.

Nemophora cupriacella   page 205   146 (349)

Very local. **FL** 6-8mm. Parthenogenetic, the male apparently unknown. Antenna a little longer than the forewing. Forewing metallic, golden green in the basal half, golden reddish or purplish beyond. **Similar species** *N. minimella, Adela cuprella*. **FS** Single-brooded, June-early August. Flies in sunshine and may be found resting on the flowers of scabious species and on Wild Teasel heads. **Hab** Downland, unimproved wet and dry grasslands, rough ground. **Fp** Field Scabious, Devil's-bit Scabious, Small Scabious, Wild Teasel; at first on seeds, later on lower leaves.

Nemophora metallica page 205

147 (346)

Very local. **FL** 7.5-10mm. Antenna 2.5 times length of forewing in the male and 1.5 times length of forewing in the female. Forewing long, uniform metallic golden greenish. **FS** Single-brooded, June-early August. Flies in sunshine and rests on flowers of the foodplant, but also flies at dusk and comes to light. **Hab** Chalk and limestone grassland. **Fp** Field Scabious, Small Scabious; at first on seeds, later on lower leaves.

Nemophora degeerella page 205

148 (338)

Common in England and Wales. Local elsewhere. **FL** 7-10mm. Head golden. Antenna nearly four times length of forewing in the male, and a little longer than forewing in the female. Forewing golden brownish, apically purplish with golden yellow streaks between the veins, particularly in the male, and a yellow cross-band edged blackish blue at two-thirds. **Similar species** *Adela croesella* **is smaller, with a reddish-brown and black head, the cross-band at just beyond one-half, and with the antenna in the male shorter. FS** Single-brooded, May-mid July. Flies by day, is sometimes seen in numbers in dappled sunshine, and is also active at dusk. **Hab** Woodland, particularly damp woodland, fens, marshes, old hedgerows. **Fp** Foodplant and early larval stage unknown, later on dead leaves on the ground.

Adela cuprella page 205

149 (366)

Local or very local. **FL** 6.5-8mm. Labial palps forward-pointing and somewhat hairy. Antenna 2.5 times length of forewing in the male, and a little longer than forewing in the female. Forewing metallic golden greenish with slight reddish or purplish reflections along the costa and at the apex. **Similar species** *Nemophora cupriacella* **has the labial palps curved upwards, with long forward-pointing hairs, and more extensive red or purple colour in the outer half of the forewing; also** *A. reaumurella*, **which is greener with a slightly rounder apex. FS** Single-brooded, April-May. Flies in sunshine, often in abundance, around the tops of tall sallow bushes. **Hab** Heathland, fens, marshes. **Fp** Sallow catkins, then fallen leaves.

Adela reaumurella page 205

150 (365)

Common only in England and Wales. Local elsewhere. **FL** 7-9mm. Antenna three times length of forewing in the male and less than 1.5 times length of forewing in the female. Forewing metallic, dark green or bluish green, sometimes with a slight golden tinge towards the costa. **Similar species** *Cauchas rufimitrella* **is smaller, the forewing golden greenish, and the antenna in the male shorter**; also *A. cuprella*. **FS** Single-brooded, mid April-June. Often seen dancing in sunshine in small swarms around the outer branches of oaks and Hazel. **Hab** Woodland, heathland, fens, marshes, scrub. **Fp** Foodplant and early larval stage unknown, later on dead leaves on the ground.

Adela croesella page 205

151 (371)

Local. Rare in the north and west. **FL** 5-6.5mm. Head reddish brown and black. Antenna slightly more than twice length of forewing in the male, and slightly longer than forewing in the female. Forewing blackish brown with a purple tinge, golden between the veins, and a yellow cross-band, edged dark purplish, at just beyond one-half. **Similar species** *Nemophora degeerella*. **FS** Single-brooded, May-June. Flies in sunshine around Wild Privet, Sea-buckthorn and Ash, and has been recorded at light. **Hab** Woodland margins, Breckland, downland, scrub, fens, marshes. **Fp** Probably on the developing seeds of plants around which adults fly, later on dead leaves on the ground.

Cauchas rufimitrella page 205   152 (382)

Local. **FL** 4.5-6mm. Antenna about twice length of forewing in the male, just less than 1.5 times length of forewing in the female. Forewing metallic, golden greenish, the costa sometimes tinged reddish or purplish. **Similar species** *Adela reaumurella*. **FS** Single-brooded, May-June. Flies in sunshine, sometimes in numbers, visiting flowers of the foodplants. **Hab** Damp grassland pastures, mosses. **Fp** Cuckooflower, Garlic Mustard; on the seeds, later on the lower leaves.

Cauchas fibulella page 205   153 (377)

Local. **FL** 3.5-5mm. Antenna slightly longer than the forewing. Forewing dark golden green, slightly purplish tinged, with an elongate whitish spot on the dorsum at about two-thirds. **FS** Single-brooded, late May-June. Flies in sunshine, visiting flowers of the foodplant. **Hab** Downland, rough ground. **Fp** Speedwells, especially Germander Speedwell; at first on the seeds, later on the lower leaves.

Nematopogon swammerdamella page 205   140 (391)

Common. Local in Scotland. **FL** 8-11mm. Antenna 2.5 times length of forewing in the male and nearly twice length of forewing in the female. Forewing pale greyish yellow, with an indistinct reticulated pattern, the base of the costa sometimes slightly darker, with the apex pointed. By far the largest of the plain-coloured longhorns. **FS** Single-brooded, late April-June. Flies by day and comes to light. **Hab** Woodland, parkland. **Fp** Dead leaves on the ground.

Nematopogon schwarziellus page 205   141 (387)

Common. **FL** 7-8mm. Antenna twice length of forewing in the male, about 1.5 times length of forewing in the female. Forewing pale greyish yellow, slightly reticulated towards the pointed apex. **Similar species N. metaxella has the forewing broader and more rounded at the apex, slightly darker, and with a dark spot or short line at two-thirds.** May also be confused with two scarcer species (neither illustrated), both found in open heathy woodland and moorland: *N. magna*, recorded from Yorkshire, parts of Scotland and western Ireland, has a densely reticulated dark greyish-brown forewing; *N. pilella*, known from south Wales northwards and very locally in western Ireland, has a dark greyish-brown forewing in the male, more variable in colour in the female, slightly reticulated towards the apex. Where *N. magna* or *N. pilella* are expected, examination of the male genitalia may be necessary; females may be very difficult to separate. **FS** Single-brooded, late April-June. Flies by day and occasionally comes to light. **Hab** Woodland, tall hedgerows, sheltered parts of moorland and heathland, other open habitats. **Fp** Dead leaves on the ground.

Nematopogon metaxella page 205   143 (390)

Scarce in the north, more common in the south. **FL** 6-8mm. Antenna three times length of forewing in the male, twice length of forewing in the female. Forewing broad with a rounded apex, greyish yellow with weak reticulation and a blurred dark spot or short line at about two-thirds. **Similar species** *N. schwarziellus*. **FS** Single-brooded, late May-July. Flies by day and is the most frequent of *Nematopogon* to come to light. **Hab** Woodland, fens, marshes, downland, gardens. **Fp** Dead and living leaves.

Incurvariidae

There are five species in this family. At rest, the wings are held roof-like, at a steep angle. The forewing length is 3-8mm. The forewings are relatively broad, dark, and with at least pale dorsal markings, which in some species are extended into full cross-bands. The hindwings are similarly broad, with narrow scales. The head is rough, with hair-scales directed upwards, giving a tufted appearance. The antennae are one-half

Incurvaria masculella

Phylloporia bistrigella

to two-thirds the length of the forewing, and in the male of two species have easily visible pectinations, appearing feathered. The labial palps are short and directed forward, and the tongue is short or rudimentary. Superficially, Incurvariidae are very similar to the next family, Prodoxidae. All species are active by day and fly in sunshine, and are occasionally attracted to light. Whilst identification of most species is straightforward, some care is required to identify a few species with certainty.

The eggs are laid on leaves and to begin with the larvae are leaf miners. Depending on species, a roughly circular, oval or violin-body-shaped case is cut from a mined portion of the leaf, the larva carrying its case and continuing to feed on living or dead leaves, usually near or on the ground. However, *Phylloporia bistrigella* pupates without additional feeding. Pupation in all species is within the case.

Further reading
British and Irish species: Heath *et al.* (1976)
European species: Bengtsson *et al.* (2008)

Incurvaria pectinea page 206   129 (423)

Common. Local in places. **FL** 6-8mm. Head pale or dark greyish brown. Antenna pectinate in the male, thread-like in the female. Forewing greyish brassy brown, with an indistinct triangular whitish spot on the dorsum before one-half, a similar smaller one at the tornus, and rarely a few indistinct spots on or near the costa. The female is paler than the male. **Similar species** *I. masculella* has the head yellowish or reddish brown, with a dark brown forewing and more distinct whitish dorsal markings. **FS** Single-brooded, April-May. Flies in sunshine by day and can be abundant; occasionally comes to light. **Hab** Woodland, parkland, heathland, scrub. **Fp** Mainly birches and Hazel, also Small-leaved Lime.

Incurvaria masculella page 206   130 (424)

Common, except in the far north. **FL** 6-8mm. Head yellowish or reddish brown. Antenna pectinate in the male, thread-like in the female. Forewing dark brown, sometimes tinged purplish, with a roughly triangular whitish spot on the dorsum before one-half, and a similar smaller one at the tornus. Sometimes there is a speckling of whitish scales in the outer half of the wing and a small white spot on the costa, almost opposite the tornal spot but placed slightly towards the base. **Similar species** *I. oehlmanniella* has a white spot or spots on the costa opposite the tornal spot or slightly towards the apex, and the male antenna is without pectinations; *I. pectinea*. **FS** Single-brooded, late April-May. Flies in sunshine by day and occasionally comes to light. **Hab** Woodland, gardens, scrub. **Fp** Hawthorn.

Incurvaria oehlmanniella page 206

Local. **FL** 6-8mm. Head pale orange or reddish brown. Antenna of both sexes thread-like. Forewing dark brown, weakly tinged purplish, with a whitish or creamy whitish spot on the dorsum before one-half, a similar smaller one at the tornus, and one or two spots on the costa opposite the tornal spot or slightly towards the apex. **Similar species** *I. masculella*. **FS** Single-brooded, May-July. Flies in sunshine by day and occasionally comes to light. **Hab** Open woodland, moorland, downland, hedgerows, suburban habitats. **Fp** Bilberry, Cloudberry, Dogwood, plums.

Incurvaria praelatella page 206

Local. **FL** 5.5-7mm. Head pale yellowish or pale reddish brown. Forewing dark purplish brown with cream markings comprising a small spot near the base, a cross-band at one-third, narrower towards the costa, a triangular spot at the tornus and a similar spot on the costa before the apex. **Similar species** *Lampronia luzella* (Prodoxidae) **does not have the pale spot near the base and has the head dark greyish brown.**
FS Single-brooded, May-mid July. Flies in sunshine by day and occasionally comes to light. **Hab** Heathland, downland, clearings in damp woodland. **Fp** Strawberry, Water Avens, other rosaceaeous herbs.

Phylloporia bistrigella page 206

Local. **FL** 3-4mm. Head whitish. Forewing brassy greyish brown, slightly tinged purplish towards the apex, with two white cross-bands, one near one-third, the other near two-thirds, the outer one occasionally interrupted about the middle; sometimes a white dot beyond. **Similar species** Male *Micropterix aruncella* (Micropterigidae). **FS** Single-brooded, May-July. Flies in sunshine around small birch trees, and comes to light. **Hab** Open woodland, heathland, scrub. **Fp** Downy and Silver Birch.

▲ Leaf mine and excised pupal case of *Phylloporia bistrigella*.

▲ Birch leaf showing mines, near-excised cases and holes left by *Incurvaria pectinea* larvae.

◀ Larva of *Incurvaria pectinea* crawling with case in which it will pupate.

Prodoxidae

There are seven species in this family. At rest, the wings are held roof-like, at a steep angle. The forewing length is 5-9mm. The forewings are relatively broad, dark, either without markings or with pale spots or cross-bands. The hindwings are similarly broad; those species without forewing markings have narrow hindwing scales, those with forewing markings

Lampronia luzella

Raspberry Moth *Lampronia corticella*

have broad-scaled hindwings. The head either has rough hair-like scales, or smooth broad scales. The antennae vary between one-third and about one-half the length of the forewing. The labial palps are short and directed forward, and the tongue is short. Superficially, Prodoxidae are very similar to the previous family, Incurvariidae. All species are active by day and fly in sunshine, and are occasionally attracted to light.

The wing pattern is variable within a species, and sometimes asymmetric between forewings, so the text and illustrations cannot match all variation found in the field, and some examples may require genitalia examination to be certain of identity.

Where known, the eggs are laid on the developing fruits of plants, but not all species are associated with fruits. The larvae are internal feeders, within the fruits, shoots or stems of plants, and in one species the larva makes a gall. Pupation normally occurs within the final feeding place.

Further reading
British and Irish species: Heath *et al.* (1976)
European species: Bengtsson *et al.* (2008)

Currant Shoot Borer *Lampronia capitella* page 206   133 (397)

Rare and much declined, but occasionally reported to be a pest. **FL** 7-9mm. Head yellowish. Forewing dark purplish brown with cream markings, a cross-band from the dorsum near the base, this sometimes indistinct towards the costa, and two triangular spots, one at the base, the other on the costa opposite the tornus. **The largest of all similar-looking Prodoxidae and Incurvariidae. FS** Single-brooded, late May-June. Flies on warm, sunny afternoons around the foodplant, and recorded at light. **Hab** Woodland, old gardens and has been recorded from a riverside. **Fp** Gooseberry, Red Currant, White Currant; in a green fruit before hibernation, then in one or more shoots in spring.

Lampronia luzella page 206   135 (398)

Local or very local. **FL** 5-6.5mm. Head dark greyish brown. Forewing dark purplish brown with creamy or pale yellow markings; near the base a slightly oblique cross-band narrowing towards the costa, sometimes formed as two separate spots, and two triangular spots, one at the tornus, the other on the costa opposite the tornus. **Similar species** *L. flavimitrella* (not illustrated), a rare species found in open woodland in Kent, Sussex and Hampshire, is larger (FL 6.5-7.5mm) and has a yellowish head, with two indistinct and broken cross-bands in the male, and two distinct complete cross-bands in the female; *Incurvaria praelatella* (Incurvariidae). **FS** Single-brooded, late May-July. Flies by day. **Hab** Woodland. **Fp** Not known, but possibly on Bramble.

Raspberry Moth *Lampronia corticella* page 206   136 (399)

Local. **FL** 4.5-6mm. Head pale yellowish brown or reddish brown. Forewing brown with pale yellow markings comprising three to five spots on the costa, two large dorsal spots and speckled with other small dots over the wing. **FS** Single-brooded, May-June. Flies in the afternoon and occasionally comes to light. **Hab** Woodland, scrub, gardens.
Fp Raspberry, Loganberry; in spring in expanding buds and shoots.

Lampronia morosa page 206   137 (400)

Local or very local. **FL** 5-6.5mm. Head brown. Forewing dark brown with an indistinct whitish dorsal spot before the tornus, sometimes a small costal spot before the apex, and scattered paler scales in outer half of wing. **FS** Single-brooded, May-June; also recorded in August. Flies by day. **Hab** Woodland, hedgerows, scrub, gardens. **Fp** Dog-rose, other wild and cultivated roses; in the shoots. The larva is bright red when small, distinguishing it from the greyish-brown larvae of *Epiblema* (Tortricidae), which feed in a similar way.

Lampronia fuscatella page 206   138 (407)

Local or very local. **FL** 6.5-9mm. Head yellow or orange. Forewing unicolorous dark brownish, sometimes with a faint purplish sheen. **FS** Single-brooded, May-June. Flies in the afternoon and comes to light. **Hab** Open woodland, heathland, moorland. **Fp** Birches, in young or small trees. The larva makes a spherical gall in thin twigs, usually at a node, sometimes with galls formed at adjacent nodes.

Lampronia pubicornis page 206   139 (409)

Rare. Recent records from Ireland and Scotland. **FL** 6-7mm. Head pale yellow. Forewing shining pale greyish brown. **FS** Single-brooded, May-June. Flies in the early morning, in afternoon sunshine, and rests on vegetation near the foodplant at other times.
Hab Sandhills, limestone grassland, where the foodplant grows in extensive stands.
Fp Burnet Rose, Sherard's Downy-rose; in spring in the shoots.

▼ Larva of *Lampronia morosa*.

▼▼ Larva of *Lampronia morosa* within a shoot of Dog-rose.

▲ Gall formed in a birch stem by *Lampronia fuscatella*.

▲ Gall opened up, showing larva in upper part of swelling.

Tischeriidae

There are six species in this family. Adults
rest with the front end raised and the wings
held roof-like, at a shallow angle. The forewing
length is 3-5.5mm. The forewings are uniformly
coloured or with a partial dark border, without
a distinct tornal angle. The hindwings are
narrower than the forewings, with long dorsal
cilia. The head has erect scales on top, directed

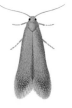

*Tischeria
ekebladella*

*Coptotriche
marginea*

forward and forming a shelf over flat scales on the face. The antennae are thread-
like, somewhat less than the length of the forewing and without a conspicuous
eye-cap. The labial palps are short and sometimes drooping, and the tongue is short.

The three yellowish species, *Tischeria ekebladella*, *T. dodonaea* and *Coptotriche
marginea*, are covered below. The three bronzy brown species are not covered.
C. heinemanni and *C. angusticollella* are scarce, and *C. gaunacella* is considered to be
extinct; all three are difficult to separate.

The larvae mine leaves of various trees and shrubs, pupating in the mine. Adults are
usually found in early summer, although *C. marginea* is typically double-brooded,
flying again later in the summer. Adults fly from dusk and come to light.

Further reading
British and Irish species: Heath *et al.* (1976)
European species: Bengtsson *et al.* (2008)

Tischeria ekebladella page 206 123 (440)

Common. More local in the far north and in Ireland. **FL** 3.5-5mm. Head and antennae
yellowish. Forewing pale brownish yellow, with the costa and apex variably speckled
darker. Hindwing grey, cilia concolorous. **Similar species** *T. dodonaea* (not
illustrated)**, a very local but widespread species on oaks and Sweet Chestnut, is
smaller (FL 3-3.5mm), with the hindwing grey and the cilia contrasting pale
golden grey.** Darker examples may also be confused with *Coptotriche marginea*, which
has a leaden-coloured head and antennae. **FS** Occasionally double-brooded in warm
summers, late April-September. **Hab** Woodland, parkland. **Fp** Oaks, Sweet Chestnut; in
a whitish blotch mine on the upperside of the leaf. The blotch mine of *T. dodonaea* is
pale brownish with darker brown concentric semicircles.

Coptotriche marginea page 206 125 (444)

Common in England and Wales. Local elsewhere. **FL** 3.5-4.5mm. Head leaden metallic on
the crown, with a yellow face and the antennae leaden metallic. Forewing brownish
yellow with the costa, the termen and occasionally the dorsum darker, and usually with a
dark spot at the tornus. **FS** Double-brooded, late April-August. May be seen at rest on the
foodplant and can be disturbed by day, and readily comes to light. **Hab** Probably wherever
the foodplants occur. **Fp** Bramble, Loganberry; in a whitish blotch mine on the upperside
of the leaf.

▲ Leaf mine of *Tischeria ekebladella* on oak. The mine is whitish or pale brown.

▲ Leaf mine of *Tischeria dodonaea* on oak. The mine is orangey brown, with concentric darker brown arcs radiating from a point along one edge.

▼ Leaf mines of *Coptotriche marginea* on Bramble. The leaf tends to contort as the larva becomes fully grown.

► Leaf mine of *Coptotriche angusticollella* on Dog-rose.

▼▼ Leaf mine of *Coptotriche marginea* on Bramble. When feeding, the larva may be visible in the early mine, or it hides within a silk-lined tunnel.

Psychidae Bagworms

There are 20 species in this family. In most species the female is wingless, whilst the male is winged and rests with the wings held roof-like, at a steep angle. A few species are found only as wingless females; their reproduction takes place without fertilisation, a process known as parthenogenesis. The forewing length is very variable, 3-14mm, the largest in the group being rather similar in shape to medium-sized macro-moths in the

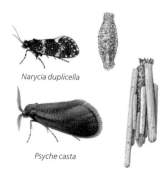

Narycia duplicella

Psyche casta

Noctuidae. The forewing of most species is not more than twice as long as wide, with a tornal angle, and is usually dull brown or grey, sometimes with a speckled pattern, and may be slightly to moderately translucent. The hindwing is as broad as the forewing. The head has erect or smooth scales and the antennae are bipectinate or thread-like, the mouthparts being much reduced. The winged adults are typically short-lived and fly by day in sunshine, often early in the morning. Only very occasionally are individuals of some species found at light.

Winged adults and larval cases of larger species are fairly easy to identify, but care is needed to distinguish the smaller species. Some species can also be separated on characters of the pupal exuviae, particularly the size and shape of the capsule of the head and appendages. The accounts below include descriptions and illustrations of adults and their larval cases. Adults of all species are rarely seen but the cases of common species, such as *Psyche casta*, are regularly encountered attached to vegetation. Cases are described from full-grown larvae; cases of immature larvae may differ in proportion and adornment.

The Psychidae can be a difficult group for a beginner; searching for cases and rearing adults is perhaps a good way to gain confidence in identifying many of the species. Some species are very local and unlikely to be encountered often, but where they do occur they can be common, and thus could be confused with more widespread species. The only species on the British list not covered below are *Canephora hirsuta*, which is doubtfully British, and *Thyridopteryx ephemeraeformis*, which may have been accidentally imported.

Larvae live within silken portable cases incorporating granular materials, adorned to a lesser or greater extent with plant and dead insect fragments. They feed on lichens, mosses, living and decaying leaves of many plants, and dead insects. The larval stage usually lasts almost one year, but in some species can last nearly two years, spread across three summers. Pupation takes place in the case. The female usually lays her eggs on or in the old larval case, although in some species they remain within the abdomen, and the cadaver forms part of the first meal for the newly emerged larvae. Young larvae may construct their first case from parts of the old larval case.

The dispersal mechanism of the larger species, *Pachythelia villosella* and *Acanthopsyche atra*, is bizarre. The female of both species is wingless, and bears no scales or appendages. It looks and behaves just like a fly maggot once free from its old larval case. It appears this behaviour is designed to attract a bird or reptile predator. Experiments have shown that larvae can hatch from eggs that have been passed out in bird faeces, suggesting that predators could be one way of helping the wingless female disperse her offspring. For most Lepidoptera, hiding from predatory birds has been an overriding evolutionary pressure, leading to the development of intricate camouflage in wing pattern and behaviour, but these two psychid species appear to have adopted quite the opposite approach!

Further reading
British and Irish species: Heath & Emmet (1985)
European species: Bengtsson *et al.* (2008)

▲ Newly emerged adult of *Narycia duplicella* with larval case and pupal exuvium.

▶ Case of *Taleporia tubulosa* fixed to a tree trunk prior to emergence of the adult.

▼ Larva and case of *Luffia ferchaultella*. Note the characteristic tilted conical shape of the case.

▼ Larva and case of *Diplodoma laichartingella*. Note the remains of insects which are commonly attached to the oral end of the case in this species.

Diplodoma laichartingella page 206
180 (747)

Local. **FL** 5-7mm, male and female winged. Head yellowish. Forewing dark brown with scattered pale yellow spots and a larger spot on the dorsum at about one-half. **FS** Single-brooded, May-June. In colder regions a generation can take up to three years. Occasionally comes to light. **Hab** Woodland, parkland. **Fp** Decaying plant matter, fungi, mosses, lichens, dead insects. **Larval case** 10-13mm, with a tough inner, triangular in cross-section, and a soft, loosely woven outer case covered with fragments of plants, insects, mineral grains and other material. The case may be found attached low down on a tree trunk or fallen log, often among spiders' webs, under bark or in hollow trees.

Narycia duplicella page 206
175 (751)

Common. More local in the north and west. **FL** 4-6mm, male and female winged. Head blackish brown. Forewing blackish brown with scattered pale yellow dots, often forming one or two ill-defined cross-bands. **FS** Single-brooded, May-July. Occasionally comes to light. **Hab** Woodland, parkland, gardens, built structures. **Fp** Probably algae. **Larval case** 5-6mm, domed in cross-section with a broad flat ventral surface, often with slight lateral flanges, and somewhat triangular at the oral end. It is covered with granular material and algae, and varies in colour depending on substrate, but is often pale greenish grey. It is found on algae and lichen-covered tree trunks, fences and walls. **Similar cases** *Dahlica inconspicuella, D. lichenella, D. triquetrella* and *Bankesia conspurcatella*, **but these are all triangular in cross-section.**

Lesser Lichen Case-bearer *Dahlica inconspicuella* page 206
177 (781)

Very local. Apparently endemic. **FL** Male 5-6mm, female wingless. Male antenna thread-like. Forewing slender, the costa slightly concave, whitish grey, variably speckled darker in a net-like pattern, with a darker dot in the middle at three-quarters and the cilia faintly chequered. **Similar species** **Male similar to** *Bankesia conspurcatella*, **which has larger blotches on the forewing and distinctly chequered cilia, and** *Luffia lapidella*, **which has the antenna with long pectinations and the forewing costa somewhat convex.** *D. inconspicuella* can be separated from *D. lichenella* (not illustrated) by rearing, as the latter species occurs only as wingless females in Britain, whereas *D. inconspicuella* has both sexes. **FS** Single-brooded, March-May. Male flies mid-morning in sunshine. **Hab** Woodland, maritime cliff and slope, coastal shingle. **Fp** Possibly *Pleurococcus* algae, lichens. **Larval case** 5-6mm, slender, triangular in cross-section, covered with fine granular material. It is usually whitish grey or greenish grey, sometimes with lichen fragments attached. The case is found on tree trunks and fences, and under scree, shingle and rubble on the ground, and is sometimes numerous. **Similar cases** *Bankesia conspurcatella*, **which is pale brownish and has minute woody fragments attached;** *D. lichenella*, **from which it is practically indistinguishable.** Cases may also be confused with *D. triquetrella* (not illustrated), a rare species recorded from a few well-scattered locations; it is a wingless parthenogenetic species in Britain (winged males occur on the Continent), found in open woodland, cliffs and quarries; the case is slightly bigger, 6-9mm, and usually adorned with plant and insect fragments at the oral end.

Lichen Case-bearer *Dahlica lichenella* page 207 (case only illustrated)
179 (762)

Very local. **FL** Parthenogenetic, female wingless; winged males occur on the Continent. **FS** Single-brooded, March-May. **Hab** Woodland, roadside trees, fences. **Fp** Possibly *Pleurococcus* algae, lichens, mosses, decaying plant matter. **Larval case** 5-7mm, triangular in cross-section and covered with fine granular materials, varying in colour depending on the materials used, but is usually pale greenish grey. The case is found attached to tree trunks, fences, rocks and stone walls, and is sometimes numerous within a very small area. **Similar cases** *D. inconspicuella, D. triquetrella, Bankesia conspurcatella*.

Taleporia tubulosa page 207 · 181 (815)

Common. Very local in the north and west. **FL** Male 7-10mm, female wingless. Male with slender wings compared with other Psychidae, the forewing greyish brown with many pale yellowish spots in a faint net-like pattern. **FS** Single-brooded, late May-June. Occasionally comes to light. **Hab** Woodland, heathland, grassland. **Fp** Lichens, decaying plant matter, dead insects. **Larval case** 14-20mm, cigar-shaped, triangular in cross-section, covered with granular material, and adorned at the oral end with fragments of plants and insects. The case is most often seen attached to tree trunks, fences and walls.

Bankesia conspurcatella page 207 · 182 (832)

Rare. Common throughout the Channel Islands. **FL** Male 6-7mm, female wingless. Forewing slender, dark greyish brown with whitish or pale grey blotches of variable size, the cilia distinctly chequered. **Similar species** *Dahlica inconspicuella, Luffia lapidella*. **FS** Single-brooded, February-March. Flies early in the morning on sunny days in late winter. **Hab** Woodland, hedgerows, mud and stone walls, gardens. **Fp** Probably leaf and insect debris. **Larval case** 6-8mm, triangular in cross-section, coated in granules of woody material, with small particles of wood along ridges, and plant and insect fragments at the oral end. The case may be found near the ground, occasionally higher in trees, and is hidden in crevices, under stones or other solid materials. Sometimes cases can be found clustered together. **Similar cases** *Dahlica inconspicuella, D. lichenella, D. triquetrella*.

Luffia lapidella page 207 · 184 (863)

Rare. Common throughout the Channel Islands. **FL** Male 5-7mm, female wingless. Male antenna with long pectinations. Forewing with the costa convex, pale brownish grey, variably speckled darker in a net-like pattern, with a darker dot in the middle at three-quarters and the cilia whitish grey, occasionally mottled darker. **Similar species** *Dahlica inconspicuella, Bankesia conspurcatella*. **FS** Single-brooded, June-July. **Hab** Coastal habitats, built structures. **Fp** Lichens. **Larval case** 6-8mm, round in cross-section, in the shape of a tilted cone, covered with fine mineral material, sometimes with flecks of lichen; it is usually greyish. **The case stands nearly at a right-angle to the substrate, distinguishing *Luffia* from other small, slender psychid cases.** The case is found in exposed places, especially on rocks and stone walls, and can be abundant. **Similar cases Very similar to *L. ferchaultella*, which is slightly smaller on average.** Where both species occur together they should be reared to help distinguish them.

Luffia ferchaultella page 207 (case only illustrated) · 185 (864)

Common. **FL** Parthenogenetic, female wingless. **FS** Single-brooded, June-July. **Hab** Woodland, urban trees, gardens, built structures. **Fp** Lichens. **Larval case** 5-7mm, round in cross-section, in the shape of a tilted cone, covered with lichen and fine mineral material, colour varying from green to black, sometimes with rings of different colours. The case is found on tree trunks, fences, rocks and walls, preferring shady places, and can be abundant. **Similar cases** *L. lapidella*.

Bacotia claustrella page 207 · 183 (866)

Very local. **FL** Male 6.5-7.5mm, female wingless. Male antenna with short pectinations without plate-like scales. Forewing slender, apex acute, brown with a darker spot at about two-thirds, scales plate-like. **Similar species** *Proutia betulina* has the antenna with long pectinations without plate-like scales and the forewing rounded only at the apex; *Psyche casta* and *P. crassiorella* (not illustrated) both have the antenna with pectinations with plate-like scales, and the forewing entirely rounded, with the hindwing dark brown in *P. casta* and greyish brown in *P. crassiorella*; and *Epichnopterix plumella* has the antenna with long pectinations without plate-like scales and forewing scales hair-like. **FS** Single-brooded, June-July. **Hab** Woodland, hedgerows, scrub. **Fp** Lichens. **Larval case** 6-7mm, stout, round in cross-section, sometimes adorned with lichen fragments; carried at a right-angle to the bark on which it sits, looking remarkably like a stubby twig. The case may be found on tree trunks, branches, posts or among thick scrub.

Proutia betulina page 207   188 (868)

Very local. **FL** Male 5.5-7mm, female wingless. Male antenna with long pectinations without plate-like scales. Forewing slender, rounded at the apex, dark brown, scales plate-like. **Similar species** *Bacotia claustrella, Psyche casta, P. crassiorella, Epichnopterix plumella*. **FS** Single-brooded, late May-July. **Hab** Woodland. **Fp** Lichens, living and decaying leaves. **Larval case** 8-10mm, somewhat pointed and covered in fragments of bark, pine needles, grass and lichen. The case may be found on tree trunks, branches and posts.

Psyche casta page 207   186 (877)

Common. **FL** Male 6-7mm, female wingless. Male antenna pectinate, the pectinations with plate-like scales. All wings broad and rounded, uniformly dark brown, scales plate-like. **Similar species** *Bacotia claustrella, Proutia betulina, Psyche crassiorella* (not illustrated), *Epichnopterix plumella*. *P. crassiorella* is a very rare species, slightly larger (FL 6.5-8.5mm) than *P. casta* with the hindwing greyish brown. **FS** Single-brooded, May-July. **Hab** Woodland, moorland, heathland, grassland, coastal habitats. **Fp** Grasses, lichens, decaying plant matter. **Larval case** 6-10mm, with blades of grass or rush attached longitudinally, often splayed out beyond the anal end. The case is usually found exposed on tree trunks, tall vegetation or posts. **Similar cases** *Epichnopterix plumella* has flat blades of grass attached longitudinally, these converging slightly beyond the anal end; and *Psyche crassiorella*, case 8-12mm, is constructed of coarser and longer blades.

Epichnopterix plumella page 207   189 (926)

Very local. **FL** Male 5-6mm, female wingless. Male antenna with long pectinations without plate-like scales. All wings broad and rounded, dull blackish brown, scales hair-like and wings somewhat translucent. **Similar species** *Bacotia claustrella, Psyche casta, P. crassiorella, Proutia betulina*. **FS** Single-brooded, April-June. Flies in sunshine around midday. **Hab** Grasslands, moorland, parkland, scrub. **Fp** Grasses. **Larval case** 8-12mm, with blades of grass attached longitudinally, converging beyond the anal end. The case is attached low down and is rarely seen. **Similar cases** *Psyche casta, P. crassiorella*.

Whittleia retiella page 207   190 (937)

Very local. **FL** Male 3-4.5mm, female wingless. All wings rounded, whitish with dark grey markings in a net-like pattern. **FS** Single-brooded, May-June. Flies in the afternoon on warm, still days and is occasionally numerous. **Hab** Saltmarsh. **Fp** Common saltmarsh-grass, cord-grasses, other grasses. **Larval case** 8-12mm, slender, with longitudinally attached grass fragments extending beyond the case. The case can be found among saltmarsh vegetation.

Acanthopsyche atra page 207   191 (954)

Rare. **FL** Male 8-10mm, female wingless. All wings broad, dull blackish grey, and thinly scaled. **FS** Single-brooded, May-June. Larval development takes nearly two years, over three summers. Males fly in sunshine from midday onwards. **Hab** Heathland, moorland. **Fp** Heather, heaths, grasses, sallows. **Larval case** 17-22mm, cylindrical, covered longitudinally with twigs and leaves of heather, and grass stems. The case is usually attached low down on vegetation or posts, and is rarely seen. **Similar cases** *Pachythelia villosella* **is bigger, with larger vegetation fragments attached.**

Pachythelia villosella page 207 page 207 192 (963)

Rare. **FL** Male 10-14mm, female wingless. All wings dull greyish brown, somewhat translucent. **FS** Single-brooded, June-early August. Larval development takes nearly two years, over three summers. Male flies in afternoon and evening sunshine. **Hab** Lowland heathland. **Fp** Heather, heaths, grasses, Bramble, Dwarf Gorse. **Larval case** 30-50mm, cylindrical, covered longitudinally with stems of heather, gorse, grass and rush. The case can be found attached to tree trunks, posts, and high up on heather bushes, and occasionally occurs in large numbers very locally. **Similar cases** *Acanthopsyche atra*.

Sterrhopterix fusca page 207 195 (1012)

Rare. Few recent records. **FL** Male 9-12mm, female wingless. All wings broad and rounded, grey and thinly scaled. **FS** Single-brooded, June-July. Larval development takes two years. Males are attracted to light. **Hab** Woodland, heathland, mosses. **Fp** Grasses, oaks, Hawthorn, sallows, birches, heather, heaths. **Larval case** 16-20mm, somewhat conical, irregularly decorated with plant matter, sometimes placed transversely. The case can be found attached to heather, tree trunks or leaves.

▲ Newly hatched larvae of *Psyche casta* constructing their first cases from the old case of their mother.

▼ Female case of *Acanthopsyche atra* fixed to a heather stem prior to emergence of the adult. Note there is no silk extension to the anal end of the case, indicating it is a female.

▼ Male case of *Pachythelia villosella* fixed to a gorse needle prior to emergence of the adult. Note the silk extension to the anal end of the case indicating it is a male.

▲ Adult female of *Acanthopsyche atra*. Note the female has no appendages or scales and is maggot-like.

▲▲ Newly emerged adult male of *Acanthopsyche atra*. Note the thinly scaled wings.

▲ Larval case of *Proutia betulina*, with the fragments of bark and needle adorning the case.

Tineidae

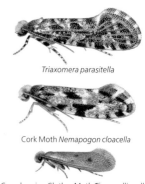

Triaxomera parasitella

There are 63 species in this family, which includes the clothes moths and several other species associated with human activities and habitation. At rest, the wings are held roof-like, at a steep angle. The forewing length is very variable, 3-16mm. Although the forewing membrane is usually without a tornal angle, the cilia at the tornus are long, producing a noticeable angle between the dorsum and termen. The forewings

Cork Moth *Nemapogon cloacella*

Case-bearing Clothes Moth *Tinea pellionella*

are frequently dull brown, although some are distinctively patterned. The hindwings are as wide as the forewings. The head is rough-scaled, although in *Psychoides* it is smooth-scaled and in *Oinophila* there is a distinct ridge between the antennae. The antennae are mostly thread-like, about three-quarters the length of the forewing. The labial palps vary, from drooping to forward-pointing or ascending, and the tongue is reduced.

Treatment of Tineidae in this guide is partial. Many species of this family are difficult to record as adults, where there may be confusion between similar-looking species (especially genera *Nemapogon* and *Tinea*). However, several species are reasonably distinctive and can be identified with care, and the most widespread species are likely to be encountered in moth traps. Twenty-three species accounts are given below.

Some species are highly restricted in distribution or are hard to find. For example, *Eudarcia richardsoni* is known from only two sites in the world, both on the Dorset coast, and is thus believed to be endemic to Britain. *Nemapogon inconditella* is known from a single example found in Devon, in 1979; it is probably resident there, but has eluded subsequent detection, despite many attempts to find it. Some previously widespread species appear to have become exceedingly rare; even into the mid-20th century Tapestry Moth *Trichophaga tapetzella* was abundant and a pest of fur and feathers, but it has hardly been seen in the past 50 years. A number of species are imported from time to time with foodstuffs, animal skins and woven natural fibres; whilst most seem not to take up residence in Britain, this remains a possibility.

Larvae feed on ferns, lichens, fungi, animal detritus such as feathers, skin and fur, and on vegetable matter. A few species feed from within portable cases. One species is associated with ants' nests.

▲ Larva of *Triaxomera parasitella*, dorsal and lateral views.

Collecting and rearing larvae is a good way of detecting a number of species, especially of the fungus-feeding group and those associated with birds' nests and owl pellets. Discarded woollen blankets in outhouses can reveal a surprising diversity of species.

Species living in warehouses or human habitations are often continually brooded, while those outside typically have fairly distinct broods and are often single-brooded. Adults of some species fly naturally at dawn, during the day, or at dusk, and several come readily to light. When disturbed, some adult moths in this family tend to run rather than fly.

Further reading
British and Irish species: Heath & Emmet (1985)
European species: Bengtsson *et al.* (2008)

▲ Larval case of the British endemic *Eudarcia richardsoni* on algae on Portland stone.

▶ Larval tube of *Infurcitinea argentimaculella* on the powdery lichen *Lepraria.*

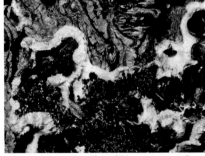

▲ Pupal exuvium and larval frass of *Triaxomera parasitella* on fungi on Silver Birch.

▲ Larva of *Morophaga choragella* exposed from within a bracket fungus.

▲ Pupal exuvium and larval frass of *Nemapogon clematella* on the hard fruiting bodies of the fungus *Diatrype disciformis* on Hazel.

▲ Loosely woven larval cases of *Psychoides verhuella* on the underside of a frond of Hart's-tongue.

Infurcitinea argentimaculella page 208 203 (572)

Local. **FL** 3.5-4mm. Forewing dark brown with a coppery gloss. Markings are variable, silvery white, with an oblique cross-band before one-half, a spot at the tornus, and spots or short streaks on the costa and towards the apex; the cilia are tipped white. **FS** Single-brooded, June-July. Flies in sunshine. **Hab** Woodland, parkland, gardens, built structures such as bridges. **Fp** *Lepraria incana*, *L. aeruginosa*, possibly other lichens, on shady and damp rocks, brickwork, old tree trunks; feeds from within thin silken tubes covered with lichen fragments. This species is most frequently detected from the larval workings.

Morophaga choragella page 208 196 (609)

Local. May be spreading northwards. **FL** 10-16mm. Head dirty whitish. Forewing sandy brown, prominent along the veins, scattered with darker brown markings; the costa has a distinct dark brown blotch near one-quarter, the middle cross-band is oblique from the dorsum at one-half to the costa at two-thirds, sometimes incomplete or obscure towards the costa, and the termen is chequered. **FS** Single-brooded, late May-August. Comes to light on warm nights, occasionally found at rest by day on tree trunks. **Hab** Woodland. **Fp** Bracket fungi, including *Piptoporus betulinus*, *Ganoderma adspersum*; possibly in rotten wood in galleries within the fungus. Frass hanging from silk threads beneath the bracket may indicate this species.

Triaxomera parasitella page 208 224 (617)

Common in the south. Very local elsewhere. **FL** 7-10.5mm. Head yellowish. Forewing brown, speckled with small whitish and darker brown spots, and orange-brown along the veins; strongly marked examples have an 'N'-shaped mark on the left wing (mirror-image on the right wing); the cilia are chequered. **FS** Single-brooded, May-July. Flies at sunrise, dusk and comes to light. **Hab** Woodland, parkland. **Fp** Bracket fungi, including *Coriolus versicolor*, also in dead wood; within the fungus. Frass-covered patches on the surface of the fungus may indicate this species or other fungus-feeding Tineidae.

Triaxomera fulvimitrella page 208 225 (613)

Local. **FL** 6-10mm. Head pale yellow. Forewing blackish brown with up to four large white spots, two on the costa and two on the dorsum, and occasionally smaller spots elsewhere; the cilia are chequered. **FS** Single-brooded, May-July. Can be found at rest on tree trunks, flies in the morning and comes to light. **Hab** Woodland. **Fp** Bracket fungi, including *Inonotus radiatus* and *Piptoporus betulinus*, perhaps preferring fungi on oaks and Beech in southern England, also in dead wood and callus-tissue around tree wounds; within the fungus. See comment under *T. parasitella*.

Archinemapogon yildizae page 208 222 (619)

Rare. **FL** 7-10mm. Forewing whitish to pale grey, darker markings variable, usually consisting of a pair of long black streaks from the base and an oblique streak from the costa at one-half, curving towards the apex. **FS** Single-brooded, May-July. Flies around birch at sunset, is active after dark and sometimes found at rest by day; occasionally abundant as larvae. **Hab** Old birch woodland. **Fp** Bracket fungi, especially *Fomes fomentarius* and *Piptoporus betulinus*, also *Inonotus radiatus*; within the fungus.

Nemaxera betulinella page 208 223 (621)

Local or very local. **FL** 5-9mm. Head whitish. Forewing whitish to pale grey, with brown spots and a characteristic orange-brown V-shaped mark in the middle of the wing and orange-brown mottling towards the termen. **FS** Possibly double-brooded, May-August. May be found at rest on trees and comes to light. **Hab** Woodland. **Fp** Bracket fungi, especially *Piptoporus betulinus* and *Coriolus versicolor*, occasionally in dead wood; within the fungus. See comment under *T. parasitella*.

Cork Moth *Nemapogon cloacella* page 208 216 (624)

Common. **FL** 5-8mm. Head yellowish orange. Forewing usually mixed pale brown, reddish brown and whitish, with a distinct white spot in the middle above the tornus and several dark brown markings, including an oblique cross-band at one-half interrupted about the middle and on the dorsum. Ground colour can vary between pale orange-brown and blackish brown, and the markings can vary in strength and extent, although the white spot is almost always present. **Similar species** *N. cloacella* is the most frequently encountered *Nemapogon* but can be confused with several local or rare species (none illustrated): Corn Moth *N. granella* (ground colour uniformly pale brown, with grey and black markings and no white spot); *N. wolffiella* (ground colour blackish brown, small whitish markings); *N. variatella* (head white); *N. ruricolella* (ground colour plain yellowish grey tinged brownish beyond the cross-band, spot small, whitish or yellowish brown). If these other species are suspected, refer to Heath & Emmet (1985) for more details on diagnosis. **FS** Probably double-brooded in the south, May-September. Flies in early morning sunshine, in late afternoon and at dusk, and is attracted to light. Adults may be found at rest on tree trunks or on the host fungus. **Hab** Woodland, parkland, gardens, occasionally indoors. **Fp** Bracket fungi, including *Piptoporus betulinus*, also in callus-tissue around wounds on birches; within the fungus, producing frass and slight silk webbing at the surface.

Nemapogon clematella page 208 220 (641)

Local. **FL** 6-8mm. Forewing white with faint yellowish-brown shading sometimes present; markings are variable in extent, dark brown, the cross-band before one-half occasionally interrupted, a basal streak along the costa, and scattered markings towards the apex and termen. **FS** Single-brooded, April-August. Flies at dawn, can be disturbed from hedgerows by day, and is attracted to light. **Hab** Woodland, hedgerows. **Fp** Fungi, including *Diatrype disciformis* and *Hypoxylon fuscum* on dead stems, and bracket fungi; within the fungus, leaving distinctive trails of dark- and light-coloured frass on the surface.

Nemapogon picarella page 208 221 (643)

Rare. Recently only from Scotland. **FL** 6-9.5mm. Head and thorax mainly white. Forewing white, markings blackish with a coppery gloss, including a broad streak from the costa at the base to the termen, more or less expanded into a cross-band at about one-half, with a short basal streak on the dorsum. **FS** Single-brooded, June-July. Occasionally flies by day and is attracted to light. **Hab** Old woodland. **Fp** Bracket fungi, especially *Piptoporus betulinus*, and in excrescences of oak; within the fungus.

Tapestry Moth *Trichophaga tapetzella* page 208 234 (661)

Rare. Formerly widespread and locally abundant; hardly seen in the past 50 years. **FL** 6.5-11mm. Forewing with the basal third dark brown with a purplish tinge, and the outer two-thirds white with scattered pale grey and yellowish-brown markings. **FS** Single-brooded, May-August. Has been recorded at light. **Hab** Unheated buildings, such as stables and outhouses, gardens, woodlands. **Fp** Animal materials, such as fur, hair, feathers, wool, birds' nests, owl pellets, wasps' nests; feeds from within a silken tube.

Common Clothes Moth *Tineola bisselliella* page 208 236 (669)

Common. **FL** 5-7mm. Head yellow or reddish yellow. Forewing pale golden, sometimes darker at the base, slightly glossy. **FS** Continuously brooded, February-September. Active in late afternoon, prefers to run rapidly rather than fly when disturbed. Infestations occur occasionally throughout the range. **Hab** Indoors, rarely found outdoors. **Fp** Animal materials, including woollen fabrics, feathers, fur, hair, infrequently outdoors in birds' nests and wasps' nests; feeds from within a flimsy silken tube and several larvae may be found together covered by silk webbing.

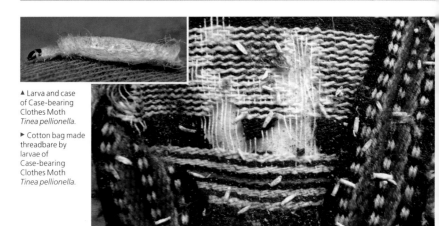

▲ Larva and case
of Case-bearing
Clothes Moth
Tinea pellionella.

▶ Cotton bag made
threadbare by
larvae of
Case-bearing
Clothes Moth
Tinea pellionella.

Case-bearing Clothes Moth *Tinea pellionella* page 208   240 (671)

Common. **FL** 4-8mm. Head orangey or reddish brown. Forewing pale greyish brown with
one to three dark brown spots, the spot at three-quarters usually the most pronounced.
Similar species *T. pellionella* is by far the most frequently encountered of the *Tinea/
Niditinea/Haplotinea* group, but it can be confused with several other local or rare resident
species (none illustrated): *T. columbariella* (ground colour uniformly pale greyish brown,
occasionally with a small spot at three-quarters); *T. dubiella* (ground colour darker greyish
brown, sprinkled with orange-brown scales); *T. flavescentella* (paler species with ground
colour whitish mixed yellow-brown); Brown-dotted Clothes Moth *N. fuscella/N.
striolella/H. insectella* (ground colour pale orange-brown, mottled with darker brown
scales). If these other species are suspected, refer to Heath & Emmet (1985) for more
details on diagnosis. **FS** Single-brooded in outbuildings, two or more generations where
heated, April-October. Flies late afternoon and occasionally comes to light. Infestations
can occur. **Hab** Indoors and in outbuildings, occasionally outdoors. **Fp** Wool, fur, hair,
feathers, sometimes stored vegetable products, occasionally in birds' nests and owl
pellets; lives in a portable case.

Large Pale Clothes Moth *Tinea pallescentella* page 208   245 (680)

Local. **FL** 6-12.5mm. Head mixed pale yellowish brown and brown; thorax usually with
dark brown tegulae. Forewing markings variable, pale greyish brown mottled with brown
scales, less so along the dorsum, and with dark brown marks at the base, forming a
sub-dorsal streak at about one-third, and usually two or three diffuse dots or marks
beyond; the apex is somewhat chequered and has a dark brown line at the base of the cilia
along the termen. **The sub-dorsal streak distinguishes small, less-well marked
forms of this species from all other resident *Tinea*. FS** Probably single-brooded, but
adults can be found at any time of year. Flies late afternoon and comes to light.
Hab Sheltered places, including outbuildings, coastal caves, tunnels. **Fp** Wool, fur, hair,
wasps' nests; makes a case in which to moult between instars.

Tinea semifulvella page 208   246 (686)

Common. **FL** 6-10mm. Head reddish orange. Forewing greyish yellow with at least the
apical quarter orange-brown, this colour extending over the rest of the wing in some
examples; a dark brown tornal spot. **FS** Probably double-brooded in the south,
May-October. Comes to light. **Hab** Woodland, hedgerows, heathland, farmland,
grassland, gardens. **Fp** Birds' nests, occasionally animal carcasses and woollen material in
the open.

Tinea trinotella page 208 247 (687)

Common. **FL** 6-8mm. Head yellowish. Forewing yellowish grey with three black spots and a fine blackish-brown stripe along the costa from the base to beyond one-third.
FS Probably double-brooded in the south, late April-September. Comes to light.
Hab Woodland, hedgerows, gardens, stables. **Fp** Birds' nests, also woollen material in the open.

Skin Moth *Monopis laevigella* page 209 227 (700)

Common. **FL** 5-10mm. Head yellowish. Forewing dark brown with a violet or coppery sheen, speckled with whitish-grey scales; a small dirty whitish translucent spot before the middle, and sometimes an obscure pale tornal spot. **Similar species *M. weaverella* has a large, usually triangular, pale yellowish spot at the tornus, the translucent spot slightly nearer the middle of the wing, and much less pale speckling. FS** Possibly double-brooded in the south, mid March-September. Flies at dusk and comes to light.
Hab Woodland, parkland, gardens, unheated buildings. **Fp** Foodstuffs of animal origin, including birds' nests, owl pellets, dry animal carcasses, woven woollen material, occasionally stored animal products, such as bird guano; feeds from within a silken tunnel.

Monopis weaverella page 209 228 (701)

Common. **FL** 5-10mm. Head yellowish. Forewing blackish brown with dark bluish or violet speckling; a large, usually triangular, pale yellowish spot at the tornus, and a whitish translucent spot almost in the middle of the wing, sometimes with whitish speckling around the spot. **Similar species** *M. laevigella*. **FS** Possibly double-brooded in the south, April-October. Flies at dusk and comes to light. **Hab** Woodland, parkland, gardens, grasslands, buildings. **Fp** Foodstuffs of animal origin, including owl pellets, faeces, dry animal carcasses.

Monopis obviella page 209 229 (704)

Common in the south. Very local elsewhere. **FL** 5-6mm. Head pale golden, thorax with a central pale yellowish band. Forewing dark brown or blackish brown with a whitish translucent spot in the middle of the wing, a speckling of whitish scales below the costa especially towards the apex, and a pale yellowish band along the dorsum. Hindwing dark grey-brown with a purplish sheen. **Similar species *M. crocicapitella* is generally slightly larger, has the forewing slightly paler with the whitish speckling much more extensive, and the hindwing pale grey; also the yellowish form of *M. imella*, which has a narrow yellowish stripe along the costa and lacks a yellowish stripe on the thorax. FS** Probably double-brooded, May-October. Flies at dusk and comes to light, and can be locally abundant. **Hab** Gardens, scrub, waste ground, rank grassland, coastal scrub, limestone cliffs, unheated buildings. **Fp** Refuse of plant or animal origin.

Monopis crocicapitella page 209 230 (705)

Local. **FL** 5-8mm. Head pale golden, thorax with a central pale yellowish band. Forewing brown or dark brown speckled with dirty whitish or pale yellowish scales, particularly towards the costa and in the outer half of the wing, with a pale yellowish or yellowish-brown dorsal streak, and a whitish translucent spot in the middle of the wing. Hindwing pale grey, darker towards the apex. **Similar species** *M. obviella, M. imella*. **FS** Probably double-brooded, May-October. Flies at dusk and comes to light, and can be locally abundant. **Hab** Coastal scrub, cliffs, gardens, unheated buildings; more frequent in maritime habitats than *M. obviella*. **Fp** Refuse of plant or animal origin, including birds' nests.

Monopis imella page 209 231 (707)

Very local. **FL** 5-7mm. Head whitish or yellowish. Forewing dark brown with a very narrow yellowish stripe along the leading edge of the costa, and a small dirty whitish translucent spot in the middle of the wing. **Similar species** An uncommon form of *M. imella* has a yellowish streak along the dorsum, and is similar to *M. obviella* and *M. crocicapitella*. **FS** Probably double-brooded, mid April-early October. Active from dusk and comes to light. **Hab** Vegetated shingle, coastal cliffs and slopes, coastal scrub; most frequent in maritime habitats. **Fp** Woollen refuse on the ground in the open, dead animal remains, birds' nests.

Monopis monachella page 209   232 (708)

Rare. Occasionally an immigrant. **FL** 6-10mm. Forewing blackish brown with a broad white blotch on the costa. **FS** Single-brooded, May-September. Comes to light. **Hab** Coastal habitats, including sand dunes and reedbeds; formerly found in fenland. **Fp** Probably animal remains.

Psychoides verhuella (not illustrated)   199 (728)

Local. **FL** 4.5-6mm. Forewing dark greyish brown with a faint purple gloss. **Similar species** *P. filicivora* has a small triangular white spot at the tornus. **FS** Single-brooded, June-July. Flies in sunshine in the early morning and late afternoon, and occasionally seen at light. **Hab** Woodland, hedgerows, gardens. **Fp** Hart's-tongue, sometimes other ferns; at first mining the frond, then feeding within sori on sporangia and later spinning empty sporangia into a loosely woven portable case in which the larva lives and moves around.

Psychoides filicivora page 209   200 (729)

Local, but spreading. This species may have originated in the Far East and has become established in Britain and Ireland. **FL** 4-6mm. Forewing dark greyish brown with a faint purple gloss and a small triangular white spot at the tornus. **Similar species** *P. verhuella* (not illustrated). **FS** In overlapping broods, April-October. Active by day, sometimes seen in large numbers around foodplants, and occasionally comes to light. **Hab** Woodland, gardens. **Fp** Hart's-tongue, Soft Shield-fern, Male-fern, spleenworts, other ferns; at first mining the frond, then feeding on sporangia and the underside of the frond, spinning sporangia into an irregular mass in which the larva hides.

▶ Spun fern sporangia concealing the larva of *Psychoides filicivora*.

Roeslerstammiidae

There are two species in this family, but *Roeslerstammia pronubella* is known only from two 19th-century specimens and may be an accidental import. The forewing length is 5.5-6.5mm. Adults rest with the wings held roof-like, at a steep angle. The forewings are elongate, with a tornal angle, and are metallic bronze. The hindwings are slightly broader than the forewings. The head has erect scales, the face being smooth-scaled. The antennae are thread-like, almost as long as the forewing, dark with a white section before the apex. The labial palps are long and curved upwards.

Roeslerstammia erxlebella

The larva begins as a leaf miner and then continues to feed externally on leaf tissues. The adult flies at night and comes to light.

Further reading
British and Irish species: Emmet (1996)
European species: Bengtsson & Johansson (2011)

Roeslerstammia erxlebella page 209   447 (1030)

Local. **FL** 5.5-6.5mm. Head yellow on the crown, labial palps white and antenna dark brown with a white band towards the apex. Forewing shining dark coppery bronze. Hindwing dark brownish bronze. **Similar species R. pronubella (not illustrated) has the hindwing yellowish with the apex and borders darker. FS** Double-brooded at least as far north as Lancashire, single-brooded in the far north, late April-September. Comes to light. **Hab** Woodland. **Fp** Limes, especially Small-leaved Lime, birches.

▼ Larva of *Roeslerstammia erxlebella*.

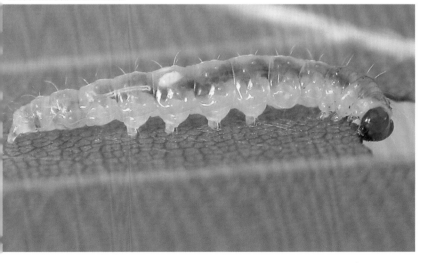

Bucculatricidae

Bucculatrix thoracella

There are 14 species in this family. The resting position of the adult varies between species, either slightly inclining or lying flat, and the wings are held roof-like, at a steep angle, or somewhat rolled around the abdomen. All species are small, with forewing length

Bucculatrix ulmella

3-4.5mm. Although the forewing membrane is without a tornal angle, the cilia at the tornus produce a noticeable angle between the dorsum and termen. The forewings are variably patterned and a few species have raised scale-tufts. The hindwings are narrower than the forewings, with very long dorsal cilia. The head typically has erect scales forming a distinct tuft, and the face is smooth. The antennae are thread-like, about two-thirds the length of the forewing, with the scape expanded to form an eye-cap. The labial palps are minute, and the tongue is short and without scales.

Several species are widely distributed, although a few are much more local. For example, *Bucculatrix maritima* is found only on saltmarshes and is often numerous where it occurs, while *B. humiliella* is found only in river valleys in the Scottish Highlands. *B. ulmifoliae* appears to have arrived recently, having been added to the British list in 2006; it is breeding on disease-resistant species of elm introduced following the decline of native elms in the 1970s from Dutch Elm Disease. Only two species are described below, both reasonably recognisable; other species are rather more difficult to identify.

Larvae begin feeding as leaf miners, later becoming external feeders. The larvae of species on broadleaved plants then eat many small windows in the underside of the leaf, the upper leaf surface remaining intact, sometimes giving the leaf a finely fenestrated appearance when viewed from above. Between instars, the larva makes a characteristic round or oval white cocoon within which moulting takes place; the presence of moulting cocoons on the host plant can help identification. The pupal cocoon is constructed almost entirely from the outside, usually in two halves, with the larva crawling inside to complete the task only when it is nearly finished. Most species

▼ Leaf mine of *Bucculatrix nigricomella* on Ox-eye Daisy.

▼ Early instar mine and larva of *Bucculatrix frangutella*.

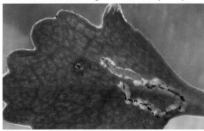

produce a characteristic ribbed, spindle-shaped cocoon. *Bucculatrix* feed on a range of herbaceous plants, shrubs and trees. Most species are single-brooded, a few are double-brooded. Adults often fly in evening sunshine, and are sometimes seen at light.

Further reading
British and Irish species: Heath & Emmet (1985)   European species: Bengtsson & Johansson (2011)

Bucculatrix thoracella page 209 — 273 (1093)

Local. Expanding its range and recently recorded in Scotland. **FL** 3-4mm. Head yellowish. Forewing yellow with a brownish cross-band near the base, a brown blotch on the costa and another on the dorsum, and a fine streak through the centre of the wing from near the base to the apex. **FS** Mainly single-brooded with a partial second brood in warm years, late April-late August. Rests by day under leaves and on tree trunks, flies in evening sunshine and occasionally comes to light. Can be very common, especially on urban trees. **Hab** Ancient woodlands, preferring open rides and woodland margins, gardens, roadside trees, parkland. **Fp** Limes, including Small-leaved Lime; at first in a short mine, later on the underside of a leaf. Fenestration by many larvae may cause leaves to turn brown, and beneath the canopy tiny green larvae may be encountered in abundance, descending on silk threads to pupate.

Bucculatrix ulmella page 209 — 274 (1094)

Common. Local in the north. **FL** 3.5-4mm. Head yellowish brown, darker in the middle, face white. Forewing yellowish white, mottled greyish brown, mottling heaviest in four patches along the costa and one on the dorsum at one-half, the last usually edged anteriorly with raised black scales. **Similar species** *B. demaryella* (not illustrated), an open-woodland and moorland species, local in the south, more common in the north and west, is slightly larger (FL 4-4.5mm), with brown mottling in three patches on the costa and one on the dorsum. **FS** Double-brooded, late April-June, late July-August. Rests by day on tree trunks and comes to light. **Hab** Woodland, scrub, roadside trees, parkland. **Fp** Oaks; at first in a mine, later on the underside of a leaf.

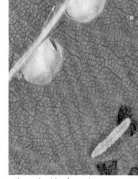

▼ Ribbed cocoon of *Bucculatrix thoracella* on the underside of a lime leaf.

► Larva and two moulting cocoons of *Bucculatrix thoracella* on the underside of a lime leaf.

▼ Larvae of *Bucculatrix ulmella* on the underside of an oak leaf. Note the characteristic feeding pattern, with the upper epidermis of the leaf left intact.

▼ Early-instar leaf mines of *Bucculatrix thoracella* on lime.

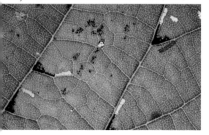

Gracillariidae

Caloptilia alchimiella

There are 95 species in this family. At rest, the wings are held roof-like, at a steep angle. All species are small or very small, with forewing length 2-8.5mm. This family includes the now well-known Horse-chestnut Leaf Miner *Cameraria ohridella*, which has colonised mainland Britain in recent years, spread rapidly, and now occurs in spectacular abundance over much of the country.

Callisto denticulella

Adults rest in an inclining posture at about 40° in *Caloptilia* and *Parornix*, raised up on extended forelegs and midlegs. The resting position is slightly raised in *Phyllonorycter*, and close to horizontal in *Phyllocnistis*. The forewings are narrow, elongate or very elongate, the membrane without a distinct tornus, although the

Firethorn Leaf Miner
Phyllonorycter leucographella

long cilia at the tornus produce a noticeable angle between the dorsum and termen. Forewing markings vary from patterned and colourful, sometimes with metallic reflections, to relatively plain. The hindwings are elongate, narrower than the forewings, with long dorsal cilia. In most genera the head has smooth scales, but in *Callisto* and *Parornix* they are erect. The antennae are thread-like, nearly as long or slightly longer than the forewing. The labial palps are slender, moderate or long, and forward-pointing or ascending; in *Caloptilia* species, segment 2 of the palp is smooth beneath, except in *C. cuculipennella* and in *Povolnya leucapennella*, where it is tufted. The tongue is not scaled. The legs in some genera are long, with bristles. In *Caloptilia* and closely-related genera the foreleg and midleg have tibiae thickened with scales, except in *Parectopa* where the tibiae are smooth-scaled; the tibial scales match the forewing colour, with the tarsi and the underside of the abdomen often whitish.

Treatment of Gracillariidae in this guide is partial and only 26 species are described, concentrating on the attractive group of the *Caloptilia* and their close relatives. *Caloptilia* can be difficult to separate. Some species have very similar wing markings, and there may be variation within a species, even between summer and autumn generations. Two commonly encountered genera, *Parornix* and *Phyllonorycter*, are not comprehensively covered as species are difficult to identify from superficial wing characters. Adults of *Parornix* are mostly mottled greyish; those of *Phyllonorycter* are well marked and beautiful, but they are very small and there are over 50 species. In both genera, some species require examination of genitalia to separate. These genera are much more easily studied by searching for the leaf mines; many species are widely distributed and straightforward to identify, based on the host plant, the shape and size of the leaf mine or leaf fold made by the larva, and time of year of appearance. Photographs of many of these larval workings and key characteristics are shown on the British Leafminers website. Rearing adults from tenanted larval workings is fairly easy for most species, and will help confirm identity.

Several species appear to be expanding their range in Britain. Since the 1970s, *Caloptilia rufipennella* has been spreading steadily and it is present in almost every county in mainland Britain, and probably continues to spread in Ireland. In the past decade, *C. semifascia* and *C. falconipennella* have become much more widespread in southern England, and the same may now be happening to *C. populetorum*. Moreover, these three species were previously local or rare, and mainly single-brooded, in the south. It appears that double-brooded populations, possibly of Continental origin, are now establishing and may become much more widespread and numerous. It may be that the same will also happen with *C. hemidactylella*, currently an exceptionally rare species in Britain, as this species has recently become unexpectedly widespread in The Netherlands. An account of *C. hemidactylella* is included below.

The larvae of all Gracillariidae start as leaf miners. In *Caloptilia* and *Parornix*, as the larva grows it quits mining to feed externally, usually on the underside of the leaf, concealing itself within successive curls or folds of a leaf, and in most species the larva pupates outside the folded leaf. In *Phyllonorycter*, the larva continues as a leaf miner, making a characteristic blister that may distort the leaf, and pupates within the mine. Most species feed on trees and shrubs; a few are found on herbaceous plants. Depending on species, adults may be active by day, fly in the evening and at dusk, and most come to light. Several species, particularly in *Caloptilia*, overwinter as an adult.

Further reading
British and Irish species: Heath & Emmet (1985)
European species: Bengtsson & Johansson (2011)
British Leafminers: http://www.leafmines.co.uk/

Parectopa ononidis page 210   299 (1100)

Local. **FL** 3.5-4.5mm. Head smooth-scaled and whitish. Forewing dark brown with shining silver costal and dorsal streaks, these edged black, at about one-quarter, one-half, three-quarters and just before the apex, and also a silver dot in the apical area. **Similar species** *Callisto denticulella* **is larger, with erect, orange-brown scales on the head. FS** Double-brooded, possibly more broods in favourable seasons, May-June, August. Rarely recorded as an adult, more often and sometimes commonly recorded as a leaf mine. **Hab** Coastal habitats, chalk downland, neutral grassland, waste ground. **Fp** Red Clover, White Clover, Strawberry Clover, possibly Common Restharrow; in a pale brown mine visible from the upperside of the leaf.

◂ Leaf mine of *Parectopa ononidis* on Red Clover.

Caloptilia cuculipennella page 210 280 (1115)

Local. Rarely found in abundance, but most frequent along the south coast. **FL** 6-7mm. Segment 2 of the labial palp somewhat tufted beneath. Forewing ground colour whitish, with brownish markings usually visible as three incomplete oblique cross-bands from the costa to the dorsum; the cilia are distinctly banded light and dark. **Similar species** *C. populetorum* **has a distinct dark spot on the costa and has a smooth underside to segment 2 of the labial palp; also pale forms of** *Povolnya leucapennella*, **which lack banding in the cilia and are usually bigger.** **FS** Single-brooded, overwinters as an adult, September-May; on the south coast adults appear in July, suggesting two generations. Flies in late afternoon and comes to light. **Hab** Cliffs and scrub on the coast, and woodland inland. **Fp** Wild Privet, Ash; constructs small, neat cones on the leaves (see also *Gracillaria syringella*).

Caloptilia populetorum page 210 281 (1125)

Local. Possibly expanding its range in the south. **FL** 6-7.5mm. Forewing variable, from whitish through yellowish brown to dark brown; paler examples have a distinctive blackish spot on the costa beyond one-half, two more near the dorsum at one-third and two-thirds, and are dark along the termen. **Similar species** *C. cuculipennella*. **FS** Single-brooded, but a double-brooded population may now be spreading in southern England; overwinters as an adult, June-May. Flies in late afternoon and comes to light. **Hab** Heathland, moorland, open woodland, parkland, gardens. **Fp** Silver Birch, Downy Birch, preferring young trees; folds and rolls the leaf in a longitudinal manner (see also *C. betulicola*, *Epinotia immundana* (Tortricidae)).

Caloptilia elongella page 210 282 (1116)

Common. **FL** 7-8.5mm. Forewing elongate, usually orange-brown or orange, occasionally pale yellowish brown, often unicolorous but sometimes with an obscure paler costal blotch and thinly speckled with darker spots on the costa in the basal half, near the dorsum, and occasionally elsewhere; the cilia are plain and the underside of the forewing grey-brown. Hindleg coxa, trochanter and femur coloured as forewing, or red-brown, sometimes yellowish. **Similar species** *C. betulicola* usually has a costal blotch, sometimes darker edged, the underside of the forewing is dark orange-brown, and the hindleg coxa, trochanter and femur are whitish; *C. elongella* and *C. betulicola* have the longest wings of any British *Caloptilia*; *C. hemidactylella* has the terminal cilia of the forewing banded; *C. rufipennella* has the forewing glossy orange-brown or red-brown with a violet sheen, and the femur of the foreleg and midleg mottled chocolate-brown and whitish.
FS Double-brooded, overwinters as an adult, June, August-May. Flies in late afternoon and comes to light. **Hab** Fens, marshes, riverbanks, woodland, gardens. **Fp** Alder; rolls the leaf in a longitudinal manner (see also *C. falconipennella*, *Epinotia immundana* (Tortricidae)).

Caloptilia betulicola page 210 283 (1112)

Common. **FL** 7-8.5mm. Forewing elongate, usually orange-brown or orange, and usually with a costal blotch. **Similar species** *C. elongella*; also *C. hemidactylella* and *C. rufipennella*. **FS** Double-brooded, overwinters as an adult, June, August-May. Flies in late afternoon and comes to light. **Hab** Woodland, heathland, moorland, gardens. **Fp** Birches; rolls the leaf in a transverse manner (see also *C. populetorum*).

▲ Cocoon of *Caloptilia elongella* on Alder. Note the elongate oval shape, and shiny semi-translucent membrane, characteristic of the cocoon of *Caloptilia*.

▲ Conical leaf roll of *Caloptilia rufipennella* on Sycamore.

Caloptilia rufipennella page 210   284 (1129)

Common. **FL** 5.5-6mm. Forewing glossy orange-brown, brown or red-brown with a violet sheen, without any markings; more rarely, the orange-brown form has small black spots in a line below the costa and one below the dorsum, with a similar spot before the termen. Femur of foreleg and midleg mottled chocolate-brown and whitish. **Similar species** *C. elongella, C. betulicola, C. hemidactylella*; also *Zelleria hepariella* (Yponomeutidae), which has a somewhat broader brown forewing, usually with a contrasting paler streak from the base, and has a declining, not inclining, posture. **FS** Single-brooded, overwinters as adult, July-early May. Comes to light. **Hab** Wherever the foodplant occurs. **Fp** Sycamore, and recorded from Silver Maple and Field Maple; constructs three neat cones on the leaves.

Azalea Leaf Miner *Caloptilia azaleella* page 210   285 (1111)

Local. A naturalised adventive, extending its range northwards. **FL** 5-5.5mm. Forewing glossy brown or reddish brown with purplish reflections, sometimes darker towards the base and the apex, with an irregular yellowish costal blotch extending from one-quarter to near the apex. **The smallest species of *Caloptilia*.** **FS** Double-brooded, occasionally a partial third brood, May-November. Flies by day and comes to light; it can sometimes be numerous. **Hab** Gardens. **Fp** Azalea cultivars; constructs two neat cones on the leaves.

Caloptilia alchimiella page 210   286 (1110)

Common. **FL** 5-7mm. Forewing pale reddish brown with purplish reflections, a sharply defined yellow basal blotch extending to about one-quarter, and a large, almost triangular, costal blotch which nearly touches the dorsum and extends almost to the apex. **Similar species** **C. robustella has a more diffused basal blotch and the yellow triangle does not extend to two-thirds along the costa; C. alchimiella is more frequent than C. robustella in the north.** If in doubt, the identity of this species-pair should be confirmed by examination of genitalia. **FS** Double-brooded in the south, single-brooded further north, May-August, early October. Can be disturbed from vegetation by day, flies in the evening and comes to light. **Hab** Woodland, parkland, hedgerows, gardens. **Fp** Oaks; constructs up to three neat cones on the leaves.

Caloptilia robustella page 210   287 (1127)

Common. **FL** 7-8.5mm. Forewing pale reddish brown with purplish reflections, a diffuse yellow basal blotch and a yellow triangular costal blotch. **Similar species** *C. alchimiella*. **FS** Double-brooded in the south, single-brooded further north, May-August, early October. Can be disturbed from vegetation by day and comes to light. **Hab** Woodland, parkland, hedgerows, gardens. **Fp** Oaks; constructs up to three neat cones on the leaves.

▲ Conical leaf roll of *Caloptilia semifascia* on Field Maple.

▲ Leaf mine and cone of *Gracillaria syringella* on Lilac.

Caloptilia stigmatella page 210   288 (1127)

Common. **FL** 6-7mm. Forewing reddish brown with a pale yellow or whitish triangular costal blotch, with an extension narrowing sharply towards the dorsum. **FS** Single-brooded, occasionally double-brooded, overwinters as an adult, late August-May, also late June-July. Easily disturbed from vegetation by day and comes to light. **Hab** Woodland, hedgerows, gardens. **Fp** Sallows, willows, poplars, Aspen, occasionally birches; constructs cones or folds on the leaves.

Caloptilia falconipennella page 210   289 (1117)

Rare, but expanding its range in the south. **FL** 6-7mm. Summer generation: forewing orange-brown to dark reddish brown, glossy with purplish reflections, and a whitish or yellowish near-triangular costal blotch. Autumn or single generation: forewing dark reddish brown from the base to one-quarter, distinctly edged, with the rest of the wing creamy whitish, heavily speckled dark brown, and a pale costal triangle sometimes discernible. **Similar species** *C. semifascia* usually has a shorter forewing; in the summer generation the costal blotch is usually quadrate, and in the autumn generation there is usually an angled white or yellowish streak from the costa at about one-quarter to the wing fold; *C. hemidactylella* could be confused with less contrasting examples of summer-generation *C. falconipennella*. There is sufficient variation is size, coloration and markings that some examples may need to be confirmed by genitalia examination. **FS** Single-brooded, but a double-brooded population may now be spreading in southern England, overwinters as an adult, July-May. Usually hard to find, but can be abundant on non-native alder trees. Can be disturbed from vegetation by day and comes to light. **Hab** Fens, marshes, riverbanks, ditches, urban areas. **Fp** Alder, Italian Alder and Grey Alder; constructs up to three folds on the margin of the leaves (see also *C. elongella*).

Caloptilia semifascia page 210   290 (1130)

Local. **FL** 5-6mm. Summer generation: forewing yellowish brown to dark reddish brown, glossy with purplish reflections, and a whitish or yellowish quadrate costal blotch. Autumn or single generation: forewing brown or dark reddish brown, glossy, with an angled white or yellowish streak from the costa at about one-quarter to the wing fold. **Similar species** *C. falconipennella*; the taxon formerly known as *Calybites hauderi* (Rebel) is now treated as the summer generation of *C. semifascia*. **FS** Single-brooded, but a double-brooded population may now be spreading in southern England, July-May. Has been found at Ivy blossom and comes to light; occasionally abundant. **Hab** Woodland, hedgerows, gardens. **Fp** Field Maple, occasionally Sycamore where the moth is abundant; constructs up to three cones on the leaves.

▲ Leaf mine of *Aspilapteryx tringipennella* on Ribwort Plantain. Note that as the larva grows, the mine expands and contorts the leaf.

▲ Three leaf coils formed by larvae of *Calybites phasianipennella* on Broad-leaved Dock. Also visible on the underside of the leaf are young blister mines close to two of the coils.

Caloptilia hemidactylella page 210   291 (1122)

Rare. May spread from Europe, where it is expanding its range. **FL** 5.5-7mm. Forewing pale yellowish brown, orange-brown or reddish orange, the costal triangle large, usually paler and conspicuous, with a few darker dots on the costa, the inner edge with a suffused darker oblique cross-band; the cilia are banded. A uniform reddish-orange form occurs. **Similar species** *C. elongella, C. betulicola, C. falconipennella* (summer generation). **FS** Single-brooded, overwinters as an adult, late September-May. **Hab** Woodland. **Fp** Sycamore, Field Maple, Norway Maple; constructs cones on the leaves.

Gracillaria syringella page 210   293 (1135)

Common. **FL** 6-7.5mm. Forewing orangey brown to dark brown, variably mottled with white marks, often with a near-complete white ring at the apex formed of opposing streaks and the cilia. **FS** At least double-brooded, March-early November. Readily disturbed from foodplants by day, and also comes to light. The larval leaf mines are often abundant. **Hab** Woodland, coastal scrub, gardens. **Fp** Wild Privet, Garden Privet, Ash, Lilac, less often White Jasmine and Snowberry; larvae mine the leaves gregariously at first, later constructing untidy cones or rolls (see also *Caloptilia cuculipennella*).

Aspilapteryx tringipennella page 210   294 (1143)

Common. **FL** 5-7mm. Head smooth-scaled. Forewing pale yellowish brown, pale greyish along the costa to near the apex, the costal half speckled with numerous small darker dots, these arranged in rows from the base. **Similar species Bedellia somnulentella (Bedelliidae) sits on midlegs and hindlegs, has the crown of the head with erect scales and the forewing with scattered blackish scales, not arranged in rows; also more uniformly marked examples of *Kessleria saxifragae* (Yponomeutidae), which has a declining not inclining posture.** **FS** At least double-brooded, April-October. Flies in afternoon and evening sunshine, and comes to light. Can be locally abundant. **Hab** Downland, rough grassland, coastal grassland, waste ground, quarries, vegetated shingle. **Fp** Ribwort Plantain; in a pale-coloured mine on the upperside of the leaf.

Euspilapteryx auroguttella page 210   297 (1145)

Common. **FL** 4.5-5mm. Forewing shining dark greyish brown with a yellow spot below the costa at one-third, another on the costa at three-quarters and two spots along the dorsum. **FS** Double-brooded, May-early June, July-August. Flies in the evening and comes to light. **Hab** Coppiced woodland, rough grassland, downland, quarries. **Fp** Perforate St John's-wort, Slender St John's-wort; at first mines a leaf, then constructs two small cones on the leaves.

▲ Leaf mine of *Acrocercops brongniardella* on oak. Note there are several larvae per leaf; mines may coalesce and larvae then feed together.

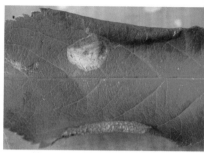

▲ Young blister mine and two leaf folds of *Callisto denticulella* on apple.

Calybites phasianipennella page 210   296 (1147)

Local. **FL** 4-5.5mm. Forewing orangey brown to greyish brown, with a narrow whitish dorsal streak from near the base to three-quarters and a whitish spot near the apex; another form has four large whitish or cream-coloured spots, two on the costa, two on the dorsum. Intermediate forms occur, and the paler markings can be somewhat obscured. **FS** Single-brooded, overwinters as an adult, mid July-May. Can be disturbed from vegetation by day and comes to light. **Hab** Heathland, acid grassland, damp woodland, fens, parkland, farmland, urban gardens. **Fp** Water-pepper, Redshank, Black-bindweed, Common Sorrel, Sheep's Sorrel, Broad-leaved Dock, Curled Dock, Water Dock, Yellow Loosestrife; makes a blister mine then cuts a strip from the leaf and rolls this into an untidy coil, making two coils to complete its feeding (see page 87).

Povolnya leucapennella page 211   292 (1150)

Local. **FL** 6-7mm. Segment 2 of the labial palp distinctly tufted beneath. Forewing ranging from very pale yellowish or very pale greenish white to whitish brown, orange-brown or red-brown, mottled with darker dots or marks which are often faint, but sometimes extensive. **Similar species** All other plain orange or brown *Caloptilia* have segment 2 of the palp smooth beneath; pale forms can be similar to *C. cuculipennella*. **FS** Single-brooded, overwinters as an adult, late June-May. Can be disturbed by day from dense or evergreen thick vegetation in woods during the winter, especially from Yew; comes to light. **Hab** Oak woodland, wooded undercliffs. **Fp** Oaks, including Holm Oak, preferring young trees; constructs at least two cones on the leaves.

Acrocercops brongniardella page 211   313 (1154)

Local. **FL** 4-5mm. Antenna longer than the forewing. Forewing orange-brown with dark-edged white marks, four along the dorsum, three on the costa, with the basal pair often joined to form a curved cross-band; the cilia are banded white and dark brown, with a dark brown pointed extension. **FS** Probably one extended brood; adult may be found in any month. Occasionally comes to light, but often found commonly as a leaf mine. **Hab** Open woodland, parkland, gardens. **Fp** Oaks, including Holm Oak; larvae feed gregariously in the upper epidermis of the leaves, making white blotch mines.

Callisto denticulella page 211   310 (1172)

Common. More local in the north and west. **FL** 5-6mm. Head with erect, orange-brown scales. Forewing dark brown with whitish near-triangular spots, two on the dorsum, three on the costa. **Similar species** *Parectopa ononidis*. **FS** Single-brooded, May-June. **Hab** Gardens, orchards, hedgerows. **Fp** Apples, including domestic cultivars; constructs two folds on the margin of the leaves.

▲ Young underside leaf mine (top left) and leaf fold (bottom right) of *Parornix anglicella* on Hawthorn.

▲ Larva of *Parornix anglicella*.

Parornix species – *Parornix anglicella* page 211   303 (1178)

Identification of *Parornix* to species requires particular care. Adults of all species are very similar to each other and are best determined by rearing from host plants or by examining genitalia. There are ten species recorded in Britain and six in Ireland. **FL** 4-5.5mm. The resting posture is inclining, in a similar manner to *Caloptilia*. Depending on species, the forewing may be dull greyish, greyish brown or blackish brown, sometimes speckled whitish and with darker grey patches in the midle of the wing; there are usually short whitish streaks along the costa to the apex, sometimes faint near the base, and a small blackish apical mark; the terminal cilia have alternating narrow dark and broad paler bands. **Similar species** For futher details on differences between *Parornix* species, and genitalia drawings, refer to Heath & Emmet (1985). **FS** Most species are double-brooded, except in the north where some are single-brooded. Adults may be found from spring to late summer, and are frequently attracted to light, sometimes in numbers.
Hab Woodland, moorland, heathland, hedgerows and gardens, depending on species.
Fp Trees and shrubs, the host plant varying between *Parornix* species; at first in a small blister mine similar to a *Phyllonorycter*, then in two or more leaf folds.

Phyllonorycter harrisella page 211   315 (1253)

Common. **FL** 3-4mm. Forewing elongate, white, pale orange-brown towards the apex with three costal and two dorsal streaks, all inwardly edged dark brown, and a black dot in the apex; the cilia have a fine dark brown basal line. **The dominant whitish ground colour and fine markings help distinguish this species from other *Phyllonorycter*.** **FS** Double-brooded, April-June, late July-early September. Readily disturbed by day from oak branches and occasionally comes to light. **Hab** Woodland, hedgerows.
Fp Pedunculate Oak, Sessile Oak; in a small oval blister mine on the underside of the leaf.

Phyllonorycter messaniella page 211   321 (1274)

Common. **FL** 3-4.5mm. Forewing shining pale golden brown with four whitish costal and four dorsal streaks, all inwardly edged dark brown, a fine whitish basal streak to about one-half, and a black dot at the apex; the cilia are pale with a fine dark brown basal line. **Similar species *P. quercifoliella* (not illustrated) has a basal streak extending to almost two-thirds.** **FS** Probably two overlapping broods, March-May, August-December. Flies freely in the early morning and on warm afternoons, readily disturbed by day and comes to light. **Hab** Woodland, parkland, hedgerows, gardens. **Fp** Holm Oak, oaks, less often Hornbeam, Sweet Chestnut, Beech; in a small, usually oval, blister mine on the underside of the leaf.

▲ Young underside leaf mine (centre) and leaf fold (left, below leaf midrib) of *Parornix scoticella* on Rowan.

▲ Underside leaf mine of *Phyllonorycter platani* on London Plane.

▲ Underside leaf mine of *Phyllonorycter sorbi* on Rowan.

▲ Underside leaf mine of *Phyllonorycter blancardella* on apple.

▲ Upperside leaf mine of *Phyllonorycter corylifoliella* on Hawthorn, before the leaf has contorted, partly obscuring the mine.

▲ Underside leaf mine of *Phyllonorycter maestingella* on Beech.

▲ Serpentine leaf mine of *Phyllocnistis xenia* in the upper epidermis of White Poplar.

▲ Upperside leaf mines of Horse-chestnut Leaf Miner *Cameraria ohridella* on Horse-chestnut.

▲ Upperside leaf mine of *Phyllonorycter leucographella* on Firethorn, before the leaf has contorted into a roll.

▲ Larva of *Phyllonorycter leucographella*.

Firethorn Leaf Miner *Phyllonorycter leucographella* page 211   332a (1269)

Common. First found in Britain in 1989, spreading rapidly north and west, reaching Scotland in 2002 and Ireland in 2005. **FL** 3.5-4.5mm. Head with erect scales, bright white centrally, orange-brown laterally. Forewing bright orange-brown with a white streak from the base to about one-third, four costal and two dorsal white streaks, with black shading in the apical area. **FS** Probably up to three broods, April-May, late July-August and October. Can be abundant, particularly the larval mine. **Hab** Urban habitats, gardens. **Fp** Firethorn, less often apples, Hawthorn, Rowan; also recorded on Beech, Wild Cherry, Cotoneaster, Laburnum; in a long, narrow mine on the upperside of the leaf, causing the leaf to curl upwards into a roll.

Phyllonorycter trifasciella page 211   361 (1319)

Common. Local in the north. **FL** 4-4.5mm. Forewing pinkish orange-brown with broad blackish cross-bands at one-fifth and one-half, the outer margin edged white with the inner margin irregular, two costal and one dorsal blackish streaks, the costal and the dorsal at about three-quarters almost joining to form a third cross-band. **Similar species** *Argyresthia trifasciata* (Argyresthiidae) often rests with head down, and has a white head. **FS** Often three generations, May-August, November. Occasionally comes to light, most often in late autumn. **Hab** Woodland, scrub. **Fp** Honeysuckle, occasionally Snowberry, Himalayan Honeysuckle; in a mine on the underside of the leaf, causing the leaf to contort.

Horse-chestnut Leaf Miner *Cameraria ohridella* page 211   366a (1330)

Common. First recorded in Britain in 2002, rapidly spreading north and west. **FL** 3.5-5mm. Forewing orangey brown with a short white basal streak, a fine white angled cross-band at one-third and another about one-half, edged darker outwardly, and with two white dark-edged short streaks on the costa towards the apex, with another on the dorsum and a more obscure one in the cilia. **Similar species The basal streak and angled cross-bands distinguish this species from similar-looking *Phyllonorycter* species and *Argyresthia trifasciata* (Argyresthiidae).** **FS** At least two overlapping broods, May-October. Flies by day and comes to light; often extremely abundant, particularly the larval mine. **Hab** Parkland, gardens, roadside trees. **Fp** Horse-chestnut, occasionally Sycamore; in a blotch mine on the upperside of the leaf.

Phyllocnistis xenia page 211   369 (1339)

Rare. First recorded in Britain in 1974, gradually spreading out from the south and east. **FL** 3-3.5mm. Forewing shining white, yellowish towards the apex, with a greyish spot on the dorsum at one-half, a prominent black dot at the base of the apical cilia and, from three-fifths, a series of silvery-grey lines along the costa, wing-tip and near the tornus, sometimes two of these joining to form a grey cross-band at three-quarters. **FS** Double-brooded, June-July and August-September. Most easily found as a leaf mine and often abundant where the moth occurs. **Hab** Scrub woodland on sand dunes and chalk downland, woodland margins, hedgerows. **Fp** Grey Poplar, White Poplar, preferring young sapling growth; in a long linear line in the upper epidermis of the leaf.

Yponomeutidae

Bird-cherry Ermine
Yponomeuta evonymella

Zelleria hepariella

Swammerdamia pyrella

There are 25 species in this family. In most species the adults rest with the wings held roof-like, at a steep angle, but in a few species the wings are wrapped around the abdomen. Species vary from small to large, with forewing length 4-13mm. This family includes the distinctive Small Ermine moths, which are white with black dots, some species of which are gregarious as larvae, producing characteristic swathes of greyish silk webbing that frequently festoon hedgerow plants in the early summer, the larvae causing defoliation.

The adult resting posture varies between species. In *Zelleria* and *Kessleria* it is declining, slightly less so in some *Swammerdamia*, but in most species the adults rest in a slightly inclining posture. The forewings are elongate, often fairly broad, with or without a distinct tornal angle. The hindwings are almost as long and broad as the forewing, or slightly narrower with a pointed apex and with moderate or long dorsal cilia. The head and face are usually smooth-scaled, but in *Swammerdamia* the head is rough-scaled and tufted between the antennae. The antennae are thread-like, two-thirds to three-quarters the length of the forewing. The labial palps are variable, from very short to moderate, and curved upwards, nearly straight or drooping. A tongue is present, although this is sometimes reduced.

The majority of British and Irish species are covered in accounts below. Several species are fairly distinctive and can be readily identified. However,

▼ Web and larvae of Hawthorn Moth *Scythropia crataegella* on Hawthorn.

e 'Swammerdamia' group
Swammerdamia, Pseudoswammerdamia,
raswammerdamia) needs particular
are, and worn examples usually require
enitalia examination to determine.
ome examples of Yponomeuta cannot be
eadily identified on external or genitalia
aracters. Rearing these from larvae
an help with identification, as most
ecies are restricted to one or a few
odplants. Two species, Euhyponomeuta
annella and Kessleria fasciapennella,
re now thought to be extinct, and Zelleria

▲ Webs of Bird-cherry Ermine Yponomeuta evonymella on Bird Cherry.

leastrella, an immigrant, is a recent addition to the British fauna. Formerly, the
rgyresthiidae, Plutellidae, Ypsolophidae and part of the Glyphipterigidae were
eated within the Yponomeutidae.

arvae of most species begin as leaf miners but progress to external feeding as they
row. Those of Scythropia, Yponomeuta, Zelleria and Kessleria all exhibit communal
ving within silk webbing, to a greater or lesser extent, and Yponomeuta larvae
ometimes occur in vast numbers. Larvae of Cedestis and Ocnerostoma feed singly,
nd Ocnerostoma remain as leaf miners throughout. Adults fly at night and come
light; some Yponomeuta species may be found in numbers in light traps on warm
ights in summer, a long distance from host plants, and it seems likely that numbers
f some species in Britain and Ireland are boosted annually by immigration from
e Continent. A few species of Yponomeuta can also be found by day, emerging in
umbers from the larval feeding sites.

rther reading
itish and Irish species: Emmet (1996)
ropean species: Bengtsson & Johansson (2011)

Hawthorn Moth *Scythropia crataegella* page 211   450 (1344)

Common in the south only. **FL** 5-7.5mm. Resting posture inclining. Forewing white or
greyish white with two brown cross-bands, brown spots along the termen, and rows of
small brown spots or streaks on the costa, dorsum and in the middle of the wing.
FS Mainly single-brooded, mid June-late July, with occasional examples mid August-late
September of a partial second generation. Comes to light. **Habitat** Woodland margins,
hedgerows, urban areas. **Fp** Hawthorns, Blackthorn, plums, Small-leaved Cotoneaster;
older larvae feed communally in a web.

The Small Ermine group

Identification of several species in this group requires care. The *Yponomeuta padella/malinellus* species pair cannot be reliably separated and it seems likely that they exist somewhere between forms of the same species and two separate species, as they show some fairly constant morphological and foodplant differences but their genitalia appear identical.

Bird-cherry Ermine *Yponomeuta evonymella*   page 211   424 (1347)

Common. Principally an immigrant in the south, but resident from the south Midlands northwards. **FL** 9.5-12.5mm. Forewing uniformly white with black dots which are **significantly more numerous than in all other *Yponomeuta* species**. **FS** Single-brooded, June-early September. Can occur in large numbers as a larva and adult. Comes to light. **Hab** Everywhere, often well away from the known foodplant in Britain. **Fp** Bird Cherry; larvae feed gregariously in a web, the webbing sometimes enveloping whole trees.

Orchard Ermine *Yponomeuta padella*   page 211   425 (1348)

Common. **FL** 9-11mm. Forewing two-toned grey and white, or uniformly grey, with black dots. It is probably reasonable to identify examples with all-grey ground colour as *Y. padella*. **Similar species** The grey-and-white form of *Y. malinellus* cannot be separated from *Y. padella* except by rearing from larvae, and even this may not be wholly reliable. Grey-and-white forms of *Y. malinellus* and *Y. padella* are similar to *Y. rorrella*, which is usually bigger, with a longer and narrower forewing, smaller-sized dots, and invariably white in the dorsal half and in a thin band along the costa. The wholly grey form of *Y. padella* is similar to *Y. sedella*, which is smaller and has a small black patch in the termen. Wholly white *Yponomeuta* could also be confused with Thistle Ermine *Myelois circumvoluta* (Pyralidae), which is larger and looks more robust, and with *Ethmia terminella* and *E. dodecea* (Elachistidae), which have fewer and larger black spots on the forewing. **FS** Single-brooded, June-August. Can occur in large numbers as a larva and adult. May be found by day at rest on the foodplant and comes to light. **Hab** Wherever the foodplants occur. **Fp** Blackthorn, hawthorn; larvae feed gregariously in a web and many metres of hedgerow can be defoliated and covered in grey webbing.

Apple Ermine *Yponomeuta malinellus*   page 211   426 (1349)

Local. **FL** 10-11.5mm. Forewing two-toned grey and white, or uniformly white, with black dots. It is probably reasonable to identify examples with all-white ground colour as *Y. malinellus*. **Similar species** **The wholly white form of *Y. malinellus* is very similar to *Y. cagnagella*, which is usually bigger, with a more arched costa and pointed apex**, but identification based on these superficial characters is not always reliable and examination of genitalia may be necessary; also the grey-and-white form of *Y. padella*. **FS** Single-brooded, June-August. Rarely occurs in large numbers as a larva, but the adult is sometimes common at light, indicating that numbers are boosted by immigration. **Hab** Orchards, hedgerows. **Fp** Apples, especially domestic cultivars; feeds gregariously in a web.

Spindle Ermine *Yponomeuta cagnagella*   page 211   427 (1350)

Common. Rare in Scotland. **FL** 9.5-13mm. Forewing uniformly white with black dots. **Similar species** The wholly white form of *Y. malinellus*. **FS** Single-brooded, late June-early September. **Hab** Scrub, hedgerows, urban areas. **Fp** Spindle, Evergreen Spindle; larvae feed gregariously in a web, the webs sometimes extensive.

▲ Larvae of *Yponomeuta padella* on Blackthorn.

▲ Webs of *Yponomeuta padella* on Hawthorn.

▲ Larvae of *Yponomeuta malinellus* on apple.

▲ Web of *Yponomeuta malinellus* on apple.

▼▶ Web and larvae of *Yponomeuta cagnagella* on Spindle.

▶ Larvae of
Willow Ermine
Yponomeuta
rorrella on willow.

▶▶ Web of
Kessleria
saxifragae on
saxifrage. Note
the larvae hide
deep within the
plant.

Willow Ermine *Yponomeuta rorrella* page 211   428 (1352)

Local. **FL** 9.5-12mm. Forewing long and narrow, two-toned grey and white with black dots which are usually smaller in size than in other *Yponomeuta*; invariably white in the dorsal half and in a thin band along the costa. **Similar species** The grey-and-white forms of *Y. padella* and *Y. malinellus*. **FS** Single-brooded, July-August, but usually only from mid July to mid August. Rarely observed as a larva in Britain, but adult numbers fluctuate annually, boosted by sometimes substantial immigration. **Hab** River valleys, plantations. **Fp** White Willow, Grey Willow; a few larvae feed together in a web.

Yponomeuta irrorella page 211   429 (1353)

Rare. Recorded recently only in Kent and Sussex. **FL** 9.5-12.5mm. Forewing white with black dots and diffuse grey patches on the wing fold at one-third and in the middle of the wing, and grey along the costa above the middle patch. **The large size and patch on the wing fold distinguish this species from other grey-and-white *Yponomeuta*.** **FS** Single-brooded, July-August. Comes to light. **Hab** Chalk downland. **Fp** Spindle; a few larvae feed together in a web but they become solitary in the final instar.

Yponomeuta plumbella page 211   430 (1354)

Common only in the south. Most frequent in chalky areas in the south, occasionally found dispersing elsewhere. **FL** 8.5-9.5mm. Forewing greyish white with black dots; a black spot in the wing fold before one-half, often a short black streak above the tornus, and a black mark along the termen towards and at the apex. **FS** Single-brooded, July-August. Comes to light. **Hab** Scrub on chalk and limestone, hedgerows. **Fp** Spindle; a few larvae feed together in a web.

Yponomeuta sedella page 211   431 (1355)

Very local. Occasionally recorded as an immigrant or transitory resident. **FL** 7.5-9mm. Forewing uniformly grey with black dots, these absent towards the apex, and a small black patch in the middle of the termen. **Similar species** Wholly grey form of *Y. padella*. **FS** Double-brooded, April-May, July-August. Comes to light. **Hab** Ancient woodland, hedge banks, gardens. **Fp** Orpine, iceplant cultivars; a few larvae feed together in a web.

Zelleria hepariella page 212   435 (1359)

Local. **FL** 5-7.5mm. Resting posture declining, sometimes steeply. Forewing slightly hooked at the apex, appearing upturned at rest; usually brown or reddish brown, less often orange- or yellow-brown, sometimes with a whitish-brown or yellowish-brown suffused streak from the base to about one-half. **Similar species** *Caloptilia rufipennella* **(Gracillariidae) has an inclining posture, and a narrower forewing without a basal streak. FS** Single-brooded, July-May, the adult hibernates. Comes to light. **Hab** Woodland, scrub, hedgerows, mainly on clayey and calcareous soils. **Fp** Ash; in a silken web among leaf-tips, sometimes containing several larvae.

Kessleria saxifragae page 212   434 (1391)

Rare. A species restricted to limestone hills and mountains, but locally common where it occurs. **FL** 5.5-7.5mm. Resting posture declining. Forewing whitish with scattered black dots, sometimes indistinct, and with irregular yellowish-brown suffusion from near the dorsum at one-quarter to the apex. **Similar species** *Aspilapteryx tringipennella* **(Gracillariidae) and** *Bedellia somnulentella* **(Bedellidae) both have an inclining posture and lack forewing suffusion. FS** Single-brooded, June-July. **Hab** Streamsides, damp rocky places. **Fp** Saxifrages; usually feeds low down among the plant in a slight web, often several larvae to a web.

Pseudoswammerdamia combinella page 212   436 (1398)

Common in the south. Very local from northern England northwards. **FL** 6.5-8mm. Resting posture usually somewhat declining. Head and thorax whitish. Forewing whitish grey with several rows of dark dots; the apex has a light coppery patch surrounded by dark scales. At rest, the adult has the appearance of a broken twig. **FS** Single-brooded, late March-early June. The moth is widespread but rarely numerous and regularly comes to light. **Hab** Wherever the foodplant occurs. **Fp** Blackthorn; older larvae feed in a thick web, sometimes several larvae in a web.

Swammerdamia caesiella page 212   437 (1400)

Common. **FL** 4.5-6.5mm. Head whitish or brownish, thorax and tegulae grey. Forewing dull grey, speckled black and white, with a white or pale grey spot, sometimes obscure, on the costa before the apex and an ill-defined oblique dark grey cross-band from the dorsum not reaching the costa. **Similar species** *Paraswammerdamia nebulella* usually has mixed grey and blackish-brown tegulae, a row of fine blackish dots on the costa near the base, and small clusters of white scales on the costa before the apex; also

▲ Larva of *Swammerdamia caesiella* under a silk web on the upperside of a birch leaf.

▲ Larva of *Swammerdamia pyrella*.

Swammerdamia compunctella, which is a larger species with a white head and pale grey thorax/tegulae, having the forewing plainer grey with a large white spot on the costa before the apex, and without costal dots near the base. Poorly marked or worn examples of the *S. caesiella*/*S. compunctella*/*P. nebulella* group require dissection to confirm identity. Male genitalia of *P. nebulella* can be determined dry by brushing away scales to reveal a distinct spine on the valva, absent in others of the group in Britain, which have a smoothly curved valva. For further details and genitalia drawings, refer to Emmet (1996). **FS** Double-brooded, May-June, July-August. Comes to light. **Hab** Heathland, scrub, open woodland. **Fp** Birches; in a slight web, typically on the upperside of a leaf.

Swammerdamia pyrella page 212 438 (1402)

Common. Rarer in the north and west. **FL** 5-6.5mm. Head whitish, thorax dark grey. Forewing grey with a darker oblique cross-band at one-half and scattered blackish scales, a small white spot on the costa before the apex, and **the termen coppery, which distinguishes this species from all others in group. Similar species** *S. passerella* (not illustrated) is a rare species found in Scotland, which is smaller (FL 4-5.5mm) and paler; this species feeds on Dwarf Birch growing on mountains above 300m and at lower altitude on blanket-bog in Sutherland, so it is very unlikely to be found with *S. pyrella*. **FS** Double-brooded, April-June, July-early September. Comes to light. **Hab** Wherever the foodplants occur. **Fp** Apples, Pear, hawthorns; under a slight web, usually on the upperside of a leaf.

Swammerdamia compunctella page 212 439 (not listed)

Scarce. Very localised in England, slightly more frequent in Scotland. **FL** 7-7.5mm. Head white, thorax pale grey. Forewing uniformly grey, with a hint of a cross-band at one-half and a few darker scales, and a distinct white spot on the costa before the apex. **Similar species** *S. caesiella*, *Paraswammerdamia nebulella*. **FS** Single-brooded, June-July. Comes to light. **Hab** Open woodland. **Fp** Rowan, hawthorns; in spring in a loose web on the leaves, usually two or three larvae to a web.

Paraswammerdamia albicapitella page 212 440 (1404)

Common. Rare in the north. **FL** 5-6mm. Head and thorax white. Forewing pale grey, with scattered white and brownish-grey scales, rows of black dots just above the wing fold and near the costa, and a blackish cross-band before one-half, broadly interrupted at the wing fold. **The white thorax and interrupted cross-band distinguish this species from all others in the group. FS** Double-brooded, late April-mid September. Comes to light. **Hab** Wherever the foodplant occurs. **Fp** Blackthorn; in a web on the leaves, occasionally with several larvae to a web.

▲ Larva of *Paraswammerdamia nebulella* in a web on Hawthorn.

Paraswammerdamia nebulella page 212   441 (1408)

Common. Local in Scotland. **FL** 5.5-7mm. Head white, thorax grey, tegulae mixed grey and blackish brown. Forewing grey, speckled dark brownish grey, with at least a row of fine blackish dots on the costa near the base, sometimes more dots in rows elsewhere, and small clusters of white scales on the costa before the apex. **Similar species** *Swammerdamia caesiella*, *S. compunctella*. **FS** Possibly one extended brood, with records confirmed by dissection from May-August. Comes to light. **Hab** Wherever the foodplants occur. **Fp** Hawthorn, Rowan, sometimes roses and Wall Cotoneaster; in spring in a light web on the leaves, occasionally with two or three larvae in a web.

Cedestis gysseleniella page 212   442 (1411)

Common. Absent from the far west and local in the north. **FL** 5.5-6.5mm. Head and thorax whitish. Forewing white with many dark-tipped scales and two golden cross-bands in the basal half; the extent of golden colour is variable and may cover much of the wing. **FS** Single-brooded, June-July. Comes to light. **Hab** Pine woodland, pine scrub on heathland. **Fp** Pines; within a needle or externally on adjacent needles in a slight web.

Cedestis subfasciella page 212   443 (1412)

Common. Local in the western half of the north. **FL** 4.5-5.5mm. Head whitish, thorax whitish to pale brown. Forewing golden brown, variably scattered with dark-tipped white scales, with a white inwardly-curved cross-band before one-half, a white spot on the costa before the apex and at the tornus, and white and brown scales, both of which may be black-tipped, extending along the termen. **FS** One extended brood, and probably a partial second brood in the south, March-mid September. Comes to light. **Hab** Pine woodland, pine scrub on heathland. **Fp** Pines; mines a needle.

Ocnerostoma piniariella page 212   444 (1416)

Local. **FL** 4-5mm. Antennae white or ringed white. Forewing of the male whitish grey, the female is whiter, especially towards the apex. **Similar species** *O. friesei* **has entirely grey antennae**; genitalia examination may be necessary to separate the two species reliably. For further detail and genitalia drawings, refer to Emmet (1996). **FS** Single-brooded, June-July. Can be tapped from branches during the day and comes to light. **Hab** Pine woodland, pine scrub on heathland. **Fp** Scots Pine; mines a needle.

Ocnerostoma friesei (not illustrated)   445 (1417)

Local. **FL** 4-5mm. Antennae grey. Forewing of the male uniformly pale grey, the female is white with greyish scales along the dorsum. **Similar species** *O. piniariella*. **FS** Up to three broods per year, March-May, July-August and occasionally November. Can be tapped from branches during the day and comes to light. **Hab** Pine woodland, pine scrub on heathland. **Fp** Scots Pine; mines a needle.

Argyresthiidae

Argyresthia brockeella

There are 25 species in this family, comprising the single genus *Argyresthia*. All species are small, with forewing length 3.5-6.5mm. The wings are held tightly appressed to the body. In almost all species the adult rests in a declining posture, with its head close to the surface and abdomen raised. The forewings are elongate, the membrane without a distinct tornal angle, although the cilia at the

Cherry Fruit Moth
Argyresthia pruniella

tornus produce a noticeable angle between the dorsum and termen. The forewings are frequently shining or bronzy and often with pale markings. The hindwings are narrower than the forewings. The head is roughly scaled above, with the face smooth. The antennae are thread-like, three-fifths to four-fifths the length of the forewing. The labial palps are moderately long, curved or forward-pointing, and the tongue is not scaled.

All but one species are covered below, with *Argyresthia illuminatella* considered to be doubtfully British. Most are distinctive and relatively straightforward to identify, although care is needed with a few species. Argyresthiidae was formerly treated as a subfamily of Yponomeutidae.

The larval habit is variable between species, but is always internal. It includes feeding within the growing shoots of woody shrubs and trees, within the thick leaves of conifers, within catkins and berries, and in the case of *A. glaucinella* in the bark of mature trees. All species are single-brooded. Adults fly about the foodplant in warm weather, and can be readily disturbed. Many fly at dusk and into the night, and come to light.

Further reading
British and Irish species: Emmet (1996)
European species: Bengtsson & Johansson (2011)

Argyresthia laevigatella page 212   401 (1433)

Common. **FL** 4.5-6.5mm. Head orange-brown. Forewing shining brownish grey without markings. **Similar species A. glabratella is smaller, has the head whitish orange and forewing paler grey**; genitalia examination may be necessary to separate the two species reliably. **FS** Single-brooded, May-July. **Hab** Plantation woodland. **Fp** Larches; mines the woody shoot.

Argyresthia glabratella page 212   403 (1435)

Local. **FL** 4-5.5mm. Head whitish orange. Forewing shining grey without markings. **Similar species** *A. laevigatella*. **FS** Single-brooded, May-July. Occasionally found commonly in Norway Spruce plantations. Can be tapped from branches during the day and comes to light. **Hab** Plantation woodland. **Fp** Norway Spruce; mines the woody shoot.

▶ Larval feeding signs of *Argyresthia glabratella*. Note the hole in the underside of the twig from which frass is ejected, beyond which the twig has died and the needles have dropped. The larva seals the hole with silk prior to pupation within the twig.

Argyresthia praecocella page 212   404 (1440)

Scarce. Recent records from southern England and northern Scotland. **FL** 4-5mm. Head yellowish white, thorax pale yellowish brown. Forewing shining pale yellowish brown, with a slight purplish reflection in the outer half of the wing. **FS** Single-brooded, May. Can be tapped from bushes during the day. **Hab** Chalk downland in the south, open woodland in the north. **Fp** Juniper; in the green berries.

Argyresthia arceuthina page 212   405 (1441)

Local. Often common where Juniper is frequent. **FL** 4-5mm. Head and thorax white. Forewing shining pale golden without markings. **FS** Single-brooded, April-June. Can be tapped from bushes during the day, flies freely on warm days and comes to light. **Hab** Chalk downland in the south, open woodland in the north. **Fp** Juniper; mines the shoot, affected shoots dying.

Argyresthia abdominalis page 212   406 (1449)

Scarce. Often common among Juniper in the south but rarer in the north. **FL** 3.5-4.5mm. Antenna slightly ringed. Head and thorax white. Forewing mixed white and pale orange, more whitish in the outer half, often with a few dark brown scales below the apex. **FS** Single-brooded, June-July. Can be tapped from bushes during the day. **Hab** Chalk downland in the south, open woodland in the north. **Fp** Juniper; mines the leaf and adjacent green bark.

Argyresthia dilectella page 212   407 (1447)

Local. Fairly widespread in the south, rarer in the north. **FL** 3.5-4.5mm. Antenna strongly ringed. Head and thorax white, tegulae golden. Forewing mottled golden brown and whitish, with a dark brown apical spot, and two brown cilia lines beyond. **Similar species** **A. cupressella is without the apical spot and cilia lines in the termen, and occurs earlier in the year. FS** Single-brooded, July-August. Occasionally numerous in gardens, although not often seen as an adult. Flies during the day, and comes to light. **Hab** Chalk downland, open woodland, gardens. **Fp** Juniper, cypress cultivars; mines the shoot, this becoming discoloured.

Argyresthia aurulentella page 212   408 (1450)

Local. Uncommon among Juniper. **FL** 3.5-4.5mm. Antenna ringed. Head and thorax whitish. Forewing obscurely marked, shining whitish grey with golden streaks. **FS** Single-brooded, July. Can be tapped from bushes during the day. **Hab** Chalk downland in the south, open woodland in the north. **Fp** Juniper; mines the leaves, affected leaves becoming discoloured.

Argyresthia ivella page 212
409 (1452)

Local. Rare in the north-west. **FL** 5-6mm. Forewing white, extensive in the basal area to one-third except along the costa, and with two brown Y-shaped cross-bands arising from the dorsum; the outer Y with one arm reaching the costa before the apex, the other arm reaching the costa beyond one-half; the inner Y with one arm meeting the outer Y on the costa, the other arm orangey white towards the base. **Similar species** *A. brockeella* **has only a small white basal blotch and is overall more golden than white.** **FS** Single-brooded, July-August. Can be disturbed from the foodplant by day and comes to light. **Hab** Orchards, hedgerows, isolated trees. **Fp** Apples, Hazel; mines the shoot.

Argyresthia trifasciata page 212
409a (1442)

Local. Spreading north and west. Often common once established. **FL** 4-5mm. Head white, thorax golden. Forewing shining golden with three narrow white cross-bands usually complete, the cross-band nearest the base often narrowing before the costa, and three white marks in the apical area. **Similar species** *Phyllonorycter trifasciella* **and** *Cameraria ohridella* **(Gracillariidae), neither of which has a white head. FS** Single-brooded, May-June. Comes to light. **Hab** Gardens. **Fp** Cypress cultivars; mines the leaf and shoot, which become discoloured.

Cypress Tip Moth *Argyresthia cupressella* page 212
409b (not listed)

Local. A North American species first found in Britain in 1997, now spreading north and west. Recorded new to Wales in 2009. Often common once established. **FL** 4-5mm. Antenna strongly ringed. Head white, thorax golden brown. Forewing distinctly mottled golden brown and whitish, with dark brown scales at and near the apex. **Similar species** *A. dilectella.* **FS** Single-brooded, June-July. Comes to light. **Hab** Gardens. **Fp** Juniper, cypress cultivars; mines the shoot, affected shoots turning brown.

Argyresthia brockeella page 212
410 (1453)

Common. **FL** 5-6mm. Forewing shining dark golden with broad white marks arising from the costa and dorsum, sometimes forming a cross-band. The extent of white markings varies and a unicolorous form occurs rarely. **Similar species** **The unicolorous form of** *A. goedartella*, **which is brassy, not dark gold;** also *A. ivella.* **FS** Single-brooded, mid May-August. Can be tapped from branches during the day, flies in afternoon sunshine and comes to light. **Hab** Wherever the foodplants occur. **Fp** Birches, Alder; in the bud in autumn, then overwinters fully fed in the male catkin, the catkin often becoming distorted.

Argyresthia goedartella page 212
411 (1454)

Common. **FL** 5-6mm. Forewing shining golden brassy, paler towards the costa, with oblique cross-bands varying from white to pale gold. The extent of white or pale markings varies, and forms which are brassy, wholly or with the cross-band pattern faintly discernible, occur frequently. **Similar species** The unicolorous form of *A. brockeella.* **FS** Single-brooded, June-September. Can be tapped from branches during the day, flies in afternoon sunshine and comes to light. **Hab** Wherever the foodplants occur. **Fp** Birches, Alder; in the bud or male catkin.

Argyresthia pygmaeella page 213
412 (1455)

Common. Local in places. **FL** 5-6.5mm. Forewing shining brassy whitish, with a golden oblique cross-band from the middle of the dorsum not reaching the costa, and further golden marks along the dorsum and near the base, the extent of all markings varying considerably. **FS** Single-brooded, June-August. Comes to light. **Hab** Wherever the foodplants occur. **Fp** Sallows, willows; mines the leaf bud at first, then the shoot.

▲ Larva of *Argyresthia brockeella*.

▼ Larva of *Argyresthia goedartella* burrowing into a catkin of birch.

▲ Larval feeding signs of *Argyresthia retinella* on birch. The larva feeds within a bud, causing it to wilt or die whilst other shoots continue to expand.

Argyresthia sorbiella page 213   413 (1456)

Local. **FL** 5.5-6.5mm. Thorax and tegulae white. Forewing white, not shining, with a golden-brown oblique cross-band from the dorsum sometimes reaching the costa, other golden-brown marks along the dorsum and short streaks along the costa. **Similar species** *A. bonnetella* has golden-brown tegulae and only one dorsal mark; also ***A. curvella*, which is smaller and has greyish-brown markings. FS** Single-brooded, June-July. Can be tapped from branches during the day and comes to light. **Hab** Open woodland. **Fp** Rowan, Whitebeam; in the shoot.

Argyresthia curvella page 213   414 (1458)

Local. **FL** 5-6mm. Forewing white with greyish-brown markings, including costal streaks from near the base to one-quarter, often extending into irregular cross-lines, and an oblique cross-band from the dorsum at one-half reaching the costa, sometimes linked to markings on the tornus and in the apical area. **Similar species** *A. sorbiella*. **FS** Single-brooded, June-July. Can be disturbed by day from apple trees and comes to light. **Hab** Orchards. **Fp** Apples; in the flowering shoot.

Argyresthia retinella page 213   415 (1459)

Common. **FL** 4.5-5mm. Forewing white with brown or greyish-brown streaks and irregular cross-lines throughout; other brown or greyish-brown markings may be present, including a diffuse streak from the base to the middle of the wing, and patches beyond one-half and in the terminal area; a tiny blackish-brown apical mark is usually present. **FS** Single-brooded, June-July. Can be tapped from branches during the day, flies in afternoon sunshine and comes to light. **Hab** Wherever the foodplant occurs. **Fp** Birches; in the male catkin or shoot. When in the shoot, causes it to wilt.

Argyresthia glaucinella page 213

Local. **FL** 4-5mm. Forewing shining brownish grey, shining greyish white along the dorsum with purplish-grey streaks, a dark brownish-grey cross-band at one-half on the dorsum extending obliquely to just beyond the middle of the wing, sometimes reaching the costa. **FS** Single-brooded, May-July. The adult may be found at rest on tree trunks and occasionally comes to light. **Hab** Open woodland, isolated trees. **Fp** Oaks, including Holm Oak, Horse-chestnut, birches in Scotland; in the bark, exuding reddish frass from crevices. This species is mostly recorded from evidence of the larvae. However, where *Dystebenna stephensi* (Elachistidae) occurs records may be unreliable, as its larval habit is similar.

Argyresthia spinosella page 213

Common. **FL** 4.5-5.5mm. Tegulae orange-brown. Forewing brownish grey with a slight purplish reflection, darker at the apex; a white streak along the dorsum from the base to the tornus, interrupted by a purplish grey-brown oblique cross-band from the dorsum at one-half to the costa at three-fifths, this often indistinct in the middle, and whitish streaks in the outer half along the costa. **Similar species** Typical form of *A. conjugella* and *A. semifusca*, both of which are larger and darker, with purple-brown or brown tegulae. **FS** Single-brooded, end May-July. May be beaten from bushes and trees, and comes to light. **Hab** Wherever the foodplants occur. **Fp** Blackthorns, plums; in the flowering shoot.

Apple Fruit Moth *Argyresthia conjugella* page 213

Common. F. *aerariella* (Midlands and north Wales) and f. *maculosa* (northern Scotland and Ireland) can be found alongside the typical form. **FL** 5-7mm. Typical form: tegulae purplish brown. Forewing greyish brown with a purplish reflection, creamy white along the dorsum, interrupted by a darker brown oblique cross-band at one-half from the dorsum to near the costa, with scattered whitish scales elsewhere, giving the forewing a mottled appearance. F. *aerariella* has the forewing shining unicolorous brown; f. *maculosa* has much of the greyish brown of the typical form replaced by creamy yellowish brown, contrasting with the cross-band and other dark brown markings. **Similar species** *A. semifusca* has a more uniform dark forewing with the cross-band hardly discernible, and is pure white on the dorsum; also *A. spinosella*. **FS** Single-brooded, May-July. May be beaten from trees, and comes to light. **Hab** Wherever the foodplant occurs. **Fp** Rowan, and apples (especially in Scotland); in the green berries of Rowan, which ripen prematurely.

Argyresthia semifusca page 213

Local. **FL** 5-7mm. Forewing dark brown with a purplish reflection and a pure white dorsal streak interrupted by a dark quadrate mark; the mark is sometimes extended towards the costa as a cross-band but is hardly darker than the ground colour. **Similar species** *A. spinosella*, *A. conjugella*. **FS** Single-brooded, June-September. Comes to light. **Hab** Open woodland. **Fp** Rowan, hawthorns; in the shoot, sometimes causing a slight swelling or the shoot to droop.

Cherry Fruit Moth *Argyresthia pruniella* page 213

Common. More local in the north. **FL** 5-6mm. Forewing mixed creamy brown and golden brown, with a white dorsal streak interrupted by an orange-brown or dark brown oblique cross-band from the dorsum to the costa. **Similar species** *A. bonnetella* has a paler, more creamy-coloured forewing, with a more oblique and narrower cross-band which does not reach the costa. **FS** Single-brooded, June-August. May be beaten from trees, flies at dusk and comes to light. **Hab** Wherever the foodplant occurs. **Fp** Cherry; at first mines the flowering shoot, then the heart of the shoot or the developing fruit.

Argyresthia bonnetella page 213

421 (1467)

Common. **FL** 4.5-6mm. Typical form: forewing mixed creamy brown and golden brown, with a white dorsal streak interrupted by a dark oblique cross-band from the dorsum not reaching the costa, the creamy brown extensive along the costa to two-thirds. F. *ossea* is almost uniformly pale yellowish white, and pure white on the dorsum. **Similar species** *A. pruniella*. **FS** Single-brooded, May-September. May be beaten from bushes and trees, flies by day and comes to light. **Hab** Wherever the foodplant occurs. **Fp** Hawthorns; in a spinning then a mine in the terminal shoot.

Argyresthia albistria page 213

422 (1468)

Common. **FL** 4.5-5.5mm. Typical form: forewing dark orangey brown or dark brown with a purplish reflection; on the dorsum a narrow white streak from the base to near one-half, a dark brown mark beyond and creamy brown before the tornus. A unicolorous form occurs with a light brown thorax, plain brown forewings and no dorsal streak. **Similar species** *A. semitestacella* **is much larger, paler orange-brown, with the white dorsal streak usually interrupted at three-fifths, and extending to the tornus.** **FS** Single-brooded, June-September. May be beaten from bushes, flies by day and comes to light. **Hab** Wherever the foodplant occurs. **Fp** Blackthorn; mines the flowering shoot.

Argyresthia semitestacella page 213

423 (1469)

Local. **FL** 6-7mm. Forewing orange-brown with a slight purplish reflection and a broad white dorsal streak extending to the tornus, often interrupted by the ground colour at about three-fifths. **Similar species** *A. albistria*. **FS** Single-brooded, August-September. This species is rarely found in numbers; it may be tapped from branches and occasionally comes to light. **Hab** Woodland. **Fp** Beech; in the shoot.

▼ Larva of *Argyresthia bonnetella* ready for pupation within a net-like cocoon.

▼ Larva of *Argyresthia albistria* on a flower of Blackthorn.

Plutellidae

There are seven species in this family. The adults are small to medium sized, with forewing length 6-11.5mm. The adult rests with the wings held roof-like, at a steep angle, with the antennae characteristically pointing forward. The forewings are elongate, often without a distinct tornus, although the cilia at the tornus may prouduce a noticeable angle between the dorsum and termen. The hindwings are as wide as the forewings. The head can be smooth-scaled or have narrow upright scales on the crown. The antennae are thread-like, except in *Eidophasia* where they are partly clothed in scales, and are about three-quarters the length of the forewing. The labial palps are long, curved and ascending.

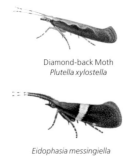

Diamond-back Moth
Plutella xylostella

Eidophasia messingiella

The family was formerly treated as a subfamily of the Yponomeutidae. All species are covered by accounts below and most species are readily identified. Several have a more northerly distribution. The family includes the familiar Diamond-back Moth *Plutella xylostella* which is a regular immigrant to Britain and Ireland. This species sometimes arrives on our shores in huge numbers, and breeds here in most summers. *P. xylostella* has a worldwide distribution; it is a pest of cabbages in sub-tropical regions, and has even been recorded in numbers within the Arctic Circle.

The larvae of British species feed on the cabbage family, Brassicaceae, typically in a slight web, and pupate in an open, net-like cocoon. Adults of most species fly at night and come to light. All *Rhigognostis* found in this country overwinter as an adult.

Further reading
British and Irish species: Emmet (1996)
European species: Bengtsson & Johansson (2011)

Diamond-back Moth *Plutella xylostella* page 213   464 (1525)

Common. **FL** 6-8.5mm. Typical form: head and thorax white; forewing greyish brown, somewhat mottled, with a whitish dorsal band with a sinuous margin, abruptly edged dark brown fading to the ground colour in the middle of the wing. Variation: head and thorax sometimes pale yellowish brown, pale orangey brown or pale grey; forewing markings may be paler or darker, with or without rows of tiny dark dots in paler examples, and occasionally the dorsal band and ground colour are unicolorous greyish. **FS** Several broods per year, recorded in every month, most frequently in the summer. Flies by day and comes to light. Numbers fluctuate annually and the moth is occasionally numerous.

Hab Found everywhere. **Fp** Many species of the cabbage family; in a slight web on the underside of the leaf or among the flowers and can cause fenestration of the leaves, with only the upper epidermis left intact as whitish patches.

▶ Larva of Diamond-back Moth *Plutella xylostella* preparing to change its skin under a slight web.

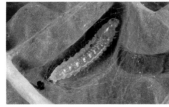

Plutella porrectella page 213

465 (1528)

Common. Local in some parts. **FL** 7-8.5mm. Forewing cream; a cream dorsal band with a sinuous margin, abruptly edged pale brownish grey, darker towards the base, and a series of pale brownish-grey streaks elsewhere, with blackish-brown dots on the dorsum and in a band along the termen. **FS** Double-brooded, late April-October. Can be disturbed from the foodplant by day and comes to light. **Hab** Wherever the foodplant occurs, including gardens, fens, marshes. **Fp** Dame's-violet; in a loose web, sometimes gregariously.

Plutella haasi page 213

465a (1529)

Rare. Known only from a very few sites in western Scotland. **FL** 6.5-8mm. Forewing pale brownish, sometimes greyish; a whitish or whitish mixed pale brownish dorsal band with a gently sinuous margin, abruptly edged dark brown fading to the ground colour in the middle of the wing, less defined towards the tornus, with a row of black marks along the dorsum and termen to the apex, and with a short row of black dots below the costa. **FS** Single-brooded, June-July. Has been found at rest on Thrift flowers and disturbed from foodplant. **Hab** Mountains. **Fp** Northern Rock-cress; in a silken spinning among the leaves.

Rhigognostis senilella page 213

466 (1533)

Local. Occasionally found in numbers. **FL** 9-11mm. Forewing mixed whitish and pale brown or greyish brown; a pale cross-band at the base from the dorsum not reaching the costa, and an oblique pale sub-triangular mark before one-half, sometimes joined to a pale area around the tornus, all pale markings more or less edged with black, with additional black marks on the termen. **FS** Single-brooded, August-April. Comes to light. **Hab** Rocky coasts, mountains. **Fp** Rock-cress, Flixweed, Dame's-violet; in a flimsy web.

Rhigognostis annulatella page 213

467 (1535)

Local. **FL** 8-11.5mm. Forewing greyish brown, or creamy mottled pale brown and blackish; a broad greyish-brown or creamy dorsal band with a sharply sinuous margin, along parts abruptly edged black, and with a few black marks along the termen. **FS** Single-brooded, July-April. Comes to light. **Hab** Mainly coastal grasslands with rocks. **Fp** Common Scurvygrass, Danish Scurvygrass, Hairy Bitter-cress; among spun flowers and leaves.

Rhigognostis incarnatella page 213

468 (1537)

Local. Recently found in west Wales. **FL** 8.5-10.5mm. Forewing dark pinkish brown; a creamy or pinkish-brown dorsal band with an irregular margin, abruptly and thickly edged black, fading to the ground colour towards the costa, with the terminal area usually paler, reddish brown or creamy brown. **FS** Single-brooded, September-June. Comes to light. **Hab** Has been recorded in woodland and a garden. **Fp** Dame's-violet, Hoary Whitlowgrass; in a thin web.

Eidophasia messingiella page 213

469 (1544)

Local. **FL** 7-8mm. Antenna clothed with black scales from the base to three-quarters, certain segments white beyond. Forewing dark brown with an inward oblique sandy-white cross-band, broadest at the dorsum, sometimes interrupted near the costa, and sometimes with a small sandy-white spot on the costa at two-thirds. **FS** Single-brooded, June-July. Comes to light. **Hab** Wasteland, embankments, rough grassland, vegetated shingle, also recorded in woodland and scrubby situations. **Fp** Hoary Cress; at first boring through the shoot then on the underside of a leaf. As an attacked shoot expands, the leaves show characteristic peppering with small holes created by the larval activity.

Glyphipterigidae

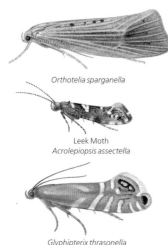

Orthotelia sparganella

Leek Moth
Acrolepiopsis assectella

Glyphipterix thrasonella

There are 14 species in this family. The adults vary between small and large micro-moths, with forewing length 3-14mm. The adult resting position is variable; in *Glyphipterix* the wings are slowly raised and lowered, unless the moth is very cold, while in all other genera the adult rests motionless with the wings held roof-like and at a steep angle. The forewings are moderately elongate, with a distinct tornus. The hindwings are broad, almost as broad as the forewings, or narrower. The head is smooth or with a tuft of scales on the crown. The antennae are thread-like, about three-fifths the length of the forewing. The labial palps are slender, upwardly curved, strongly so in *Orthotelia*, and the tongue is well developed, although it is rudimentary in *Orthotelia*.

Digitivalva, *Acrolepiopsis* and *Acrolepia* were formerly included within the Yponomeutidae. All species are covered here. Most *Glyphipterix* have a series of pale costal streaks and a pale dorsal hook-shaped streak that may lead to confusion with some *Cydia* species (Tortricidae), and most are superficially similar to each other, so care is needed with identification. Other species are relatively straightforward to identify. *Orthotelia sparganella* is included in the Glyphipterigidae, although it does not fit well on external morphology; it is much larger than the others, resembling one of the 'wainscot' macro-moths in the Noctuidae.

The most widespread and abundant species of *Glyphipterix* is the tiny Cocksfoot Moth *G. simpliciella*, which is often seen flying by day or at rest on tall grasses and on flowers.

Larvae feed internally on seeds, within stems, flowers and berries, or mine leaves. *Digitivalva pulicariae*, *Acrolepiopsis* and *Acrolepia* hibernate as adults. Adults fly by day or at night, and most species are attracted to light.

Further reading
British and Irish species: Heath & Emmet (1985) for *Glyphipterix* species; Emmet (1996) for others
European species: Bengtsson & Johansson (2011)

◄ Feeding damage caused by Leek Moth *Acrolepiopsis assectella* on onion. On thick leaf tissue the larva feeds within the hollow leaf, creating pale patches below the epidermis and depositing frass; on thin leaf tissue, such as in young leeks, the larva eats the whole shoot.

Orthotelia sparganella page 214 {470 (1576)}

Local. **FL** 9-14mm. Forewing brown with veins creamy, giving a streaked appearance, and often a few dark brown dots between veins near the dorsum. The female is much larger than the male, and usually paler. **Similar species *Chilo phragmitella* and *Calamatropha paludella* (Crambidae), and macro-moths such as Silky Wainscot *Chilodes maritima* (Noctuidae), which all have a similar blunt apex but rest with the wings flat, not roof-like. FS** Single-brooded, end June-August. Flies at dusk and sometimes comes to light. **Hab** Wetlands. **Fp** Bur-reeds, irises, Reed Sweet-grass; mines the leaves and stems.

Digitivalva perlepidella page 214 {471 (1555)}

Scarce. **FL** 5-6mm. Forewing purplish; pale brownish creamy markings comprising a narrow cross-band before one-half, with one dorsal and two costal spots beyond, sometimes fused and with the markings mixed purplish, and broadly edged orangey brown except at the tornus. **FS** Single-brooded, May-June. Flies by day. **Hab** Chalk downland, limestone hills. **Fp** Ploughman's-spikenard; mines the lower leaves.

Digitivalva pulicariae page 214 {472 (1556)}

Local. **FL** 6-7mm. Forewing greyish brown sometimes mixed orangey brown towards the dorsum, with scattered blackish, brownish and whitish scales; two rather indistinct whitish marks on the dorsum, at one-quarter and one-half, the latter subquadrate and sometimes extended faintly towards the costa, containing wavy greyish-brown lines perpendicular to the dorsum. **Similar species *Acrolepiopsis assectella* is usually slightly larger, with the dorsal mark before one-half more triangular, clear white or containing only a short greyish-brown line. FS** Single-brooded, and can be found in any month of the year, but most frequent in spring/summer; the adult hibernates. Flies by day in warm weather, in the evening and comes to light. **Hab** Wherever the foodplants occur. **Fp** Fleabane, occasionally Hemp-agrimony; mines the leaves.

Leek Moth *Acrolepiopsis assectella* page 214 {473 (1565)}

Local, but spreading north. Can be abundant in the south and an occasional pest of leeks. **FL** 5.5-7.5mm. Forewing greyish brown with scattered dark brown and white scales, with heavy mottling of white in the terminal area; a distinct clear white triangular mark from the dorsum before one-half, sometimes containing a short brown line or a few brown scales, and occasionally a small oblique white mark nearer the base and one or two small white marks near the tornus. **Similar species *A. betulella* is brown, not greyish brown, and brown near the termen with a smaller white mark on the dorsum**; also *D. pulicariae*. **FS** Double-brooded, and can be found in any month of the year, but most frequent in late summer/autumn; the adult hibernates. Flies in the evening and comes to light. **Hab** Gardens, allotments. **Fp** Onions, leeks, garlic; mines and feeds within the leaves and shoots.

Acrolepiopsis betulella page 214 {474 (1567)}

Very local. **FL** 6-7mm. Forewing brown, marbled darker brown, with a small white triangular spot on the dorsum before one-half, sometimes containing one or two blackish dots, white dots scattered elsewhere in the wing and sometimes with a black streak in the middle of the wing before the apex. **Similar species** *A. assectella*. **FS** Single-brooded, July-May; the adult hibernates. Rarely seen as an adult. **Hab** Shaded woodland. **Fp** Ramsons; within the flowers and seeds.

Acrolepiopsis marcidella page 214 475 (1569

Very local. **FL** 6.5-7.5mm. Forewing pale orangey brown with a small white spot on the dorsum before one-half, sometimes edged blackish, usually with an oblique black streak from the costa beyond one-half to the middle of the wing, and whitish dots scattered elsewhere; the termen and cilia are somewhat sinuate. **FS** Single-brooded, October-early July; the adult hibernates, although some adults emerge from pupae in early spring. Rests by day deep in bushes and can be disturbed from them, flies around the foodplant at dusk, and very occasionally comes to light. **Hab** Deciduous woodland, hedgerows, hedge banks. **Fp** Butcher's-broom; mines the bark, cladodes and later the berries.

Acrolepia autumnitella page 214 476 (1572

Local. Seemingly becoming more widespread. Recorded new to Ireland in 2009. **FL** 5-6mm. Forewing somewhat coppery brown mottled with blackish-brown marks, a small whitish- or yellowish-brown triangular mark on the dorsum before one-half, and a scattering of white scales towards the termen, sometimes forming white cross-lines, and a short black streak at about four-fifths. **FS** Double-brooded, and can be found in any month of the year; the adult hibernates. Flies freely on warm, still evenings and comes to light. **Hab** Wherever the foodplants occur. **Fp** Bittersweet, occasionally Deadly Nightshade and Tomato; mines the leaves.

Cocksfoot Moth *Glyphipterix simpliciella* page 214 391 (1594

Common. Often abundant. **FL** 3-4.5mm. Forewing dark leaden grey with narrow silvery-white streaks, a single silvery spot before the termen and a black spot at the apex. Hindwing with cilia grey throughout. **Simlar species** Very similar to *G. schoenicolella*, **which has the forewing with a bolder dorsal streak at one-half, and the hindwing with white cilia towards the base; *G. forsterella* and *G. equitella* are both rather larger species with two silvery spots before the termen, and *G. forsterella* has a silver dot in the black apical spot.** **FS** Single-brooded, May-early July. Flies in sunshsine and occasionally comes to light. **Hab** Unmanaged grassland. **Fp** Cock's-foot, Tall Fescue; within the seeds. The larva enters the stem to pupate and infested stems can be detected by the presence of small holes along their length.

Glyphipterix schoenicolella page 214 392 (1595

Local. **FL** 3-4mm. Forewing dark leaden grey with silvery-white streaks, a single silvery spot before the termen and a black spot at the apex. Hindwing with white cilia towards the base. **Similar species** *G. simpliciella, G. equitella, G. forsterella.* **FS** Single-brooded, May-September. Can be found in abundance by day at rest on or flying around the foodplants and occasionally comes to light. **Hab** Wet heaths, moors, less often neutral grasslands. **Fp** Black Bog-rush, Toad Rush, Sea Club-rush, probably sedges; in the seeds.

Glyphipterix equitella page 214 393 (1587

Local or very local. **FL** 4-5mm. Forewing pale greyish brown near the base, darker brown in the rest of the wing, with bold silvery streaks and two silvery spots before the termen. **Similar species** *G. simpliciella, G. schoenicolella, G. forsterella.* **FS** Single-brooded, June-July. Flies by day and comes to light. **Hab** Coastal cliffs, stone walls. **Fp** Biting Stonecrop, English Stonecrop; mines the leaves and stem.

Glyphipterix forsterella page 214 394 (1592

Local. **FL** 4-6.5mm. Forewing greyish dark brown, a prominent silvery streak on the dorsum at one-half, with narrower streaks elsewhere, two silvery spots before the termen, and a silver dot within the black spot at the apex. **Similar species** *G. simpliciella, G. schoenicolella, G. equitella.* **FS** Single-brooded, May-June. Rests by day on the flowers of sedges and comes to light. **Hab** Woodland rides. **Fp** Sedges; in the seeds.

► Cocksfoot Moth *Glyphipterix simpliciella*. Above shows an entrance hole in a grass stem. Below shows the grass stem and cocoon split open to reveal the hibernating larva within.

►▼ Larva and leaf mines of *Acrolepia autumnitella*. Note the larva frequently changes mines, making several irregular blotches per leaf.

▼ Butcher's-broom berry containing a larva of *Acrolepiopsis marcidella*. Note the differential ripening of the berry, dark mines beneath the rind, and the entrance hole, caused by the larva.

Glyphipterix haworthana page 214   395 (1591)

Local. **FL** 6.5-7.5mm. Forewing greyish dark brown, a prominent silvery streak on the dorsum at one-half, with narrower streaks elsewhere, two silvery spots before the termen, and a black spot at the apex. **The largest species of *Glyphipterix*, with dark forewings and contrasting silvery-white streaks. Similar species *Grapholita lunulana* and *G. orobana* (Tortricidae) have no streak at the tornus and short streaks along the costa.** **FS** Late April-May. Rarely seen as an adult. **Hab** Bogs. **Fp** Cottongrass; in the seeds, spinning the tenanted seedhead to the stem or other vegetation so that it remains above ground over winter.

Glyphipterix fuscoviridella page 214   396 (1584)

Common. **FL** 5-8mm. Forewing brassy, unmarked. **Similar species Unmarked forms of *G. thrasonella*, which have a pronounced lobe at the apex.** **FS** Single-brooded, end April-June. Flies weakly in sunshine just above the foodplant, at dusk, and comes to light. **Hab** Short acid or calcareous grassland. **Fp** Field Wood-rush; in the stems, mining down into the centre of the plant.

Glyphipterix thrasonella page 214   397 (1580)

Common. Often abundant where it occurs. **FL** 5-7.5mm. Forewing with a pronounced lobe at the apex, brassy, with distinctive streaks which appear shining white or blue depending upon orientation of the light; these markings may be much reduced or virtually absent. The female has a narrower forewing than the male, with a more arched costa. **Similar species** Unmarked forms are similar to *G. fuscoviridella*. **FS** Single-brooded, May-August. Flies by day in sunshine and comes to light. **Hab** Wet grasslands, moorland, bogs, mires. **Fp** Rushes; possibly in the stem or seeds.

Ypsolophidae

There are 16 species in this family, split
between two genera in Britain and Ireland,
Ypsolopha and *Ochsenheimeria*. Adults of
most species rest with the wings held tightly
appressed to the body, but are wrapped around
the body in a few, particularly *Ochsenheimeria*.
Ypsolopha are all medium-sized micro-moths,

Honeysuckle Moth
Ypsolopha dentella

Ochsenheimeria taurella

forewing length 7-16mm, whereas *Ochsenheimeria* are small, forewing length
4.5-6mm. In some species of *Ypsolopha* the resting position is declining, but in
others and in *Ochsenheimeria* the adults lie almost horizontal or slightly inclining.
The forewings of *Ypsolopha* are elongate with a distinct tornal angle, and the apex
of some species is hooked. The forewings of *Ochsenheimeria* are short compared
with the abdomen, which may extend slightly beyond them, and are without a tornal
angle. The hindwings are about as broad as the forewings. The head of *Ypsolopha* has
flattened or loose scales, while in *Ochsenheimeria* it is densely covered in long, spiky
scales, with a smooth face. The length of the antennae is between one-half and three-
quarters the length of the forewing; in *Ochsenheimeria* the antennae may be clothed
in scales and are usually held at a wide angle or pointing forward. The labial palps
are moderately long and curved, forward pointing or upwardly curved; **in *Ypsolopha***
there is a scale-tuft on the underside of the second segment of the labial palp
which is long and forward-pointing, giving species a characteristic furry snout,
similar to that in the genus *Crambus* (Crambidae). The tongue is present.

Ypsolophidae was formerly treated as part of the Yponomeutidae. All but one species
of *Ypsolopha* are covered below, with *Ypsolopha asperella* now being considered
extinct; most are distinctive and readily identified. One of the three *Ochsenheimeria*
species is covered; adults are active during the day, flying in sunshine around
midday, but are otherwise difficult to detect. They can be swept from grassland, and
aggregations of *O. vacculella* are sometimes found under loose bark on trees. Since
the adults are rarely encountered, all species may be more numerous than records
suggest.

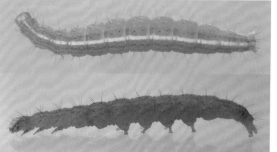

The larvae of *Ypsolopha*
feed on various trees and
shrubs. They are long
and narrow, and tapered
at both ends. Larvae of
all species feed in the
same way. They rest on

◄ Larva of *Ypsolopha scabrella*, dorsal
and lateral views.

a slight web on the underside of a leaf and when disturbed become very active and drop readily to the ground. They pupate in a papery cocoon shaped like the hull of a boat, usually near the ground. All *Ypsolopha* species fly at night and come to light; three species hibernate as adults. The larvae of *Ochsenheimeria* all feed within grass stems. While seemingly difficult to find in Britain and Ireland, *Ochsenheimeria* can become occasional pests of cereals crops in eastern Europe.

Further reading
British and Irish species: Heath & Emmet (1985) for *Ochsenheimeria*; Emmet (1996) for *Ypsolopha*
European species: Bengtsson & Johansson (2011)

Ypsolopha mucronella page 215   451 (1480)

Local. **FL** 13-16mm. Forewing long and narrow, pointed at the apex, buff coloured, veins paler, with a narrow brownish or blackish streak from the base sometimes extending to the tornus, small blackish scale-tufts towards the dorsum, and a few blackish dots elsewhere. **FS** Single-brooded, August-June; the adult hibernates. Comes to light. **Hab** Woodland and scrub on calcareous soils. **Fp** Spindle.

Ypsolopha nemorella page 215   452 (1481)

Local. Very local in north-west England. **FL** 10-12mm. Forewing broad and hooked at the apex, buff coloured with veins darker, sometimes with a brownish streak extending from the base to the tornus, a black spot towards the dorsum at one-half, and scattered black scales elsewhere. **FS** Single-brooded, June-August. Comes to light. **Hab** Woodland, scrub. **Fp** Honeysuckle.

Honeysuckle Moth *Ypsolopha dentella* page 215   453 (1482)

Common. **FL** 10-11mm. Forewing broad and hooked at the apex, chestnut-brown; a broad band along the dorsum from the base to the tornus, varying from white to pale brownish yellow, with a thin white line extending obliquely into the middle of the wing. **FS** Single-brooded, mid June-early September. Comes to light. **Hab** Woodland, scrub, gardens. **Fp** Honeysuckle.

Ypsolopha scabrella page 215   455 (1486)

Common in much of England. More local in the west. **FL** 7-11mm. Forewing broad, forming a lobe beyond the tornus and pointed at the apex, grey, heavily streaked whitish brown in the costal half, and streaked or marked brownish or blackish in the dorsal half, with several blackish scale-tufts towards the dorsum. **FS** Single-brooded, June-October. Comes to light. **Hab** Woodland, scrub, gardens. **Fp** Hawthorns, apples, occasionally Cotoneaster.

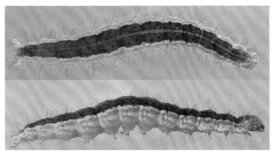

► Larva of Honeysuckle Moth *Ypsolopha dentella*, dorsal and lateral views.

Ypsolopha horridella page 215
456 (1488)

Local. Scarce in northern England. **FL** 7-10mm. Forewing extended to a broad point at the apex, brown, variably mottled darker brown and somewhat greyish in a band along the dorsum containing three small black scale-tufts. **FS** Single-brooded, July-September. Comes to light. **Hab** Woodland, scrub, gardens. **Fp** Blackthorn, apples.

Ypsolopha lucella page 215
457 (1489)

Scarce. **FL** 8-9mm. Resting posture steeply declining. Head and thorax white. Forewing very broad with a slightly concave termen, giving the appearance of a hook at the apex, dull yellowish brown, often reticulated darker, sometimes with a narrow white band along the dorsum. **Similar species Y. alpella and Y. sylvella, both of which rest horizontally and have the thorax the same colour as the forewing, not white.** FS Single-brooded, July-August. Comes to light. **Hab** Woodland, scrubby heathland. **Fp** Oaks.

Ypsolopha alpella page 215
458 (1491)

Local in the south. Very local in the north. **FL** 7-9mm. Thorax straw coloured, head paler. Forewing straw coloured, reticulated darker, with two oblique cross-bands extending from the dorsum to near the middle of the wing, not beyond. **Similar species Y. sylvella has cross-bands which extend to the costa, sometimes interrupted about the middle of the wing**; see also Y. lucella. **FS** Single-brooded, July-October. Comes to light, sometimes found away from woodland. **Hab** Woodland. **Fp** Oaks.

Ypsolopha sylvella page 215
459 (1492)

Local. **FL** 8-9.5mm. Head and thorax straw coloured. Forewing straw coloured, weakly reticulated darker, with two oblique cross-bands extending from the dorsum to the costa, often interrupted in the middle of the wing. **Similar species** Y. alpella and Y. lucella. **FS** July-November. Comes to light, sometimes found away from woodland. **Hab** Woodland. **Fp** Oaks.

Ypsolopha parenthesella page 215
460 (1493)

Common. **FL** 7-9mm. Head white or pale yellowish brown, thorax white, yellowish brown, brown or greyish brown. Forewing orangey brown, occasionally pale brown, sometimes unicolorous, but usually with a white patch in the costal half from the base to at least one-half. **Similar species Y. ustella does not have a form with a white costa contrasting with an orangey-brown forewing; unicolorous forms of Y. parenthesella have a white or pale yellowish-brown head; unicolorous forms of Y. ustella have the head concolorous with the forewing; Y. ustella overwinters as an adult, Y. parenthesella does not**. FS Single-brooded, July-early October. Comes to light. **Hab** Woodland. **Fp** Oaks, Hazel, Hornbeam, birches.

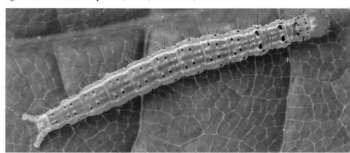

▶ Larva of
Ypsolopha
parenthesella.

Ypsolopha ustella page 215
461 (1494)

Common. **FL** 7-10mm. Forewing very variable in colour, from whitish, through light greyish to several shades of brown, sometimes unicolorous and almost unmarked, or with variable mottling of brown or black, or streaked whitish along the veins with a black band from the base to near the apex. **Similar species** *Y. parenthesella*. **FS** Single-brooded, July-April. Can be tapped from dense vegetation in the winter and comes to light. **Hab** Oak woodland. **Fp** Oaks.

Ypsolopha sequella page 215
462 (1495)

Common. Spreading in Scotland. **FL** 8-10mm. Head and thorax white. Forewing white, with several black spots along the dorsum and smaller black marks along the costa and in the middle of the wing. **FS** Single-brooded, July-October. Comes to light. Rarely seen in numbers. **Hab** Woodland, scrub. **Fp** Field Maple, sometimes Sycamore, especially in Scotland.

Ypsolopha vittella page 215
463 (1496)

Local. **FL** 8-10mm. The typical form has the forewing whitish grey, mottled with darker grey or brownish dots and short oblique lines, with black or brown markings along the dorsum, sometimes joined to form a narrow band. A darker form has the forewing almost entirely blackish, with a brownish-grey band along the costa. **FS** Single-brooded, July-September. Comes to light. **Hab** Woodland. **Fp** Elms, Beech.

Ochsenheimeria taurella page 215
251 (1520)

Local. **FL** 5.5-6mm. Head with erect, spade-shaped scales which are split at the apex. Antenna with over half its length heavily thickened with long scales, those near the middle three times the width of the antenna. Forewing yellowish brown or greyish brown, with loose, raised scales in the middle, and in paler examples often scattered with darker scales; cilia dark tipped along the termen, especially noticeable on the underside of the forewing. **Similar species** *O. urella* (not illustrated), a widespread species more frequent in the north, has the antenna with scales near the middle less than three times the width, with a more uniform pale yellowish-brown forewing with terminal cilia not or hardly dark-tipped; Cereal Stem Moth *O. vacculella* (not illustrated; see photograph below), a very local species in England, has thread-like, dark-ringed antennae, and scales on the forewing hardly raised. **FS** Single-brooded, July-September. Flies in sunshine for a short while around midday. **Hab** Grasslands. **Fp** Various grasses; mines the lower part of the stem, causing upper parts to wilt, the larva changing stems frequently.

▶ Adult of Cereal Stem Moth *Ochsenheimeria vacculella*. The characteristic thread-like antenna is just visible adjacent to the forewing costa.

Praydidae

There are currently five species listed in this family, split between two genera, *Prays* and *Atemelia*. At rest, the wings are held roof-like, at a steep angle, with the moth in a slightly inclining posture and the antennae along the

Ash Bud Moth
Prays fraxinella

side of the body. *Prays* are all medium-sized micro-moths, forewing length 6-8mm, while *Atemelia*, represented by a single species, is small, forewing length 4-5mm. The forewings are elongate, moderately broad, the membrane without a distinct tornal angle, although the cilia at the tornus produce a noticeable angle between the dorsum and termen. The hindwings are similar in breadth to the forewings. The antennae are thread-like, about one-half the length of the forewing. The labial palps are curved upwards.

Praydidae was formerly treated as a subfamily of the Yponomeutidae. Of the species within *Prays*, Ash Bud Moth *P. fraxinella* is native and the most widespread; the small, uniformly dark form is recognised as a separate species in Europe, *P. ruficeps*, and in Britain and Ireland the same taxonomic distinction should probably be followed, although, as yet, there are no published records. The other species, *P. citri*, *P. peregrina* and *P. oleae*, are probably accidental imports. *P. peregrina* appears to be resident, and the other two are occasionally imported with nursery stock, *P. citri* on *Citrus* trees and *P. oleae* on Olive trees. *Atemelia torquatella* is native and a northern species in Britain. The resident species are covered below.

Larvae feed from within a bud or mine the leaves and bark of the foodplant. Adults fly at night and come to light.

Further reading
British and Irish species: Emmet (1996)
European species: Bengtsson & Johansson (2011)

▼ Larva of Ash Bud Moth *Prays fraxinella* feeding on Ash.

▲ Feeding signs of Ash Bud Moth *Prays fraxinella* on Ash.

Atemelia torquatella page 216   448 (1422)

Very local. A northern species. **FL** 4-5mm. Forewing dark brown, with a creamy white spot on the costa before the apex and one at the tornus, these occasionally almost joined, with another on the wing at about one-third. **The slightly inclining posture distinguishes this species from similar-looking species in the Elachistidae. FS** Single-brooded, June-July. Adults have been found feeding on flowers of Mountain Everlasting.
Hab Mosses and lower slopes of mountains, among areas of regenerating birch.
Fp Birches, especially Downy Birch seedlings, Bog-myrtle; forms blotch mines, sometimes with several larvae to a mine; a slight web is spun beneath the mine, and frass ejected from the mine collects in the web.

Ash Bud Moth *Prays fraxinella* page 216   449 (1424)

Common. **FL** 7-8mm. Head and thorax predominantly white. Forewing white, with a large triangular blackish-brown blotch on the costa almost reaching the dorsum, and blackish markings along the termen. A melanic form occurs which has white replaced with dark grey-brown except on the head; other forewing markings can usually be discerned faintly.
Similar species *P. ruficeps* (not illustrated) is slightly smaller, FL 6-7.5mm, has the head orangey brown and thorax and forewings unicolorous dark brown. **FS** Probably double-brooded, May-October. Comes to light. **Hab** Woodland, hedgerow trees, parks, gardens.
Fp Ash; initially mining a leaf then the bark of a twig, and in spring and summer feeds externally in a spinning, especially among Ash flower galls. *P. ruficeps* feeds similarly.

Prays peregrina page 216   449b (not listed)

Rare. Presumed adventive, probably now resident. First recorded in this country in London, in 2003, and described new to science from Britain in 2007. **FL** 6mm. Head pale grey, thorax brown. Forewing white, speckled with dark dots, with a brownish curved triangle on the dorsum at about one-half and a brownish mark at the tornus. **FS** Number of broods not known, June-October. Comes to light. **Hab** Suburban gardens.
Fp Unknown.

Bedelliidae

Bedellia somnulentella is the sole representative of this family found in Britain and Ireland. The forewing length is 3.5-4.5mm. The adult rests in an inclining posture, the tip of the abdomen and wings on the ground, standing only on the midlegs and hindlegs, and the wings held rolled around the abdomen. The forewings are narrow and tapering, without a tornal angle. The hindwings are elongate, narrow, the apex pointed, with long dorsal cilia. The crown of the head has erect hair-like scales. The antennae are thread-like, about as long as the forewing, with a thickened scape. The labial palps are short.

Bedellia somnulentella

The larva of *B. somnulentella* is a leaf miner. It pupates outside the mine, attached to the underside of a leaf; the pupa has a keel and a pronounced beak. The adult is sometimes abundant near the southern coastline, and it seems likely that its numbers may be boosted by immigration from Europe during warm summers.

Further reading
British and Irish species: Heath & Emmet (1985)
European species: Bengtsson & Johansson (2011)

Bedellia somnulentella page 216   264 (1602)

Common in the south. Local further north. **FL** 3.5-4.5mm. Forewing pale brownish, speckled darker, sometimes with a small dark spot on the dorsum at one-quarter, at one-half and at the tornus. **Similar species** *Aspilapteryx tringipennella* **(Gracillariidae) sits on its forelegs and midlegs, has the head with smooth scales, and has dots on the forewing arranged in rows. FS** Double-brooded, August-early September, October-May; the adult hibernates. Flies in warm sunshine, may be disturbed during the day by tapping vegetation and comes to light. **Hab** Woodland margins, hedgerows, wasteland, urban habitats, coastal habitats. **Fp** Field Bindweed, Hedge Bindweed, Morning-glory; mines the leaf, making irregular pale brown blotches from which frass is expelled. It frequently changes leaves; a network of loose silken threads under the leaf may help the larva to move between mines.

▶ Larvae and leaf mines of *Bedellia somnulentella* on bindweed.

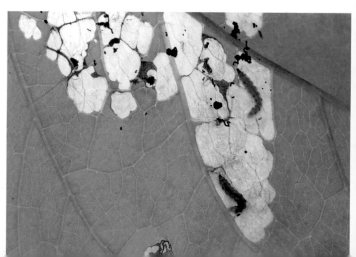

Lyonetiidae

Laburnum Leaf Miner *Leucoptera laburnella*

Apple Leaf Miner *Lyonetia clerkella*

There are nine species in this family. Adults rest with the front end raised, except in *Lyonetia*, the wings held in a roof-like position, at a steep angle. Members of this family are all very small, with the forewing length 2.5-4.5mm. The forewings are elongate and mostly shining white with short dark lines, not unlike some white Gracillariidae. The hindwings are elongate and narrow, the apex pointed, with long dorsal cilia. The head has erect scales on the crown. The antennae are long, thread-like, between two-thirds and as long as the length of the forewing, with the scape dilated to form an eye-cap. The labial palps are variable in their development, from rudimentary to long. The mouthparts are weakly developed.

Two species are covered below, Apple Leaf Miner *Lyonetia clerkella*, which is an extremely widespread and often abundant moth, regularly seen as a leaf mine and as an adult in moth traps, and Laburnum Leaf Miner *Leucoptera laburnella*, which is commonly seen wherever Laburnum trees are planted. *Leucoptera* is the main genus; most species are bright white with thin costal and apical lines, orange or yellow markings between, and a patch of leaden metallic scales at the tornus. They are superficially similar and care is needed in the identification of adults. Seven are recognised in Britain and Ireland overall, although this number has changed from time to time as one or another taxon has been raised to specific status, or sunk to a subspecies. One of the seven, *Leucoptera sinuella*, is considered to be extinct.

▲ Tenanted leaf mine of Apple Leaf Miner *Lyonetia clerkella* on cherry. Note the long, thin mine with linear frass; the larva is also longer and thinner than any in the Nepticulidae.

▶ Cocoon of Apple Leaf Miner *Lyonetia clerkella*.

Larvae of Lyonetiidae form linear or blotch mines on herb, shrub and tree species mainly in the Pea family (Fabaceae) or the Rosaceae. The combination of mine shape and host plant are usually sufficient to determine species, which is especially helpful in identifying *Leucoptera*. The cocoon of *Lyonetia* is suspended by two or three silken threads. Pupae can overwinter more than once, which may explain the decline of certain species in some years, in particular Pear Leaf Blister Moth *Leucoptera malifoliella*, which can be apparently absent in one year, then widespread the next. Adults of *Leucoptera* tend to rest motionless during the day on leaf tips, where they are conspicuous, which has led some to consider that they may be mimicking tiny wasps, the thin lines on the forewing representing the legs, the orange marks the thorax, and the metallic tornal spot the eye. Adults fly mainly in evening sunshine and occasionally come to light.

Further reading
British and Irish species: Heath & Emmet (1985)
European species: Bengtsson & Johansson (2011)

Laburnum Leaf Miner *Leucoptera laburnella* page 216 254 (1610)

Common. **FL** 3.5-4.5mm. Forewing shining white, two oblique dark-edged yellow bars on the costa at two-thirds and four-fifths, three straight dark grey lines in the white cilia at the apex, and a patch of leaden metallic scales at the tornus. **Similar species** All other *Leucoptera* (not illustrated); if found away from Laburnum, examples should be carefully checked. Species with forewing white: *Leucoptera laburnella* f. *wailesella* is smaller (FL 2.5-3.5mm) and very local in England, on Dyer's Greenweed; *L. spartifoliella* is common, on Broom and Dyer's Greenweed; *L. lathyrifoliella* is rare, on Narrow-leaved Everlasting-pea; *L. orobi* is rare, on Bitter-vetch. Species with forewing grey: *L. lotella* is local on downs and in marshes, on Common Bird's-foot-trefoil and Greater Bird's-foot-trefoil; Pear Leaf Blister Moth *L. malifoliella* is widespread, on apple and pear. **FS** Double-brooded, May, July-August. Sometimes abundant around Laburnum trees, flies at dusk and occasionally comes to light. **Hab** Mainly urban, wherever Laburnum trees are planted; f. *wailesella* on chalk and neutral grassland. **Fp** Laburnums, Dyer's Greenweed, occasionally Garden Lupin; initially in a blotch, then in an irregular mine and later in a large blotch.

Apple Leaf Miner *Lyonetia clerkella* page 216 263 (1627)

Common. **FL** 4-4.5mm. Typical form: head white or whitish, forewing elongate, shining white with an orangey brown spot at two-thirds towards the costa, an angular greyish-brown cross-band beyond, and a brownish blotch in the apical area; the apex has a distinct black dot, from which a few dark lines radiate out into the cilia. Melanic form: head dark brown, forewing unicolorous, metallic dark brown, tinged golden. Intermediate forms also occur. **Similar species** *L. prunifoliella* (not illustrated), a rare former resident, now seen as an occasional immigrant but possibly overlooked. It is slightly larger (FL 4.5-5mm) and has rather variable white and reddish-brown forewing markings; the most similar form has the forewing shining white, with a brown outwardly curved streak from the dorsum at one-half. It is single-brooded, September-May, and the adult hibernates. **FS** Up to three generations a year, mainly May-October, adults of the autumn brood hibernating until spring, although rarely seen. Can be abundant. Comes to light. **Hab** Woodland, scrub, heathland, moorland, urban parks, gardens, waste ground. **Fp** Hawthorns, birches, apples, pears, plums, cherries, Blackthorn, Whitebeam, Rowan, Cockspurthorn, cotoneasters, occasionally sallows and hops; in a very long, linear mine containing a narrow line of frass.

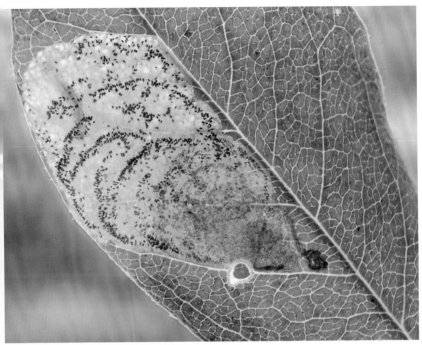

▲ Tenanted leaf mine of Laburnum Leaf Miner *Leucoptera laburnella* on Laburnum.

▼ Two leaf mines of Pear Leaf Blister Moth *Leucoptera malifoliella* on apple. Note the rather round mine and circular frass pattern which distinguish this species from the mine of *Bohemannia pulverosella* (Nepticulidae).

Douglasiidae

Tinagma ocnerostomella

There are two species in this family, both belonging to the genus *Tinagma*. Adults rest with the front end raised, the wings held in a roof-like position. They are very small or small, with forewing length 3.5-4mm. The forewings are narrow, the membrane without a tornus, although the cilia at the tornus produce a noticeable angle between the dorsum and termen. The hindwings are elongate and very narrow, the apex pointed, with long dorsal cilia. The head has smooth scales. The antennae are thread-like, about two-thirds the length of the forewing and without an eye-cap. The labial palps are short and point forwards, and the tongue is not scaled.

Both species are associated with Viper's-bugloss. Adults are found in early to mid summer, and can be found flying in sunshine around the foodplant.

Further reading
British and Irish species: Heath & Emmet (1985)
European species: Bengtsson & Johansson (2011)

Tinagma ocnerostomella page 216 398 (1037)

Local in the south and east. **FL** 3.5-4mm. Forewing grey, finely speckled whitish. **Similar species** *T. balteolella* (not illustrated), a rare species confined to the coast of south-east England, is slightly larger with a pale spot or cross-band. Could also be confused with a dark *Elachista* (Elachistidae). **FS** Single-brooded, late May-mid July. Can be numerous where it occurs. May be found flying around the foodplant on warm days and comes to light. **Hab** In sparsely vegetated places where the foodplant occurs, including coastal cliffs, vegetated shingle, chalk downland, waste ground. **Fp** Viper's-bugloss; feeds and pupates in the pith of a stem. Easily reared by collecting dead stems in winter.

▼ Feeding signs of *Tinagma ocnerostomella* within a dead stem of Viper's-bugloss. Note the pale-coloured frass leading up to the cocoon and the exit chamber leading to the outer surface of the stem, prepared in advance by the larva.

Autostichidae

There are four species in this family. Adults rest almost flat to the ground, with wings partly overlapping and antennae over the forewing, curving towards or above the costa. The forewing length is 5-7.5mm. The forewings of *Oegoconia* are quite narrow and rounded at the apex, mainly blackish, with pale yellow cross-bands. The hindwings are similar in width to the forewings. The head has smooth scales. The antennae are thread-like, about two-thirds the length of the forewing. The labial palps are long, curving upwards in front of the face, and the tongue is scaled at the base.

Oegoconia quadripuncta

The three superficially similar *Oegoconia* species are difficult to identify with certainty. The remaining species, *Symmoca signatella*, is a rare accidental importation. The larvae are associated with leaf litter or dried vegetable matter. Adults fly at night and come to light.

Further reading
British and Irish species: Emmet & Langmaid (2002a)

Oegoconia quadripuncta page 216 870 (2941)

Local. **FL** 5.5-7.5mm. Thorax pale yellow, dark brown towards the head. Forewing dark brownish, with a pale yellow spot near the base, a broad pale yellow cross-band at about one-half and another at about two-thirds. **Similar species** *O. deauratella* and *O. caradjai* are sufficiently similar, and variable in wing pattern, that reliable identification of all three species can be achieved only by examination of genitalia (see male genitalia on p.389). However, certain characters may be helpful in separating typical forms: *O. deauratella* is the darkest of the three species, with the central cross-band generally narrower than in *O. quadripuncta*; *O. caradjai* has a slightly narrower forewing than the other two species, and is the palest of the three, usually with the broadest middle cross-band, and a more-or-less distinct and broad cross-band at one-quarter which is rarely present in *O. quadripuncta*. *O. quadripuncta* is the most widespread species, *O. caradjai* is local in the south and east, and *O. deauratella* appears to be spreading north and west from the south-east. **FS** Single-brooded, late June-September. Comes to light. **Hab** A wide range of habitats, sometimes plentiful at the bases of hedges.

▶ Larva of *Oegoconia caradjai* in leaf litter.

Fp Decaying leaves and other vegetable matter; in a slight web.

Blastobasidae

There are seven species in this family. Adults rest flat to the ground with the wings overlapping and usually wrapped around the abdomen, sometimes more splayed, and the antennae are held alongside the body. The forewing length is 6-11mm. The forewings

Blastobasis adustella

are elongate, without a tornal angle, dull grey, brown or yellowish brown and with weak markings. The hindwings are nearly the width of the forewings. The head is smooth, with flattened scales. The antennae are thread-like, two-thirds the length of the forewing, the scape with a pecten, this partly concealed by scales; the second segment of the antenna in the male is notched (see p.39). The labial palps are long, strongly curved upwards, with the third segment pointed, and the tongue is developed.

The family is widespread, being most diverse in the tropics; the majority of the Palaearctic species, about 20, occur on the island of Madeira. All species in Britain and Ireland are thought to be adventives, possibly having been accidentally introduced through the horticultural trade, and the two covered below have spread rapidly. A further two, *Blastobasis rebeli* and *B. vittata*, appear now to be resident, although rare, but it is likely these will spread in time. *B. phycidella* may still be resident in Guernsey but has not been observed there recently. The occurrence of *B. normalis* in Britain is open to question. More species may arrive in due course, so unusual-looking *Blastobasis* should be retained for critical examination.

Adults are obscurely marked and can be difficult to separate on wing pattern. Larvae feed on a diverse range of vegetable matter. Adults fly at night, mostly during the summer months, although sometimes later, and come to light. They can also be readily disturbed from their resting place by day, when they typically drop to the ground.

Further reading
British and Irish species: Emmet & Langmaid (2002a)

▶ Larva of *Blastobasis lacticolella*, dorsal and lateral views.

▲ Larva of *Blastobasis adustella*, exposed from within its silken tube.

Blastobasis adustella page 216   873 (2905)

Common. **FL** 6-9mm. Forewing elongate, cream to buff, variably marked dark brownish, with a pale oblique streak from the dorsum at about one-quarter. **Similar species** *B. vittata* (not illustrated), recorded only from Hampshire and Sussex, is usually smaller and more obscurely marked, often has a warmer, orangey brown ground colour, or is brownish grey, and sometimes has a broad dark cross-band at about one-third. This species can be separated reliably only by examination of genitalia, which are extremely similar to *B. adustella*. Also *B. rebeli* (not illustrated), recorded from Hampshire, Sussex, Lancashire and south Wales, which has a greyish ground colour and a broad blackish cross-band at one-third. **FS** Single-brooded, late June-October. Can be abundant in Yew woodland. Most often seen at light, but can be disturbed from cover by day. **Hab** Woodland, scrub, heathland, urban parks, gardens. **Fp** A variety of vegetable matter, including spongy oak-galls, dead juniper leaves and the empty seed-pods of Gorse; in a silken gallery.

Blastobasis lacticolella page 216   874 (2904)

Common. Expanding its range. Recorded new to Ireland in 2002. **FL** 6-11mm. Forewing cream to straw yellow, variably shaded brown, with a darker oblique streak from the dorsum at about one-third, this sometimes reduced to a spot, with two spots, placed vertically together, at about two-thirds. Some forms are rather plain with markings obscured, and completely unmarked forms occur. **FS** Possibly double-brooded, late May-November, but has also been recorded in late December. Can be numerous where found. Readily disturbed by day and comes to light. **Hab** Woodland, coastal scrub, urban parks, gardens. **Fp** Recorded on a range of unrelated foodstuffs, including shoots of Tamarisk, spongy oak-galls, rosehips, hawthorn berries, moss, dead insects and the dried skin of apples; in a silken gallery.

Oecophoridae

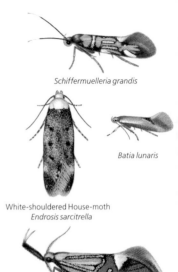

Twenty-seven species have been recorded in this family. At rest, the body is held near to horizontal in most species, slightly inclining in a few, e.g. *Pleurota bicostella*, and distinctively declining in *Crassa unitella*. The wings of most species are held in a steep roof-like position, but in a few, such as *Hofmannophila* and *Endrosis*, are nearly flat and overlapping, and in *Pleurota* are wrapped around the body. The forewing length varies from 3-11mm. The forewings are elongate, often fairly broad, and several species have a tornal angle. The hindwings can be as broad as the forewings, although they are narrower in some, with the apex more pointed. The head is smooth-scaled, sometimes rough on the crown or tufted at the sides. The antennae are between two-fifths and three-quarters the length of the forewing. They are often held along the sides of the body, but point forwards in some species, e.g. *Esperia*. The labial palps are often strongly upwardly curved, but sometimes point forward or are slightly drooping, and are occasionally long or very long, e.g. *Alabonia* and *Pleurota*.

Schiffermuelleria grandis

Batia lunaris

White-shouldered House-moth
Endrosis sarcitrella

Alabonia geoffrella

Most species are covered below. The adults of several species are brightly coloured or distinctively marked and are readily identified, although care may be needed to identify a few species. Some species are exceedingly local and possibly declining, e.g. *Schiffermuelleria grandis*, whereas others are widely distributed and may be found commonly, at least in the south of Britain, e.g. *Esperia sulphurella*. Two species are frequently found in houses, White-shouldered House-moth *Endrosis sarcitrella* and Brown House-moth *Hofmannophila pseudospretella*. Two species are thought to be extinct, *Denisia augustella* and *Epicallima formosella*, while *Borkhausenia minutella*, which declined to the point of extinction by the 1960s, was recorded once again in England in the 2000s, and it is now found locally in the Channel Islands, on Guernsey and Jersey. Two species have become established since 2000 and are very locally numerous: *Metalampra italica*, which is otherwise known only from Italy, and *Barea asbolaea*, a native of Tasmania.

The taxonomy of the Oecophoridae has been subject to recent change. Formerly, the family included members of the Lypusidae, Chimabachidae, Peleopodidae, Stathmopodidae, and the major genera *Depressaria* and *Agonopterix*, which are now included within the Elachistidae.

The larvae of most species feed on fungi on dead wood, on dead leaves, or on dry animal or vegetable matter. When associated with dead wood, larvae are often found underneath loose bark in slight silk tubes. Adults of most species are found in the summer months, with a few also found at other times of the year.

▲ Larva of *Schiffermuelleria grandis* on dead wood.

Many fly at night and come to light, others flying during the day in sunshine or on warm, still, overcast afternoons, and a few are active at or around sunrise.

Further reading
British and Irish species: Emmet & Langmaid (2002a)
European species: Palm (1989); Tokár *et al.* (2005)

Bisigna procerella page 217   639 (2242)

Rare. First recorded in Britain in 1976. Known only from a few sites in Kent and recorded once in East Sussex. **FL** 5-6mm. Forewing orange, with an oblique whitish line from the base of the costa to the dorsum, two others from the dorsum, not reaching the costa, at one-quarter and one-half, and a dark, somewhat irregular and suffused leaden metallic cross-band from the costa at two-thirds towards the tornus. **FS** Single-brooded, July. Comes to light. **Hab** Woodland. **Fp** Continental authors give lichens on tree trunks, but not known in Britain.

Schiffermuelleria grandis page 217   634 (2248)

Rare. Perhaps declining, possibly overlooked. **FL** 6-8mm. Forewing shining yellowish orange, tinted coppery, edged darker along the costa, termen and dorsum, with three silvery cream blotches, two metallic silvery lines, black-edged, near the middle of the wing and a black-edged silvery line forming a cross-band at one-quarter, this not quite reaching the dorsum or costa. **FS** Single-brooded, late May-June. Flies at dawn and in the first few hours of morning sunshine. Occasionally seen at light. Has been disturbed from a very ancient Ivy-covered stump. **Hab** Ancient woodland, scrub, hedgerows. **Fp** Under bark and within dead wood, including standing dead branches of oaks, Beech and Gorse, trunks of Ivy, rotten fence posts; in a slight web.

Denisia subaquilea page 217   635 (2266)

Local. Predominantly northern. **FL** 5-6.5mm. Forewing cream, suffused brown or blackish brown with ill-defined darker blotches at one-third, two-thirds and the tornus, the first two sometimes joined. **FS** Single-brooded, late May-July. Males fly soon after sunrise, in late afternoon and early evening. **Hab** Heathland, upland moorland to 1,000m; also limestone grassland in North Wales and cliffs in south-west Britain. **Fp** Dead leaves of Bilberry; in a slight silk spinning incorporating frass.

Denisia similella page 217   636 (2262)

Local. A northern species, found patchily in southern Scotland and northern England. **FL** 6-7.5mm. Forewing dark greyish brown, darker in the female, thinly scattered with yellowish scales, and with four variably shaped yellow blotches, the smallest at the tornus. **FS** Single-brooded, June-July. Has been found sitting on tree trunks. Flies at dawn, late evening and dusk, and comes to light. **Hab** Parkland, coniferous and deciduous woodland, especially those containing old or decaying birch trees. **Fp** Fungi, including *Daedalea quercina* on an oak stump, *Piptoporus betulinus* on dying birch, also *Fomes fomentarius*; within the fungus or dead wood.

▲ Larva of *Crassa unitella* under dead bark.

▲ Larva of Brown House-moth *Hofmannophila pseudospretella*, dorsal and lateral views.

Denisia albimaculea page 217   638a (2261)

Very local. **FL** 4-5.5mm. Forewing dark greyish brown with whitish or yellowish-white markings, comprising a cross-band at one-quarter and one-half, the latter sometimes broken, a spot on the costa at about three-quarters, with another at the tornus, this sometimes merging with the central cross-band. **Similar species** *D. augustella* (not illustrated), little known and believed extinct, has a purplish tinge to the forewing, the cross-bands broader and more yellow, and the tornal spot confluent with the central cross-band. **FS** Single-brooded, late May-June. Has been found resting on tree trunks by day and comes to light. **Hab** Woodland, parks, gardens. **Fp** Under the bark of trees; probably in a silken gallery.

Metalampra italica page 217   642a (2279)

Very local. First found in Britain in 2003, otherwise known only in Italy. **FL** 7mm. Forewing orange-brown, the costa finely brown, with a yellowish spot on the costa at about three-quarters and a pale streak from the base to the dorsum before one-half, enclosing a brownish patch. Has been tapped from Yew. Comes to light. **FS** Single-brooded, late June-August. **Hab** Gardens, scrub, woodland. **Fp** Under the bark of dead trees and shrubs; in a slight web.

White-shouldered House-moth *Endrosis sarcitrella* page 217   648 (2282)

Common. **FL** 6-9mm. **Head and thorax white.** Forewing whitish, coarsely speckled dark greyish brown, giving the wing a silvery appearance, with an indistinct whitish spot on the costa at three-quarters and darkish spots at one-quarter, one-half and two-thirds. **FS** A succession of broods, any month of the year. Found at rest on walls in buildings by day, and comes to light. **Hab** In buildings, sheds and barns, and many outdoor habitats. **Fp** Dead animal and vegetable matter, including stored cereals, rotten wood, dead insects and detritus in birds' nests; in a silken gallery.

Brown House-moth *Hofmannophila pseudospretella* page 217   647 (2284)

Common. **FL** 7-11mm. Forewing pale brown variably speckled darker, with two dark spots or short dashes at one-third, a more circular spot at two-thirds and series of short dark dashes, sometimes contiguous, from just before the apex and along the termen. **FS** Double-brooded, possibly continuously brooded in warmer conditions. Found at rest on walls in buildings by day, flies freely in late evening, and comes to light. **Hab** In buildings, sheds and barns, and many outdoor habitats. **Fp** On a wide variety of dead animal and vegetable matter, including cotton, seeds, wool, fur and dried skins, books and has even been noted feeding on slug-bait; in a slight silk web.

Borkhausenia fuscescens page 217   644 (2287)

Common. **FL** 3-5mm. A tiny, rather obscure-looking species. Forewing greyish brown speckled dark greyish brown, shaded darker at the base, apex, tornus and along the costa, with dark spots at about one-third, just before one-half, and a generally larger spot just before two-thirds. **Similar species** *B. minutella* is rare and seen only in the Channel Islands and once recently in south-east England. It is larger (FL 5-7mm), plain dark purplish brown with a creamy spot on the tornus and another on the costa at three-quarters. **FS** Single-brooded, June-early September. Flies freely on warm evenings, frequently alighting on glass windows, and occasionally comes to light. **Hab** Woodland, parkland, hedgerows, farmland, gardens; also inside barns and outhouses. **Fp** Leaf litter, also bred from birds' nests and decaying coniferous wood; in a frass-covered silken tube.

Crassa tinctella page 217   637 (2298)

Local. Not seen recently in northern England and may be declining. **FL** 6-7.5mm. Adult rests horizontally. Thorax greyish yellow. Forewing bronzy or greyish brown, slightly darker towards the apex. **Similar species** **C. unitella rests in a declining posture, is a darker species, with the yellowish-orange head and labial palps strongly contrasting with the brown thorax and forewing. Any colour contrast in C. tinctella is slight. FS** Single-brooded, May-June. Flies in the evening and comes to light. **Hab** Ancient woodland. **Fp** Continental authors have given the larva found on lichen on rotting willows. It has also been bred from dead elm.

Crassa unitella page 217   642 (2299)

Common. More local from the Midlands northwards but may be spreading in the north. **FL** 6-8mm. Head and labial palps yellowish orange. Forewing unmarked, golden brown. **Similar species** *Crassa tinctella*. **FS** Single-brooded, June-early September. **Hab** Woodland, parkland, wooded heathland, scrubby grassland, farmland. **Fp** On fungus and under dead bark of a range of trees; in a slight silk web.

Batia lunaris page 217   640 (2303)

▲ *Batia internella*

Common. **FL** 4-5mm. Forewing yellowish, with the base of the dorsum, the costa to about three-quarters, and the terminal area distinctly suffused reddish brown; an oblique triangular blackish mark near the tornus, the tip in the middle of the wing pointed and curving towards the termen. **Similar species** *B. lambdella* **is much larger and paler, with the base of the dorsum hardly darker than the yellowish ground colour, the triangular mark broader at the base with the tip rounded. Also B. internella, a rare species, though perhaps overlooked, which is intermediate in size between B. lunaris and B. lambdella; B. internella has the base of the dorsum not darker than the yellowish ground colour, the costal suffusion narrower, and the triangular mark broader at the base, with the tip pointed, not rounded. FS** Single-brooded, mid June-early September. Comes to light. **Hab** A range of habitats, including woodland, vegetated shingle, gardens. **Fp** The grey-green crustose lichen *Lecanora conizaeoides* on tree trunks and old wooden posts, also bred from midge-galls on sallow; in a frass-covered silk tube on the surface or in a crack.

Batia lambdella page 217   641 (2301)

Common. Perhaps more frequent in coastal localities. **FL** 6-9mm. Forewing yellowish, with the costa to about three-quarters and terminal area somewhat suffused reddish brown; an oblique triangular blackish mark near the tornus, the upper process rounded. **Similar species** *B. lunaris, B. internella*. **FS** Single-brooded, June-mid August and recorded in early September. Comes to light. **Hab** Scrubby areas with gorse, especially mature gorse. **Fp** Under dead bark of gorse; in a slight silk web.

Esperia sulphurella page 217 649 (2312)

Common. More local in Scotland. **FL** 5.5-8.5mm. Forewing bronzy greyish brown with scattered yellow scales; the male has two short yellow longitudinal streaks from the base, a triangular yellow spot at the tornus and a small vertical yellow mark, sometimes indistinct, on the costa at three-quarters; the female has a straight yellowish streak from the base, almost reaching the triangular tornal mark. In both sexes the hindwing is pale yellow, mostly edged dark greyish brown. **FS** Single-brooded, April-June. Flies in morning sunshine and occasionally comes to light. **Hab** Woodland, scrubby heathland, hedgerows, gardens. **Fp** Under bark of decaying wood of most deciduous and coniferous trees and shrubs, in rotting fence posts, and on the decaying hard fungus *Daldinia concentrica*, preferring dry areas; in a slight silk web, producing copious frass.

Dasycera oliviella page 217 650 (2314)

Very local. **FL** 6-8mm. Forewing blackish, with variable purple and copper iridescence, a metallic cross-band at about one-third and pale yellow markings comprising a blotch near the base and a straight cross-band at about one-half, and with pale yellow speckling beyond the cross-band. **FS** Single-brooded, June-July. Flies at sunrise, in daytime sunshine and occasionally comes to light. **Hab** Ancient broadleaved woodland, especially recent clearings and coppice, also orchards. **Fp** Under bark of decaying wood, especially standing dead wood such as branches and neglected coppice of oak and Hazel, also Blackthorn and Pear; in a slight silk web.

Oecophora bractella page 217 651 (2317)

Rare. **FL** 6-8mm. Head and thorax yellow. Forewing yellow to about one-third, the costa edged blackish blue, with the basal patch bordered by an iridescent bluish or purplish cross-band, the rest of the wing blackish, with a yellow spot on the costa at two-thirds and the tips of the cilia white at the apex. **FS** Single-brooded, late May-July. Flies from dawn to sunrise and comes to light. **Hab** Ancient semi-natural woodland, native or replanted, and usually confined to large woodlands. Has been found in neglected Sweet Chestnut coppice. **Fp** Under bark of decaying wood in dry conditions, including oaks, Sweet Chestnut, birches, Ash, Hazel, Blackthorn and various species of conifer; sometimes in close association with the mycelia of honey fungus *Armillaria*; in a slight silk web.

Alabonia geoffrella page 217 652 (2321)

Common. More local in Wales and northern England. **FL** 8-10mm. Labial palps long. Forewing yellowish in the basal third, merging to orangey brown in the mid-wing, darker in the outer third, the veins finely highlighted greyish brown, with two black-edged metallic silver streaks from the base, the upper one angled so that they meet at about one-half; another metallic streak from the costa at about one-half, with two large triangular pale yellow spots, one on the costa at two-thirds and one almost opposite at the tornus. **FS** Single-brooded, May-June. Flies at sunrise and in morning sunshine, occasionally later in the day; sometimes recorded at light. **Hab** Deciduous and plantation woodland, woodland-edge habitat, scrub, including Blackthorn thicket, hedgerows. **Fp** Within soft decaying twigs of Blackthorn lying on the ground, and in the base of stems of dead Bramble in hedges; also bred from dead Hazel and sallow.

Tachystola acroxantha page 217 656 (2330)

Local. Probably an Australian species. Expanding its range. Recorded new to Wales in 2002. **FL** 6-9mm. Forewing pinkish brown, shaded darker towards the apex, with three somewhat indistinct dark brown spots; the cilia along the termen largely pinkish orange, contrasting with the forewing colour. **FS** Possibly in a succession of broods, mid April-December. Found at rest indoors and in sheds, and readily comes to light. **Hab** Gardens, especially under hedges and in dry, leafy compost heaps. **Fp** Among leaf litter; in a slight silk tube.

▲ Larva of *Oecophora bractella*.

▲ Larva of *Esperia sulphurella*.

◀ Larva of *Tachystola acroxantha* on leaf litter.

Pleurota bicostella page 218   654 (2348)

Common. Local in the Midlands. **FL** 7.5-11mm. Labial palps long, pointing forward. Forewing whitish, variably shaded with greyish-brown scales, with a broad white streak along the costa nearly reaching the apex, this broadly edged darker dorsally, and a black spot or dash at one-third, with another spot at two-thirds. **Similar species Grass moths in *Agriphila* and *Crambus* (Crambidae), which all have shorter palps without an erect terminal segment**; *Pleurota aristella*, which has only been found on the Channel Islands, has a white costal streak with another from the base through the middle of the wing to the apex, and does not have the two distinctive dark spots in the centre of the wing. *P. bicostella* is not known from the Channel Islands. **FS** Single-brooded, June-July. Can be numerous where found. Males fly between sunset and dusk, and again from dawn to sunrise. Occasionally recorded at light. **Hab** Heathland, moorland, often with a preference for damp heathland. Also heathy vegetated shingle. **Fp** Cross-leaved Heath, Bell Heather; in a silken web among shoots.

Aplota palpella page 218   653 (2384)

Rare. Possibly overlooked. **FL** 5-6.5mm. **Labial palps prominent in the male.** Forewing blackish brown with yellowish-brown scales, except along the costa, and blackish spots at two-fifths, just below at one-half, at two-thirds and another between this and the tornus, these two spots sometimes joined. **FS** Single-brooded, July-August. Rests by day on moss and occasionally comes to light. **Hab** Parkland, preferring isolated veteran trees; woodland, on standard trees at the edge or in rides; also old walls. Has been recorded in a conifer plantation. **Fp** On mosses such as *Hypnum cupressiforme*, *Homalothecium sericeum* and *Orthotrichum* spp. on trunks of oaks, elms, Beech, or on rocks or walls; in a silk tube in a fissure in the bark and under moss, where the moss grows sparsely.

Lypusidae

There are four species in this family. The adult rests slightly inclining and the wings are held in a slight to moderate roof-like position. The forewing length is 4.5-10.5mm. The forewings have a gently arched costa, and

Pseudatemelia josephinae

are elongate and moderately broad without a tornal angle, and all species are plainly coloured with few or no markings. The hindwings are broad, nearly as broad as the forewings. The head has erect scales on the crown, and the face is smooth or has forward projecting scales. The long, thread-like antennae are three-quarters to four-fifths the length of the forewing. The labial palps are moderate, slightly curved upwards.

The four species were formerly included in the Oecophoridae. There are three species of *Pseudatemelia*, all superficially similar, being pale brownish grey, and all typically occur in woodlands. *Amphisbatis incongruella* is widely distributed but probably overlooked as it is small and dark greyish, and flies in sunshine by day from March to early May; it is found on heathland, moorland, downland and coastal grassland.

Larvae of all four species live within a portable case, *Pseudatemelia* feeding on leaf litter, and *Amphisbatis* has been observed feeding on dead flowers. All are single-brooded.

Further reading
British and Irish species: Emmet & Langmaid (2002a)
European species: Palm (1989); Tokár *et al.* (2005)

Pseudatemelia josephinae page 218   660 (3055)

Local. **FL** 9-10mm. Forewing pale brownish grey with scattered paler scales, with two small, rather indistinct darker spots at one-third, and another, larger, at two-thirds. Abdomen brownish grey dorsally. **Similar species** *P. flavifrontella* (not illustrated), a local species in England and Wales, is similar but usually slightly larger, has the abdomen yellowish brown dorsally, and flies earlier in the year (late May-early July). *P. subochreella* (not illustrated), a local species in England and Wales, has the forewing pale yellowish brown, sometimes with a greyish tinge, but is otherwise unmarked. It flies late May-early

July. Where there is overlap in time of appearance and geographical range, examination of genitalia is likely to be necessary to confirm identification (see male and female genitalia on p.390). **FS** Single-brooded, late June-early August. Flies in the evening and comes to light. **Hab** Deciduous woodland. **Fp** Decaying leaves; the larval case can be found in early spring by sifting through leaf litter at the base of trees.

◀ Larva of *Pseudatemelia josephinae* removed from its portable case.

Chimabachidae

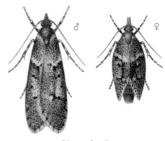

Diurnea fagella

There are three species in this family. All show sexual dimorphism, the male with full wings and the female having much reduced wings and incapable of flight. The male sits slightly inclining, with the wings held in a shallow, roof-like position, slightly overlapping or curved over the abdomen. Male forewing length is 8-14mm; the forewings are broad, with slight scale-tufts, and the hindwings are as broad or broader than the forewings. The head has narrow or hair-like scales. The antennae are from one-half to almost two-thirds the length of the forewing. The labial palps are rather short, forward pointing or slightly curved upwards, and the tongue is rudimentary.

The three species, which were formerly included within the Oecophoridae, are covered below and are straightforward to identify. The larvae feed from within spinnings on leaves of trees and shrubs. Characteristically, they have the pair of true legs on the third thoracic segment modified into paddle-like projections, which are easy to spot with a hand lens, and the species are easy to separate on the colour of the head and prothoracic plate. Pupation takes place in a silken cocoon in detritus or under the soil surface. All species are single-brooded, and adults are found in spring or late autumn.

Further reading
British and Irish species: Emmet & Langmaid (2002a)
European species: Palm (1989); Tokár *et al.* (2005)

Diurnea fagella page 218   663 (2231)

Common. **FL** Male 13-14mm, female 8-11mm. Forewing ground colour variable, pale grey, speckled darker, sometimes tinged yellowish grey, ranging through darker grey to brownish black; an outwardly angled black cross-line, sometimes Y-shaped, at one-third, often interrupted or absent before the dorsum, the inner edge creamy, this more noticeable in darker examples; usually several spots between two-fifths and two-thirds, the outer pair sometimes joined into a transverse black mark. Female forewing short and pointed, with markings similar to the male, often more defined. **FS** Single-brooded, March-May. Frequently seen in numbers. The male comes to light. **Hab** Deciduous woodland, particularly oak woodland, parkland, gardens; also birch woodland in parts of Scotland. **Fp** Oaks, Beech, Sweet Chestnut, Hornbeam, Aspen, Hazel, Blackthorn, willows, birches, probably other deciduous trees and shrubs; feeds slowly between spun flat leaves.

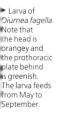

► Larva of *Diurnea fagella*. Note that the head is orangey and the prothoracic plate behind is greenish. The larva feeds from May to September.

▲ Larva of *Diurnea lipsiella*. Note the head is brown and the prothoracic plate is yellowish. The larva feeds from June to early August.

▲ Larva of *Dasystoma salicella*. Note the head is dark brownish, sometimes blackish, the prothoracic plate is pale greenish and brown, and the pinaculae (body spots) are more noticeable than in the other two species. The larva feeds from May to September.

▲ Adult pair of *Dasystoma salicella*, the male on the left.

▲ Adult female *Dasystoma salicella*. Note the much reduced wings.

Diurnea lipsiella page 218   664 (2232)

Local. **FL** Male 10-12mm, female 7-8mm. Antennae of the male are finely pectinate. Male forewing light brown, variably shaded grey and yellowish brown, with an obscure dark spot usually present at two-thirds; **a diagonal line from the costa at about three-fifths to the tornus, sometimes prominent**, beyond which the wing is darker; sometimes with a pale streak from the base to about one-half. Female forewing is elongate and pointed, whitish grey with contrasting short black streaks or shading. **FS** Single-brooded, October-early November. The male flies in the afternoon on warm, windless days, and comes to light. **Hab** Oak woodland. **Fp** Oaks, particularly Sessile Oak, also Bilberry, Aspen, Small-leaved Lime; feeds slowly between spun flat leaves.

Dasystoma salicella page 218   665 (2234)

Local. Scarce in some areas, with a single record in Ireland from Co. Tyrone. **FL** Male 8-9mm, female with much reduced wings, 3-4.5mm. Male forewing brownish, with a creamy oblique cross-line from the costa at one-quarter not reaching the dorsum, the outer edge often dark brown, a creamy suffusion in the central area edged at two-thirds with a dark brown curved line, and with scattered creamy scales towards the apex. The female has a whitish forewing, with a blackish-brown cross-band just beyond one-half and another before the apex. **FS** Single-brooded, April. The male flies in warm sunshine around midday. **Hab** Woodland, bogs, wet heathland, coastal scrub. **Fp** On a range of plants, including Bog-myrtle, Blackthorn, Meadowsweet, Silverweed, Buckthorn, Dogwood, sallow, Bramble; feeds slowly in a spinning.

Peleopodidae

There is just one species in this family in Britain and Ireland, *Carcina quercana*. The adult rests with the wings flat and overlapping. The forewing length is 8-10mm. The forewings are broad with a curved costa, and the hindwings are almost as broad as the forewings. The head has smooth scales. The thread-like antennae are hidden along the side of the body and are almost the length of the forewing.

Carcina quercana

The labial palps are moderately long, slender and curved upwards. The species was formerly included in the Oecophoridae.

Further reading
British and Irish species: Emmet & Langmaid (2002a)
European species: Palm (1989); Tokár *et al.* (2005)

Carcina quercana page 218   658 (2328)

Common. More local in Scotland. **FL** 8-10mm. Forewing variable in colour, ranging from cream to orangey brown or blackish pink, with an indistinct dark cross-band at one-quarter; a narrow yellow blotch on the costa near the base, and another, usually quadrate, at about three-fifths, and with two dark spots, sometimes obscure, at about one-third and two-thirds; cilia yellow, the termen narrowly edged darker. **FS** Mainly single-brooded, July-August, but has been recorded May-early December, suggesting a second brood. Readily disturbed by day from trees and shrubs, and comes to light. **Hab** Deciduous woodland, hedgerows, scrub, gardens. **Fp** A wide range of trees and shrubs, including oaks, Beech, Sweet Chestnut, Sycamore, apples, roses, Bramble; under a flat silken web on the underside of a leaf.

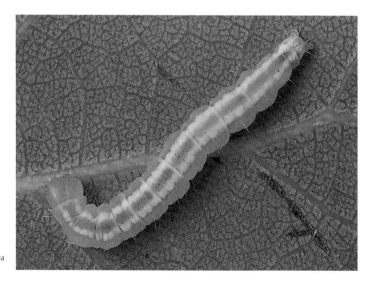

▶ Larva of *Carcina quercana*.

Elachistidae

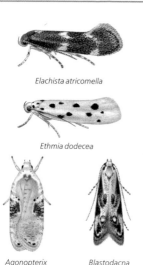

Elachista atricomella

Ethmia dodecea

Agonopterix
alstromeriana

Blastodacna
hellerella

There are 113 species in this family. Adult resting positions are variable; they are horizontal with wings overlapping in *Agonopterix* and *Depressaria*, but roof-like in most others, at a steep angle in *Elachista*, moderate to steep in *Ethmia*, and in *Blastodacna* the wings are held somewhat roof-like, with the tips of the wings flat to the resting surface. The forewing length varies considerably, from 3-17mm. The forewings are elongate, in *Elachista* being about three times as long as broad, with a tornal angle, although this is sometimes indistinct. There is no consistent wing pattern or coloration between genera in the family. For example, *Agonopterix* and *Depressaria* are obscurely shaded white, grey, blackish or brownish, markings mostly consisting of black dots and dashes; *Elachista* are mostly white with dots and shading, or black with a pale cross-band with or without paired costal and tornal spots; most *Ethmia* are distinctively patterned black and white; and raised scale-tufts are present in *Chrysoclista*, *Blastodacna* and *Spuleria*. The hindwings are narrower (e.g. *Elachista*), as broad as (e.g. *Agonopterix*) or slightly broader than the forewings (e.g. *Ethmia*), with the dorsal cilia moderate to long, as in *Elachista*. The hindwing apex in *Elachista* is pointed. The head is smooth, some with forward-projecting scales on the face, rough scales on the crown or slight tufts on the neck. The thread-like antennae are two-thirds to seven-eighths the length of the forewing and are directed backwards. The labial palps are usually long and slender, and may be drooping, forward pointing or curved upwards.

A recent taxonomic review has led to a substantial change within the Elachistidae. Experienced moth recorders may find considerable difficulty in understanding how a once discrete entity of very similar species can now incorporate such diversity and yet remain within the same family. This family now includes *Agonopterix*, *Depressaria* and related species (formerly part of the Oecophoridae), the *Ethmia* species (formerly treated as Ethmiidae), and those genera formerly treated either as the Agonoxenidae or part of the Cosmopterigidae, i.e. *Chrysoclista*, *Spuleria*, *Blastodacna* and *Dystebenna*.

◄ Larva and leaf mine of *Elachista maculicerusella* on a reed leaf.

▲ Larva and leaf mine of *Elachista argentella* on a grass leaf. Note the larva is dark coloured and the frass is packed at the top of the mine.

▲ Larvae and leaf mines of *Elachista gangabella* on False-brome. Note the dark-coloured larvae and that there are sometimes several larvae to a leaf.

Treatment of the Elachistidae in this guide is partial. Only a small number of *Elachista* and *Depressaria* are included, together with several *Agonopterix*. Many *Elachista*, which now incorporates species formerly in *Biselachista* and *Cosmiotes*, are very small and superficially similar and thus can be difficult to identify, requiring examination of the genitalia; several species are also only rarely seen at light. Similarly, many *Depressaria* and some *Agonopterix* can prove difficult to determine, especially as examples in light traps are often worn. The species covered include regular visitors to light traps, such as *A. heracliana* and the obscurely marked male of *Elachista canapennella*, and others which are considered relatively easy to identify. Also included are a few widespread species, such as *E. argentella*, which are readily encountered by day. Coverage of *Ethmia* and other smaller genera is more complete. Four species of Elachistidae are considered extinct in Britain, including the colourful *Hypercallia citrinalis*, although this can still be found in parts of Ireland, while *Elachista nobilella* was discovered in Britain for the first time in 2003, in Surrey.

The larvae of *Elachista* are leaf miners on grasses or sedges; their mines can be easily confused with Diptera (true flies), although the larvae differ in their morphology, the moth larvae having chewing mouthparts and those of the flies sucking mouthparts. Many *Depressaria* and *Agonopterix* feed within slight tubular spinnings among leaves, flowers or seeds on Apiaceae, Asteraceae or Fabaceae; the larvae are often very active and when disturbed eject themselves remarkably rapidly from the spinning. Larvae of *Chrysoclista* feed underneath living bark, exuding reddish frass which reveals their presence. *Ethmia* larvae are typically brightly coloured, feed externally and are mainly associated with the Boraginaceae.

Adults fly at dusk and into the night, or fly by day, although a few are known to be active at dawn. Many *Depressaria* and *Agonopterix* species hibernate as adults and a few are often recorded at light traps on milder winter nights.

Further reading
British and Irish species: Emmet (1996); Emmet & Langmaid (2002a)
European species: Traugott-Olsen & Schmidt Nielsen (1977) – covers most British and Irish *Elachista* species; Palm (1989) – covers most species other than *Elachista*;Tokár et al. (2005) (part)

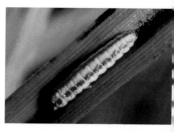
▲ Larva of *Elachista cinereopunctella* ready to pupate near the base of a leaf of Glaucous Sedge. Note the bright colour of the larva which distinguishes it from the whitish-grey larva of *E. gleichenella*, which may use the same foodplant.

Elachista atricomella page 219   597 (1865)

Common. **FL** 5.5-6.5mm. Male head dark grey, thorax blackish brown. Forewing blackish brown, somewhat mottled paler, with an oblique creamy cross-band from the costa at one-third to the dorsum before one-half, often interrupted in the middle, a creamy white triangular tornal spot, another opposite on the costa and sometimes a small patch of whitish scales beyond. The female has a yellowish-brown head, with broader white markings on the forewing. **The relatively large size, dark head and interrupted cross-band of the male are diagnostic. FS** Single-brooded, late May-late September. Can be disturbed by day, flies in the evening and regularly comes to light. **Hab** Woodland rides, unmanaged grassland, roadside verges, waste ground. **Fp** Cock's-foot; feeds within a long, narrow, whitish mine, the larva changing leaves as it grows.

Elachista albifrontella page 219   601 (1856)

Common. **FL** 4-4.5mm. **Head and neck tufts white**, thorax blackish brown. Forewing blackish brown with a slightly oblique white cross-band from the costa at one-third to the dorsum before one-half, a small white tornal spot, and a larger one on the costa beyond. **Similar species** *E. luticomella* (not illustrated), a common species, is slightly larger (FL 5-5.5mm) with a yellowish head and neck tufts; *E. apicipunctella* (not illustrated), a local species, also slightly larger (FL 5-5.5mm), has a white head and greyish-brown neck tufts, and shining white markings usually including a streak or spot before the apex. **FS** Single-brooded, early June-late July. Can be seen at rest or swept from grasses by day, and comes to light. **Hab** Woodland rides, unmanaged grassland, gardens. **Fp** Cock's-foot, Yorkshire-fog, Creeping Soft-grass, Tufted Hair-grass, probably other grasses; feeds within a whitish blotch and may change leaves.

Elachista canapennella page 219   607 (1883)

Common. **FL** 4-5mm. Male forewing dark greyish brown, mottled paler, the markings usually faint, with hints of a whitish spot at the tornus and a cross-band just before one-half. Female smaller, the forewing blackish brown with white markings, including a cross-band just before one-half, a spot on the costa at three-quarters and another almost opposite on the dorsum. **Similar species** (none illustrated) Male *E. subnigrella*, is smaller (FL 3.5-4mm) and more speckled in appearance; the female is similar to female *E. freyerella* and *E. stabilella*, both of which have a white patch in the terminal area extending into the cilia. Examination of genitalia may be required to confirm identification. **FS** Double-brooded in the south, April-September, single-brooded in the north, June-August. The male comes to light, sometimes in numbers; the female is rarely seen. **Hab** Found in a wide range of habitats. **Fp** Creeping Soft-grass, False Oat-grass, Creeping Bent, probably other grasses; feeds within a linear mine which broadens as the larva grows.

Elachista rufocinerea page 219

608 (2009)

Common. **FL** 5-6mm. Forewing pale cream, the base of the costa reddish brown and the whole wing sprinkled with pale reddish-brown scales, but with a few longitudinal stripes of the ground colour showing through. **FS** Single-brooded, mid April-early July, occurring earlier in the south than in the north. Flies at sunset, but readily disturbed from vegetation by day. Occasionally comes to light. **Hab** Grassland, including woodland, parkland, bogs, heaths and vegetated shingle. **Fp** Creeping Soft-grass, False Oat-grass, Tall Fescue, Smaller Cat's-tail; feeds within a broad, rather transparent mine.

Elachista maculicerusella page 219

609 (1974)

Common. **FL** 4.5-6mm. Forewing white, variably shaded greyish brown, with an irregular dark brownish cross-band at about one-half, a blotch at about two-thirds, and a blackish spot on the dorsum at about one-half. **FS** Double-brooded at least to Lancashire, late April-August, single-brooded further north, July-August. Comes to light. **Hab** Damp grassland, fens, marshes, river and canal banks, wet woodland. **Fp** Reed Canary-grass, Common Reed, less frequently on other grasses; feeds within a large, whitish mine (see p.136).

Elachista argentella page 219

610 (1863)

Common. Perhaps most frequent near the coast. Occasionally abundant. **FL** 5-6mm. Head and forewing white, sometimes slightly darker near the base. **Similar species Opostega salaciella (Opostegidae) rests with wings held roof-like at a shallow angle, and has a distinctive eye-cap at the base of the antenna**; *Elachista subalbidella* (not illustrated), a common species on heaths and moors, has a pale yellowish or yellowish-brown unmarked forewing. **FS** Single-brooded, early May-early August, and has been recorded in April. Can be disturbed from vegetation by day, is frequently seen sitting on grass stems, flies in late afternoon and evening, and comes to light. **Hab** A wide range of grassland habitats, including coastal cliffs, saltmarshes, vegetated shingle, downland, parkland. **Fp** A variety of grasses, including Cock's-foot, Yorkshire-fog, Red Fescue, Sheep's-fescue; the fat, greyish larva can readily be observed within the mine on fine-bladed grasses in early spring.

Elachista subocellea page 219

613 (2022)

Local in the south, rare in the north. **FL** 4-5.5mm. Forewing white, shaded yellowish brown, faintly speckled blackish, and greyish or greyish brown on the costa, particularly from the base, with whitish cross-bands at one-third and two-thirds. **FS** Single-brooded, mid May-early August. Can be found by day, especially early in the morning, and comes to light. **Hab** Woodland edges, clearings, coastal grassland, scrubby downland. **Fp** False Brome; feeds within a blister mine in the leaf.

Elachista bisulcella page 219

623 (1877)

Local. **FL** 4-5mm. Forewing dark brown, with a creamy white cross-band at one-half which broadens towards the dorsum, the inner margin sharply defined, the outer margin suffused with yellow, with a yellowish patch in the cilia below the apex. **The yellowish outer margin to the cross-band is diagnostic. Similar species** Some species of *Syncopacma* (Gelechiidae) are similarly patterned, although they are bigger, sit flat, not roof-like, and have a white cross-band nearer to two-thirds on the forewing. **FS** Possibly double-brooded in the south, May-September, single-brooded in the north, July-August. Occasionally comes to light. **Hab** Woodland, coastal grassland. **Fp** Tufted Hair-grass, Tall Fescue; feeds within a slightly inflated blotch mine, the mine later becoming long and broad.

Semioscopis avellanella page 219 666 (1668)

Local. **FL** 10-13mm. Forewing pale grey, lightly speckled blackish, with a longitudinal black streak from the base to one-quarter then bending towards the costa, almost reaching one-half, a black chevron at two-thirds and a series of black spots along the termen. **Similar species *S. steinkellneriana* has a broader, slightly darker forewing, with the black streak from one-fifth to about one-half, this not starting from the base.** **FS** Single-brooded, March-mid May. Can be found at rest on tree trunks and comes to light. **Hab** Mature birch woodland, ancient Small-leaved Lime woodland. **Fp** Birches, Small-leaved Lime, occasionally Hornbeam; in a long, narrow tube along the leaf edge.

Semioscopis steinkellneriana page 219 667 (1670)

Local. Very local in Scotland. **FL** 9-11mm. Forewing grey, sometimes tinged brownish, lightly speckled black, with a broad curved black streak from one-fifth to about one-half, another streak at two-thirds, this resembling an inverted L, and with black spots along the termen. **Similar species** *S. avellanella*. **FS** Single-brooded, late March-May. Flies at dawn and comes to light. **Hab** Old hedgerows, thickets, woodland. **Fp** Blackthorn, Rowan, occasionally Hawthorn; within a downward-folded leaf.

Luquetia lobella page 219 668 (1674)

Local in southern England. **FL** 8-10mm. Forewing greyish, tinged brownish when freshly emerged, and mottled darker grey, with a series of three small raised tufts of black scales just before one-third and two more at just after one-half; these tufts can become rubbed off with age. **FS** Single-brooded, May-early July. Occasionally disturbed from hedgerows by day and comes to light. **Hab** Hedgerows, woodland edges, scrubby situations. **Fp** Blackthorn, preferring to feed on the lower leaves and low-growing suckers; under a white silk web on the underside of a leaf.

Agonopterix heracliana page 219 688 (1736)

Common. An abundant species throughout Britain. **FL** 8-11mm. Forewing pale greyish brown to greyish brown, variably speckled darker, sometimes tinged reddish, the base paler, with a pair of blackish dots at one-third, the lower dot partly edged white, a white dot at just before one-half and another just after one-half, both circled darker, sometimes with an obscure dark greyish blotch above, with an obscure, pale, angled cross-band at three-quarters, and a series of dark brown dots along the termen. **Similar species** **A. ciliella (not illustrated), a common species, is larger (FL 9.5-12mm), with about five distinct dark lines in the hindwing cilia, and the cilia also pink-tinged; the underside of the moth is often strongly tinged with pink.** *A. heracliana* may have one distinct and up to four faint cilia lines. If there is doubt, determination should be by

► Larva of *Agonopterix heracliana* outside its spinning.

genitalia examination. **FS** Single-brooded, can be found all year round. Hibernates as an adult. Feeds at night at Ivy blossom and on over-ripe blackberries. Readily comes to house and moth-trap lights. **Hab** Found in most habitats, although not montane. **Fp** Cow Parsley, Rough Chervil, Hogweed, other umbellifers (Apiaceae); in a slight tube in the leaves, sometimes also the flowers.

Agonopterix purpurea page 219 691 (1732)

Local. Scarce in the north. Recorded new to Scotland in 2007. **FL** 6-7mm. **A small, dark purplish species.** Forewing dark purplish brown, variably speckled blackish towards the termen, broadly speckled whitish and blackish along the costa, more distinctly whitish at the base to the dorsum, with a blackish blotch on the costa at one-half, and with a small white dot, narrowly edged black, just below the blotch. **FS** Single-brooded, can be found all year round. Hibernates as an adult. Has been found flying in the afternoon sun and occasionally comes to light. **Hab** Calcareous and neutral grassland, coastal grasslands, vegetated shingle, road verges, open woodland. **Fp** Wild Carrot, Rough Chervil, Cow Parsley, Upright Hedge-parsley; in a slight tube in the leaves.

Agonopterix subpropinquella page 219 692 (1722)

Common in the south, scarce in the north. More coastal in Scotland and Ireland. **FL** 8-11mm. Head and thorax pale brown, dark brown or black. Forewing pale brown, sometimes with a slight reddish tinge, variably speckled darker, with a pair of blackish dots at one-third and an obscure greyish-brown blotch at one-half. **Similar species** *A. scopariella* (not illustrated) has two or three white dots centrally; *A. propinquella* (not illustrated), a slightly smaller species (FL 8-9.5mm) has a more arched costa and rounded apex, and a pale greyish or brownish ground colour with a dark-edged, creamy brown patch at the base, extending a little way along the costa. *A. scopariella* is local in England and Wales, widespread in Scotland, and the larva feeds on Broom and Tree Lupin; *A. propinquella* is scarce in the south, widespread in the north and regularly comes to light, and the larva feeds on Spear Thistle and Creeping Thistle. **FS** Single-brooded, can be found all year round. Hibernates as an adult. Comes to light. **Hab** Grasslands, open woodland, waste ground, roadside verges, vegetated shingle, hedgerows, gardens. **Fp** Common Knapweed, Greater Knapweed, Spear Thistle, Greater Burdock, Globe Artichoke, Cardoon; in a slight tube in the leaves.

Agonopterix nanatella page 219 694 (1712)

Very local. Only historically recorded in Ireland. **FL** 7-9mm. Forewing sandy brown, lightly mottled brown, with a black dot at about two-fifths and another, sometimes slightly obscure, at three-fifths, usually with an obscure greyish blotch above and between the two. **Similar species A. assimilella is usually larger and has a row of dots along the termen**; *A. carduella* (not illustrated), a very local, mainly coastal species, has a more arched costa and is orangey or pinkish. **FS** Single-brooded, July-October. Rarely seen as an adult, but occasionally comes to light. More readily found in the larval stage. **Hab** Chalk and limestone grassland. **Fp** Carline Thistle; mines the leaves of rosette plants, then rolls a leaf into a tube when larger; often there are several larvae to a plant.

▶ Larva of *Agonopterix carduella* outside its spinning.

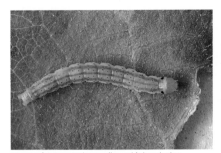

▲ Larva of *Agonopterix arenella* outside its spinning.

▲ Larva of *Agonopterix atomella* (not described) appearing from its spinning on Dyer's Greenweed.

Agonopterix alstromeriana page 219   695 (1730

Common. More local from north-west England northwards. **FL** 8-10mm. Head and thorax white, sometimes tinged pale brownish. Forewing white, variably mottled greyish, with a pair of black spots at one-third, a blackish blotch on the costa at one-half, bordered below by a short thick rusty-red streak, and with a row of black dots along the termen. **FS** Single-brooded, can be found all year round. Hibernates as an adult. Comes to light. **Hab** A wide range of habitats, including roadside verges, waste ground, agricultural headlands, hedgerows, marshes, riverbanks, woodland. **Fp** Hemlock; in a slight tube in the leaves, sometimes also the flowers.

Agonopterix arenella page 219   697 (1719

Common. **FL** 9-11mm. Forewing pale sandy brown, variably shaded reddish brown and finely speckled blackish, with a pair of oblique black dots at one-third, another at just after one-half, a dark greyish blotch between these dots, several dark greyish marks along the costa and a series of blackish dots along the termen. **FS** Single-brooded, can be found all year round. Hibernates as an adult. Comes to light. **Hab** A wide range of habitats, including downland, open grassland, gardens, waste ground. **Fp** Thistles, knapweeds, burdocks, Saw-wort; under a web on the underside of a leaf.

Agonopterix kaekeritziana page 220   698 (1757

Common. **FL** 9.5-12.5mm. Forewing pale yellowish brown, more or less suffused with pale orangey brown shading, with a black dot at one-third and at two-thirds, and obliquely towards the tornus is a somewhat indistinct greyish or orangey brown blotch. **Similar species** *A. pallorella* is without orangey brown suffusion and has a long dark brown streak below the dorsum; *A. bipunctosa* (not illustrated), a rare species on Saw-wort, also without orangey brown suffusion, has the costa edged black at the base. **FS** Single-brooded, July-September. Can be found at night on flowers, and seldom comes to light. **Hab** Grasslands. **Fp** Greater Knapweed, Common Knapweed; in a rolled leaf.

Agonopterix pallorella page 220   700 (1756

Very local. Now confined to the south, south Wales and western Ireland, formerly with a few scattered records further north. **FL** 9-11mm. Forewing pale sandy brown, veins streaked darker, with a black dot near the base, at one-third and at two-thirds, and with a strong dark brown streak below these dots from one-fifth to about three-quarters, and a row of black dots along the termen. **Similar species** *A. kaekeritziana*, *A. bipunctosa* (not illustrated). **FS** Single-brooded, July-May. Hibernates as an adult. Occasionally comes to light. **Hab** Chalk and limestone grassland, especially near the coast. **Fp** Greater Knapweed, occasionally Common Knapweed, Saw-wort; in a rolled leaf.

Agonopterix ocellana page 220 701 (1691)

Common. **FL** 9-11mm. Forewing pale sandy brown, variably speckled darker, with a pair of black or greyish-brown dots at one-third, often joined and mixed with red, a greyish-brown blotch at one-half bordered below by a short red streak, a red-circled white dot at three-fifths, this sometimes slightly obscure, and a row of greyish-brown dots along the termen. **FS** Single-brooded, can be found all year round. Hibernates as an adult. Comes to light. **Hab** A wide range of habitats, including woodland, scrub, fens, marshes. **Fp** Various species of willow, including White Willow, Purple Willow, Osier, Goat Willow, Creeping Willow; among spun leaves.

Agonopterix assimilella page 220 702 (1707)

Common. Local in the north-west. **FL** 8-10mm. Forewing pale sandy brown, variably mottled darker and speckled greyish brown and black, except near the base, with a pair of black dots at two-fifths, although one is sometimes indistinct, a rusty brown elongate mark at about one-half, this sometimes partly obscure, merging into a greyish-brown blotch of variable intensity, and with a row of brownish dots along the termen. **Similar species Pale forms may be confused with *A. kuznetzovi* (not illustrated), a rare species on Saw-wort known only from south-west England, which has a black dot at about three-fifths**; also *A. nanatella*. **FS** Single-brooded, April-September. Occasionally comes to light. **Hab** A wide range of habitats, including waste ground, scrub, heathland, hedgerows. **Fp** Broom; between spun green stems in early spring.

Agonopterix umbellana page 220 705 (1764)

Local. Perhaps more frequent in coastal localities. **FL** 9-12mm. Forewing pale sandy brown, dorsum and veins variably lined brownish, with a series of blackish dots along the termen. **FS** Single-brooded, August-May. Hibernates as an adult. Occasionally comes to light. **Hab** Heathland, scrub, waste ground, coastal cliffs, dunes. **Fp** Gorse, Hairy Greenweed; in spun shoots.

Agonopterix nervosa page 220 706 (1763)

Common. **FL** 8-10mm. Forewing pale sandy brown, lightly speckled blackish and variably mottled reddish brown or greyish brown except at the base, the veins often streaked with these colours, with a black dot, sometimes two, at one-third followed at one-half by a greyish blotch of variable intensity, below which is a short red streak, and a red-circled white dot just after one-half, this sometimes slightly obscure; **the slightly pointed apex and brownish termen distinguish this species from other *Agonopterix*. FS** Single-brooded, June-September, and has been found in January. Comes to light. **Hab** A wide range of habitats, including woodland, waste ground, heathland, hedgerows, scrubby downland. **Fp** Broom, gorses, Dyer's Greenweed, Petty Whin, Tree Lupin; in spun shoots.

▼ Larva of *Agonopterix nervosa* outside its spinning.

Agonopteris liturosa   page 220

Local. Very local in Ireland, very scarce in northern Scotland. **FL** 8-10mm. Thorax creamy brown, contrasting with dark purplish-brown tegulae. Forewing dark purplish brown to reddish brown, shaded black in the basal third and broadly along the costa to three-quarters, this variably speckled whitish, with a curved black mark at one-third and a smaller, sometimes rudimentary, black mark at just after one-half, the termen bordered by a black line. **Similar species** *A. conterminella* **(not illustrated), a common species on willows, is more brownish, with the thorax and tegulae pale or dark, but not contrasting with each other. FS** Single-brooded, June-early September. Occasionally comes to light. **Hab** Grassland, downland, scrub, waste ground, open woodland. **Fp** Various species of St John's-wort; in spun shoots.

Agonopteris yeatiana   page 220

Local. More frequent near the coast, especially in northern England, Scotland and Ireland. **FL** 9-11mm. Forewing pale sandy brown to light greyish brown, lightly speckled black, the base paler, this edged blackish towards the dorsum, with a pair of black dots at one-third and a white dot, edged darker, at three-fifths, between these an obscure greyish blotch, the termen with a row of black dots. **FS** Single-brooded, can be found all year round. Hibernates as an adult. Comes to light. **Hab** Coastal and inland grassland, marshes, open scrub and woodland. **Fp** Wild Carrot, Pepper-saxifrage, Rough Chervil, Hemlock Water-dropwort, Milk-parsley, Wild Celery, possibly other umbellifers (Apiaceae); in a slight tube in the leaves, sometimes also the flowers.

Agonopteris rotundella   page 220

Local. Primarily coastal. **FL** 7-8mm. Forewing creamy, tinged brownish, sometimes greyish towards the termen, variably lightly speckled with darker scales, with a black dot at one-third, another at three-fifths and a series of dark dots bordering the termen. **FS** Single-brooded, can be found all year round. Hibernates as an adult. Occasionally comes to light. **Hab** Downland, coastal cliffs. **Fp** Wild Carrot; in a slight tube in the leaves, sometimes also the flowers.

Depressaria daucella   page 220

Common. **FL** 11-12mm. Forewing brown, reddish brown or greyish brown, speckled black forming numerous short longitudinal black streaks, with a small, often obscure, black blotch at the base near the dorsum and a faint, paler brown, acutely-angled cross-band at three-quarters; the black streaks may be weak or absent in some forms. **Similar species** *D. ultimella* **(not illustrated), a local species in the south on Fool's-water-cress, is much smaller (FL 8-9mm), mottled with creamy and black scales, and with a distinct creamy longitudinal streak in the central area;** *D. radiella* **is usually large and has a series of blackish dots along the termen. FS** Single-brooded, late July-April. Hibernates as an adult. Comes to light. **Hab** Margins of watercourses, ditches, marshes, wet meadows, waste ground, damp moorland, wet woodland. **Fp** Water-dropworts, Whorled Caraway, Stone Parsley; in a silk spinning in the flowers and seeds.

Parsnip Moth *Depressaria radiella*   page 220

Common. **FL** 12-14mm. Forewing pale greyish brown, speckled sandy brown and greyish brown, with a series of longitudinal blackish streaks, mostly from about one-half, a paler acutely-angled cross-band at three-quarters and a series of blackish dots along the termen. **Similar species** *D. daucella*. **FS** Single-brooded, late July-early June. Hibernates as an adult. Occasionally disturbed by day and comes to light. **Hab** Grassland, field margins, woodland edge, gardens, waste ground. **Fp** Hogweed, Wild Parsnip. On the Channel Islands reported on Fool's-water-cress; in a silk spinning in the flowers and seeds

▲ Larva of *Depressaria daucella* outside its spinning.

▲ Larva of Parsnip Moth *Depressaria radiella*, dorsal and lateral views.

Depressaria pulcherrimella page 220   676 (1798)

Local. **FL** 8-10mm. Head and thorax creamy brown. Forewing pale pinkish brown, with a scattering of greyish-brown and blackish scales, an oblique black mark at one-third with a cream mark beyond, two short black streaks before one-half, a creamy dot at about two-thirds with an obscure pale cross-band beyond at a right-angle around this dot, and a row of black spots along the termen. **Similar species** *D. douglasella* has a slightly broader, more greyish-brown and mottled forewing, and the rather obscure pale cross-band is obtusely angled. **FS** Single-brooded, June-September. Flies in the evening and comes to light. **Hab** Grasslands with dense growth of the foodplants. **Fp** Mainly Pignut, less often Wild Carrot, Burnet-saxifrage; in the flowers and developing seeds, distorting the umbel.

Depressaria douglasella page 220   677 (1799)

Local. Perhaps more frequent in coastal localities. Very local or scarce in northern England, Wales and Ireland. **FL** 8-10mm. Head and thorax predominantly whitish, sometimes tinted pale sandy brown. Forewing greyish brown, variably mottled greyish white, brown and black, with an obscure oblique blackish V-shaped mark at about one-third, an obscure greyish-white dot at about two-thirds and an obscure pale cross-band beyond, obtusely angled around this dot. **Similar species** *D. pulcherrimella*. **FS** Single-brooded, July-September. Comes to light. **Hab** Grassland, rough disturbed ground, chalk pits, farmland. **Fp** Wild Carrot, Wild Parsnip, occasionally Upright Hedge-parsley; in a tube in the leaves.

Telechrysis tripuncta page 220   646 (2398)

Local. Scarce in northern England. **FL** 5.5-7mm. Forewing dark greyish brown, with two pale cream spots on the costa and one above the dorsum near the tornus. **FS** Single-brooded, late May-early July. Flies at dawn and dusk, and occasionally comes to light. Usually seen only in small numbers. **Hab** Woodland, hedgerows, usually at low altitudes but has been recorded from a hawthorn hedge at 300m. **Fp** Has been bred from a rotten stump of dead Hazel.

Hypercallia citrinalis page 220   657 (3078)

Rare. Currently found only in the Burren, Co. Clare. Formerly Kent, with 19th-century records from Essex and Co. Durham. **FL** 6-9mm. Forewing bright yellow, patterned with red markings. **FS** Single-brooded, late June-July. Flies at dawn and dusk and occasionally comes to light. **Hab** Limestone pavement, short-turfed calcareous grassland, possibly also grassy woodland rides. **Fp** Common Milkwort, Chalk Milkwort; in an inconspicuous spinning among flowerbuds and leaves.

Ethmia terminella page 221

Rare. Resident in East Sussex , Kent and possibly Suffolk; probably an immigrant or wanderer elsewhere. **FL** 8-10mm. Forewing whitish, the costa and dorsum greyish, with two rows of three small black dots and a series of black dots along the termen. **Similar species Could be mistaken for a small *Yponomeuta* species (Yponomeutidae), but readily separated by the series of dots along the termen.** **FS** Single-brooded, late May-mid July. Has been found on the foodplant, fence posts and telegraph poles by day, and comes to light. **Hab** Vegetated shingle, coastal chalk downland; also recorded from a sand dune. **Fp** Viper's-bugloss; on the flowers and unripe seeds.

Ethmia dodecea page 221

Very local. An occasional immigrant or wanderer. **FL** 8-10mm. Forewing whitish, with about ten blackish dots of variable size. **FS** Single-brooded, May-August. Comes to light. **Hab** Woodland rides, wooded fenland, coastal grassland, scrub. **Fp** Common Gromwell; usually feeds gregariously within a slight web, the larvae often stripping the plant of its leaves.

Ethmia quadrillella page 221

Very local. Predominantly eastern. Possibly an occasional immigrant or wanderer. **FL** 7-9mm. Forewing patterned black and white, the apex black, with another strong black blotch from the costa to the middle of the wing, and two smaller black dots towards the base. **FS** Single-brooded, May-August. Has been found flying by day, but most often comes to light. **Hab** Fenland, wetlands, damp open woodland, waste ground, riverbanks, gardens. **Fp** Common Comfrey, Tuberous Comfrey, Lungwort, Common Gromwell, also recorded on Wood Forget-me-knot; feeds from under a slight web.

Ethmia bipunctella page 221

Rare. Resident only in East Sussex, Kent and possibly north to Suffolk. An occasional immigrant or wanderer inland. **FL** 9-13mm. Forewing blackish brown costally and white dorsally, the apex pale grey, with a series of black dots along the termen. Abdomen orange-yellow. **FS** Possibly double-brooded, late April-early October. Can be found by day at rest on the foodplant, fence posts and telegraph poles, and comes to light. **Hab** Vegetated shingle, coastal chalk cliffs. **Fp** Viper's-bugloss; on the flowers and leaves under a slight web.

Ethmia pyrausta page 221

Rare. Known from a single example in the 19th century until rediscovered in 1996, since when a few examples have been recorded. Probably overlooked. **FL** 7.5-10.5mm. Forewing greyish black with black spots at about one-quarter, one-half and three-quarters. Abdomen with a yellowish tip. Has been found in a water trap and caught in a spider's web. **FS** Single-brooded, May-mid June. **Hab** Montane situations and has been recorded above 800m and at about 1,000m. **Fp** On the Continent, on Alpine Meadow-rue; the larva has not yet been found in Scotland.

Blastodacna hellerella page 221

Common. Very local in Scotland. **FL** 4.5-6mm. Forewing dark or blackish brown; a white streak from the base to the tornus, narrowed at about one-third by dark brown, the outer part of this streak often indistinct, with an angled white line towards the apex usually present; there are two patches of raised scales, above the dorsum before one-half, and in the middle of the wing before the tornus. Sometimes the white is more extensive across the wing. **Similar species *B. atra* is often slightly larger and darker, and has yellowish brown towards the base.** **FS** Single-brooded, May-August. Rests by day on tree trunks and fences, and comes to light. **Hab** Woodland, parkland, farmland, scrub, gardens. **Fp** Hawthorn, Midland Hawthorn, Whitebeam; in a berry.

▲ Larva of *Ethmia bipunctella* on Viper's-bugloss.

▲ Larva of *Blastodacna hellerella* burrowing into a hawthorn berry.

Apple Pith Moth *Blastodacna atra* page 221   906 (2056)

Local. **FL** 5-6mm. Forewing blackish brown; pale examples have an indistinct creamy white dorsal streak from the base to the tornus, with at least a yellowish-brown streak towards the base and often another at about two-thirds; two patches of dark raised scales, above the dorsum before one-half, and in the middle of the wing before the tornus; dark examples may be almost without white scaling. **Similar species** *B. hellerella*. **FS** Single-brooded, late May-early September. Flies at dusk and comes to light. **Hab** Gardens, orchards, woodland. **Fp** Apples; mines a twig, causing the shoot and blossom-trusses to die.

Spuleria flavicaput page 221   904 (2060)

Local. Rare in Scotland. **FL** 6-7mm. Head bright yellow. Forewing shining dark brown, with two small patches of raised darker scales, on the dorsum at one-third and above the tornus. **FS** Single-brooded, May-June. Flies in morning sunshine and has been recorded at light. **Hab** Woodland margins, scrub, hedgerows. **Fp** Hawthorn, Midland Hawthorn; lives in a twig.

Dystebenna stephensi page 221   907 (2062)

Rare. Found mainly in south-eastern England. **FL** 4-5mm. Head and thorax predominantly white. Forewing whitish, variably shaded orange-brown and greyish brown, with a dark greyish-brown mark on the costa at the base and at about one-half, with a more elongate dark greyish-brown mark at the apex; there are two patches of raised scales, above the dorsum before one-half, and in the middle of the wing before the tornus. **FS** Single-brooded, late June-early September. Can sometimes be found in numbers on the trunks of old oak trees. **Hab** Parkland, open woodland. **Fp** Oaks; in the living bark, exuding reddish frass in a similar manner to the larva of *Argyresthia glaucinella* (Argyresthiidae).

Chrysoclista lathamella page 221   902 (2049)

Rare. Infrequently recorded and little known. Recently discovered new to Ireland. **FL** 5-6mm. Forewing orange, bordered shining blackish brown, this broader from the base of the costa to almost one-half where it joins the dorsal border, separating the orange into two patches, with three round metallic silvery white slightly raised spots, the first at one-third, the second slightly beyond and the third at two-thirds. **FS** Single-brooded, June-August. Flies by day around the host tree, has been captured in a malaise trap, and is occasionally recorded at light. **Hab** Woodland, bogs, commons, coastal landslips. **Fp** Willows and sallows; lives in bark.

Chrysoclista linneella page 221   903 (2048)

Very local. **FL** 5-6mm. Antennae greyish brown, tipped white. Forewing orange, bordered shining blackish brown, with three slightly raised silver spots, each edged blackish brown. **FS** Single-brooded, late May-August. Can be found resting on tree trunks of the host tree by day, waving its antennae. Comes to light. **Hab** Urban and suburban roadsides, parkland, woodland. **Fp** Limes; feeds under bark, exuding reddish frass.

Stathmopodidae

Stathmopoda pedella

There are three species in this family on the British list. Adults rest in a characteristic fashion, horizontal but standing on the forelegs and midlegs, with the hindlegs held almost at right-angles to the body and slightly angled upwards, and with the wings rolled around the abdomen. The forewing length is 4-7mm. The forewings are elongate and taper towards the apex. The hindwings are narrow and pointed, with the dorsal cilia very long. The head has smooth scales. The antennae are thread-like, about three-quarters the length of the forewing. The labial palps are long, slender and curved upwards.

Stathmopodidae was formerly treated as a subfamily of Oecophoridae. There are three species of *Stathmopoda* recorded: *S. pedella* is native, while *S. diplaspis* and *S. auriferella* are adventives, both added recently to the British list, having been reared from the calyx of imported Pomegranates bought from supermarkets. A fourth species, possibly in the genus *Calicotis* and not yet identified to a species, has been found recently in two formal gardens, one in England and one in Wales. It is associated with ferns and is likely to have been imported from the southern hemisphere; it has become established and may well spread.

Further reading
British and Irish species: Emmet & Langmaid (2002a)
European species: Palm (1989)

Stathmopoda pedella   page 221   877 (2403)

Local. Very local in northern England. **FL** 5-7mm. Hindlegs yellowish, banded dark brown. Head yellowish. Forewing yellowish, the base greyish brown, with the costa greyish brown, this sometimes interrupted at about one-half and near the apex, and the dorsum with greyish-brown blotches at about one-third and two-thirds, these extending across the wing. **FS** Single-brooded, late June-early August. Can be tapped from the underside of leaves of the foodplant by day, and occasionally comes to light. **Hab** Alder carr, fens, marshes, woodland, parkland. **Fp** Alder, Grey Alder; in the green fruits, exuding orange frass.

► Feeding signs of the larva of *Stathmopoda pedella*. Note the bright orange frass on the surface of a green cone of Alder.

Batrachedridae

There are three species in this family. The adult rests with the front end slightly raised, the wings rolled around the abdomen and the antennae held over the back. The forewing length is 4-8mm. The forewings are especially narrow and elongate, without a distinct tornal angle. The hindwings are similarly narrow, with long dorsal cilia. The head has smooth scales. The antennae are four-fifths the length of the forewing and are held over the back. The labial palps are curved upwards, and the tongue is scaled.

Batrachedra praeangusta

Of the three species recorded in Britain, *Batrachedra praeangusta* and *B. pinicolella* are native, and *B. parvulipunctella* has been recorded only once, in England. The two native species are covered below. They are mainly associated with woodlands and mature trees. Adults can be disturbed from the foodplants by day, and are attracted to light.

Further reading
British and Irish species: Emmet & Langmaid (2002a)

Batrachedra praeangusta page 221   878 (2428)

Common. Local in Scotland. **FL** 7-8mm. Forewing elongate, appearing greyish, variably shaded yellowish white, with a short longitudinal blackish streak at around one-third and another from about two-fifths to three-quarters, interrupted at about two-thirds by a yellowish-white spot. **FS** Single-brooded, mid June-early October. Can be disturbed by day and comes to light. **Hab** Woodland, parkland, gardens. **Fp** White Poplar, Aspen, White Willow, Goat Willow; feeds in the female catkins and can move onto the buds.

Batrachedra pinicolella page 221   879 (2429)

Common in England. **FL** 4-6mm. Forewing elongate, pale yellowish with fine dark brown speckling, stronger on the costa, and with a small dark brown spot at about four-fifths. **Similar species** Resembles *Coleophora salicorniae* and *C. clypeiferella* (Coleophoridae), which have a broader forewing and diffuse dots, with the antennae pointing forward; also *B. parvulipunctella*, which is generally larger and paler, with two short dark streaks on the wing. **FS** Single-brooded, mid June-August. Can be tapped from branches during the day and comes to light. **Hab** Coniferous woodland, gardens. **Fp** Norway Spruce; mines the needles.

▼ Larva of *Batrachedra praeangusta*.

Coleophoridae

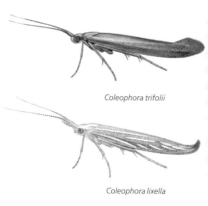

Coleophora trifolii

Coleophora lixella

There are 109 species in this family, with virtually all belonging to the genus *Coleophora*. The adults rest in a slightly inclining position. The wings are held variably between species, in a shallow to moderately steep roof-like position, or more closely wrapped around the abdomen; the apex of the wings of some species rests on the surface. The antennae point forwards or are diverging. The forewing length is 3-11mm. The forewings are narrow and elongate, without a tornal angle. They are generally plainly coloured, or lightly speckled, or with pale streaks. The hindwings are narrower than the forewings, with very long dorsal cilia. The head is smooth-scaled. The long antennae are thread-like, the base in a few species thickened with scales or hairs, and two-thirds to three-quarters the length of the forewing. The labial palps are slender and moderately curved upwards. The tongue is functional, but short.

Many species are very widely distributed, for example *C. serratella, C. gryphipennella* and *C. alticolella*. Others are extremely restricted. *C. galbulipennella* is resident only in two coastal sites in Kent, although it can be extremely abundant there, the larvae causing leaves of the foodplant, Nottingham Catchfly, to wilt and curl. Four species are now thought to be extinct, including *Augasma aeratella*, which forms a gall, not a case, in Knotgrass, and *C. albella*, believed to be the only British moth with a larviparous female, that is, she lays young larvae, not eggs, in the flowers of Ragged-robin. New species to our shores are found from time to time: *C. calycotomella* was discovered on Broom in Surrey, in 2004, and Clover Case-bearer *C. frischella* is now acknowledged to occur locally in Britain and Ireland, having been recognised among the very similar and more widespread species, *C. alcyonipennella*.

Comprehensive coverage of Coleophoridae is not possible in this guide. Only a few species can readily be identified from wing and other external characters, and examination of genitalia is required to confirm identification of most examples caught in moth traps. It happens that most of the easily recognisable species are the larger and rarer ones. Conversely, those regularly encountered are the smaller and more widespread ones, and there are so many species that they can be confused with each other. This situation is reflected in the accounts below. A small number of the most striking species are described and illustrated, even though they are unlikely to be found often, together with the few widespread ones that are fairly easy to identify. In addition, there is a longer list of species for which there are short identification notes and an illustration of the forewing of each. The intention is to enable the reader to get to a group of closely-related species, and perhaps to a species, although it is unlikely that records for most of these species would be accepted without additional verification, particularly by genitalia examination. The last three species in the notes are those that have a dot in the wing, and can be separated fairly easily.

Rearing adults from the larval stage is probably the most satisfactory way to become familiar with the Coleophoridae. Photographs of a range of cases and feeding signs are shown on pp.154-157. The larvae

▲ Case and feeding pattern of *Coleophora serratella* on birch. The case is made from a leaf margin which has been mined and then excised and lined inside with silk.

construct portable cases in characteristic ways, and with knowledge of the foodplant, feeding pattern or how the case is made from the leaves or seeds, most species can be determined from the larval stage. As larvae grow, the case is lengthened with additional leaf or seed fragments, or with silk. All species overwinter in the larval stage, a few species having a two-year life cycle. However, only a small number can easily be found over the winter months, such as *C. alticolella* and *C. glaucicolella* among the seedheads of rushes, or *C. argentula* on the old flowerheads of Yarrow. A number of species change cases between autumn and spring, and the shape of the case may be radically different. For example, the larva of *C. lixella* uses a single calyx from the flower of thyme as its case while feeding on thyme flowers in the autumn, abandoning this case after resuming feeding in spring, when it feeds on a range of grasses, constructing a case from part of a hollowed-out blade.

Larvae of many of the leaf-mining species characteristically feed in short bouts, moving to new parts of the same leaf, or to a different leaf, fixing the case with silk, usually to the underside, and then mining the leaf for a short while. This leaves telltale signs of feeding, and the size and frequency of small mines can help to determine the species, even if the case cannot be found. Normally the upper epidermis remains intact but the larva of *C. siccifolia* severs this wholly or partially when it has finished mining. A few species only form a case at the end of their final instar, for example *Metriotes lutarea*, which makes its case from a hollowed-out seed capsule of Greater Stitchwort, a huge structure in comparison with the diminutive larva. Most species pupate within the case, except for example *M. lutarea*, which makes a cocoon in bark, or *C. salicorniae* and *C. clypeiferella*, which make a cocoon in the ground. In most species, adults are active at dusk and fly at night, and some species come readily to light; a few species are active during the day, such as *C. alticolella* and *C. glaucicolella*, which can be observed in numbers flying around rush flowers on a warm summer afternoon.

Further reading
British and Irish species: Emmet (1996)

Metriotes lutarea page 222   487 (2438)

Very local. **FL** 4.5-6mm. Forewing elongate, shining grey with a yellowish-brown tinge.
Similar species Several *Coleophora* are superficially similar, but behaviour, time
of year and presence of the foodplant are a good indication of this species.
FS Single-brooded, late April-early June. Can be numerous. In sunshine, flies over the
foodplant and rests in the flowers of the foodplant. Occasionally comes to light.
Hab Woodland, old hedgerows. **Fp** Greater Stitchwort; on the ripening seeds, when full-
grown making a case from the seed capsule.

Goniodoma limoniella page 222   488 (2442)

Local. **FL** 5-6mm. Forewing elongate, curved at the tip, reddish brown, with a series of
irregularly darker-edged leaden metallic markings, including streaks in the basal half,
along the costa, dorsum and in the middle of the wing, with spots beyond and a line along
the termen to the apex. **FS** Single-brooded, July-August. Can be numerous where found.
Flies in afternoon sunshine, but can be hard to see, and comes to light. **Hab** Drier parts of
saltmarshes where the foodplant occurs. **Fp** Common Sea-lavender; on the seeds in the
flowers, using a calyx as a case and pupating in a stem.

Coleophora albitarsella page 222   515 (2496)

Common. Much more local in the north. **FL** 5-6.5mm. Antenna white, ringed darker in the
basal half. Forewing elongate, shining greyish brown. All tarsi of the male and the middle
of the hind tarsi of the female are whitish. **FS** Single-brooded, late May-mid August.
Sometimes flies in sunshine and comes to light. **Hab** Hedgerows, scrub, gardens, open
woodland, fens, downland. **Fp** Calamint, Wild Basil, Ground-ivy, various mints, Wild
Clary, Selfheal, Marjoram; in a case made of leaf and silk, this about 9mm in length.

Coleophora trifolii page 222   516 (2498)

Common. More local in northern England. **FL** 7-10mm. **Upper margin of the eye edged
yellow.** Antenna blackish, whitish towards the apex. Forewing elongate, metallic golden
green, with a purplish sheen towards the apex, this more prominent in the female.
Similar species There are five other metallic greenish or bronzy *Coleophora*: *C. mayrella*
(not illustrated), a common species, is small (FL 5-6mm), with the basal two-fifths of the
antenna thickly clothed in purplish reflecting dark scales; *C. deauratella*, *C. frischella* and
C. alcyonipennella (all not illustrated) are small but very similar and often require genitalia
examination to separate; *C. amethystinella* (not illustrated), a scarce species, is large,
without a pale tip to the antenna; also *C. paripennella*, which is smaller and shining greyish
brown with a metallic bronze reflection, not green. **FS** Single-brooded, May-August. Flies
in sunshine and comes to light. **Hab** Rough grassland, waste ground, gardens, marshes,
dunes, downland, roadside verges. **Fp** Ribbed Melilot, Tall Melilot; in a somewhat hairy
case made of a seed-pod or pods, this about 8mm in length.

Coleophora lixella page 222   530 (2654)

Local. Very local in Scotland. **FL** 8-10mm. Antenna white on the upperside towards the
apex. Forewing elongate, curved at the tip, pale yellow or yellowish brown with a few
longitudinal silvery white streaks, these finely and partially edged blackish brown;
blackish-brown scaling is more prevalent towards the apex. **Similar species** *C. tricolor*
(not illustrated), a rare species on Basil Thyme, resident only in Breckland in East Anglia,
has the antenna white, ringed greyish brown throughout; trap-caught examples will

normally need to be confirmed by examination of genitalia. **FS** Single-brooded, mid June-August. Readily disturbed from vegetation by day, flies freely in afternoon sunshine and comes to light. **Hab** Chalk and limestone grassland, sandhills, rocky slopes. **Fp** Thyme flowers in autumn, forming a case from the calyx; in spring on various grasses, constructing a case from a grass blade, this up to 11mm in length.

Coleophora ochrea page 222   531 (2647)

Rare. **FL** 7-9mm. Forewing elongate, pale yellowish brown with up to four pale silvery streaks, usually with one on the dorsum to about one-half and another from the base to beyond the middle, approaching the dorsum at about two-thirds; darker, reddish-brown scales form indistinct streaks over the wing. **FS** Single-brooded, July-August. Perhaps most readily found in the larval stage in spring, when the larva causes pale brown blotching of leaves of the foodplant. The larval cases can be numerous. **Hab** Chalk and limestone habitats, favouring rocky, barren situations, particularly where the foodplant overhangs ledges. **Fp** Common Rock-rose; in a tubular case formed of leaves, this up to 15mm in length.

Coleophora vibicella page 222   538 (2639)

Rare. Declining, with a few sites currently known in Dorset, Hampshire, Isle of Wight and Sussex. **FL** 8-11mm. Forewing elongate, pale yellowish brown, darker brown towards the costa, with a silvery white streak from the base to just over one-half, another just below the costa from about one-third to two-thirds and one from about one-half angled towards the apex. **FS** Single-brooded, July-August. The larval cases can be numerous. Recorded at light. **Hab** Neutral grassland by the coast, woodland rides. **Fp** Dyer's Greenweed; in a black case formed entirely of silk, this up to 19mm in length.

Coleophora paripennella page 222   560 (2850)

Common. **FL** 5-6.5mm. Antenna dark greyish brown, white towards the apex. Labial palps dark greyish brown. Forewing elongate, shining greyish brown with a bronzy metallic reflection. Legs shining greyish brown. **Similar species** *C. violacea* (not illustrated) has mainly whitish-brown labial palps, with the legs matt, not shining; *C. fuscocuprella* (not illustrated) has the antenna somewhat ringed towards the apex; also the *C. trifolii* group. **FS** Single-brooded, end June-early August. Comes to light. **Hab** Grasslands, rough ground, roadside verges, woodland rides. **Fp** Common Knapweed, Creeping Thistle, sometimes other knapweeds and thistles; in a narrow black tubular case formed almost entirely of silk, this up to 8mm in length.

Coleophora wockeella page 222   527 (2866)

Rare. Much declined, currently known only from Surrey. **FL** 9-11mm. Antenna clothed in scales to just over one-half. Forewing elongate, curved slightly at the tip, the dorsal half pale reddish brown, slightly darker towards the costa, with a fine white costal streak from the base to nearly one-half and another from the base approaching the dorsum at about one-half. **FS** Single-brooded, June-July. The larval cases can be numerous. Recorded at light. **Hab** Edges of woodland rides. **Fp** Betony; in a tubular case formed of leaf sections, this up to 10mm in length.

▲ Case of *Metriotes lutarea* in seed-pod of Greater Stitchwort.

▲ Case of *Coleophora ochrea* on Common Rock-Rose.

▲ Case of *Coleophora betulella* on birch.

▲ Case of *Coleophora vibicella* on Dyer's Greenweed.

▲ Case of *Coleophora gryphipennella* on a rose shoot.

▲ Larval feeding pattern of *Coleophora gryphipennella* on ro

▲ Case of *Coleophora alnifoliae* on Alder.

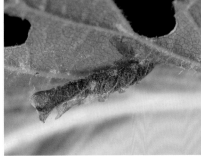

▲ Case of *Coleophora limosipennella* on English Elm.

▲ Case of *Coleophora binderella* on Alder.

▲ Case of *Coleophora lineolea* on Black Horehound.

▲ Case of *Coleophora genistae* on Petty Whin.

▲ Case of Larch Case-bearer *Coleophora laricella* on Larch.

▲ Case of *Coleophora conyzae* on Fleabane.

▲ Case of *Coleophora discordella* on bird's-foot-trefoil.

▲ Case of *Coleophora pennella*.

▲ Case of *Coleophora galbulipennella* on Nottingham Catchfly.

▲ Feeding damage by *Coleophora galbulipennella* on Nottingham Catchfly.

▲ Feeding marks of *Coleophora artemisicolella* on Mugwort seed-pods.

▲ Case of *Coleophora lithargyrinella* on Greater Stitchwort.

▲ Case of *Coleophora peribenanderi* on Creeping Thistle.

▲ Case of *Coleophora aestuariella* on Annual Sea-blite.

▲ Case of *Coleophora striatipennella* on Common Mouse-ea

▲ Case of *Coleophora alticolella*.

▲ Case of *Coleophora caespititiella*.

▲ Case of *Coleophora salicorniae* on glasswort.

▲ Case of *Coleophora albitarsella* on Ground Ivy.

▲ Early case of *Coleophora hemerobiella* on hawthorn.

▲ Final case of *Coleophora hemerobiella* on hawthorn.

▶ Final case of
Coleophora lixella
on grass.

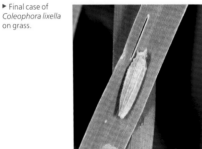

▲ Case of *Coleophora albicosta* on a seed-pod of gorse.

▲ Case of *Coleophora lusciniapennella* case on Bog-myrtle.

▲ Cases of *Coleophora trifolii* on seed-pods of mellilot.

Notes on common species of *Coleophora* likely to be encountered which are likely to require further investigation to confirm identity (forewings illustrated on page 223)

C. lutipennella 490 (2453) **FL** 5-6mm. Almost identical to *C. flavipennella*, which is similarly common in oak woodland. Forewing pale yellowish brown. **FS** June-August. **Hab** Woodland. **Fp** Oaks.

C. gryphipennella 491 (2456) **FL** 5-6.5mm. Similar to the greyish-brown form of *C. serratella* and the *C. spinella* group. Antenna wholly ringed white and greyish brown. Forewing yellowish brown, tinged greyish. **FS** June-July. **Hab** Woodland, hedgerows, scrub, gardens. **Fp** Roses.

C. serratella 493 (2468) **FL** 5-7mm. Antenna with white and brown or greyish-brown bars along the inner half only. Forewing dark brown or greyish brown. Greyish-brown form similar to those in the *C. spinella* group. **FS** July-August. **Hab** Woodland, scrub, hedgerows. **Fp** Mainly birches, Alder, Hazel.

Apple & Plum Case-bearer *C. spinella* 495 (2469) **FL** 5-6mm. The most widespread of the three very similar species in the *C. spinella, C. coracipennella* and *C. prunifoliae* group. Antenna barred and ringed white and greyish brown. Forewing greyish brown, scales darker tipped. **FS** June-July. **Hab** Woodland, hedgerows, scrub, gardens. **Fp** Mainly hawthorns, apples, less often plums, Blackthorn.

C. juncicolella 510 (2492) **FL** 3-4mm. A very small species; large examples could be confused with *C. laricella*. Antenna ringed whitish and dark grey. Forewing narrow, greyish, tinged pale brown. **FS** June-July. **Hab** Heathland, moorland. **Fp** Heather, Bell Heather.

C. lineolea 522 (2518) **FL** 5.5-7mm. The all-white antennae help separate this from similar species. Forewing whitish, with most veins highlighted yellowish brown. **FS** June-August. **Hab** Dry grassland, roadside verges, waste ground, gardens. **Fp** Mainly Black Horehound, Hedge Woundwort, Lamb's-ear.

Larch Case-bearer *C. laricella* 526 (2683) **FL** 4-5.5mm. Small examples could be confused with *C. juncicolella*. Antenna grey, slightly ringed. Forewing grey. **FS** June-July. **Hab** Plantation woodland, gardens. **Fp** European Larch.

Pistol Case-bearer *C. anatipennella* 533 (2592) **FL** 6-8.5mm. One of five very similar species. Forewing white, with scattered dark greyish-brown scales. **FS** June-July. **Hab** Woodland, scrub, hedgerows. **Fp** Mainly Blackthorn, also plums, hawthorns, apples.

C. pyrrhulipennella 541 (2601) **FL** 5.5-7mm. One of several similar species with white streaks on a darker background. Forewing pale yellowish grey, tinged brownish, with broad white streaks. **FS** May-August. **Hab** Heathland, moorland, bogs. **Fp** Heather, Bell Heather.

C. albicosta 544 (2662) **FL** 6-7.5mm. One of several similar species with white streaks on a darker background. Forewing yellowish brown with scattered light greyish-brown scales, with a broad white streak along the costa, a narrow white streak from the base to near the tornus and another in the middle of the wing angled above the tornus. **FS** May-July. **Hab** Wherever the foodplant is frequent. **Fp** Gorse.

C. discordella 547 (2572) **FL** 5.5-7mm. One of several similar species with white streaks on a darker background. Antenna ringed white and greyish brown. Forewing brown, tinged yellowish, darker towards the apex, with a broad white streak along the costa and narrow white streaks on the dorsum and two in the middle. **FS** June-August. **Hab** Grasslands, sand dunes, shingle. **Fp** Common Bird's-foot-trefoil, Greater Bird's-foot-trefoil.

C. striatipennella 553 (2809) **FL** 5.5-6.5mm. One of many species in the *C. alticolella* group. Forewing whitish, veins yellowish brown, tinged grey. **FS** May-August. **Hab** Woodland rides, grasslands, fens. **Fp** Lesser Stitchwort, Common Chickweed, Common Mouse-ear.

C. peribenanderi 559 (2786) **FL** 6-7.5mm. One of many species in the *C. alticolella* group. Forewing yellowish brown, with whitish veins. **FS** June-August. **Hab** Woodland rides, grasslands, fens. **Fp** Mainly Creeping Thistle.

C. otidipennella 578 (2690) **FL** 5-6.5mm. The earliest of the *C. alticolella* group on the wing. Forewing greyish, with whitish veins. Easily disturbed by day, rarely seen at light. **FS** May-June. **Hab** Grasslands, heathland, moorland, woodland rides. **Fp** Field Wood-rush, Heath Wood-rush.

C. alticolella 584 (2692) **FL** 5-6mm. Most similar to the common *C. glaucicolella*; one of many species in the *C. alticolella* group. Forewing greyish brown usually with a narrow, dirty whitish streak along the costa. **FS** June-July. **Hab** Wet grasslands, fens, marshes, wet heathland, moorland. **Fp** Rushes.

Coleophora species with a dark dot or dots on a pale ground colour

C. hemerobiella 523 (2524) **FL** 6-7.5mm. Forewing whitish, heavily speckled grey, with a grey dot at four-fifths. **FS** July. **Hab** Woodlands, hedgerows, scrub, gardens. **Fp** Mainly hawthorns, apples, cherries, plums.

C. salicorniae 588 (2858) **FL** 6-7mm. Forewing pale brownish yellow, somewhat darker speckled, with a weakly defined round dark dot at three-quarters. **FS** July-August. **Hab** Saltmarshes, but disperses widely. **Fp** Glassworts.

C. clypeiferella 589 (2854) **FL** 6.7.5mm. Forewing pale brownish yellow, somewhat darker speckled, with a weakly defined dark mark at three-quarters, and sometimes a similar mark at one-third. **The first abdominal segment has a clypeus* on the dorsal surface, which distinguishes this species from all others. FS** July-August. **Hab** Waste ground, arable field margins, but disperses widely. **Fp** Fat-hen.

*The clypeus is a shield of blunt, peg-like spines that are used by the adult to open the cocoon and push up to the surface of the ground. The clypeus cannot be seen unless the forewings are parted to show the abdomen.

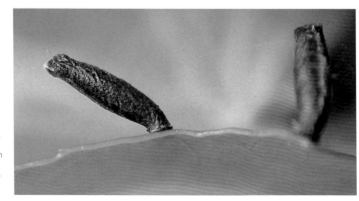

▶ Larval cases of Apple & Plum Case-bearer *Coleophora spinella*. The case is made from a leaf margin which has been mined and then excised, and lined inside with silk.

Momphidae

Mompha locupletella

Mompha subbistrigella

There are 15 species in this family, all of which are covered below. The adults rest in a horizontal position and the wings are held roof-like at a moderate to steep angle, curling slightly around the body, giving a distinctive tapered appearance to this small group when

Mompha propinquella

viewed from above. The forewing length is 3-8.5mm. The forewings are elongate, the membrane without a distinct tornal angle, although the cilia at the tornus may produce an angle between the dorsum and termen. The forewings usually have scale-tufts and sometimes metallic markings. The hindwings are narrower than the forewings. The head is smooth. The antennae are thread-like, three-quarters the length of the forewing and are held along the side the body. The labial palps are ascending, segment three sometimes reaching above the head, and the tongue is scaled.

The majority of species are associated with open habitats, such as grassland, marshland, open woodland and along hedgerows. Several species are found widely throughout Britain and, to a lesser extent, Ireland, although a few, such as *Mompha bradleyi*, are localised; this species was recognised in Britain in the early 1990s and may now be spreading. Providing adults are in good condition, many species can be identified on external characters, although care is needed with a few superficially similar species pairs, especially *M. lacteella* and *M. propinquella*, *M. divisella* and *M. bradleyi*, and *M. sturnipennella* and *M. subbistrigella*. If examples are worn, identification by examination of genitalia is likely to be necessary.

All but one species is associated with the plant family Onagraceae, the willowherbs. Several species are easiest to find in the larval stage. The larval habits are variable

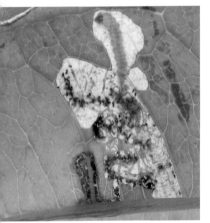

between species and include leaf-mining, feeding within spun shoots or seed-pods, forming a gall within a small stem, or living within a larger stem of the foodplant. Adults of many species can be found at light. In southern Britain *M. subbistrigella* and *M. epilobiella* are probably the most frequently encountered at light, and the former is also regularly seen indoors and in sheds, hibernating in late autumn and winter.

Further reading
British and Irish species: Emmet & Langmaid (2002a)
European species: Koster & Sinev (2003)

◄ Tenanted mine of *Mompha raschkiella* on Rosebay Willowherb.

► Tenanted mine of
Mompha locupletella
on a willowherb.

Mompha miscella page 224   884 (2872)

Local. **FL** 3.5-4.5mm. Forewing brownish, greyish at the base, with greyish cross-bands from the costa at one-fifth to the dorsum at one-third, sometimes merged with the greyish base, and at two-thirds, these outwardly edged with patches of black raised scales; a whitish spot on the costa at about three-quarters and often another at about one-half; the apical area is greyish. **FS** Double-brooded, late April-October. Found by day and comes to light. **Hab** Chalk and limestone grassland, locally on acid soils in Scotland. **Fp** Common Rock-rose, Hoary Rock-rose, White Rock-rose; mines the leaves.

Mompha langiella page 224   880 (2876)

Very local. More frequent in the west. **FL** 4.5-5mm. Forewing shining dark brown, with an irregular white spot beyond one-half and a scatter of white scales along the costa in the apical half and towards the apex. **FS** Single-brooded, August-April. Hibernates as an adult. **Hab** Woodland, shaded rides. **Fp** Enchanter's-nightshade, Great Willowherb, sometimes Broad-leaved Willowherb, Hoary Willowherb, Rosebay Willowherb; mines the leaves. The mine starts as an irregular gallery, widening into a blotch, in June and July (cf. *M. terminella*).

Mompha terminella page 224   881 (2878)

Very local. Perhaps more frequent in the west. **FL** 3-4.5mm. Forewing dark orange, with markings including a leaden grey blotch enclosing a blackish spot near the base, a white costal spot at three-quarters and a leaden grey dorsal blotch with two small black scale-tufts. **FS** Single-brooded, late June-August. Can be swept by day and occasionally comes to light. **Hab** Woodland, shaded woodland rides. **Fp** Enchanter's-nightshade; mines the leaves. The mine starts as a gallery in circles or semicircles, widening into a blotch, in August and September (cf. *M. langiella*).

Mompha locupletella page 224   882 (2879)

Common. Local in Scotland. **FL** 4.5-5.5mm. Forewing bright orange, with shining grey leaden markings, including a basal blotch edged blackish, costal spots at one-quarter and one-half, a dorsal spot at one-half edged with a black scale-tuft, and another spot at the tornus with a whitish spot on the costa opposite; the apex is dark brown. **FS** Double-brooded in England and Wales, possibly single-brooded further north, late May-early September. May be found at rest by day on leaves and occasionally comes to light. **Hab** Damp situations. **Fp** Chickweed Willowherb, Marsh Willowherb, Broad-leaved Willowherb, Spear-leaved Willowherb; mines the leaves.

Mompha raschkiella page 224   883 (2880)

Common. Local in central and western Ireland. **FL** 3.5-5mm. Forewing shaded blackish brown and orange-brown, the basal area shining blackish brown, the costa blackish brown with shining leaden grey spots near the costa at one-third and two-thirds, near the dorsum at one-half and three-quarters, edged with black scale-tufts, with a white costal spot at three-quarters. **FS** Double-brooded, possibly with a partial third brood, early May-October. Can be found by day at rest on leaves and occasionally comes to light. **Hab** Waste ground, woodland clearings, heaths, roadside verges. **Fp** Rosebay Willowherb; mines the leaves.

Mompha conturbatella page 224   885 (2883)

Common. May have declined to local in northern England. **FL** 7-8.5mm. **Typically the largest *Mompha* species.** Forewing dark grey, mottled with brownish and blackish spots, with a greyish basal patch, two black scale-tufts on the dorsum, an almost square white mark on the costa at about four-fifths and a smaller white scale-tuft on the dorsum at about two-thirds. **FS** Single-brooded, late May-late September. Can be disturbed by day and comes to light. **Hab** Waste ground, woodland clearings, scrubby downland, roadside verges. **Fp** Rosebay Willowherb, Broad-leaved Willowherb; within a tight spinning in the terminal shoot.

Mompha ochraceella page 224   886 (2884)

Common. **FL** 6.5-7mm. Forewing yellowish orange, mottled darker, with an indistinct yellowish costal spot at three-quarters and with small tufts of dark brown scales at one-fifth and one-half, although these can become obscure in worn examples. **Similar species *M. epilobiella* is duller and smaller, with the forewing yellowish brown, not orange.** **FS** Single-brooded, late May-early August. Comes to light. **Hab** Damp situations, streambanks, ditches. **Fp** Great Willowherb; hibernates in the root, mining the lower stem and leaves in spring.

Mompha lacteella page 224   887 (2885)

Rare. **FL** 4.5-6mm. Head and thorax pale yellowish brown. Forewing dark greyish brown to blackish brown, with a pale yellowish-brown basal blotch and several orange-brown marks, one beyond the basal blotch, others in the middle and apical half, and with a large black scale-tuft on the dorsum at one-half, a white costal spot at three-quarters and a white scale-tuft almost opposite; dark brownish grey towards the termen and apex. **Similar species *M. propinquella* has a white head, thorax and basal blotch, and is reddish brown towards the apex.** Genitalia examination may be needed to confirm identification. **FS** Single-brooded, May-July. Occasionally comes to light. **Hab** Woodland, waste ground. **Fp** Broad-leaved Willowherb, possibly Great Willowherb; mines the leaves.

Mompha propinquella page 224   888 (2886)

Common. **FL** 5-6mm. Head and thorax white. Forewing brownish grey to blackish brown, the base whitish, this almost reaching the costa, with reddish-brown blotches beyond the whitish basal area, near the middle and in the apical half, and with a large black scale-tuft on the dorsum at one-half, a whitish costal spot at three-quarters and a white scale-tuft almost opposite; reddish brown towards the termen and apex. **Similar species** *M. lacteella*. **FS** Single-brooded, late June-mid September. Regularly comes to light and has been found flying at dawn. **Hab** Open woodland, waste ground, gardens. **Fp** Great Willowherb, Broad-leaved Willowherb, possibly other small species of willowherb; mines the leaves.

Mompha divisella page 224   889 (2887)

Very local. Now spreading after a period of considerable scarcity. **FL** 5-6mm. Forewing somewhat long and narrow; dark greyish brown, finely speckled whitish, with the dorsal area bright white, yellowish brown towards the base, and with a white cross-band at three-quarters, this with a pale brown streak on either side. **Similar species *M. bradleyi* usually has a slightly shorter and broader forewing, and the dorsal area is a duller white, with a peppering of brown and grey scales.** Identity of trap-caught examples is best confirmed by examination of genitalia. **FS** Single-brooded, August-May. Hibernates as an adult. Occasionally comes to light. **Hab** Waste ground, uncultivated field margins, roadsides, pavements, cemeteries, other very open and dry areas; also damp woodland and some shady situations. **Fp** Broad-leaved Willowherb, Marsh Willowherb, Spear-leaved Willowherb, Hoary Willowherb, Great Willowherb; feeds within a stem, forming a gall, sometimes causing branching of the plant above the gall.

▲ The terminal shoot of a willowherb opened up to show the larva of *Mompha conturbatella* within.

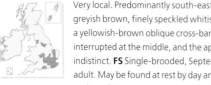

▶ A gall formed by the larva of *Mompha divisella* in a small willowherb species. Note the whitish silken exit hole created by the larva before it pupated.

Mompha bradleyi page 224 — 889a (2887)

Very local. May be spreading. **FL** 5-6mm. Forewing somewhat short and broad; dark greyish brown, finely speckled whitish, with the dorsal area white, peppered with brown and grey scales; a white cross-band at three-quarters is often interrupted. **Similar species** *M. divisella*. **FS** Single-brooded, August-May. Hibernates as an adult. **Hab** Marshes, ditches, roadside verges, rough ground in urban areas. **Fp** Great Willowherb; feeds within a stem, forming a gall, preferring the stalks just below the flowers.

Mompha jurassicella page 224 — 890 (2888)

Very local. Predominantly south-eastern. May be spreading. **FL** 5-6.5mm. Forewing dark greyish brown, finely speckled whitish, with the basal half of the dorsum yellowish brown, a yellowish-brown oblique cross-band at three-quarters, this sometimes narrowed or interrupted at the middle, and the apex with a yellowish-brown spot, but this can be indistinct. **FS** Single-brooded, September-April; also recorded in June. Hibernates as an adult. May be found at rest by day and occasionally comes to light. **Hab** Dry situations, such as allotments, waste ground, dry ditches. **Fp** Great Willowherb; mines the stem.

Mompha sturnipennella page 224 — 891 (2890)

Very local. Well established in the south-east. Seems to occur in discrete colonies, and occasionally numerous where found. **FL** 4-6mm. Forewing somewhat long; grey with a whitish cross-band before one-half not reaching the costa and broadening towards the dorsum, and a whitish oblique cross-band at three-quarters, with dark brown streaks in the middle of the wing and short white costal streaks towards the apex. **Similar species** *M. subbistrigella* **has a shorter forewing, without streaks in the middle of the wing and without white costal streaks near the apex. FS** Double-brooded, July-May. Hibernates as an adult. Can be found flying around the foodplant in sunshine and sometimes comes to light. **Hab** Waste ground, heaths, urban situations. **Fp** Rosebay Willowherb; in the first generation the larva feeds in a stem, forming a gall, and in the second generation it feeds within a seed-pod.

▲ Larval working of *Mompha epilobiella* in a side shoot of Great Willowherb.

◄ Larva of *Mompha subbistrigella*, dorsal and lateral views.

Mompha subbistrigella page 224   892 (2891)

Common. More local in the north, Scotland and Ireland. **FL** 3.5-5.5mm. Forewing somewhat short; dark grey with an irregular white cross-band before one-half, this broadening towards the dorsum, and another whitish cross-band at three-quarters, which is oblique and sometimes interrupted in the middle. **Similar species** *M. sturnipennella*. **FS** Single-brooded, recorded throughout the year. Hibernates as an adult. Can be found indoors and in sheds, and comes to light. **Hab** Margins of ponds, ditches, woodland clearings, waste ground, gardens. **Fp** Broad-leaved Willowherb, Marsh Willowherb, Hoary Willowherb, Square-stalked Willowherb; within a seed-pod.

Mompha epilobiella page 224   893 (2892)

Common. More local in the north, Scotland and Ireland. **FL** 5-6mm. Forewing yellowish brown, the costa more greyish, with indistinct yellowish markings, including a faint costal spot at three-quarters and up to three very small dark brown scale-tufts along the dorsum. **Similar species** *M. ochraceella*. **FS** Double-brooded, found throughout the year. Hibernates as an adult. Comes to light. **Hab** Occurs in damp and dry habitats, including woodland rides, coastal grasslands, marshes, waste ground, gardens. **Fp** Great Willowherb, occasionally Evening-primrose, possibly Broad-leaved Willowherb, Marsh Willowherb; in a spun shoot, often causing the shoot to grow sideways.

Scythrididae

Scythris picaepennis

Twelve species have been recorded in this family. The adults rest in an inclining position and the wings are held roof-like at a steep angle and wrapped close to the abdomen, with the antennae along the side of the body. The forewing length is 3.5-9mm. The forewings are elongate, the membrane with the tornal angle indistinct, although the long cilia at the tornus produce an angle between the dorsum and termen. The forewings are typically dull brown or glossy bronze, often without markings. The pointed hindwings are slightly narrower than the forewings. The head is smooth, with long, flattened scales. The antennae are thread-like, one-half to two-thirds the length of the forewing, the scape having a pecten. The labial palps are curved, ascending, not reaching above the head, and the tongue is covered in scales at the base.

This is a group of rather obscure micro-moths. They are more prevalent in the southern half of Britain, with *Scythris picaepennis* probably being the most widespread. Several species are scarce or rare, including *S. fallacella*, *S. siccella*, *S. empetrella* and *S. potentillella*. *S. siccella* is confined to one site in Britain, and within that to just a few hundred square metres of habitat. *S. fuscoaenea* is probably now extinct. On the other hand, *S. inspersella*, although still scarce, may now be spreading. *S. sinensis* is an adventive and has been recorded only twice, from the same shopping store. Coverage of this family below is partial and includes a few of the more recognisable species that may be encountered.

Care is needed to identify to family and species levels, as superficially some could be confused with *Coleophora* (Coleophoridae) or *Elachista* (Elachistidae) species. The smaller, more uniformly coloured species may require examination of genitalia to confirm identification.

◄ Larval workings of *Scythris siccella*. The larva lives in a sand-covered silken tube and mines leaves of a variety of plants from beneath.

▲ Larval workings of *Scythris empetrella*. The larva lives in a sand-covered silken tube, drawing this up into heather shoots growing close to the ground.

Larvae live in a web or silk tube on or close to the foodplant, typically low-growing plants, although *S. inspersella* is found high up in spun shoots and flowering stalks of various willowherbs. Larvae of several species live in delicate silk tubes adorned with sand that lie on the surface, attaching the tube to the vegetation on which they feed. Adults are usually active in sunshine, but tend to run and hop rather than fly, especially when disturbed; some species can be found at rest on flowers. A few species are very occasionally seen at light. Typically members of the Scythrididae inhabit warm, dry, well-drained habitats. Perhaps the best way to record them, other than by searching for the larval stage, is with the aid of a sweep net.

Further reading
British and Irish species: Emmet & Langmaid (2002a)
European species: Bengtsson (1997)

Scythris grandipennis page 225   911 (2088)

Local. Scarce in northern England. **FL** 5-9mm. Forewing bronzy dark grey or brown, sometimes greenish tinged, occasionally with a thin scattering of paler scales, particularly towards the apex. **FS** Single-brooded, June-early July. Flies in sunshine and occasionally recorded at light. Perhaps more readily seen in the larval stage. **Hab** Heathland, downland, scrubby situations, waste ground; predominantly on dry soils. **Fp** Gorses, preferring Dwarf Gorse; the larvae are gregarious and live in a tightly spun web, sometimes covering much of the plant.

Scythris crassiuscula page 225   914 (2110)

Local. **FL** 4-6mm, the female usually smaller. Forewing unicolorous dark bronzy green or bronzy brown. In the female, the penultimate two or three segments on the underside of the abdomen are yellowish brown or dirty whitish. **Similar species** *S. picaepennis* **appears rougher scaled, with a scattering of pale brown scales towards the apex; the female has the two penultimate segments of the underside of the abdomen whitish.** Examination of genitalia may be needed to confirm identification. **FS** Double-brooded in at least parts of its range, May-September. Rests on the foodplant and flowers, especially yellow Compositae, by day and is active in sunshine. **Hab** Chalk and limestone grassland. **Fp** Common Rock-rose, possibly other rock-roses; in a loose web extended over several shoots of the plant.

Scythris picaepennis page 225   915 (2105)

Local. Very local in Scotland, more frequent in the southern half of England. **FL** 4-6mm. Forewing appearing somewhat roughly scaled, dark greyish brown, sometimes with a purplish tinge, usually with a thin scattering of pale brown scales, especially in the apical half of the wing. Examples from the west coast of Scotland have a much more vivid purplish tinge. **Similar species** *S. crassiuscula*. **FS** Single-brooded, June-mid Spetember. Often found resting on yellow flowers, and very occasionally seen at light. **Hab** Grasslands on chalky and sandy soils, also soft-rock cliffs, railway embankments. **Fp** Common Bird's-foot-trefoil, Wild Thyme; in a web covering the plant, extending into the ground, where the larva hides when disturbed.

Scythris limbella page 225   918 (2169)

Rare. Possibly declining. **FL** 6-8mm. Forewing brownish, tinged yellowish grey, with cream markings of variable intensity, including a streak from the dorsum at about one-quarter and a distinct blotch near the tornus. **FS** Double-brooded or with a long flight period, depending on the weather, June-September. Occasionally found on walls or fences and comes to light. **Hab**. Dry pastures, arable fields, waste ground. **Fp** Goosefoots, oraches; among spun flowers and shoots.

Cosmopterigidae

There are 16 species in this family. The adults rest variably; it is in a slightly declining posture in *Cosmopterix*, moderately declining in *Anatrachyntis*, whilst in *Limnaecia, Pancalia* and *Sorhagenia* it is almost horizontal, and in *Pyroderces* can be either declining or inclining. Most species rest with the wings held steeply roof-like and slightly rolled around the abdomen, although *Limnaecia, Pancalia* and *Sorhagenia* hold their wings in a moderately shallow roof-like position. The forewing length is

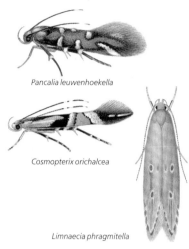

Pancalia leuwenhoekella

Cosmopterix orichalcea

Limnaecia phragmitella

3.5-11mm. The forewings are elongate or very elongate and sometimes very narrow, the membrane without an obvious tornal angle, although the long cilia at the tornus produce a noticeable angle between the dorsum and termen. The forewings are smooth or with scale-tufts and sometimes the cross-bands may be metallic. The hindwings are usually very narrow. The head is smooth. The antennae are thread-like, three-quarters to four-fifths the length of the forewing. The labial palps are long, thin and sharply curved upwards.

This family used to contain more genera, but *Blastodacna, Spuleria, Dystebenna* and *Chrysoclista* have been transferred to the Elachistidae. Most of those that remain are attractive, if small, micro-moths. *Cosmopterix pulchrimella* and *C. scribaiella* have been added to the British list recently, as have two *Anatrachyntis* species, *A. badia* and *A. simplex*, which have been reared from the calyx of imported Pomegranates. The distinctive *Pyroderces argyrogrammos*, a southern European species, is known

▶ Adult of *Sorhagenia rhamniella*.

from a single example found in a light trap on Guernsey, in the Channel Islands, in May 2002. Two species are now considered to be extinct, *Euclemensia woodiella* and *Cosmopterix schmidiella*.

Coverage of this family below is partial, with *Sorhagenia* and the two extinct species omitted. The three *Sorhagenia* species are all small, brownish or greyish moths with three bands of prominent scale-tufts (see photograph previous page), and they are very difficult to separate reliably on external characters, requiring examination of genitalia to confirm identification. They are all associated with buckthorns. Although superficially very similar, *Cosmopterix* species are usually readily identifiable. Both *Pancalia* species need particular care to distinguish them, with *P. leuwenhoekella* being the more widespread of the two.

The larvae of *Cosmopterix* species are leaf miners, and are much easier to find in this stage than as adults, typically feeding in late summer. The mines of *C. pulchrimella* can be found in winter and may be numerous. The larvae of *Sorhagenia* feed in young twigs, buds and spun terminal leaves, from late spring to early summer. Evidence of the presence of larvae of *Limnaecia phragmitella* can be spotted in spring from the large amounts of down hanging out of the previous year's flower spike of Lesser Bulrush or Bulrush; the down and seeds in uninfested flower spikes disperse much earlier in the year. Adults of some species fly by day in sunshine, for example *Pancalia* which fly low among their foodplants, whilst others are nocturnal and occasionally come to light.

Further reading
British and Irish species: Emmet & Langmaid (2002a)
European species: Koster & Sinev (2003)

Pancalia leuwenhoekella page 225   899 (3104)

Local. **FL** 4.5-5.5mm. Antenna in both sexes thread-like, with a white section below the tip. Forewing orange-brown, variably edged blackish brown; a raised silver cross-band at one-fifth, not reaching the dorsum and sometimes interrupted in the middle, two costal and two dorsal raised silver spots beyond and a silver mark along the termen; the silver spot at the tornus is perpendicular to the dorsum. **Similar species** *P. schwarzella* is typically larger, with the silver spot at the tornus oblique; the antenna in the male is unicolorous, and in the female thickens towards the middle below a white section. Examination of genitalia may be necessary to confirm identification. **FS** Single-brooded, late April-June. Also noted in July and early August. Flies in sunshine. **Hab** Chalk and limestone grassland, railway embankments, woodland clearings. **Fp** Dog-violet, Hairy Violet; in a web among the roots, the larva eating the bark.

Pancalia schwarzella page 225   900 (3105)

Very local. Distribution poorly understood owing to confusion with *P. leuwenhoekella*; rare in the south, possibly more widespread in Scotland. **FL** 5-7mm. Antenna in the male is unicolorous, and in the female thickens towards the middle. Forewing with the silver spot at the tornus oblique. **Similar species** *P. leuwenhoekella*. **FS** Single-brooded, April-mid July; possibly double-brooded in southern England, and may start flying earlier in April than *P. leuwenhoekella*. Flies in sunshine, visiting yellow flowers such as dandelions, hawkweeds and bird's-foot-trefoils. **Hab** Grasslands, sand dunes. **Fp** Dog-violet, Hairy Violet, Marsh Violet; in a web in the lower leaves, where the plants grow among mosses.

▲ A seedhead of Bulrush with seeds adhering to silk from larvae of *Limnaecia phragmitella*.

▲ Larva of *Limnaecia phragmitella* extracted from the seedhead of Bulrush.

Limnaecia phragmitella page 225 898 (3154)

Common. More local in the north. **FL** 8-11mm. Forewing pale yellowish brown, with an indistinct brownish line from about one-third to just beyond two-thirds, this encompassing a white-edged darker spot at nearly one-half and another at about two-thirds. **FS** Single-brooded, late June-August. Comes to light. Larvae can be abundant. **Hab** Lake and river margins, freshwater ponds, ditches; occasionally wanders from breeding sites. **Fp** Lesser Bulrush, Bulrush; in the seeds among the flowering spike, the larval silk preventing seeds and down from dispersing. Also found in the stems.

Cosmopterix zieglerella page 225 894 (3163)

▶ Leaf mine of *Cosmopterix zieglerella* on Hop. The finger-like projections and silk within the mine distinguish it from mines made by flies (Diptera).

Very local. **FL** 4-5mm. Forewing shining dark brown, with a golden metallic cross-band near the base and an orange cross-band just beyond one-half, this bordered each side by a raised pale golden metallic cross-band, with a silver spot towards the apex and a shining white spot extending into the cilia. **Similar species** The *Cosmopterix* species are superficially similar, with the exception of *C. lienigiella*. *C. orichalcea* has a large brassy basal patch on the forewing.

C. scribaiella has white streaks in the basal area, the orange cross-band extended towards the apex, and has an unbroken apical line. *C. pulchrimella* is typically the smallest *Cosmopterix*, with white streaks in the basal area and the apical line broken into two short dashes. *C. schmidiella* is an extinct species which is most similar to *C. zieglerella* but has an uninterrupted apical line and the antenna with a broader white ring below the tip. **FS** Single-brooded, late May-July. Can be disturbed by day from the foodplant and comes to light. **Hab** Hedgerows, scrub, open woodland. **Fp** Hop; mines the leaf.

Cosmopterix orichalcea page 225 896 (3165)

Very local. Although the most widespread species of the group, it is infrequently encountered. **FL** 4-5mm. **Similar species** Other *Cosmopterix*, except *C. lienigiella*. **FS** Single-brooded, late May-August. Can be swept from the foodplant by day and occasionally comes to light. **Hab** Fens, damp woodland, ditches, streambanks. **Fp** Reed Canary-grass, Sweet Vernal-grass, Tall Fescue, Common Reed, Millet, Holy-grass; mines the leaf.

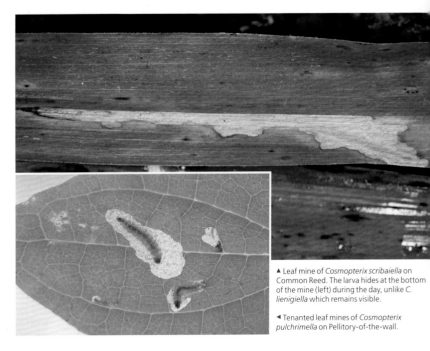

▲ Leaf mine of *Cosmopterix scribaiella* on Common Reed. The larva hides at the bottom of the mine (left) during the day, unlike *C. lienigiella* which remains visible.

◄ Tenanted leaf mines of *Cosmopterix pulchrimella* on Pellitory-of-the-wall.

Cosmopterix scribaiella page 225   896a (3166)

Rare. First discovered in Britain in 1996. **FL** 5-5.5mm. **Similar species** Other *Cosmopterix*, except *C. lienigiella*. **FS** Single-brooded, June-October. Can be swept from the foodplant and occasionally comes to light. **Hab** Ponds, ditches, alder carr, river margins, usually where the foodplant grows in drier situations. **Fp** Common Reed; mines the leaf, the larva is highly mobile in the mine. Often there are many mines on a single plant.

Cosmopterix pulchrimella page 225   896b (3167)

Very local. First discovered in Britain in 2001 and expanding its range. Predominantly coastal, although occasionally found inland. **FL** 3.5-4mm. **Similar species** Other *Cosmopterix*, except *C. lienigiella*. **FS** A single extended generation from autumn to spring, or possibly double-brooded; adults mostly appear October-November, and have been reared in February from mines found in January. Sometimes found in numbers. Can be found flying over the foodplant by day and occasionally comes to light. **Hab** Coastal locations, including cliffs, stone walls, caves, roadside verges, path edges, woodland, scrub. **Fp** Pellitory-of-the-wall; mines the leaf, preferring broad leaves on plants growing in deep shade.

Cosmopterix lienigiella page 225   897 (3170)

Very local, with a single record in Ireland. **FL** 4.5-6mm. Forewing yellowish brown, with a series of narrow white lines near the base, a raised silver metallic cross-band at about one-half, and another just beyond, with a white line beyond to the apex. **FS** Single-brooded, mid May-mid October. Can be swept from the foodplant by day and has been found at night on the leaves; occasionally comes to light. **Hab** Reedbeds in both fresh and brackish water, soft-rock cliffs, and elsewhere where the foodplant grows. **Fp** Common Reed; mines the leaf, the larva hardly moves in the mine. Usually only one mine per plant.

Gelechiidae

There are 163 species in this family. It is often considered to be a difficult group to get to grips with, since it contains so many species, lots of which are small and superficially similar, and there is a good number that fit the colloquial description of micro-moths as 'small brown jobs'. However, the family contains considerable diversity in morphology and life history. In this guide, full coverage of the trickiest genera has been avoided. Accounts are restricted to a combination of distinctive, easily recognised species, together with some of the most frequently seen, even if they are difficult to identify.

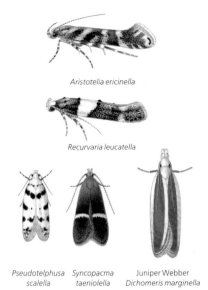

Aristotelia ericinella

Recurvaria leucatella

Pseudotelphusa   Syncopacma   Juniper Webber
scalella   taeniolella   *Dichomeris marginella*

Adult resting positions are variable; they are almost horizontal to slightly inclining in most genera, strongly inclining in a few, such as *Metzneria* and *Chrysoesthia*, and a few species seem to prefer to sit horizontally but with the wings above the ground, such as *Dichomeris alacella*. The wings may be held roof-like, as in *Aristotelia ericinella*, slightly rolled around the body, as in *Sophronia semicostella*, or held flat, as in *Helcystogramma rufescens*. The antennae tend to lie along the dorsum, along the costa, or frequently are held at a backward-pointing angle from the head, above and to the side of the forewings. The forewing length is 3-11mm. The forewings are moderately to strongly elongate, generally white, grey, brown or blackish, the coloration and markings being very varied. General characters include the presence of a distinct fold, or crease, in the wing membrane, which runs from the base nearly to the tornus, often with a spot part way along the fold. Also, in the disc, the flat area of wing membrane in the middle of the forewing without veins, there are often two spots. A number of species have scale-tufts on the wing, such as *Recurvaria leucatella*, and a few have metallic markings, such as *Chrysoesthia drurella*. **The hindwings may be narrower or wider than the forewings and have the termen concave and the apex often extended to a point or finger-like projection, this being characteristic of the Gelechiidae,** although this feature is less pronounced in a few species, such as *Anacampsis*. Only when the wings are fully spread does the hindwing character become apparent. The head is smooth-scaled. The antennae are thread-like, about two-thirds to three-quarters the length of the forewing. The labial palps strongly curve upwards, often reaching above the head, the second segment frequently with a conspicuous tuft of scales. Occasionally these palps are elongate and point straight forwards. The tongue is long and mostly scaled.

Many species of gelechiid are localised, some exceedingly so. *Syncopacma albifrontella* has been added to the British list recently from a single example from Aberdeenshire; it is probably resident in Scotland, but the nearest known location elsewhere in Europe is Switzerland. *S. suecicella* is known only from one small part of the Lizard Peninsula, in Cornwall, and *Helcystogramma lutatella* from the coastal cliffs of Dorset, between Lulworth and Portland. Other species are more widely distributed and are regularly encountered in light traps, such as *Brachmia blandella* and *Scrobipalpa costella*.

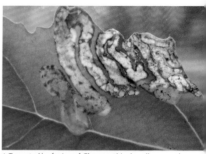

▲ Tenanted leaf mine of *Chrysoesthia drurella* on orache. Note the serpentine form of the mine.

There are a few adventive species, such as Peach Twig Borer *Anarsia lineatella* which is associated with imported plums, peaches and apricots, although this species now appears to be established in the wild in Sussex. *Gelechia senticetella*, thought to have been introduced in the horticultural trade, has now become established in the south-east of Britain. *Syncopacma polychromella* has been seen on a few occasions, most recently as an immigrant, but earlier examples appear to have been accidental imports. *Tuta absoluta* is a South American species associated with tomato crops, and has been found in the Channel Islands, Isles of Scilly, Gloucestershire, Worcestershire and Yorkshire; it could appear anywhere and may become established. Eleven species are thought to have become extinct, including *Dichomeris derasella* and *Syncopacma vinella*.

There are several genera or groups of species that require care in identification, for example *Metzneria*, *Bryotropha*, *Scrobipalpa*, *Caryocolum* and *Syncopacma*, and reference to the genitalia is often required. Many species are perhaps best found in the larval stage, the foodplant often being a useful guide to a species' identification.

The larvae of a number of species feed internally in seedheads, stems or roots, a few being leaf miners, with many others feeding from within spinnings amongst the leaves of herbaceous plants, bushes or trees. A few species are associated with mosses or lichens. Most species are restricted to one or a very few foodplants. Pupation typically takes place within a cocoon. Adults of most species are nocturnal and some are attracted readily to light, whereas a few fly by day. Some species can be disturbed from their resting place by day, but they are more likely to run or drop to the ground than fly. Most are single-brooded, a few are double-brooded and a very few overwinter as adults.

Further reading
British and Irish species: Emmet & Langmaid (2002b)
European species: Elsner *et al.* (1999); Huemer & Karsholt (1999 & 2010) (Gelechiidae part)

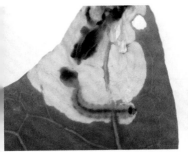

▲ Tenanted leaf mine of *Chrysoesthia sexguttella* on goosefoot. Note the mine is a white blotch.

▲ Leaf roll of the larva of *Anacampsis blattariella* on Silver Birch.

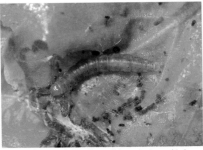

▲ Larva of *Carpatolechia alburnella* in a spinning on birch.

▲ Spinning of *Teleiodes sequax* in the terminal shoot of Common Rock-rose.

▲ Larva of *Bryotropha terrella* among moss and grass.

▲ Larva of *Mirificarma lentiginosella* emerging from its spinning in a shoot of Dyer's Greenweed.

◄ Larva of *Gelechia senticetella* exposed within its spinning on cypress.

▲ Larva of *Athrips mouffetella*.

◄ Larva of *Neofaculta ericetella*.

▲ Leaf mine of Beet Moth *Scrobipalpa ocellatella* on the upperside of a leaf of Sea Beet.

▲ Larva of *Scrobipalpa costella* ready to pupate.

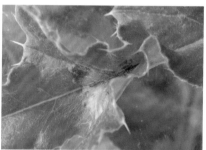

▲ Leaf mine of *Scrobipalpa acuminatella* on the upperside of a leaf of Creeping Thistle.

▲ Leaf mine of *Scrobipalpula tussilaginis* on Colt's-foot. The presence of silk within the mine distinguishes it from mines made by flies (Diptera).

▲ Larval working of *Caryocolum fraternella* on stitchwort. The larva causes the shoot to grow abnormally.

▲ Larval case of *Thiotricha subocellea* on the flowers of Marjoram. The case is formed from hollowed-out calyxes stacked end to end.

▲ Larva of *Dichomeris ustalella* on a lime leaf.

▲ Larva of *Helcystogramma rufescens* exposed from its leaf roll on grass.

Aristotelia ericinella page 226   752 (3230)

Local. Very local in Scotland. **FL** 6-6.5mm. Forewing shining dark brown, reddish brown towards the dorsum, with silvery white markings, including four oblique, sometimes partly obscure, cross-bands, the middle two meeting to form a V-shape, and other small irregular spots in the apical area. **FS** Single-brooded, mid June-early September. Can be numerous. Flies in the afternoon, early evening and comes to light. **Hab** Heathland, moorland. **Fp** Heather; feeds on the leaves, hiding in a slender silk tube among twigs of the plant.

Aristotelia brizella page 226   753 (3237)

Very local. Coastal. **FL** 4-5.5mm. Forewing yellowish brown to darker brown, particularly in the costal half, with narrow, irregular, transverse silvery grey lines, these sometimes obscure, and up to three black spots, one at about one-half, the others towards the tornus, but which may be absent. **FS** Double-brooded, mid May-early June, late July-August. Flies at sunrise, in the evening, and also recorded at light. **Hab** Saltmarshes, cliffs, vegetated shingle. **Fp** Thrift, occasionally Common Sea-lavender; in a flowerhead.

Metzneria lappella page 226   724 (3273)

Common. Local in the west and north. **FL** 8-10mm. Forewing sandy brown, variably mottled darker, especially towards the base, with the veins greyish brown, and blackish dots at one-third, about one-half, and another just beyond, although these vary in intensity and are sometimes almost obsolete. **Similar species *M. metzneriella* is generally darker, with orangey or reddish brown in the middle of the wing and towards the costa**; *M. aestivella* (not illustrated) lacks the black spots in the middle of the wing and is typically smaller; *M. neuropterella* (not illustrated), a rare species, is the largest of the genus, has the veins broadly edged dark grey-brown and lacks distinctive black spots; *M. aprilella* is shaded reddish orange, without distinctive black spots. Worn examples of any *Metzneria* may need to be identified by examination of genitalia. **FS** Single-brooded, June-July. Flies at night and comes to light. **Hab** Waste ground, open woodland, downland. **Fp** Greater Burdock, Lesser Burdock; feeds on the seeds, pupating within the seedhead.

Metzneria metzneriella page 226   726 (3280)

Common. Very local in northern Scotland. **FL** 7-9.5mm. Forewing yellowish brown, with reddish-brown shading and the veins highlighted greyish and whitish; blackish dots at one-third, about one-half, and another just beyond. **Similar species** Other *Metzneria* species, particularly *M. lappella*. **FS** Single-brooded, May-August. Flies at night and comes to light. **Hab** Grassland, roadside verges, waste ground. **Fp** Common Knapweed, Saw-wort; feeds on the seeds, pupating within the seedhead.

Metzneria aprilella page 226   727a (3285)

Local. **FL** 7-9mm. Forewing sandy brown, heavily shaded reddish orange, the veins greyish brown and with yellow streaks near the base, on the costa at one-quarter and at the tornus. **Similar species** Other *Metzneria* species, but only when worn. **FS** Single-brooded, May-August. Flies at night and comes to light. **Hab** Grassland, roadside verges, waste ground, particularly on chalky soils. **Fp** Greater Knapweed; feeds on the seeds, pupating within the seedhead.

Apodia bifractella page 226   730 (3292)

Common in the south. **FL** 4.5-6mm. Head orange-yellow. Forewing blackish brown, lightly speckled grey, with an orangey zigzag cross-line from the tornus to the costa at four-fifths, this often obsolete in the middle, and sometimes visible only as a pale mark on the costa. **FS** Single-brooded, July-August. Active from late afternoon and can be found resting on flowers of the foodplants. Comes to light. **Hab** Damp meadows, ditches, fens, marshes, open downland, saltmarshes, woodland rides, coastal landslips. **Fp** Common Fleabane, Ploughman's-spikenard, Sea Aster; in the seedhead, with no external sign of presence.

Ptocheuusa paupella page 226   748 (3295)

Common in the south. **FL** 4.5-6mm. Forewing pale yellowish brown, finely speckled darker, with variable white markings, including a costal streak and an oblique streak towards the apex at about two-thirds, and with a darker spot at about one-half, this sometimes obsolete. **FS** Double-brooded, late May-early July, late August-September. Readily disturbed from vegetation by day, can be found on the flowers of the foodplants and comes to light. **Hab** Damp grassland, ditches, woodland rides, edges of saltmarshes, coastal landslips. **Fp** Common Fleabane, Golden-samphire, sometimes Common Knapweed and mints; in the seedhead, a raised section of florets indicating the presence of a larva.

Argolamprotes micella page 226   734 (3308)

Very local. **FL** 5-7mm. Forewing shining dark brown, with an oblique whitish streak from the costa near the base, not reaching the dorsum and sometimes reduced, and a few scattered whitish spots beyond, including one at the tornus and another opposite on the costa. **FS** Single-brooded, June-July. Easily disturbed by day, flies at night and comes to light. **Hab** Hedgerows, woodlands, shady tracks, gardens, coastal landslips. **Fp** Raspberry, Bramble; in a shoot.

Monochroa cytisella page 226   728 (3312)

Local. Recorded new to Scotland in 2000. **FL** 5-6mm. Forewing yellowish brown, darker in the costal half, sometimes more uniformly darker brown throughout, with a fine oblique whitish streak from the costa at about three-quarters. **FS** Single-brooded, late June-August. Readily disturbed from the foodplant by day and comes to light. **Hab** Woodland, parkland, heathland, hillsides, coastal cliffs, possibly preferring warm, sheltered places. **Fp** Bracken; within a slight swelling in the stem or in a side shoot, causing it to wilt.

Monochroa palustrellus page 226   737 (3327)

Very local. Mainly found in south-east England. Recorded new to Ireland in 2006. **FL** 7-10mm. Forewing pale yellowish brown, with blackish streaks between the veins and an elongated, sometimes oval, black spot at about one-half, and a round black spot at about two-thirds, both spots outlined paler. **FS** Single-brooded, late June-August, also recorded in September. Comes to light. **Hab** Waste ground, vegetated shingle, sand dunes, dry grassland, also fens and marshes. **Fp** Curled Dock, possibly other docks; feeds in the rootstock, stem and leaf petiole.

Bryotropha affinis page 226   779 (3389)

Common. Local in Scotland. **FL** 4.5-6.5mm. **Forewing dark brown with yellowish bases to the scales, giving the moth a speckled appearance**; four blackish-brown spots, often with small patches of yellowish scales adjacent, one spot at about one-quarter, one towards the costa at one-third, one towards the dorsum at one-half and another at two-thirds; there are patches of yellow scales at the tornus and opposite on the costa, sometimes joined to form an angled cross-band. Examples from coastal locations can be much paler. **Similar species** *B. umbrosella* and *B. similis* (neither illustrated) lack yellowish speckling, *B. umbrosella* has white markings, *B. similis* is glossy in appearance; *B. basaltinella* and *B. dryadella* (neither illustrated) have a pair of blackish dots at one-third, one above the other, and do not have one at one-half. If there is any doubt, determination should be made by examination of genitalia. **FS** Single-brooded, May-August. Flies at dusk and regularly comes to light. **Hab** Urban areas, parkland, rural buildings. **Fp** Mosses, including *Tortula muralis*, growing on walls; in a silk tube in the moss.

Bryotropha senectella page 226   782 (3372)

Common. Local in Scotland. **FL** 4.5-6mm. **Sides of the head and labial palps yellowish brown.** Forewing brown, mottled with darker brown scales; four dark brown spots, sometimes ill-defined, often with small patches of yellowish-brown scales adjacent, one spot at about one-quarter, one towards the costa at one-third, one towards the dorsum at one-half and another at two-thirds; paler patches of scales are usually discernible at the tornus and opposite on the costa and are sometimes joined to form an angled weak cross-band. **FS** Single-brooded, mid June-early September. Flies at dusk and comes to light. **Hab** Open habitats, including grasslands, coastal slopes, sand dunes, open woodland. **Fp** Mosses, including *Homalothecium lutescens*, growing amongst grasses; in a delicate silk tube.

Bryotropha terrella page 226   787 (3373)

Common. **FL** 7-8mm. Forewing light, dark or greyish brown, usually with one black dot on the fold at one-third, another obliquely above in the disc, and a further spot in the disc at two-thirds, these spots sometimes pronounced, sometimes obsolete; a faint acutely angled paler cross-band is often discernible from the tornus to the costa, with darker brown speckling towards the apex. **Similar species** *B. desertella* (not illustrated), a local species most often found in sand dunes and sandy heaths, is usually smaller (FL 5.5-7.5mm) and has a slightly narrower forewing, often with the impression of a faint streak in the middle of the wing, and the forewing is often yellowish brown and silkier in appearance. *B. politella* ground colour is uniformly greyish brown and does not normally have an angled cross-band, and there is no darker speckling towards the apex. However, larger and darker forms of *B. desertella* occur which are similar to *B. politella*; both may be confused with paler forms of *B. terrella*. If there is any doubt, determination should be made by examination of genitalia. **FS** Single-brooded, mid May-August. Flies from dusk and frequently comes to light. **Hab** A wide range of open habitats. Can be abundant in dry, sandy grasslands. **Fp** Has been observed eating Common Bent and the moss *Rhytidiadelphus squarrosus;* feeds from within a silken gallery at stem bases or amongst the moss.

Bryotropha politella page 226   788 (3372)

Common in the north, very local in the south. **FL** 6-8mm. Forewing glossy, brownish grey, usually with one black dot on the fold at one-third, another obliquely above in the disc, and a further spot in the disc at two-thirds, these spots often very small or obsolete; a faint acutely angled paler cross-band is sometimes discernible from the tornus to the costa, and there are usually dark brown scales along the termen. **Similar species** *B. terrella*, *B. desertella*. **FS** Single-brooded, mid May-July. Comes to light. **Hab** Dry grassland, chalk downland in the south. Can be abundant in dry grasslands in the north. **Fp** Likely to feed on mosses and grasses.

Bryotropha domestica page 226 789 (3383)

Common. More local in Scotland, with scattered records north to the Glasgow area. **FL** 5-6.5mm. Forewing yellowish brown, finely speckled darker, with several prominent black spots, usually including one near the base, a pair at about one-third, one above the other and sometimes fused, and another at two-thirds, and with a paler angled cross-band at about three-quarters, sometimes obscure. **FS** Single-brooded, mid May-early September. A common species in urban areas and regularly comes to light. **Hab** Gardens, parkland, less often woodland. **Fp** Mosses, including *Tortula muralis*, growing on walls; in a silk tube in the moss.

Recurvaria leucatella page 226 758 (3400)

Local. **FL** 6-7mm. Head white. Forewing blackish, with a broad white cross-band at one-quarter, broader on the dorsum than the costa, a white tornal spot and a similar, smaller spot opposite on the costa, with white speckling towards the apex. **FS** Single-brooded, late June-August. Can be disturbed from hedgerows by day and comes to light. **Hab** Hedgerows, gardens. **Fp** Hawthorn, apples, occasionally Rowan; in spun leaves.

Stenolechia gemmella page 226 755 (3407)

Local. **FL** 5-5.5mm. Forewing white, variably and finely speckled darker, with a darker dorsal and costal spot at the base, an almost complete blackish oblique cross-band at one-half and a small grey spot on the tornus. **FS** Single-brooded, July-early October. Has been found resting in crevices on oak trunks and occasionally comes to light. **Hab** Woodland. **Fp** Oaks; bores into buds and shoots.

Parachronistis albiceps page 226 756 (3410)

Local. **FL** 5-6mm. Head white. Forewing dark greyish brown to blackish, with scattered white spots, including three along the costa, the one near the base often extended to form a narrow oblique cross-band, joining an elongate white streak on the dorsum to near the tornus; weak scale-tufts are usually present at one-third and two-thirds. **FS** Single-brooded, July-August, and has been recorded in late May. Comes to light. **Hab** Woodland, gardens. **Fp** Hazel; in a bud.

Teleiodes vulgella page 226 765 (3415)

Common in England and Wales. **FL** 5.5-6.5mm. Forewing grey, mottled with dark greyish brown, with a variable number of black marks, including a raised scale-tuft at one-third and a pair of raised tufts, usually joined into a bar at two-thirds, with one or two spots in the middle, and blackish shading along the costa. **FS** Single-brooded, mid May-August. Rests on tree trunks and fences by day and comes to light. **Hab** Woodland, scrub, hedgerows. **Fp** Mainly hawthorns, Blackthorn; in spun leaves.

Teleiodes luculella page 227 774 (3419)

Common. Local in the north. **FL** 5-6mm. Forewing dark grey to blackish, with a large semicircular white costal blotch from one-fifth to one-half, not reaching the dorsum, this often with a yellowish tinge, particularly away from the costa, and a white costal spot at about two-thirds. **Similar species** ***T. flavimaculella*** **(not illustrated), a scarce species in south-east England, has the mark near the costa orange, not whitish or yellowish, and the ends of this mark do not reach the costa. FS** Single-brooded, May-mid August. Rests on tree trunks by day and comes to light. **Hab** Woodland, parkland, scrub. **Fp** Oaks; in spun leaves.

Teleiopsis diffinis page 227 776 (3448)

Common. Local in Ireland. **FL** 7-9mm. A long-winged species. Forewing greyish brown or greyish, variably spotted blackish and faintly speckled whitish; at one-fifth are three obliquely placed raised black spots, sometimes joined into a cross-line, with two raised spots at about one-half in the middle of the wing, and another pair at three-quarters, beyond which a whitish cross-band is sometimes discernible. **FS** At least double-brooded in most years, May-October. Also recorded in March. Comes to light. **Hab** Dry grassland, heathland, vegetated shingle. **Fp** Sheep's Sorrel; feeds from a silken tube at the base of the stem.

Carpatolechia proximella page 227 770 (3430)

Local. **FL** 6.5-8mm. **Segment 2 of the labial palp white, speckled greyish brown.** Forewing whitish, speckled black and grey, with an oblique line of black spots from the costa at one-fifth, a pair of black spots at one-third and another pair at two-thirds, and with a whitish-grey angled cross-band from the tornus opposite to the costa. Examples from the Scottish Highlands have more blackish scaling in the central part of the wing. **Similar species** C. alburnella is somewhat shorter winged and more whitish in appearance, with black spots less distinct in the central part of the wing. **FS** Single-brooded, April-July. Comes to light. Usually occurs at low density. **Hab** Woodland, river valleys, heathland, moorland. **Fp** Birches, alders; in a folded leaf.

Carpatolechia alburnella page 227 771 (3428)

Local. **FL** 6-7mm. **Segment 2 of the labial palp pure white.** Forewing whitish, speckled grey, and grey and blackish towards the apex, with pale yellowish-brown or greyish marks in the central part of the wing, sometimes visible as a raised tuft of scales at about one-quarter and three-quarters; black spots of varying intensity on the costa, the one at about two-thirds typically the largest. Examples from the Scottish Highlands have more blackish scaling in the central part of the wing. **Similar species** C. decorella (not illustrated), a widespread but rather uncommon species on oak, is very variable but has a whitish almost unmarked form with a distinct black bar at the base of the costa, longer than the equivalent mark in C. alburnella; C. proximella . **FS** Single-brooded, late June-September. Comes to light. Usually occurs at low density. **Hab** Woodland, parkland, heathland. **Fp** Birches; in spun leaves or a folded leaf.

Pseudotelphusa scalella page 227 764 (3453)

Local. **FL** 5-7mm. Head white. Forewing white with blackish markings, including a basal spot, an oblique cross-band at about one-fifth, a triangular costal spot at two-fifths, a spot at three-fifths, and a triangular spot on the dorsum near the tornus, with smaller spots in the apical area. **FS** Single-brooded, May-June. Rests on tree trunks by day and comes to light. **Hab** Woodland, parkland. **Fp** Possibly on moss on tree trunks.

Pseudotelphusa paripunctella page 227 773 (3432)

Local. **FL** 5-7mm. Forewing grey, brown or pale yellowish brown, with a blackish spot or patch near the base of the costa and two similar patches at two-fifths and three-fifths, these sometimes obscure; there are three pairs of fine black dots at one-fifth, two-fifths and three-fifths, a pale, rather angular cross-band beyond, and black dots around the termen. The ground colour varies according to the foodplant: from oak, it tends to be yellowish brown, from Bog-myrtle, grey-brown. **Similar species** *Teleiodes wagae* **(not illustrated), a very local species on Hazel, has a greyish-brown forewing with four dots in a row at two-fifths, not two.** **FS** Single-brooded, May-June. Comes to light. **Hab** Woodland margins, heathland, hedgerows, preferring young trees, and moorland, fens, bogs. **Fp** In southern England and Wales mainly on oaks, associated with Bog-myrtle in northern England and Scotland, also bred from Dwarf Birch in Scotland; between flatly spun leaves.

Altenia scriptella page 227   766 (3461)

Very local. Declining, recently found only south of a line from Worcestershire to Norfolk. **FL** 6-7mm. Head white. Forewing white, in part tinged greyish and finely speckled black, with a yellowish-brown costal spot at the base and a dark greyish blotch from one-fifth to three-fifths, extending from the dorsum over much of the wing and meeting the costa as a black mark. **FS** Single-brooded, June-July. Often rests on a tree trunk or fence paling by day and comes to light. **Hab** Hedgerows, woodland margins, open country. **Fp** Field Maple, preferring saplings; in a folded leaf.

Gelechia rhombella page 227   800 (3469)

Local. Very local in northern England, rare in Scotland. **FL** 6-8mm. Forewing pale grey, finely speckled blackish, with a distinct black mark at the base on the costa, two blackish spots, one oblique at about one-third, the other beyond one-half, and a small blackish mark on the costa at about two-thirds. **Similar species** *G. hippophaella* (not illustrated), a scarce species on Sea-buckthorn on sandhills in south-east and eastern England, is slightly larger (FL 7.5-9mm), with pale yellowish-grey forewings and small black spots in the centre of the wing. **FS** Single-brooded, July-August. Can be disturbed from the foodplant by day and comes to light. **Hab** Old orchards, gardens, parkland, open woodland. **Fp** Apples, pears; in a flat spinning between two leaves or a turned-down leaf-edge.

Gelechia senticetella page 227   801a (3471)

Local. A naturalised adventive, spreading out from south-east Engalnd. **FL** 6-7.5mm. Forewing greyish, tinged pale yellowish brown to reddish brown and densely speckled blackish, with veins variably outlined blackish and dark spots or streaks in the centre of the wing, sometimes joined to form a longitudinal streak. **FS** Single-brooded, July-August. Comes to light. **Hab** Gardens, parks. **Fp** Juniper, cypresses, including Lawson's Cypress; older larvae feed from a conspicuous silken tube between leaflets, turning these brown.

Gelechia sororculella page 227   802a (3474)

Common. Local in the north. **FL** 6.5-7.5mm. Thorax dark reddish brown. Forewing dark brown or blackish brown, with a longitudinal black band in the middle of the wing and **a white-ringed black dot just before the middle**, and with scattered whitish scales forming a pale angled cross-band from the tornus to the costa opposite. Markings may be more defined or obscure. **FS** Single-brooded, July-August. Rests on tree trunks, may be tapped from bushes by day and occasionally comes light. **Hab** Woodland, fens, scrub, wet heathland, riverbanks. **Fp** Mainly Goat Willow, Grey Willow; in a spinning in the leaves or female catkins.

Gelechia nigra page 227   806 (3482)

Very local. Recent records are mainly from south-east England, north-west to Worcestershire. **FL** 6-9mm. Forewing blackish brown to blackish, with a few whitish scales, a black longitudinal streak at about one-half, and a faint whitish angled cross-band at about three-quarters. **FS** Single-brooded, June-August. Rests on tree trunks by day, although often flies off when approached, and comes to light. **Hab** Woodland, riverbanks, suburban habitats. **Fp** Aspen, White Poplar, Grey Poplar; between flatly spun leaves.

Gelechia turpella page 227 807 (3483)

Rare. Declining and now extremely local in south-east England. **FL** 8-10mm. Forewing dark greyish brown with scattered paler scales, a blackish mark at the base and at one-half on the costa, two blackish spots in the middle of the wing and a paler, slightly obscure, angled cross-band at about three-quarters. **FS** Single-brooded, late June-July. Can be found by day resting on the trunks of old poplars (although it is difficult to see as it runs up the trunk when approached), and comes to light. **Hab** Parkland, woodland, woodland margins. **Fp** Black Poplar, Lombardy Poplar, possibly also sallows; feeds between spun leaves.

Mirificarma mulinella page 227 792 (3507)

Common. **FL** 6-7mm. Variable. Forewing brown, sometimes slightly paler along the costa and in the dorsal half, with a blackish spot at one-third and two others at about two-thirds, though these can be obscure; often a longitudinal dark streak from one-third to the apex and another from the base to beyond one-third, these sometimes merging. **FS** Single-brooded, July-September. Comes to light. **Hab** Heathland, farmland, waste ground. **Fp** Gorse, Broom, possibly Tree Lupin and a cultivar of Dyer's Greenweed; feeds in the flowers.

Sophronia semicostella page 227 841 (3749)

Common. Rare in north-west England. **FL** 8-9mm. **Labial palps distinct, segment two strongly tufted.** Forewing elongate, greyish brown, faintly streaked reddish brown and variably speckled whitish, with a white streak along the costa from the base to about two-thirds, a pair of darker spots, one above the other, at two-thirds, and distinctly banded cilia. **FS** Single-brooded, June-July. Flies in early afternoon, at dusk and sometimes comes to light. Seldom seen in numbers. **Hab** Dry grassland, heathland, vegetated shingle. **Fp** Continental authors give Sweet Vernal-grass.

Aroga velocella page 227 796 (3530)

Local. **FL** 7-8.5mm. Variable. Forewing greyish brown to blackish brown, sometimes tinged yellowish brown or reddish brown, the veins often highlighted paler, with a tapered pale greyish or creamy streak on the dorsum and a whitish spot on the tornus with another opposite on the costa; sometimes there is a black spot on the fold at about one-third, with another obliquely above, a further spot at two-thirds, and black dots around the termen. All markings can be obscure. **FS** Double-brooded, May, July-August. Can be found in numbers where it occurs. Flies in the afternoon in warm weather, and comes to light. **Hab** Heathland, moorland, woodland clearings, hedgerows, parkland, acid grassland, vegetated shingle. **Fp** Sheep's Sorrel; feeding in a silken tube at the base of the stem.

Prolita sexpunctella page 227 794 (3549)

Local. Has declined in some areas. **FL** 8-10.5mm. Forewing dark greyish brown mixed reddish brown, with several irregular whitish cross-bands, and black marks between in the middle of the wing; cilia silver-grey. **FS** Single-brooded, May-June. Flies in sunshine. Can be abundant where Heather is short, especially in areas regenerating after fire. **Hab** Heathland, moorland, mosses. **Fp** Heather; in spun leaves.

Prolita solutella page 227 795 (3550)

Rare. Found in west Cornwall and central and north-east Scotland. No recent records from northern England. **FL** 8-10.5mm. Forewing pale to dark greyish brown, speckled greyish white from the base to about four-fifths, with prominent black spots at one-quarter and one-third, these sometimes joined, and at just beyond one-half, and with a whitish cross-band at three-quarters, this sometimes broken and indistinct. Examples from Cornwall tend to be paler than those from Scotland. **FS** Single-brooded, May-July. Flies in sunshine, can be disturbed from the foodplants in dull weather and comes to light. **Hab** Dry, herb-rich, cattle-grazed pasture, dry grassy heathland, coastal cliffs. **Fp** Petty Whin, Hairy Greenweed; from within a silken tube.

Athrips mouffetella page 227 762 (3559)

Local. **FL** 7-8.5mm. Forewing greyish, slightly darker along the costa, with two pairs of blackish transversely placed spots, at about one-quarter and at about two-thirds; other smaller blackish spots variably present. **FS** Single-brooded, late June-early September. Comes to light. **Hab** Woodland, hedgerows, gardens. **Fp** Honeysuckle, Fly Honeysuckle, Snowberry; in a dense white silk spinning in a shoot.

Scrobipalpa acuminatella page 227 822 (3580)

Common. **FL** 5-7mm. Variable. Dark yellowish brown to greyish brown, sometimes mixed orangey brown along the veins, with three black dots in the middle of the wing. Markings are often obscure or obsolete, and the ground colour can be mixed or unicolorous. A form occurs in Scotland which has a pale greyish-brown forewing with orangey streaks towards the base. The female is usually smaller than the male and has a more pointed forewing. **FS** Double-brooded across much of its range, April-September, possibly single-brooded in the far north. Regularly comes to light. **Hab** Grasslands, parkland, field margins, waste ground, fens, coastal slopes. **Fp** Thistles; on the lower leaves in a brownish leaf mine.

Scrobipalpa costella page 227 819 (3592)

Common. More local in northern parts of its range. **FL** 6-7mm. Forewing yellowish brown to reddish brown, variably shaded greyish, with an almost triangular blotch extending from one-fifth to about two-thirds along the costa, not reaching the dorsum; the costal triangle is obscure in darker examples but usually retains its outline, at least in part. **FS** Possibly more than one generation, emerging over a long period, and recorded in every month of the year. Comes to light. **Hab** Hedgerows, gardens, marshes, vegetated shingle. **Fp** Bittersweet; feeds in a blotch mine, spun leaves or in a berry or stem.

Syncopacma larseniella page 227 844 (3781)

Local. **FL** 5.5-7mm. Forewing black, faintly speckled greyish, with a straight, whitish or yellowish-white cross-band at two-thirds, although rarely this is obsolete; the underside of the forewing has a small yellowish costal spot. **Similar species** *S. taeniolella* **has a slightly inwardly curved, or sometimes straight, cross-band, and the underside of the forewing has a narrow yellowish-white cross-band and a similarly coloured costal spot on the underside of the hindwing.** *S. cinctella* (not illustrated) has a forewing which broadens slightly in the outer half, with a slightly inwardly curved cross-band; however, the superficial distinctions between *S. larseniella* and *S. cinctella* are subtle and genitalia must be examined to confirm identification. See also *Elachista bisulcella* (Elachistidae). **FS** Single-brooded, June-July. Can be disturbed in the late afternoon and comes to light. **Hab** Rough ground, woodland margins, fens, marshes, damp areas. **Fp** Greater Bird's-foot-trefoil, Common Bird's-foot-trefoil; between spun leaves.

Syncopacma taeniolella page 227 847 (3785)

Local. Very local from the Midlands northwards, scarce in Scotland. **FL** 5.5-7mm. **Similar species** *S. larseniella*, *S. cinctella*. **FS** Single-brooded, end June-early August. Can be disturbed in the late afternoon and comes to light. **Hab** Chalk and limestone grassland, rough ground, coastal slopes. **Fp** Mainly Common Bird's-foot-trefoil; between spun leaves.

Aproaerema anthyllidella page 227 843 (3708)

Common. More local in Scotland. **FL** 5-6mm. Forewing blackish or dark greyish brown, faintly speckled greyish, usually with a small yellowish spot on the costa at two-thirds and sometimes a smaller one opposite on the dorsum; often there are a few yellowish-brown scales or a small yellowish spot on the fold before one-half, and always a thin blackish-brown line in the cilia. **Similar species** *Eulamprotes immaculatella* (not illustrated), a very local species, does not have yellowish-brown scales on the fold, and does not have a line in the forewing cilia. *Syncopacma albipalpella* (not illustrated), a rare species on Petty Whin, also does not have yellowish-brown scales in the middle of the wing, and rests with the wings roof-like at a shallow angle, whereas *A. anthyllidella* rests roof-like at a steep angle. However, genitalia should be examined to confirm identification if *E. immaculatella* or *S. albipalpella* is suspected. **FS** Probably at least double-brooded, late April-October. Flies from late afternoon and comes to light. **Hab** Grassland, downland, waste ground, coastal areas. **Fp** Usually Kidney Vetch, but also restharrows, Sainfoin, Lucerne, clovers; in spring forms a blotch mine in leaves, causing them to fold upwards, and in summer feeds in the flowers and seeds, although on restharrows feeds within spun leaves.

Anacampsis populella page 228 853 (3804)

Common. More local in the north. **FL** 7-9mm. Forewing variable, from almost unmarked pale greyish to contrasting black and whitish grey. The comparatively unmarked form has ill-defined blackish spots along the fold at one-fifth and one-third, another obliquely above in the disc, and a further spot at two-thirds, with others along the termen. The more strongly-marked form is blackish in the dorsal half to two-thirds, speckled whitish grey elsewhere, with a creamy grey band along the costa from the base to beyond one-half. Both forms usually have an angled whitish-grey cross-band from the tornus to the costa. The abdomen of the female has segments 2-4 pale yellowish brown. **Similar species *A. blattariella* is, on average, slightly smaller, has the cross-band more abruptly angled, and some examples are more strongly marked than *A. populella*. However, examination of the genitalia is needed to confirm identity of either species, unless they have been reared. FS** Single-brooded, late June-early September. Can be disturbed from vegetation by day and comes to light. **Hab** Woodland, hedgerows, scrub. **Fp** Poplar, Aspen, willows, including Creeping Willow; within a rolled leaf.

Anacampsis blattariella page 228 854 (3805)

Common in much of England. **FL** 7-9mm. **Similar species** *A. populella*. **FS** Single-brooded, July-September. Rests by day on the trunks of large birches and comes to light. **Hab** Woodland, parkland, scrub, heathland. **Fp** Birches; within a rolled leaf.

Anarsia spartiella page 228 856 (3823)

Local. Very local in Scotland. **FL** 6-8mm. Forewing variable, grey to dark grey, sometimes paler along the costa and between the veins, often with several short, dark, oblique streaks or spots on the costa towards the apex, the largest and most distinctive at about one-half. **FS** Single-brooded, June-August. Can be disturbed from vegetation by day and comes to light. **Hab** Heathland, downland, scrub, vegetated shingle, waste ground. **Fp** Gorse, Broom, Dyer's Greenweed; in spun shoots.

Hypatima rhomboidella page 228 — 858 (3827)

Common. **FL** 7-9mm. Forewing elongate, pale greyish, mottled darker, with a large blackish, almost triangular blotch on the costa at about one-half, this preceded by one or two small costal spots, and with a small blackish streak just below the apex. The wings are folded closely together at the wing tip when at rest. **FS** Single-brooded, July-September, has been recorded in mid June and October. Flies at night and comes to light.
Hab Woodland, heathland, scrub. **Fp** Birches, Hazel; in a spun shoot or rolled leaf.

Nothris congressariella page 228 — 839 (3830)

Rare. **FL** 7-10mm. Forewing brown, yellowish brown towards the costa, with a blackish-brown streak in the middle of the wing, interrupted by two yellowish-brown spots, black speckling edging the veins towards the apex and a series of blackish spots near the base of the cilia. **FS** Double-brooded, May-early July, September-October. **Hab** Coastal; low cliffs, scree slopes, waste ground, sandhills. **Fp** Balm-leaved Figwort; larger larvae feed under a folded leaf.

Neofaculta ericetella page 228 — 797 (3833)

Common. **FL** 7-9mm. Forewing pale to dark grey, veins variably speckled white and black, especially in the outer third, with an indistinct black spot along the fold at one-third, another obliquely above in the disc, often linked to a further spot at two-thirds by small black and white dashes or dots, and with black spots along the termen. **FS** Single-brooded, late April-July. Easily disturbed from Heather by day and comes to light. Can be abundant. **Hab** Heathland, moorland, gardens. **Fp** Heather, Bell Heather, Cross-leaved Heath; feeds at night and hides by day in a slight spinning made between grass blades.

Juniper Webber *Dichomeris marginella* page 228 — 862 (3849)

Local. **FL** 7-8mm. Head white. Forewing reddish brown, with a white costal streak from the base to near the apex and a white dorsal streak from the base to the termen. **FS** Single-brooded, June-August. Can be disturbed from the foodplant by day and comes to light. **Hab** Chalk and limestone grassland, gardens. **Fp** Common Juniper, including garden varieties; after hibernation, feeds from within a web among the needles.

Dichomeris juniperella page 228 — 863 (3848)

Rare. Central and eastern Highlands of Scotland only. **FL** 9-11mm. Forewing grey, finely speckled darker, with blackish spots, sometimes indistinct, on the fold at one-third, another obliquely above in the disc, a further spot at two-thirds, and blackish dots along the termen; there is a paler, slightly obscure cross-band at three-quarters and a series of black spots at the base of the cilia. **FS** Single-brooded, late June-July. Can be disturbed from the foodplant by day and comes to light. **Hab** Open woodland on the lower slopes of mountains. **Fp** Common Juniper; feeds from within a web among the needles.

Dichomeris ustalella page 228 — 864 (3850)

Rare. **FL** 7-10.5mm. Forewing dark reddish brown, darker towards the termen, and variably yellowish brown in the costal half from about one-fifth to near three-quarters; the cilia are yellowish brown, tinged reddish. **FS** Single-brooded, May-June. Comes to light. **Hab** Woodland. **Fp** Small-leaved Lime; between flatly spun leaves.

Brachmia blandella page 228 866 (3863)

Common in much of England. Local in Wales. **FL** 5-6.5mm. Forewing with the termen slightly sinuate, yellowish brown, variably shaded darker, with an obscure cross-band from the costa at about three-fifths to the tornus; there are usually a few black marks, comprising a small raised scale-tuft above the dorsum at three-fifths, a dot obliquely above in the disc, and a further spot at two-thirds, with a pale line at the base of the cilia. **FS** Single-brooded, late June-early August. Can be disturbed from vegetation by day and comes to light. **Hab** Grassland, parkland, woodland, scrub, gardens. **Fp** Gorse; in a slight silk web amongst loose bark and dead spines. Has been reared from the seedheads of Marsh Thistle, and an insect gall on Giant Fir.

Helcystogramma rufescens page 228 868 (3870)

Common. **FL** 7-9mm. Outer margin of labial palps with brown scales. Forewing yellowish brown, often unicolorous, sometimes darker between the veins, occasionally with an obscure angled paler cross-band at two-thirds and up to three fine dark brown or black dots in the middle of the wing, these often obsolete. **Similar species *H. lutatella* (not illustrated), a rare species confined to the Dorset coast, has the labial palps white above and below, with no brown scales on the outer margin, and the forewing is unicolorous grey-brown or brown, and usually has three fine dots. FS** Single-brooded, mid June-August. Can be disturbed from vegetation by day and comes to light. **Hab** Rank grassland, downland, woodland rides, gardens, vegetated shingle. **Fp** A range of grasses, including False Oat-grass, meadow grasses, Cock's-foot and Tor-grass; feeds from within a spun leaf roll.

Acompsia schmidtiellus page 228 861 (3883)

Very local, with old records from Derbyshire and Lincolnshire. **FL** 7-8mm. **A broad-winged species.** Forewing yellowish brown, partially suffused reddish brown in the outer third, the costa finely edged black at the base, usually with two black spots at about one-third, one above the other, another just before two-thirds and one near the tornus. **FS** Single-brooded, July-August. The adult is rarely seen but sometimes comes to light. Can be numerous in the larval stage. **Hab** Chalk and limestone grassland, roadside verges. **Fp** Marjoram; in a folded or rolled leaf.

Hollyhock Seed Moth *Pexicopia malvella* page 228 809 (3888)

Very local. Possibly declining. Most frequent in the south-east. **FL** 9-10mm. Forewing yellowish brown, speckled darker, with a broad, sometimes faint, brownish cross-band from the tornus to the costa and similar shading before the apex, and with an ill-defined dark spot on the fold at about one-third, another obliquely above in the disc, and a further spot at about two-thirds. **Similar species *Platyedra subcinerea* is paler and has a narrower forewing, with blackish spots at the base of the dorsum and at one-third, the latter usually ringed paler. FS** Single-brooded, June-August. Comes to light. **Hab** Gardens, grazing marshes, river margins. **Fp** Marsh-mallow, garden Hollyhock; feeds in the seeds.

Platyedra subcinerea page 228 808 (3890)

Very local. Possibly declining and recently found only south of a line from Dorset and Wiltshire to Essex. **FL** 8-9mm. Forewing pale greyish brown, mixed darker, with blackish spots at the base of the dorsum and at one-third, usually ringed paler, and another, sometimes obsolete, at two-thirds. **Similar species** *Pexicopia malvella*. **FS** Single-brooded, potentially found all year round. Hibernates as an adult. Comes to light. **Hab** Gardens, waste ground, coastal marshes and grasslands. **Fp** Common Mallow, sometimes garden Hollyhock; feeds in the flowers and seeds.

Alucitidae

There is one species in this family. The forewing length is 7.5-9mm. All wings of the sole British species are divided into six 'fingers', or 'plumes'. At rest the wings may be spread out, revealing the plumes, or closed together over the dorsum, so the plumes are not obvious. The antennae are thread-like and about one-half the length of the forewing. The tongue base is not scaled. Adults fly from dusk and come to light.

Twenty-plume Moth
Alucita hexadactyla

Further reading
British and Irish species: Beirne (1952); Hart (2011)

Twenty-plume Moth (Many-plumed Moth) *Alucita hexadactyla* page 228   1288 (5323)

Common. **FL** 7.5-9mm. Forewing and hindwing deeply divided into six feathery plumes, greyish brown, with the appearance of a dark brown cross-band at about one-half, followed by a dark bar from the costa at about three-quarters. **FS** Single-brooded, recorded in every month of the year. Can be disturbed easily from its foodplant by day, is often found hibernating in sheds and dense vegetation, and comes to light.
Hab Woodland, hedgerows, gardens. **Fp** Honeysuckle; feeds in the flower buds and flowers, and mines the leaves.

▼ Larval workings of Twenty-plume Moth *Alucita hexadactyla* in Honeysuckle flowers. Holes and dark coloration in the flower buds indicate larval feeding.

Pterophoridae
Plume moths

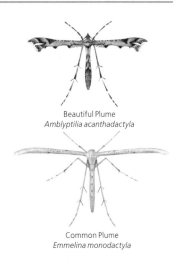

Beautiful Plume
Amblyptilia acanthadactyla

Common Plume
Emmelina monodactyla

There are 44 species in this family, referred to as the plume moths, which contains the familiar and frequently encountered Common Plume *Emmelina monodactyla* and White Plume Moth *Pterophorus pentadactyla*. All species stand up on their legs, resting the wings horizontally above the substrate, extended at or nearly at right-angles to the body; in some species, the forewing and hindwing are visible in this position, in others the forewing completely overlaps the hindwing. *Agdistis* species rest with the wings rolled into a tube and extended upwards and forwards, well above horizontal, and stand high up on long, slender legs. The forewing length is 5-16mm. Only in *Agdistis* are the forewings and hindwings undivided. In all other genera the forewings are narrow and cleft into two lobes, referred to here as costal and dorsal lobes, towards the apex. The position of this cleft and shape of the lobes are important features for identification. The hindwings are divided into three lobes, and sometimes have specialised scales on the third, or dorsal, lobe, known as a scale-tooth. The head is smooth, with a variable number of erect scales around the collar. The antennae are thread-like, and are one-half to about three-quarters the length of the forewing. The labial palps point forward or are gently curved upwards, and the tongue is well developed. The legs and abdomen are typically long and slender.

A number of species have been added to the British list in recent years as a result of detailed fieldwork and analysis. Mountain Plume *Stenoptilia islandicus* has been found on saxifrages in one mountainous part of the Scottish Highlands, and Small Scabious Plume *S. annadactyla* has very recently been recorded breeding in Breckland in East Anglia. There may be additional members of *Stenoptilia* yet to be added to the British list; the genus contains many species in Europe, and they are hard to separate, even on genitalia characters, and the taxonomic status of some requires clarification. Reedbed Plume *Emmelina argoteles*, a very close relative of *E. monodactyla*, has been found in the fens in Cambridgeshire, and the Tamarisk Plume *Agdistis tamaricis* is resident in Jersey, in the Channel Islands. One species, Scarce Light Plume *Crombrugghia laetus*, is a rare immigrant and has recently been found new to Ireland. Several species are very restricted in distribution or have a disjunct distribution, and even some of the more widespread species appear to occur in discrete colonies. Two species are now considered extinct in this country, Gentian Plume *Stenoptilia pneumonanthes* and Downland Plume *Oxyptilus pilosellae*.

The larvae feed on herbaceous plants, although one species, Rose Plume *Cnaemidophorus rhododactyla*, is associated with Dog-rose. Several feed internally in flowerheads or in growing stems. A few others, such as Wood Sage Plume *Capperia britanniodactylus* and Dingy White Plume *Merrifieldia baliodactylus*, bite the stem of the foodplant on which they feed, causing the section above to wilt, the larva preferring to feed on the wilted leaves. Perhaps the most remarkable feeding habit is that of Sundew Plume *Buckleria paludum*, the larva of which feeds on the leaves and petioles of the insectivorous plant Round-leaved Sundew; how the moth, as an adult and a larva, avoids being entrapped in the sticky hairs on the leaves is not understood.

The larvae of plume moths tend to be louse-shaped and slightly or moderately hairy,

resembling the larva of the blue butterflies in the Lycaenidae; they are often similarly coloured to the leaves of the foodplant and are well camouflaged in this respect. Adults of some species fly naturally by day, although the majority fly from dusk and into the night, with only a few species seen regularly at light, and even then typically in small numbers. Many species are perhaps best found in the larval stage, knowledge of the foodplant aiding species' identity. Adult moths of some species can be difficult to identify, particularly in *Stenoptilia*, *Merrifieldia* and *Hellinsia*, but many species can be readily identified.

▲ Larva of Cliff Plume *Agdistis meridionalis* eating the flowers of Rock Sea-lavender.

Further reading
British and Irish species: Beirne (1952); Hart (2011)
European species: Gielis (1996)

Cliff Plume *Agdistis meridionalis* page 229   1487 (5349)

Rare. Coastal. **FL** 10-12mm. Forewing with a pointed apex and straight termen, not cleft into lobes; pale greyish brown, whitish or creamy, heavily speckled with dark brown scales in a band along the costa and dorsum, with a small dark spot just below the costa at three-quarters and a series of three or four obscure dark spots just above the dorsum. **Similar species** *A. bennetii* **is usually larger, rather more greyish white and less speckled along the costa and dorsum. FS** Double-brooded, June-October. Rests by day with the wings rolled, pointing forward and upward. **Hab** Grassy coastal slopes, cliffs, undercliffs. **Fp** Rock Sea-lavender; on the leaves and flowers, preferring plants on cliff ledges and in crevices.

Saltmarsh Plume *Agdistis bennetii* page 229   1488 (5348)

Local. Coastal, but occasionally wanders inland. Occasionally common. **FL** 12-14mm. Forewing with the apex and termen somewhat rounded, not cleft into lobes; pale greyish brown, usually with an indistinct dark spot at three-quarters just below the costa, and a series of three or four dark spots just above the dorsum. **Similar species** *A. meridionalis*. **FS** Double-brooded, June-July and August-September. Can be disturbed on warm days, flies at dusk and comes to light. **Hab** Saltmarsh, chalk cliffs in Kent. **Fp** Common Sea-lavender, Sea Aster, at one site Rock Sea-lavender; on the leaves, often resting by day along the midrib of a leaf.

rish Plume *Platyptilia tesseradactyla* page 229   1499 (5374)

Scarce. Ireland only. The single Kent specimen is correctly identified but considered to have been mislabelled rather than a wanderer. **FL** 8.5-10mm. Forewing cleft into lobes from about three-quarters; greyish brown, with whitish scales most prominent along the costa and somewhat arranged in very fine oblique cross-lines in the middle of the wing, a darker brown triangular costal mark just before the cleft and a pale transverse line across both wing lobes. **Similar species** *P. gonodactyla*, *P. calodactyla*. **FS** Single-brooded, end May-June. Lives in discrete colonies. Easily disturbed on sunny days and flies in the evening around the foodplant. **Hab** Limestone grassland, embankments and pavement. **Fp** Mountain Everlasting; in the flower stem and shoot, distorting growth or causing rosette leaves to wilt.

Goldenrod Plume *Platyptilia calodactyla* page 229   1500 (5369)

Very local. Mainly coastal. **FL** 9-12.5mm. Forewing cleft into lobes from about three-quarters; pale orangey brown with whitish scales scattered in the middle of the wing, usually richer reddish brown in the outer half, especially on the costal lobe, a dark brown triangular costal mark just before the cleft and a pale transverse line across the wing lobes, most obvious on the costal lobe. **Similar species** *P. gonodactyla*, *P. tesseradactyla*. **FS** Single-brooded, June-July. Recorded rarely at light. **Hab** Grasslands near the coast, grassy areas in open woodland. **Fp** Goldenrod; in spring feeds in the base of the stems, sometimes causing the leaves to wilt.

Triangle Plume *Platyptilia gonodactyla* page 229   1501 (5368)

► Larva of Triangle Plume *Platyptilia gonodactyla* in the seedhead of Colt's-foot.

Common. **FL** 9-13mm. Forewing cleft into lobes from about three-quarters; pale brown, with whitish scales often arranged in very fine oblique cross-lines in the middle of the wing, a darker triangular costal mark just before the cleft and a paler transverse line across the wing lobes, most obvious on the costal lobe. **Similar species P. tesseradactyla** is a small, pale species; **P. calodactyla** is a darker, richer coloured species, with the forewing costal lobe relatively short and blunt, and the hindwing dorsal lobe with a wide scale-tooth at one-half; **P. gonodactyla** has the forewing costal lobe relatively long and pointed, and the hindwing dorsal lobe has a narrow scale-tooth before one-half. **FS** Double-brooded, late May-July, August-September. Flies at dusk and comes to light. **Hab** Embankments, quarries, waste ground, rough grassland. **Fp** Colt's-foot, possibly butterburs; autumn larvae overwinter in the stems, in spring spinning the seedheads, causing drooping.

Hoary Plume *Platyptilia isodactylus* page 229   1502 (5371)

Local. Very local in western and northern Scotland. **FL** 9-14mm. **An obscurely marked species with a blunt forewing apex.** Forewing cleft into lobes from about three-quarters, the costal lobe short and blunt; ground colour greyish brown or yellowish brown, with indistinct brown marks on the costa before the cleft and at one-third. Examples of the second generation are often smaller, more greyish and plainer than those of the first, with a pair of dark dots at the base of the lobes. **FS** Double-brooded, late May-September. Occasionally flies in late afternoon but usually from dusk, and comes to light. **Hab** Damp, deciduous woodland, fens, coastal marshes, grazed lawns (New Forest), water meadows, riverbanks, ditches. **Fp** Marsh Ragwort; in the stem, exuding frass from the leaf axil, the summer brood also in the flowerhead and on leaves.

Tansy Plume *Gillmeria ochrodactyla* page 229   1503 (5378

Very local. **FL** 12-14mm. Forewing cleft into lobes from about three-quarters; yellowish brown, with a small pale brown blotch on the dorsum at about one-quarter and a pale mark on the costa just before the cleft. **Similar species** *G. pallidactyla* **has the labial palps visible beyond a tuft of hairs on the face, the hindleg tibia between the spurs is uniformly grey or pale brownish,** the forewing markings are more suffused and the dorsal lobe of hindwing has a relatively narrow scale-tooth; *G. ochrodactyla* **has a longer, thicker tuft of hairs on the face, obscuring the palps, the hindleg tibia between the spurs is distinctly banded whitish and brownish,** the forewing markings are more defined, and the hindwing dorsal lobe has a relatively wide scale-tooth. **FS** Single-brooded, late June-August. Rests on the flowers of the hostplant after dark and comes to light. **Hab** Riverbanks, dry grassland, roadsides, embankments, vegetated shingle, quarries. **Fp** Tansy; overwinters in the root, in spring feeds in a new shoot, causing it to wilt.

Yarrow Plume *Gillmeria pallidactyla* page 229   1504 (5377

Common. **FL** 11-13mm. Forewing cleft into lobes from about three-quarters; pale yellowish or whitish brown, darker along the costa from the base to an indistinct triangular costal spot just before the cleft. **Similar species** *G. ochrodactyla*. **FS** Single-brooded, June-August. Readily disturbed by day and comes to light. **Hab** Embankments, waste ground, quarries, hedgerows, dry grassland. **Fp** Yarrow, Sneezewort, occasionally Tansy; overwinters in the root, in spring feeds on a succession of shoots, causing them to wilt.

Beautiful Plume *Amblyptilia acanthadactyla* page 229   1497 (5381

▶ Larva of Beautiful Plume *Amblyptilia acanthadactyla* ready for pupation on a stem of Hedge Woundwort.

Common. **FL** 9-11mm. Forewing cleft into lobes from just beyond two-thirds; reddish brown lightly speckled with whitish scales, with a darker triangular mark on the costa just before the cleft and a pale transverse line across both lobes; the scale-tooth on the hindwing dorsal lobe is wide, and short in height. **Similar species** *A. punctidactyla* **is darker, appearing greyish brown and more variegated with distinct whitish speckling; the scale-tooth on the hindwing dorsal lobe is narrow, and tall in height.** **FS** Double-brooded, July and September-early June. Overwinters as an adult. Adults of the autumn generation have been found at Ivy blossom. Flies from dusk and comes to light. **Hab** Woodland, parkland, hedgerows, heathland, moorland, gardens, rough ground. **Fp** A range of plants, including Hedge Woundwort, restharrows, mints, goosefoots, crane's-bills, heathers; feeds on the flowers and unripe seeds.

Brindled Plume *Amblyptilia punctidactyla* page 229   1498 (5382

Local. **FL** 9-11mm. Forewing cleft into lobes from just beyond two-thirds; greyish brown with whitish speckling, with a darker triangular mark on the costa just before the cleft and a pale transverse line across both lobes. Examples from Scotland may be paler, with a whitish thorax. **Similar species** *A. acanthadactyla*. **FS** Double-brooded, July and September-early June. Overwinters as an adult. Adult visits blossoms of ragworts, Hedge Woundwort, sallows and Ivy, and comes to light. **Hab** Woodland, hedgerows, preferring damp, shady places. **Fp** Mainly Hedge Woundwort, also columbines, Common Stork's-bill, crane's-bills, Primrose; feeds on the flowers and unripe seeds, but found only on shaded plants.

Saxifrage Plume *Stenoptilia millieridactyla* page 230   1506 (5405)

Very local. Native in Ireland; introduced accidentally to Britain in the late 1960s and expanding its range. **FL** 8-10mm. Forewing cleft into lobes from just beyond two-thirds; greyish brown becoming reddish or orange-brown towards the dorsum, with a dark fused double spot at the base of the cleft, a small dark spot just beyond one-third and a longitudinal spot in the dorsal half of the costal lobe. Hindwings tinged bronze or reddish. **FS** Mainly single-brooded, June-July, with a partial second brood late August-early September. Can be numerous in gardens. Readily disturbed by day, flies from dusk and comes to light. **Hab** Rockeries in gardens, limestone pavement. **Fp** Mossy Saxifrage, including garden cultivars; in spring feeds externally on the leaves.

Dowdy Plume *Stenoptilia zophodactylus* page 230   1507 (5426)

Common. More local in Wales and scarce in Scotland. **FL** 9-11mm. Forewing cleft into lobes, relatively widely separated, from about two-thirds; pale greyish brown, fading to orange-brown or sandy brown along the dorsum, a dark brown spot at one-third and either one spot just before the cleft, or if two, then the second fainter and obliquely displaced towards the costa; the cilia of the costal lobe have at least two patches of dark scales. Legs whitish and a whitish band on the 'saddle' (the upper end of the abdomen adjacent to the thorax). **Similar species The 'S. bipunctidactyla' group (see below), which are more robust moths with a fused double spot at the cleft base and only one spot in the cilia of the costal lobe, brown legs and the saddle with a yellowish-brown band edged whitish; the wing cleft is narrower than in S. zophodactylus.** **FS** Double-brooded, late May-early November. Occasionally found in numbers. Easily disturbed by day, flies from dusk and comes to light. **Hab** Dry grassland, chalk downland, embankments, quarries, open woodland, sand dunes, vegetated coastal shingle, coastal cliffs and grassland, wet heathland. **Fp** Centaury, Yellow-wort, Autumn Gentian (Felwort), occasionally Marsh Gentian; among the flowers and developing seeds.

Stenoptilia bipunctidactyla group

There are four very similar species in this group: Twin-spot Plume *S. bipunctidactyla*, a widespread species on Devil's-bit Scabious and Field Scabious; Small Scabious Plume *S. annadactyla* (not illustrated), known only from Breckland in East Anglia, on Small Scabious; Scarce Plume *S. inopinata* (not illustrated), a very rare immigrant, and Gregson's Plume *S. scabiodactylus* (not illustrated), a very local species in the Midlands, Wales and northern England. Another related species is Mountain Plume *S. islandicus* (not illustrated), currently known only from one mountain in Scotland, on Mossy Saxifrage. The account below is for the most widespread species, *S. bipunctidactyla*. If any of the other species is suspected, detailed examination, including genitalia, will be necessary.

Twin-spot Plume *Stenoptilia bipunctidactyla* page 230   1508 (5397)

Common. More local in the north. **FL** 9-12mm. Forewing cleft into lobes from just beyond two-thirds; greyish brown fading to orange-brown along the dorsum, with a dark spot at one-third, a fused double spot at the base of the cleft and a small black spot in the cilia of the costal lobe. **Similar species** *S. zophodactylus*, *S. pterodactyla*. **FS** Two overlapping generations, late May-early October. Easily disturbed by day, flies in the afternoon, at dusk and comes to light. **Hab** Grasslands, woodland rides, roadside verges; may occur wherever the foodplants grow. **Fp** Field Scabious, Devil's-bit Scabious; in the flowers in summer and autumn, in the young shoots in spring.

► Larva of
Rose Plume
*Cnaemidophorus
rhododactyla*
feeding in the flower
bud of Dog-rose.

Brown Plume *Stenoptilia pterodactyla* page 230   1509 (5390)

Common. More local in the north. **FL** 10-13mm. Forewing cleft into lobes from about two-thirds; pale reddish brown to yellowish brown, with two isolated, rarely fused, small darker spots at the base of the cleft; the costal lobe has a small dot in the terminal cilia and white cilia along the costal margin. **Similar species** The **'S. bipunctidactyla' group have greyish-brown forewings, and the costal lobe is grey with dark cilia on the costal margin. FS** Single-brooded, late May-early August. Easily disturbed by day, flies from dusk, visits flowers at night and comes to light. **Hab** Woodland margins, hedgerows, dry grassland, roadside verges. **Fp** Germander Speedwell; hibernates in a stem, feeds on flowers in spring.

Rose Plume *Cnaemidophorus rhododactyla* page 230   1496 (5434)

Rare. Probably an occasional immigrant. **FL** 9-12mm. Legs white, joints banded reddish brown. Forewing cleft into lobes from about two-thirds; orangey brown with a transverse white line at the base of the cleft and a whitish blotch just below the costa at about one-half. Hindwing with a scale-tooth at the apex of the lobe. **FS** Single-brooded, late June-August. Comes to light. Usually occurs in small colonies, often short-lived. **Hab** Woodland margins, hedgerows, scrub on downland. **Fp** Dog-rose; in spring on the flower buds and shoots.

Crescent Plume *Marasmarcha lunaedactyla* page 230   1495 (5436)

Common in southern England. Very local further north. **FL** 9-10mm. Forewing cleft into lobes from about two-thirds; orangey dark brown in the male, orangey brown in the female, with an incomplete pale crescent-shaped transverse band at the base of the cleft. **FS** Single-brooded, mid June-early August. Easily disturbed by day and comes to light. **Hab** Downland, sand dunes, sea cliffs, vegetated coastal shingle, quarries. **Fp** Common Restharrow, Spiny Restharrow; on the buds and leaves.

Small Plume *Oxyptilus parvidactyla* page 230   1490 (5443)

Local. Scarce away from southern England. **FL** 6-8mm. Forewing cleft into lobes from near one-half; dark reddish to blackish brown with two transverse white lines on the lobes, most obvious on the costal lobe. The hindwing has a scale-tooth in the cilia at the apex of the dorsal lobe. **FS** Single-brooded, late May-August. Flies in afternoon sunshine and at dusk, visiting thyme flowers, and very occasionally comes to light. **Hab** Chalk and limestone grassland, dry grassland, heathland, coastal grassland, vegetated shingle. **Fp** Mouse-ear-hawkweed; in the centre of a rosette.

▲ Larva of Wood Sage Plume
Capperia britanniodactylus on
a leaf of Wood Sage.

► Larval working of Wood
Sage Plume *Capperia
britanniodactylus* on Wood
Sage. Note the larva has bitten
part way through the stem,
causing the shoot to wilt.

Breckland Plume *Crombrugghia distans* page 230   1491 (5445)

Very local. Probably an occasional immigrant. **FL** 7-10mm. Forewing cleft into lobes from near one-half; pale to dark reddish brown, with two pale transverse lines on the costal lobe. The hindwing has a scale-tooth at about two-thirds. **Similar species Scarce Light Plume C. *laetus* (not illustrated), a rare immigrant, is often slightly paler and has the scale-tooth in the hindwing less well developed; however, if C. *laetus* is suspected examination of genitalia is necessary to confirm identification.** **FS** Double-brooded, May-June, July-August, with a possible occasional third brood in October. Easily disturbed by day, particularly in hot weather, and flies freely from dusk. Occasionally comes to light. **Hab** Sand dunes, Breckland grassland, chalk downland, coastal vegetated shingle. **Fp** Mainly Smooth Hawk's-beard, also Mouse-ear-hawkweed; spring larvae on the leaves, summer larvae on the flowers.

Wood Sage Plume *Capperia britanniodactylus* page 230   1494 (5459)

Local. Very local in northern England. **FL** 8-10mm. Forewing cleft into lobes from near one-half; dark brown with two transverse white lines, these most obvious on the costal lobe. The hindwing dorsal lobe has a distinct dark brown scale-tooth towards the apex which extends into the dorsal and costal margins, with white scaling at the apex and along the dorsal margin towards the base. **FS** Single-brooded, May-August. Sometimes abundant around the foodplant. Can be disturbed by day, flies from dusk and comes to light. **Hab** Chalk downland, open woodland, parkland, dry heathland, acid grassland, sand dunes, coastal vegetated shingle. **Fp** Wood Sage; large larvae eat through part of the stem causing wilting, then feed on both wilted and unaffected leaves.

Sundew Plume *Buckleria paludum* page 230   1493 (5478)

Rare. **FL** 5-7mm. Forewing cleft into lobes from just before one-half; dark brown, with two transverse white lines across the costal lobe. Hindwing without an obvious scale-tooth. **FS** Double-brooded, June-August. Flies low down on warm, still afternoons and is easily disturbed from vegetation; also flies at dusk and comes to light. **Hab** Boggy heathland. **Fp** Round-leaved Sundew; in spring on the leaves, in summer on the leaves, flowers and developing seeds.

▲ Larvae and feeding pattern of Spotted White Plume *Porrittia galactodactyla*. Note the holes in the leaf and the white down which has been scraped into a semicircle.

▲ Peppershot feeding damage caused by larvae of Spotted White Plume *Porrittia galactodactyla* on a leaf of burdock.

White Plume Moth *Pterophorus pentadactyla* page 230   1513 (5485)

Common. Very local north of southern Scotland. **FL** 12-16mm. Forewing cleft into lobes from about one-half, the lobes tapering; white. Hindwing white. **FS** Double-brooded in the far south, March-August, single-brooded further north, with an occasional partial second brood, June-early August and September. Easily disturbed by day, flies from dusk and comes to light. Can be found at night on flowers such as valerians. **Hab** Dry grassland, gardens, hedgerows, waste ground. **Fp** Hedge Bindweed, Field Bindweed, probably Sea Bindweed; on the leaves and flowers.

Spotted White Plume *Porrittia galactodactyla* page 231   1514 (5488)

Local. **FL** 10-12mm. Forewing cleft into lobes from about one-half, the lobes tapering; whitish with dark markings, including a blackish dot at one-third, a pair of black spots at the base of the cleft, appearing V-shaped, and a dark spot on the costa at about two-thirds. **FS** Single-brooded, June-July. Rarely seen as an adult. Flies at dusk and after dark and occasionally comes to light. **Hab** Woodland, dry grassland. **Fp** Lesser Burdock, possibly Greater Burdock; larvae are rather gregarious and make distinctive peppershot holes in the leaf from the underside, scraping away the white down into a semicircle at the edge of the hole.

Thyme Plume *Merrifieldia leucodactyla* page 231   1510 (5501)

Local. **FL** 9-12mm. Basal third of the antenna with a line of dark brown scales on the upper surface, bounded either side by rows of pale yellow scales. Forewing cleft into lobes from about one-half, the lobes tapering; pale yellowish in the dorsal half, yellowish brown in the costal half, the extent of each variable, and edged darker brown along much of the costa, and with dark cilia. **Similar species Western Thyme Plume *M. tridactyla* (not illustrated), a rare species on thyme, confined to western Ireland and west Cornwall, has the basal third of the antenna ringed pale yellow and dark brown; the forewing is very similar, although the yellow is more variable, often pale sulphur yellow, and with less contrasting cilia.** Scales on the antenna rub off easily. **FS** Single-brooded, June-August, with an occasional second brood into September. Easily disturbed by day, flies from dusk and comes to light. **Hab** Chalk and limestone grassland, rocky coasts. **Fp** Wild Thyme; on the leaves and developing seeds.

Dingy White Plume *Merrifieldia baliodactylus* page 231   1512 (5506)

Local. Rare away from southern England. **FL** 10-13mm. Forewing cleft into lobes from about one-half, the lobes tapering; pale yellowish, with an indistinct fine dark brown line along the costa joined to a more prominent narrow mark or spot at about or just beyond one-half. **FS** Single-brooded, possibly with an occasional partial second brood, May-August. Easily disturbed by day from the foodplant, flies from dusk and comes to light. **Hab** Chalk and limestone grassland, coastal cliffs, quarries, waste ground. **Fp** Marjoram; after hibernation the larva eats through a stem near the top causing leaves to wilt, feeding on the wilted leaves.

Horehound Plume *Wheeleria spilodactylus* page 231   1515 (5512)

Rare. **FL** 9-12mm. Forewing cleft into lobes from about one-half, the lobes tapering; pale yellowish white, tinged greyish, with greyish-brown markings, including a small blotch at the cleft and a costal spot just beyond the cleft. **FS** Single-brooded, July-August, with an occasional partial second generation, September-October. Occurs in discrete colonies. Can be disturbed by day in favourable weather and occasionally comes to light. **Hab** Calcareous downland, Breckland field margins, coastal cliffs, suburban habitats, also found on vegetated coastal shingle. **Fp** White Horehound; on the leaves.

Short-winged Plume *Pselnophorus heterodactyla* page 231   1516 (5521)

Rare. **FL** 9-11mm. Forewing cleft into lobes from just beyond one-half; reddish brown, darker towards the apex, around the cleft and towards the base, with white costal spots, including one at about one-half, another just beyond the cleft and one at the apex; the cilia of the dorsal lobe are chequered white and dark brown. **FS** Single-brooded, June-July. **Hab** Partly shaded areas of large woods in the Cotswolds, also shaded mountain gullies in Scotland. **Fp** Wall Lettuce, Marsh Hawk's-beard, Nipplewort; after hibernation the larva bites through the midrib of the leaf and rests on the wilted part.

Dusky Plume *Oidaematophorus lithodactyla* page 231   1523 (5528)

Local. Very local further north towards the Borders. **FL** 12-14mm. The tibia of the midleg is thickened with scales in the middle, and the hindleg is banded dark brown above the joints. Forewing cleft into lobes from about two-thirds; greyish brown to reddish brown, sometimes variably shaded greyish, with a dark brown oblique mark before the base of the cleft, the upper part of which curves away towards a longitudinal blackish-brown spot on the costa above the cleft base. **Similar species Strongly marked examples of *Emmelina monodactyla*, a less robust-looking species, do not have banded markings on the legs.** **FS** Single-brooded, late June-August. Flies at dusk and occasionally comes to light. **Hab** Dry grassland, chalk downland, hedgerows, roadside verges, open woodland, ditches, sheltered riverbanks. **Fp** Common Fleabane, Ploughman's Spikenard; at first in the shoots, then openly on the leaves, making many small holes.

▶ Two larvae of Dusky Plume *Oidaematophorus lithodactyla* on a leaf of Ploughman's Spikenard.

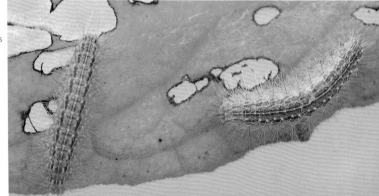

Mugwort Plume *Hellinsia lienigianus* page 231 1518 (5543)

Very local. Scarce in the northern part of its range. **FL** 8-10mm. Forewing cleft into lobes from about two-thirds; creamy white, tinged greyish and sometimes slightly reddish, scattered with dark brown scales, a large arched or oblique dark spot before the base of the cleft, one dark spot on the dorsal edge of the costal lobe and two dark spots on the costal edge, the one nearest the cleft being longer. **FS** Single-brooded, June-September. Flies at dusk and occasionally recorded at light. **Hab** Woodland, dry grassland, waste ground. **Fp** Mugwort; feeds from within a tent-like spun leaf, making up to four successive tents in the spring.

Small Goldenrod Plume *Hellinsia osteodactylus* page 231 1520 (5545)

Local. **FL** 8-11mm. Forewing cleft into lobes from just beyond one-half; yellowish white to bright yellow, an ill-defined and faint brownish streak from the base extending towards the cleft but not reaching it, with a small dark brown spot at the base of the cleft and a brown dash, sometimes indistinct, extending half way along the costa of the costal lobe. **Similar species Scarce Goldenrod Plume *H. chrysocomae* (not illustrated), a rare species in south-east England on Goldenrod, is usually brownish yellow, has a larger, more diffuse spot centred just below the cleft base, and a faint streak from the forewing base extending to the cleft or beyond.** Examination of genitalia may be needed if *H. chrysocomae* is suspected. **FS** Single-brooded, June-August. Easily disturbed by day, flies from dusk and comes to light. **Hab** Open, broadleaved woodland, restored mine-waste sites, coastal grasslands and cliffs. **Fp** Goldenrod, occasionally on Common Ragwort; on the flowers and seedheads.

Plain Plume *Hellinsia tephradactyla* page 231 1522 (5534)

▶ Larva and feeding pattern of Plain Plume *Hellinsia tephradactyla* on Goldenrod.

Local. Very local in Scotland.
FL 9-11mm. Forewing cleft into lobes from just beyond one-half; greyish or pale brownish white, sometimes darker, speckled with darker scales, these loosely grouped into three or four longitudinal rows from the base to the lobes, with two dots at the base of the cleft and more defined dots at the margins of the lobes. Examples from western Ireland are not as speckled. **FS** Single-brooded, June-July. Easily disturbed by day, flies from dusk and comes to light. **Hab** Rocky coasts, woodland, preferring the foodplants growing in shade or on north-facing slopes. **Fp** Goldenrod; in spring, on the leaves, making holes.

Citron Plume *Hellinsia carphodactyla* page 231 1519 (5537)

Very local. **FL** 8-11mm. Forewing cleft into lobes from just beyond one-half; yellow or pale yellow, with scattered dark brown scaling, especially in the basal half, and with dark spots usually at one-third, at the base of the cleft, and on the costa just beyond the base of the cleft; **there are small dark marks, one on the dorsal edge of the costal lobe and three on the dorsal edge of the dorsal lobe.** **FS** Double-brooded, late May-September. Flies from dusk and occasionally comes to light. **Hab** Chalk and limestone grassland, embankments, rocky cliffs, coastal landslips, occasionally open woodland. **Fp** Ploughman's Spikenard; summer-generation larvae feed in the flowers, those in spring in the heart of the plant rosette.

◀ Gall and exit hole at the base of the flowering stem of Hemp-agrimony, caused by the larva of Hemp-agrimony Plume *Adaina microdactyla*.

Hemp-agrimony Plume *Adaina microdactyla* page 231   1517 (5550)

Common. More local further north. **FL** 6-8mm. **A small, pale species.** Forewing cleft into lobes from just beyond one-half; pale yellowish, sometimes yellowish white, pale brown or pale grey, with a small brown spot at the base of the cleft and two spots on the costa of the costal lobe. **FS** Double-brooded, May-early September. Easily disturbed by day, flies from dusk and comes to light. **Hab** Downland, damp marshy localities, wet flushes on soft-rock cliffs. **Fp** Hemp-agrimony; makes a gall at the base of the flower stalks and hibernates fully fed in the stem. Small holes in dead stems seen in winter indicate the presence of larvae.

Common Plume *Emmelina monodactyla* page 231   1524 (5552)

Common. More local from the Borders northwards. **FL** 9-13mm. Rests with its wings tightly rolled, resembling a small cross. Forewing cleft into lobes from about two-thirds; pale greyish white to dark reddish brown; the dark brown markings usually including a small spot at one-third, a larger spot before the cleft, a streak on the costa of the costal lobe, and three spots on the dorsal margin of the dorsal lobe. **Similar species Reedbed Plume *E. argoteles* (not illustrated), a rare species on Hedge Bindweed recognised recently from the fens of Cambridgeshire, is on average slightly smaller than *E. monodactyla*, usually plainer in appearance and with the costa of the costal lobe straighter.** Examination of genitalia will be needed if *E. argoteles* is suspected; also *Oidaematophorus lithodactyla*. **FS** Single-brooded, September-May. Also found in late June-early August, and these may be long-lived hibernated examples or possibly another generation. Often seen on fence posts and walls, and indoors. Flies at dusk, comes to light and can be found at Ivy blossom, ripe blackberries and, in spring, at sallow blossom. **Hab** A wide range of habitats, including woodland, scrub, hedgerows, gardens. **Fp** Bindweeds, including Hedge Bindweed and Field Bindweed; on small leaves, flowers and developing seeds.

Schreckensteiniidae

There is one species in this family. The adult rests with the body horizontal and the wings flat and slightly diverging, with the hindlegs held at an angle to the body and pointing slightly upwards, somewhat similar to the Stathmopodidae. The forewing length is 4.5-5.5mm. The forewings are narrow. The hindwings are narrow and pointed, not as broad as the forewings, the dorsal cilia moderately long. The head is smooth-scaled. The antennae are thread-like, three-fifths the length of the forewing. The labial palps are slightly upwardly curved.

Schreckensteinia festaliella

Further reading
British and Irish species: Emmet (1996)

Schreckensteinia festaliella page 232   485 (5291)

Common. Local in Scotland. **FL** 4.5-5.5mm. Rests only on the forelegs and midlegs, the hindlegs held obliquely to the abdomen. Forewing whitish to yellowish brown, the costa, dorsum and termen darker brown, with a similarly coloured streak from near the base almost to the apex, this expanded at about two-thirds. **FS** More or less continually brooded, March-September. Flies in the afternoon and occasionally comes to light.
Hab Found in a range of habitats, perhaps most frequently in open woodland.
Fp Bramble, Raspberry, Hazel; beneath a slight web on the underside of the foodplant, skeletonising leaves. Pupates in an open network cocoon attached to a stem.

▼ Larva of *Schreckensteinia festaliella* ready for pupation within a net-like cocoon.

Epermeniidae

There are eight species in this family. Adults rest in a slightly inclining posture, with the wings held roof-like at a steep angle over the body, the antennae mostly along the side of the wings. The forewing length is 3.5-7.5mm. The forewings are elongate, sometimes with one or more scale-tufts along the dorsal margin which appear as a conspicuous crest, or crests, when at rest and viewed side-on. Some species have a

Phaulernis fulviguttella

Epermenia chaerophyllella

hooked apex to the wing. The hindwings are narrower than the forewings, with the dorsal cilia usually long. The head is smooth-scaled. The antennae are thread-like, three-fifths the length of the forewing. The labial palps are curved upwards, and the tongue base is not scaled.

Two species, *Phaulernis fulviguttella* and *Epermenia chaerophyllella*, are widely distributed; the rest are found predominantly in the southern half of the country and two are rare. The larvae are mostly associated with umbellifers, with one found on Bastard-toadflax; they feed externally under silken webbing, in some cases gregariously, or within spun-together seeds of the foodplant, except *E. falciformis*, which feeds in leaves spun together and in the flower stems. Pupation takes place within an open network cocoon. Adults of some species can be found by day on umbellifer flowers, others fly at night and occasionally come to light. The species are all fairly straightforward to identify.

Further reading
British and Irish species: Emmet (1996)

Phaulernis dentella page 232   477 (5298)

Local. **FL** 4-4.5mm. Forewing nearly unicolorous, blackish brown with a scattering of white scales, sometimes with obscure paler patches above the tornus and on the dorsum at one-quarter; there is a distinct scale-tuft on the dorsum at one-third. **FS** Single-brooded, June. Found on sunny days resting briefly on the flowers of umbellifers. **Hab** Dry grassland, chalk pits, waste ground, hedgerows. **Fp** Rough Chervil, Burnet-saxifrage, Ground-elder; in a slight spinning among developing seeds.

Phaulernis fulviguttella page 232   478 (5296)

Common. More local in north-west England and Ireland. **FL** 4.5-5mm, slightly larger in Shetland. Forewing dark brown with a yellowish-orange spot before the tornus, another above the tornus at about two-thirds and with other more or less distinct spots near the base and on the dorsum. **Similar species** *Pammene aurana* (Tortricidae) **has a shorter and broader forewing with larger spots.** **FS** Single-brooded, July-August. Can be found at rest on the flowers of the foodplant. **Hab** Damp grassland, woodland rides, fens, marshes. **Fp** Hogweed, Wild Angelica; in a spinning among the developing seeds.

Epermenia farreni page 232   479 (5310)

Rare. The disjunct distribution suggests this species is overlooked. **FL** 4-4.5mm. Forewing greyish brown with scattered white scales, a whitish spot on the dorsum at one-third and on the costa before the middle, these sometimes joined, and with a whitish spot at the tornus and a whitish patch in the apex. **FS** Single-brooded, June-July. Can be found at rest on the foodplant or surrounding vegetation. **Hab** Dry grassland, chalk downland, waste ground. **Fp** Wild Parsnip in England, Hogweed in eastern Scotland; feeds within an individual seed, usually with no external sign of its presence.

Epermenia profugella page 232   480 (5312)

Local. Scarce in northern England. **FL** 3.5-4.5mm. Forewing unicolorous greyish brown with a bronze sheen when fresh. **FS** Single-brooded, July-August. Can be found flying around the foodplant in early evening sunshine. **Hab** Chalk and limestone grassland, dry grassland, waste ground. **Fp** Wild Carrot, Burnet-saxifrage, Wild Angelica, Ground-elder; feeds within two or three of the developing seeds spun together.

Epermenia falciformis page 232   481 (5305)

Common. Local in northern England, also found on Islay in Scotland. Recently recorded from a few sites in eastern Ireland. **FL** 6-7.5mm. Forewing yellowish brown, mottled darker at the base, along the costa and towards the apex, with an oblique dark brown cross-band from the dorsum at one-third to the costa at two-thirds, the apex hooked and edged blackish brown; the dorsum has two, sometimes three, black scale-tufts, the largest at one-third. **FS** Double-brooded, May-September. Occasionally comes to light. **Hab** Damp woodland, damp grassland, marshland. **Fp** Wild Angelica, Ground-elder; first-generation larvae feed within spun leaflets, those of the second mining a branch stem below the inflorescence, causing it to wilt.

▼ Larva and feeding pattern of *Epermenia chaerophyllella* on carrot. Note the larvae feeding externally and the sparse strands of silk with droplets and frass attached.

▼ (Inset) Larva of *Epermenia chaerophyllella* on the underside of a leaf.

permenia insecurella page 232 482 (5300)

Rare. Possibly declining. **FL** 4-5mm. Forewing variable, usually whitish and mottled greyish especially along the costa and towards the apex, with scattered black dots, and sometimes with orangey brown marks at and beyond one-half; the dorsum has two scale-tufts, the largest at about one-half, the second sometimes indistinct. **FS** Double-brooded, May-August. Flies in late afternoon sunshine and occasionally comes to light. **Hab** Chalk and limestone grassland. **Fp** Bastard-toadflax; mines the shoots and leaves when young, later feeding in a slight web among the leaves and flowers.

permenia chaerophyllella page 232 483 (5303)

Common. **FL** 5.5-6mm. Forewing variable, usually blackish brown with most markings obscured, sometimes lighter; paler examples are greyish near the base, with a dark oblique cross-band at one-half, and brownish and blackish beyond, often with patches of whitish scales; the dorsum has a large mixed black and brown scale-tuft at one-third, another, slightly smaller, at one-half, and two small ones beyond; **the wing apex is slightly hooked. Similar species *E. aequidentellus* has a narrower forewing without a hooked apex. FS** Double- or occasionally triple-brooded, can be found all year round, and hibernates as an adult. Occasionally flies in the evening and regularly comes to light. **Hab** Found in a wide range of habitats, including gardens, waste ground, grasslands, hedgerows, scrub, woodland. **Fp** Hogweed, Wild Parsnip, Cow Parsley, Wild Angelica, Wild Carrot; young larvae begin as leaf miners, older larvae feed in a slight web on the underside of the leaf.

Epermenia aequidentellus page 232 484 (5302)

Local. Recently recorded new to Wales, Scotland and Ireland. **FL** 4.5-5.5mm. Forewing white, speckled blackish or, in darker examples, shaded yellowish brown or blackish brown, with a blackish spot or dash at three-quarters; the dorsum has up to four small black scale-tufts, the largest at one-third. **Similar species** *E. chaerophyllella*. **FS** Double-brooded, June-October. Occasionally comes to light. **Hab** Often found near the coast. Limestone or chalk downland, also recorded from woodland and gardens. **Fp** Wild Carrot, Burnet-saxifrage, also reported from Moon Carrot; often a leaf miner throughout, but large larvae may feed in a slight web on the underside of the leaf.

▼ Larva and feeding pattern of *Epermenia aequidentellus* on Burnet-saxifrage. Note the larva leaf mining, even though it is too big to fit within the leaf, and the sparse strands of silk with frass attached.

Choreutidae

There are six species in this family. Adults rest with the body horizontal, standing well above the substrate on the legs, with most species holding the wings flat and slightly diverging; in *Tebenna*, the wings are held

Nettle-tap
Anthophila fabriciana

Tebenna micalis

somewhat roof-like, the apices tucked down and pressed together, as if protecting the back end of the body. The general shape of the Choreutidae is similar to the Tortricidae, especially the broad hindwing, but the resting position in the latter is quite different and all species rest on the substrate. The forewing length is 4-8.5mm. The forewing and hindwing are broad, a few species having metallic scales. The head is smooth-scaled. The antennae are thread-like, one-half to two-thirds the length of the forewing. The labial palps point forwards or are ascending, and in some genera they have scale-tufts beneath the second segment; the tongue is well developed and scaled at the base.

This family includes one of the most widespread and common micro-moths, the Nettle-tap *Anthophila fabriciana*. The other species are much more restricted in range, and *Choreutis diana* is confined to a single site in the Highlands of Scotland. Most species are straightforward to identify, but the two *Prochoreutis* species require careful attention.

Larvae feed externally on leaves from within a web. They pupate in dense, spindle-shaped cocoons. Adults fly by day in sunshine and sometimes settle on flowerheads and leaves, walking to new positions in erratic bouts, pausing momentarily between movements. Most species have been recorded at light.

Further reading
British and Irish species: Heath & Emmet (1985)

Nettle-tap *Anthophila fabriciana* page 232   385 (5269)

Common. **FL** 5-7mm. Forewing dark brown, variably speckled white, with dark cross-bands and whitish spots on the costa, one at about one-third and another at two-thirds which is usually extended, forming a fine line to the tornus. **FS** At least double-brooded, April-November. Flies by day, is frequently seen in numbers over the foodplant, and rests on flowers such as Common Ragwort. Occasionally comes to light. **Hab** Rough ground, hedgerows, woodland margins and rides, gardens. **Fp** Nettle, occasionally Pellitory-of-the-wall; in a web on the upperside of the leaf, drawing the edges together.

Prochoreutis sehestediana page 232   387 (5272)

Very local. **FL** 4-5.5mm. Forewing brown, variably shaded darker with several white spots, and white speckling from the dorsum at about one-half extending towards the costa; scattered bluish-silver spots, some forming a cross-line at about three-quarters, beyond which is a cross-band of paler ground colour; the apex is somewhat pointed. **Similar species** *P. myllerana* has the white spots typically larger and more rounded, the patch of white speckling at most reaching half way across the wing from the dorsum; beyond the silvery blue outer cross-line the ground colour is usually lighter but does not form a distinct cross-band; the apex is somewhat rounded. The with-or-without cross-band difference is best seen by observing the wings at an angle in bright daylight. Worn examples are likely to require examination of genitalia to confirm identification. **FS** At least double-brooded, late May-September. Flies by day, especially around the foodplant. Has been recorded at light. **Hab** Fens, marshes, wet heathland, woodland rides. **Fp** Skullcap, Lesser Skullcap; probably in a silken web.

Prochoreutis myllerana page 232 388 (5271)

Local. **FL** 4.5-6mm. **Similar species** *P. sehestediana*. **FS** At least double-brooded, late May-September. Flies by day, especially around the foodplant, and occasionally comes to light. **Hab** Fens, marshes, wet ditches, edges of ponds, wet heathland, woodland rides. Flies over or rests on the foodplant by day and has been recorded at light. **Fp** Skullcap, Lesser Skullcap; initially mines a leaf, later in a loose silken web.

Tebenna micalis page 232 386 (5279)

Scarce, with records mainly from southern coastal counties. An immigrant that occasionally becomes temporarily established, but it appears not to survive the winter here. **FL** 4.5-6mm. **The resting posture is characteristic.** Forewing brown with two speckled white broad cross-bands, these sometimes linked in the middle, the outer one curving around a black-outlined silver spot in the middle of the wing. **FS** Probably single-brooded in Britain, late July-October. Flies in sunshine and visits flowers of the foodplant. Occasionally comes to light. **Hab** Coastal landslips, wet meadows, margins of ponds, ditches, occasionally gardens. **Fp** Common Fleabane; on the underside of the leaf, within a silk web spun between the stem and the leaf, preferring lower leaves, the feeding causing the upper epidermis to turn brown.

Apple Leaf Skeletoniser *Choreutis pariana* page 232 389 (5282)

Local. Very local in some areas, especially further north and in Wales. **FL** 5-7mm. Forewing variable, shaded bright orangey or reddish brown to greyish brown, often with a wavy blackish line near the base and a blackish or reddish-brown line at five-sixths. **Similar species** *C. diana* (p.232), a rare species on birch, known from a single site in Scotland, is larger (FL 7-8.5mm), with forewing greyish, yellowish grey or tinged brownish grey, and an irregular whitish cross-line at one-third and a similar cross-bar at two-thirds. **FS** Double-brooded, June-July and late August-March; has been recorded in mid May. Hibernates as an adult. Rests by day on leaves and comes to light. **Hab** Woodland, scrub, isolated trees in urban areas, gardens. **Fp** Apples, occasionally hawthorns, Wild Pear, Rowan; in a silk web on the upper leaf surface, skeletonising the leaf.

▲ Larva and feeding pattern of *Prochoreutis myllerana* on Skullcap.

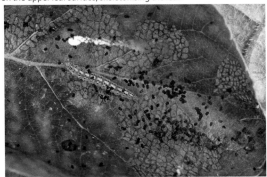

▲ Larva and feeding pattern of Apple Leaf Skeletoniser *Choreutis pariana* on the upperside of an apple leaf.

Micropterigidae x 4

Micropterix tunbergella
(p.47)

Micropterix mansuetella
(p.47)

Micropterix aureatella
(p.47)

Micropterix aruncella
(p.47)

Micropterix calthella
(p.47)

Eriocraniidae x 4

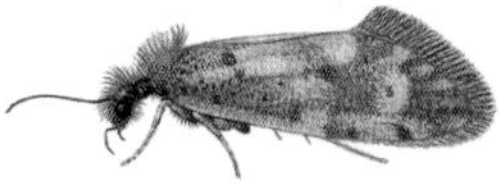

Dyseriocrania subpurpurella
(p.48)

Heringocrania unimaculella
(p.48)

Nepticulidae x 6 # Opostegidae x 4

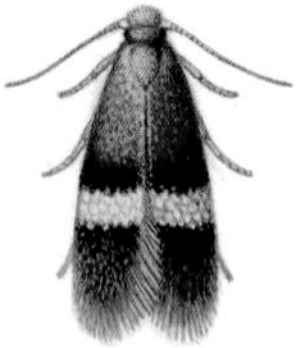

Stigmella aurella
(p.51)

Ectoedemia decentella
(p.51)

Opostega salaciella
(p.54)

Pseudopostega crepusculella
(p.54)

Heliozelidae x 6

Antispila metallella
(p.55)

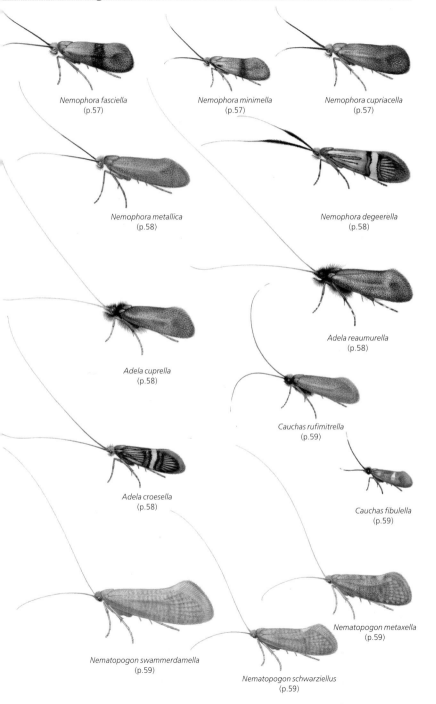

Nemophora fasciella
(p.57)

Nemophora minimella
(p.57)

Nemophora cupriacella
(p.57)

Nemophora metallica
(p.58)

Nemophora degeerella
(p.58)

Adela reaumurella
(p.58)

Adela cuprella
(p.58)

Cauchas rufimitrella
(p.59)

Adela croesella
(p.58)

Cauchas fibulella
(p.59)

Nematopogon metaxella
(p.59)

Nematopogon swammerdamella
(p.59)

Nematopogon schwarziellus
(p.59)

Incurvariidae
X

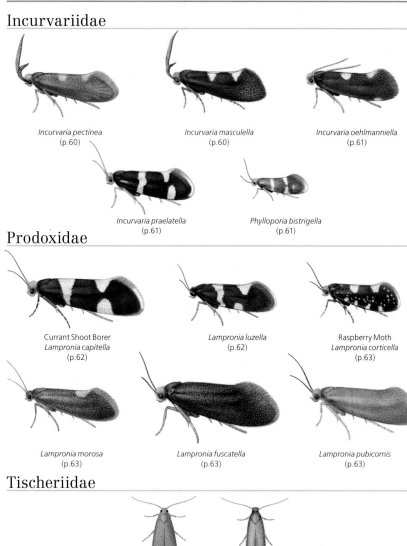

Incurvaria pectinea
(p.60)

Incurvaria masculella
(p.60)

Incurvaria oehlmanniella
(p.61)

Incurvaria praelatella
(p.61)

Phylloporia bistrigella
(p.61)

Prodoxidae
X

Currant Shoot Borer
Lampronia capitella
(p.62)

Lampronia luzella
(p.62)

Raspberry Moth
Lampronia corticella
(p.63)

Lampronia morosa
(p.63)

Lampronia fuscatella
(p.63)

Lampronia pubicornis
(p.63)

Tischeriidae
X

Tischeria ekebladella
(p.64)

Coptotriche marginea
(p.64)

Psychidae Bagworms
moths x 3 cases x 2.

Diplodoma laichartingella
(p.68)

Narycia duplicella
(p.68)

Lesser Lichen Case-bearer
Dahlica inconspicuella
(p.68)

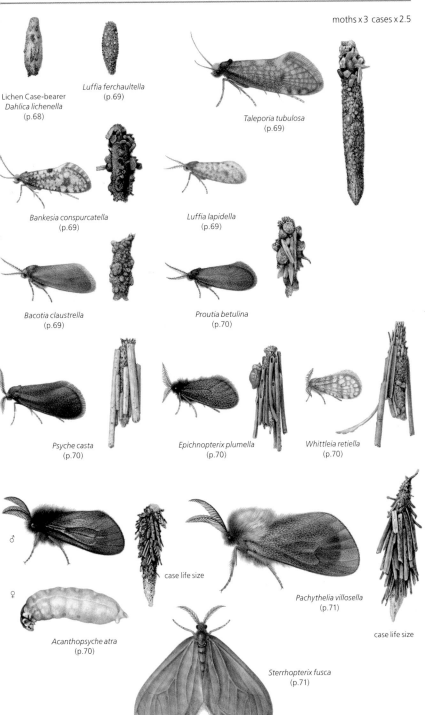

moths x 3 cases x 2.5

Lichen Case-bearer
Dahlica lichenella
(p.68)

Luffia ferchaultella
(p.69)

Taleporia tubulosa
(p.69)

Bankesia conspurcatella
(p.69)

Luffia lapidella
(p.69)

Bacotia claustrella
(p.69)

Proutia betulina
(p.70)

Psyche casta
(p.70)

Epichnopterix plumella
(p.70)

Whittleia retiella
(p.70)

♂

♀

case life size

Pachythelia villosella
(p.71)

Acanthopsyche atra
(p.70)

case life size

Sterrhopterix fusca
(p.71)

Tineidae

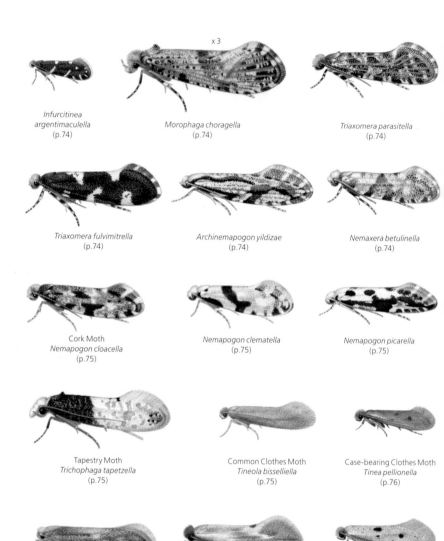

x 3

Infurcitinea
argentimaculella
(p.74)

Morophaga choragella
(p.74)

Triaxomera parasitella
(p.74)

Triaxomera fulvimitrella
(p.74)

Archinemapogon yildizae
(p.74)

Nemaxera betulinella
(p.74)

Cork Moth
Nemapogon cloacella
(p.75)

Nemapogon clematella
(p.75)

Nemapogon picarella
(p.75)

Tapestry Moth
Trichophaga tapetzella
(p.75)

Common Clothes Moth
Tineola bisselliella
(p.75)

Case-bearing Clothes Moth
Tinea pellionella
(p.76)

Large Pale Clothes Moth
Tinea pallescentella
(p.76)

Tinea semifulvella
(p.76)

Tinea trinotella
(p.77)

x 4

Skin Moth
Monopis laevigella
(p.77)

Monopis weaverella
(p.77)

Monopis obviella
(p.77)

Monopis crocicapitella
(p.77)

Monopis imella
(p.78)

Monopis monachella
(p.78)

Psychoides filicivora
(p.78)

Roeslerstammiidae

x 4

Roeslerstammia erxlebella
(p.79)

Bucculatricidae

x 5

Bucculatrix thoracella
(p.81)

Bucculatrix ulmella
(p.81)

Gracillariidae

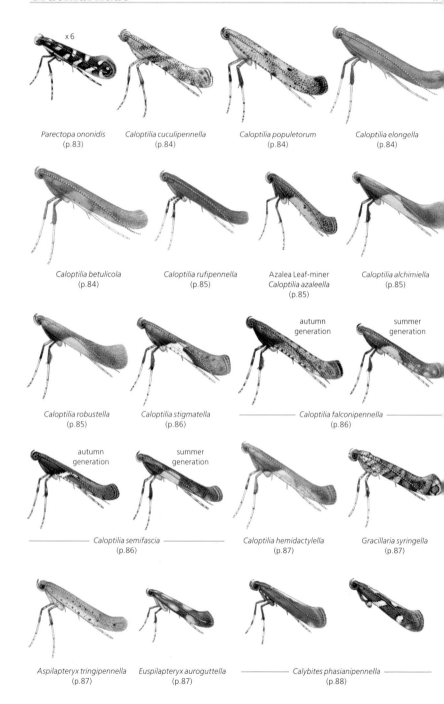

x 6

Parectopa ononidis
(p.83)

Caloptilia cuculipennella
(p.84)

Caloptilia populetorum
(p.84)

Caloptilia elongella
(p.84)

Caloptilia betulicola
(p.84)

Caloptilia rufipennella
(p.85)

Azalea Leaf-miner
Caloptilia azaleella
(p.85)

Caloptilia alchimiella
(p.85)

Caloptilia robustella
(p.85)

Caloptilia stigmatella
(p.86)

autumn
generation

summer
generation

Caloptilia falconipennella
(p.86)

autumn
generation

summer
generation

Caloptilia semifascia
(p.86)

Caloptilia hemidactylella
(p.87)

Gracillaria syringella
(p.87)

Aspilapteryx tringipennella
(p.87)

Euspilapteryx auroguttella
(p.87)

Calybites phasianipennella
(p.88)

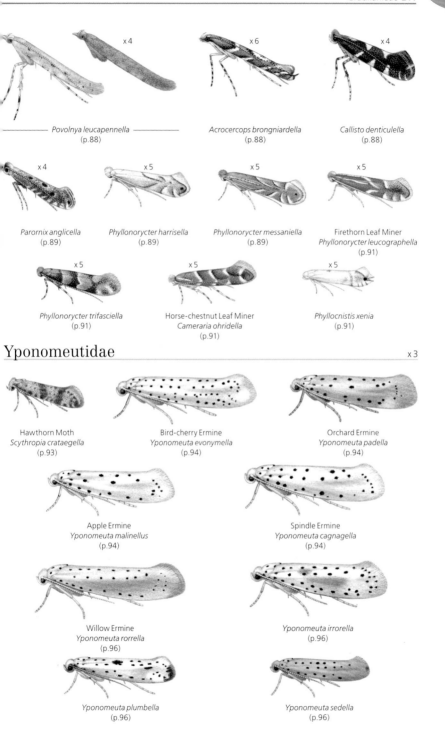

Povolnya leucapennella
(p.88)

x 4

Acrocercops brongniardella
(p.88)

x 6

Callisto denticulella
(p.88)

x 4

Parornix anglicella
(p.89)

x 4

Phyllonorycter harrisella
(p.89)

x 5

Phyllonorycter messaniella
(p.89)

x 5

Firethorn Leaf Miner
Phyllonorycter leucographella
(p.91)

x 5

Phyllonorycter trifasciella
(p.91)

x 5

Horse-chestnut Leaf Miner
Cameraria ohridella
(p.91)

x 5

Phyllocnistis xenia
(p.91)

x 5

Yponomeutidae

x 3

Hawthorn Moth
Scythropia crataegella
(p.93)

Bird-cherry Ermine
Yponomeuta evonymella
(p.94)

Orchard Ermine
Yponomeuta padella
(p.94)

Apple Ermine
Yponomeuta malinellus
(p.94)

Spindle Ermine
Yponomeuta cagnagella
(p.94)

Willow Ermine
Yponomeuta rorrella
(p.96)

Yponomeuta irrorella
(p.96)

Yponomeuta plumbella
(p.96)

Yponomeuta sedella
(p.96)

x 3

Zelleria hepariella
(p.97)

Kesslaria saxifragae
(p.97)

Pseudoswammerdamia combinella
(p.97)

Swammerdamia caesiella
(p.97)

Swammerdamia pyrella
(p.98)

*Swammerdamia
compunctella*
(p.98)

*Paraswammerdamia
albicapitella*
(p.98)

Paraswammerdamia nebulella
(p.99)

Cedestis gysseleniella
(p.99)

Cedestis subfasciella
(p.99)

Ocnerostoma piniariella
(p.99)

Argyresthiidae

x 4

Argyresthia laevigatella
(p.100)

Argyresthia glabratella
(p.100)

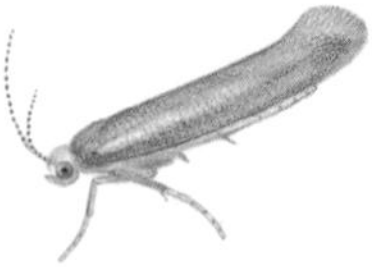

Argyresthia praecocella
(p.101)

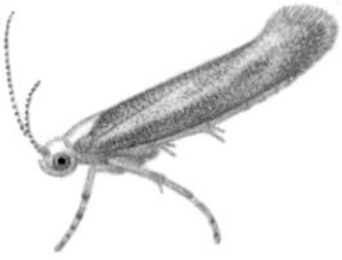

Argyresthia arceuthina
(p.101)

Argyresthia abdominalis
(p.101)

Argyresthia dilectella
(p.101)

Argyresthia aurulentella
(p.101)

Argyresthia ivella
(p.102)

Argyresthia trifasciata
(p.102)

Cypress Tip Moth
Argyresthia cupressella
(p.102)

Argyresthia brockeella
(p.102)

Argyresthia goedartella
(p.102)

x4

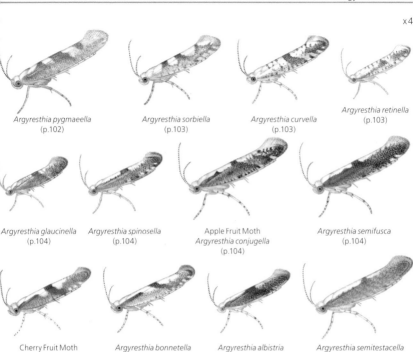

Argyresthia pygmaeella
(p.102)

Argyresthia sorbiella
(p.103)

Argyresthia curvella
(p.103)

Argyresthia retinella
(p.103)

Argyresthia glaucinella
(p.104)

Argyresthia spinosella
(p.104)

Apple Fruit Moth
Argyresthia conjugella
(p.104)

Argyresthia semifusca
(p.104)

Cherry Fruit Moth
Argyresthia pruniella
(p.104)

Argyresthia bonnetella
(p.105)

Argyresthia albistria
(p.105)

Argyresthia semitestacella
(p.105)

Plutellidae

x2.5

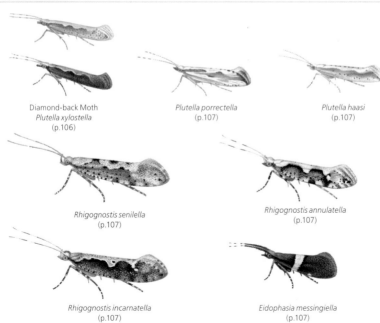

Diamond-back Moth
Plutella xylostella
(p.106)

Plutella porrectella
(p.107)

Plutella haasi
(p.107)

Rhigognostis senilella
(p.107)

Rhigognostis annulatella
(p.107)

Rhigognostis incarnatella
(p.107)

Eidophasia messingiella
(p.107)

Glyphipterigidae

Orthotelia sparganella
(p.109)

Digitivalva perlepidella
(p.109)

Digitivalva pulicariae
(p.109)

Leek Moth
Acrolepiopsis assectella
(p.109)

Acrolepiopsis betulella
(p.109)

Acrolepiopsis marcidella
(p.110)

Acrolepia autumnitella
(p.110)

Cocksfoot Moth
Glyphipterix simpliciella
(p.110)

Glyphipterix schoenicolella
(p.110)

Glyphipterix equitella
(p.110)

Glyphipterix forsterella
(p.110)

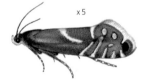

Glyphipterix haworthana
(p.111)

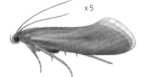

Glyphipterix fuscoviridella
(p.111)

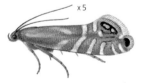

Glyphipterix thrasonella
(p.111)

Ypsolophidae

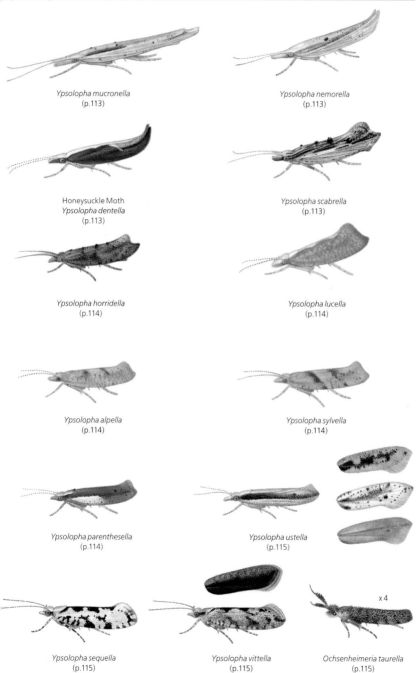

Ypsolopha mucronella
(p.113)

Ypsolopha nemorella
(p.113)

Honeysuckle Moth
Ypsolopha dentella
(p.113)

Ypsolopha scabrella
(p.113)

Ypsolopha horridella
(p.114)

Ypsolopha lucella
(p.114)

Ypsolopha alpella
(p.114)

Ypsolopha sylvella
(p.114)

Ypsolopha parenthesella
(p.114)

Ypsolopha ustella
(p.115)

Ypsolopha sequella
(p.115)

Ypsolopha vittella
(p.115)

x 4

Ochsenheimeria taurella
(p.115)

Praydidae
x 3

Atemelia torquatella
(p.117)

Ash Bud Moth
Prays fraxinella
(p.117)

Prays peregrina
(p.117)

Bedelliidae
x 4

Bedellia somnulentella
(p.118)

Lyonetiidae
x 5

Laburnum Leaf Miner
Leucoptera laburnella
(p.120)

———————— Apple Leaf Miner ————————
Lyonetia clerkella
(p.120)

Douglasiidae
x 5

Tinagma ocnerostomella
(p.122)

Autostichidae
x 3

Oegoconia quadripuncta
(p.123)

Oegoconia deauratella
(p.123)

Oegoconia caradjai
(p.123)

Blastobasidae
x 3

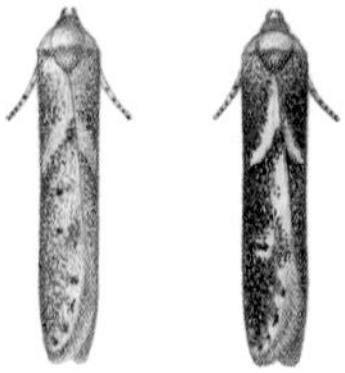

Blastobasis lacticolella
(p.125)

— *Blastobasis adustella* —
(p.125)

Oecophoridae

x 3

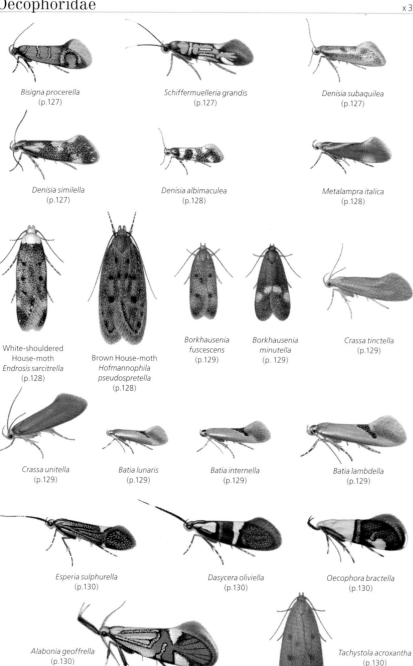

Bisigna procerella
(p.127)

Schiffermuelleria grandis
(p.127)

Denisia subaquilea
(p.127)

Denisia similella
(p.127)

Denisia albimaculea
(p.128)

Metalampra italica
(p.128)

White-shouldered
House-moth
Endrosis sarcitrella
(p.128)

Brown House-moth
Hofmannophila
pseudospretella
(p.128)

Borkhausenia
fuscescens
(p.129)

Borkhausenia
minutella
(p. 129)

Crassa tinctella
(p.129)

Crassa unitella
(p.129)

Batia lunaris
(p.129)

Batia internella
(p.129)

Batia lambdella
(p.129)

Esperia sulphurella
(p.130)

Dasycera oliviella
(p.130)

Oecophora bractella
(p.130)

Alabonia geoffrella
(p.130)

Tachystola acroxantha
(p.130)

x 3

Pleurota bicostella
(p.131)

Pleurota aristella
(p.131)

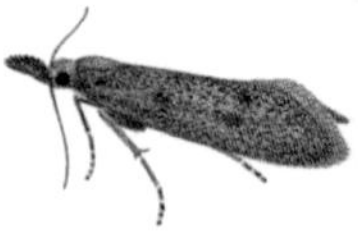

Aplota palpella
(p.131)

Lypusidae

x 3

Pseudatemelia josephinae
(p.132)

Chimabachidae

x 2

——————— *Diurnea fagella* ———————
(p.133)

——— *Diurnea lipsiella* ———
(p.134)

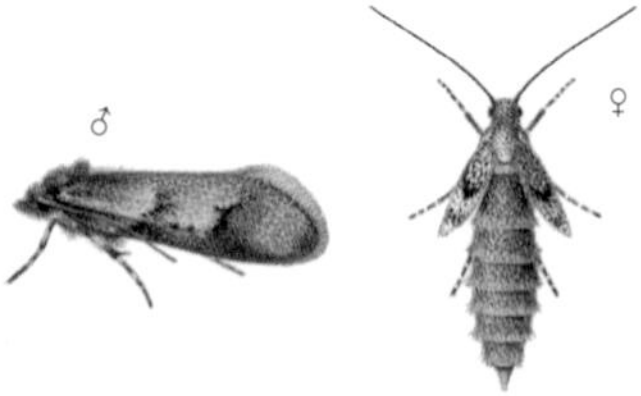

——— *Dasystoma salicella* ———
(p.134)

Peleopodidae

x 3

Carcina quercana
(p.135)

Elachistidae

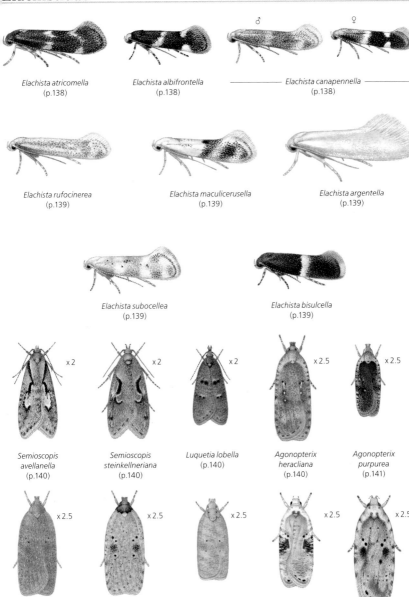

Elachista atricomella
(p.138)

Elachista albifrontella
(p.138)

♂ ♀
————— Elachista canapennella —————
(p.138)

Elachista rufocinerea
(p.139)

Elachista maculicerusella
(p.139)

Elachista argentella
(p.139)

Elachista subocellea
(p.139)

Elachista bisulcella
(p.139)

x 2
Semioscopis
avellanella
(p.140)

x 2
Semioscopis
steinkellneriana
(p.140)

x 2
Luquetia lobella
(p.140)

x 2.5
Agonopterix
heracliana
(p.140)

x 2.5
Agonopterix
purpurea
(p.141)

x 2.5 x 2.5
——— Agonopterix subpropinquella ———
(p.141)

x 2.5
Agonopterix
nanatella
(p.141)

x 2.5
Agonopterix
alstromeriana
(p.142)

x 2.5
Agonopterix
arenella
(p.142)

x2.5

*Agonopterix
kaekeritziana*
(p.142)

*Agonopterix
pallorella*
(p.142)

*Agonopterix
ocellana*
(p.143)

*Agonopterix
assimilella*
(p.143)

*Agonopterix
umbellana*
(p.143)

*Agonopterix
nervosa*
(p.143)

*Agonopterix
liturosa*
(p.144)

*Agonopterix
yeatiana*
(p.144)

*Agonopterix
rotundella*
(p.144)

*Depressaria
pulcherrimella*
(p.145)

*Depressaria
douglasella*
(p.145)

Telechrysis tripuncta
(p.145)

*Depressaria
daucella*
(p.144)

Parsnip Moth
*Depressaria
radiella*
(p.144)

x 3

Hypercallia citrinalis
(p.145)

x 2.5

Ethmia terminella
(p.146)

Ethmia dodecea
(p.146)

Ethmia quadrillella
(p.146)

Ethmia bipunctella
(p.146)

Ethmia pyrausta
(p.146)

x 4

*Blastodacna
hellerella*
(p.146)

x 4

Apple Pith Moth
Blastodacna atra
(p.147)

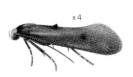

x 4

Spuleria flavicaput
(p.147)

x 4

*Dystebenna
stephensi*
(p.147)

x 4

Chrysoclista lathamella
(p.147)

x 4

Chrysoclista linneella
(p.147)

Stathmopodidae

x 4

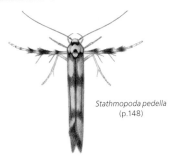

Stathmopoda pedella
(p.148)

Batrachedridae

x 4

Batrachedra praeangusta
(p.149)

Batrachedra pinicolella
(p.149)

Coleophoridae

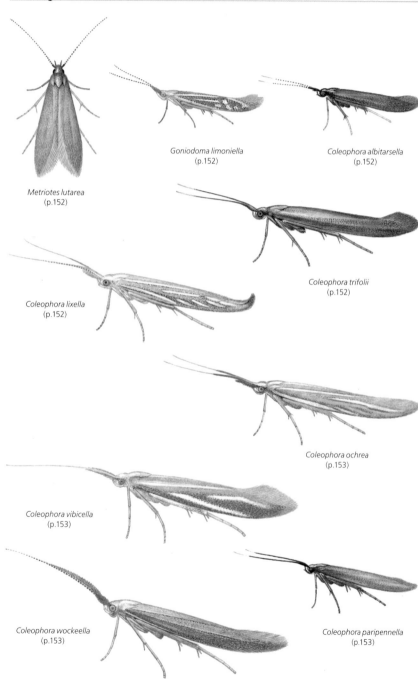

Metriotes lutarea
(p.152)

Goniodoma limoniella
(p.152)

Coleophora albitarsella
(p.152)

Coleophora trifolii
(p.152)

Coleophora lixella
(p.152)

Coleophora ochrea
(p.153)

Coleophora vibicella
(p.153)

Coleophora wockeella
(p.153)

Coleophora paripennella
(p.153)

x4

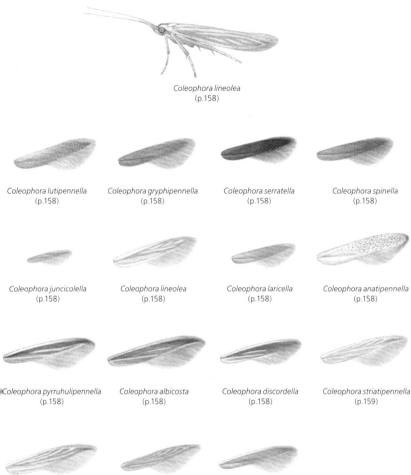

Coleophora lineolea
(p.158)

Coleophora lutipennella
(p.158)

Coleophora gryphipennella
(p.158)

Coleophora serratella
(p.158)

Coleophora spinella
(p.158)

Coleophora juncicolella
(p.158)

Coleophora lineolea
(p.158)

Coleophora laricella
(p.158)

Coleophora anatipennella
(p.158)

Coleophora pyrruhulipennella
(p.158)

Coleophora albicosta
(p.158)

Coleophora discordella
(p.158)

Coleophora striatipennella
(p.159)

Coleophora peribenanderi
(p.159)

Coleophora otidipennella
(p.159)

Coleophora alticolella
(p.159)

Coleophora hemerobiella
(p.159)

Coleophora salicorniae
(p.159)

Coleophora clypeiferella
(p.159)

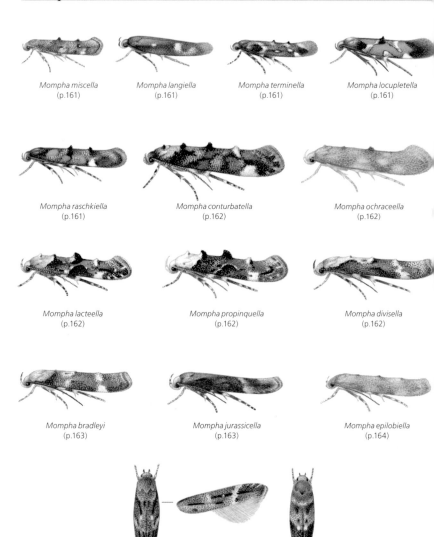

Mompha miscella
(p.161)

Mompha langiella
(p.161)

Mompha terminella
(p.161)

Mompha locupletella
(p.161)

Mompha raschkiella
(p.161)

Mompha conturbatella
(p.162)

Mompha ochraceella
(p.162)

Mompha lacteella
(p.162)

Mompha propinquella
(p.162)

Mompha divisella
(p.162)

Mompha bradleyi
(p.163)

Mompha jurassicella
(p.163)

Mompha epilobiella
(p.164)

Mompha sturnipennella
(p.163)

Mompha subbistrigella
(p.164)

Scythrididae

x 4

Scythris grandipennis
(p.166)

Scythris crassiuscula
(p.166)

Scythris picaepennis
(p.166)

Scythris limbella
(p.166)

Cosmopterigidae

x 6

Pancalia leuwenhoekella
(p.168)

Pancalia schwarzella
(p.168)

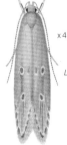

x 4

Limnaecia phragmitella
(p.169)

Cosmopterix zieglerella
(p.169)

Cosmopterix orichalcea
(p.169)

Cosmopterix scribaiella
(p.170)

Cosmopterix pulchrimella
(p.170)

Cosmopterix lienigiella
(p.170)

Gelechiidae

Aristotelia ericinella
(p.175)

Aristotelia brizella
(p.175)

*Metzneria
lappella*
(p.175)

*Metzneria
metzneriella*
(p.175)

*Metzneria
aprilella*
(p.175)

Apodia bifractella
(p.176)

Ptocheuusa paupella
(p.176)

Argolamprotes micella
(p.176)

*Monochroa
palustrellus*
(p.176)

*Bryotropha
affinis*
(p.176)

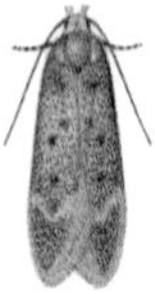

*Bryotropha
senectella*
(p.177)

*Bryotropha
terrella*
(p.177)

*Bryotropha
politella*
(p.177)

*Bryotropha
domestica*
(p.178)

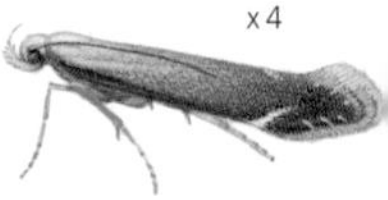

Monochroa cytisella
(p.176)

Recurvaria leucatella
(p.178)

Stenolechia gemmella
(p.178)

Parachronistis albiceps
(p.178)

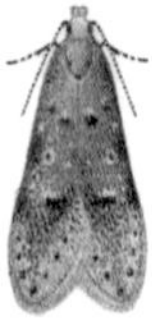

Teleiodes vulgella
(p.178)

×3

Teleiodes luculella
(p.178)

Teleiopsis diffinis
(p.179)

*Carpatolechia
proximella*
(p.179)

*Carpatolechia
alburnella*
(p.179)

*Pseudotelphusa
scalella*
(p.179)

*Pseudotelphusa
paripunctella*
(p.179)

Altenia scriptella
(p.180)

Gelechia rhombella
(p.180)

*Gelechia
senticetella*
(p.180)

*Gelechia
sororculella*
(p.180)

*Gelechia
nigra*
(p.180)

*Gelechia
turpella*
(p.181)

*Mirificarma
mulinella*
(p.181)

Sophronia semicostella
(p.181)

*Aroga
velocella*
(p.181)

*Prolita
sexpunctella*
(p.181)

*Prolita
solutella*
(p.181)

*Athrips
mouffetella*
(p.182)

*Scrobipalpa
acuminatella*
(p.182)

*Scrobipalpa
costella*
(p.182)

*Syncopacma
larseniella*
(p.182)

*Syncopacma
taeniolella*
(p.182)

*Aproaerema
anthyllidella*
(p.183)

x ?

*Anacampsis
populella*
(p.183)

*Anacampsis
blattariella*
(p.183)

*Anarsia
spartiella*
(p.183)

*Hypatima
rhomboidella*
(p.184)

*Nothris
congressariella*
(p.184)

*Neofaculta
ericetella*
(p.184)

Juniper Webber
*Dichomeris
marginella*
(p.184)

*Dichomeris
juniperella*
(p.184)

*Dichomeris
ustalella*
(p.184)

Brachmia blandella
(p.185)

*Helcystogramma
rufescens*
(p.185)

*Acompsia
schmidtiellus*
(p.185)

Hollyhock Seed Moth
Pexicopia malvella
(p.185)

*Platyedra
subcinerea*
(p.185)

Alucitidae

x 2.5

Twenty-plume Moth
Alucita hexadactyla
(p.186)

Pterophoridae Plume moths

x 1.5

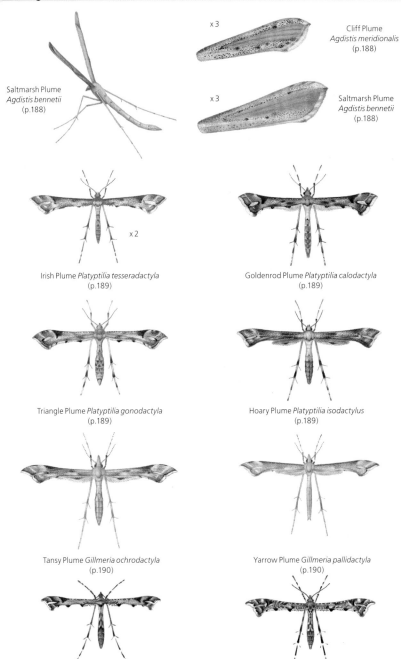

x 3

Cliff Plume
Agdistis meridionalis
(p.188)

Saltmarsh Plume
Agdistis bennetii
(p.188)

x 3

Saltmarsh Plume
Agdistis bennetii
(p.188)

x 2

Irish Plume *Platyptilia tesseradactyla*
(p.189)

Goldenrod Plume *Platyptilia calodactyla*
(p.189)

Triangle Plume *Platyptilia gonodactyla*
(p.189)

Hoary Plume *Platyptilia isodactylus*
(p.189)

Tansy Plume *Gillmeria ochrodactyla*
(p.190)

Yarrow Plume *Gillmeria pallidactyla*
(p.190)

Beautiful Plume *Amblyptilia acanthadactyla*
(p.190)

Brindled Plume *Amblyptilia punctidactyla*
(p.190)

x 1.

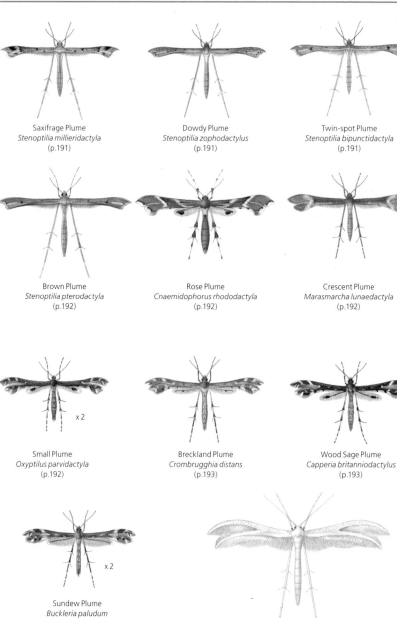

Saxifrage Plume
Stenoptilia millieridactyla
(p.191)

Dowdy Plume
Stenoptilia zophodactylus
(p.191)

Twin-spot Plume
Stenoptilia bipunctidactyla
(p.191)

Brown Plume
Stenoptilia pterodactyla
(p.192)

Rose Plume
Cnaemidophorus rhododactyla
(p.192)

Crescent Plume
Marasmarcha lunaedactyla
(p.192)

Small Plume
Oxyptilus parvidactyla
(p.192)

Breckland Plume
Crombrugghia distans
(p.193)

Wood Sage Plume
Capperia britanniodactylus
(p.193)

Sundew Plume
Buckleria paludum
(p.193)

White Plume Moth
Pterophorus pentadactyla
(p.194)

x 1.5

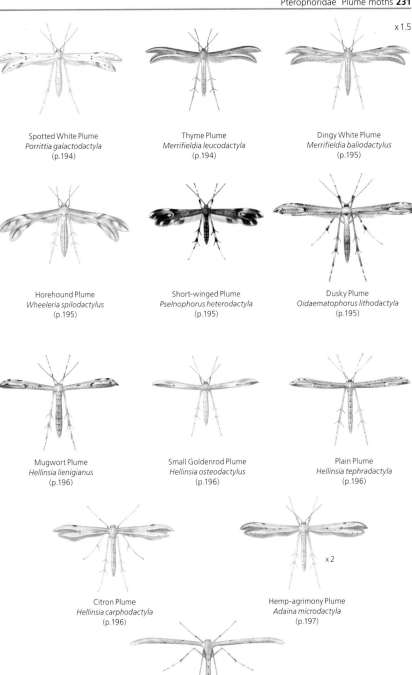

Spotted White Plume
Porrittia galactodactyla
(p.194)

Thyme Plume
Merrifieldia leucodactyla
(p.194)

Dingy White Plume
Merrifieldia baliodactylus
(p.195)

Horehound Plume
Wheeleria spilodactylus
(p.195)

Short-winged Plume
Pselnophorus heterodactyla
(p.195)

Dusky Plume
Oidaematophorus lithodactyla
(p.195)

Mugwort Plume
Hellinsia lienigianus
(p.196)

Small Goldenrod Plume
Hellinsia osteodactylus
(p.196)

Plain Plume
Hellinsia tephradactyla
(p.196)

Citron Plume
Hellinsia carphodactyla
(p.196)

Hemp-agrimony Plume
Adaina microdactyla
(p.197)

x 2

Common Plume
Emmelina monodactyla
(p.197)

Schreckensteiniidae

x 4

Schreckensteinia festaliella
(p.198)

Epermeniidae

x 4

Phaulernis dentella
(p.199)

Phaulernis fulviguttella
(p.199)

Epermenia farreni
(p.200)

Epermenia profugella
(p.200)

Epermenia falciformis
(p.200)

Epermenia insecurella
(p.201)

Epermenia chaerophyllella
(p.201)

Epermenia aequidentellus
(p.201)

Choreutidae

x 3

Nettle-tap
Anthophila fabriciana
(p.202)

*Prochoreutis
sehestediana*
(p.202)

*Prochoreutis
myllerana*
(p.203)

Tebenna micalis
(p.203)

Apple Leaf Skeletoniser
Choreutis pariana
(p.203)

Choreutis diana
(p.203)

Tortricidae

Phtheochroa inopiana
(p.261)

Phtheochroa sodaliana
(p.261)

Phtheochroa rugosana
(p.262)

Hysterophora maculosana
(p.262)

Cochylimorpha alternana
(p.262)

Cochylimorpha straminea
(p.262)

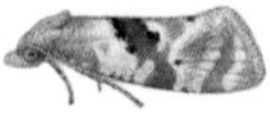

Phalonidia manniana
(p.262)

Phalonidia affinitana
(p.263)

Phalonidia gilvicomana
(p.263)

Phalonidia curvistrigana
(p.263)

Gynnidomorpha minimana
(p.263)

Gynnidomorpha permixtana
(p.263)

Gynnidomorpha vectisana
(p.264)

Gynnidomorpha alismana
(p.264)

Gynnidomorpha luridana
(p.264)

f. *fasciella*

Agapeta hamana
(p.264)

Agapeta zoegana
(p.264)

———— *Eupoecilia angustana* ————
(p.265)

Vine Moth *Eupoecilia ambiguella*
(p.265)

Commophila aeneana
(p.265)

x 2.5

——————— *Aethes tesserana* ———————
(p.265)

Aethes hartmanniana
(p.265)

Aethes piercei
(p.266)

Aethes williana
(p.266)

Aethes cnicana
(p.266)

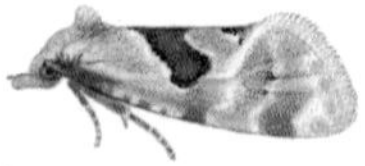

Aethes rubigana
(p.266)

Aethes smeathmanniana
(p.266)

Aethes margaritana
(p.267)

Aethes dilucidana
(p.267)

Aethes francillana
(p.267)

Aethes beatricella
(p.267)

Cochylidia implicitana
(p.267)

Cochylidia heydeniana
(p.268)

Cochylidia subroseana
(p.268)

Cochylidia rupicola
(p.268)

Cochylis roseana
(p.268)

Cochylis flaviciliana
(p.269)

Cochylis dubitana
(p.269)

Cochylis molliculana
(p.269)

Cochylis hybridella
(p.270)

Cochylis atricapitana
(p.270)

Cochylis pallidana
(p.270)

Cochylis nana
(p.270)

Falseuncaria ruficiliana
(p.270)

Falseuncaria degreyana
(p.271)

x2.5

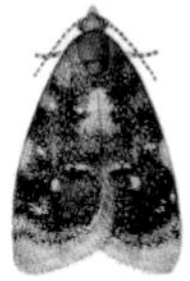

Spatalistis bifasciana
(p.271)

Green Oak Tortrix
Tortrix viridana
(p.271)

——— *Aleimma loeflingiana* ———
(p.271)

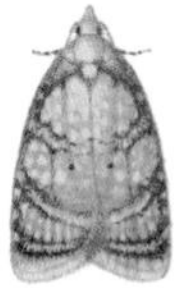

Acleris bergmanniana
(p.271)

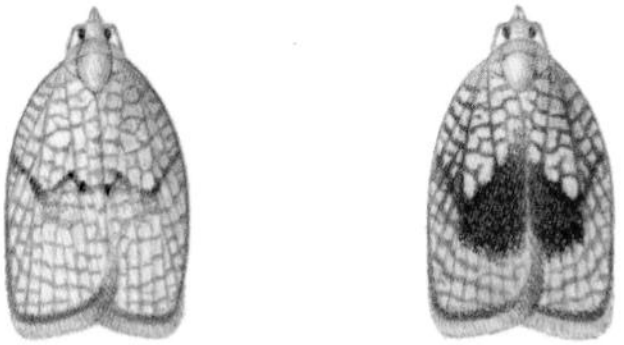

——————— *Acleris forsskaleana* ———————
(p.272)

Acleris holmiana
(p.272)

Acleris comariana forewing

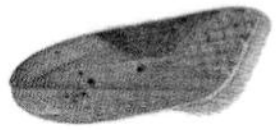

Acleris caledoniana forewing
(p.273)

Acleris laterana
(p.272)

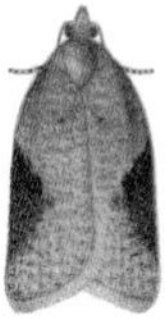

Strawberry Tortrix
Acleris comariana
(p.272)

Acleris sparsana
(p.273)

♂   ♀

——— Rhomboid Tortrix ———
Acleris rhombana
(p.273)

——— *Acleris aspersana* ———
(p.273)

Acleris ferrugana
(p.274)

Acleris notana
(p.274)

Acleris shepherdana
(p.274)

——— *Acleris schalleriana* ———
(p.274)

x 2.5

——————— Garden Rose Tortrix ———————
Acleris variegana
(p.275)

Acleris permutana
(p.275)

——— *Acleris kochiella* ———
(p.275)

Acleris logiana
(p.276)

——— *Acleris umbrana* ———
(p.276)

——————— *Acleris hastiana* ———————
(p.276)

——————————— *Acleris cristana* ———————————
(p.277)

——— *Acleris hyemana* ———
(p.277)

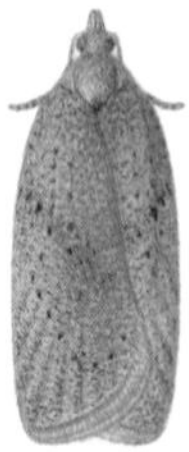

Acleris lipsiana
(p.277)

——— *Acleris rufana* ———
(p.277)

x 2.5

Acleris lorquiniana
(p.278)

Acleris abietana
(p.278)

Acleris maccana
(p.278)

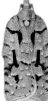

——— *Acleris literana* ———
(p.278)

——— *Acleris emargana* ———
(p.279)

Acleris effractana
(p.279)

Neosphaleroptera nubilana
(p.279)

Exapate congelatella
(p.279)

Tortricodes alternella
(p.280)

x 2.5

Eana osseana
(p.280)

Eana incanana
(p.280)

f. *bellana*

ssp. *colquhounana*

—— *Eana penziana* ——
(p.280)

Cnephasia sp.
(p.281)

Grey Tortrix
Cnephasia stephensiana
dark form
(p.282)

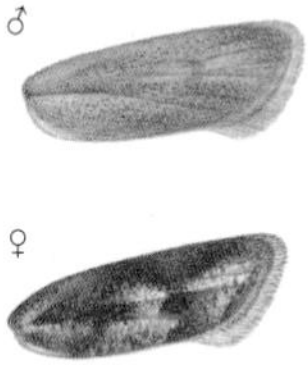

♂

♀

Cnephasia longana
(p.281)

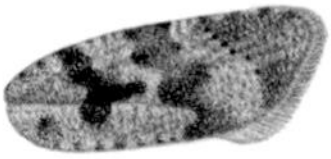

Cnephasia communana
(p.282)

Cnephasia conspersana
(p.282)

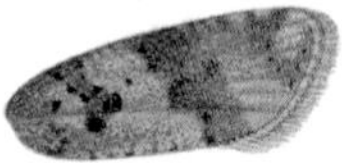

Cnephasia stephensiana
(p.282)

Flax Tortrix
Cnephasia asseclana
(p.282)

Cnephasia pasiuana
(p.282)

Cnephasia pumicana
(p.283)

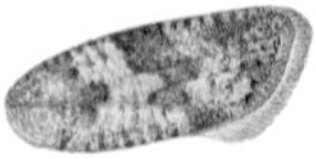

Cnephasia genitalana
(p.283)

Light Grey Tortrix
Cnephasia incertana
(p.283)

x 2.5

Eulia ministrana
(p.284)

Sparganothis pilleriana
(p.283)

*Pseudargyrotoza
conwagana*
(p.284)

────── Red-barred Tortrix ──────
Ditula angustiorana
(p.284)

Epagoge grotiana
(p.284)

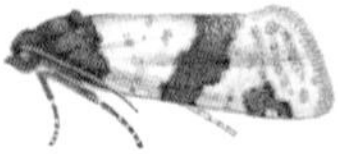

Periclepsis cinctana
(p.284)

────── *Philedone gerningana* ──────
(p.285)

Capua vulgana
(p.285)

── *Philedonides lunana* ──
(p.285)

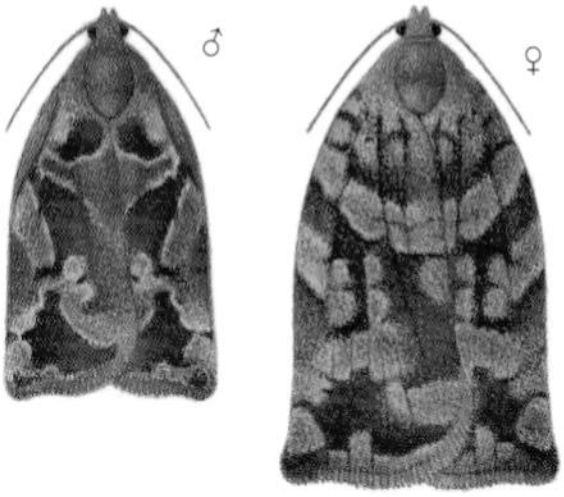

────── *Archips oporana* ──────
(p.286)

────── Large Fruit-tree Tortrix ──────
Archips podana
(p.286)

Brown Oak Tortrix
Archips crataegana
(p.286)

Variegated Golden Tortrix
Archips xylosteana
(p.286)

────── Rose Tortrix ──────
Archips rosana
(p.287)

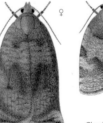

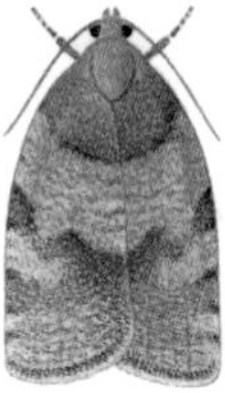

*Choristoneura
diversana*
(p.287)

x 2.5

Argyrotaenia ljungiana
(p.287)

Choristoneura hebenstreitella
(p.287)

Ptycholomoides aeriferanus
(p.288)

Ptycholoma lecheana
(p.288)

Chequered Fruit-tree Tortrix
Pandemis corylana
(p.288)

————— Barred Fruit-tree Tortrix —————
Pandemis cerasana
(p.288)

Pandemis cinnamomeana
(p.289)

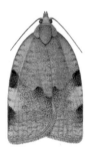

Dark Fruit-tree Tortrix
Pandemis heparana
(p.289)

Pandemis dumetana
(p.289)

Syndemis musculana
(p.289)

Lozotaenia forsterana
(p.290)

——————— Carnation Tortrix ———————
Cacoecimorpha pronubana
(p.290)

Bilberry Tortrix
Aphelia viburnana
(p.290)

x2.5

Timothy Tortrix
Aphelia paleana
(p.291)

Aphelia unitana
(p.291)

Clepsis senecionana
(p.291)

Clepsis rurinana
(p.291)

─────── *Clepsis consimilana* ───────
(p.292)

Cyclamen Tortrix
Clepsis spectrana
(p.292)

Lozotaeniodes formosana
(p.292)

─────── Light Brown Apple Moth ───────
Epiphyas postvittana
(p.292)

Summer Fruit Tortrix
Adoxophyes orana
(p.293)

Olindia schumacherana
(p.293)

Isotrias rectifasciana
(p.293)

x 2.

Bactra furfurana
(p.293)

Bactra lancealana
(p.293)

Bactra lacteana
(p.294)

Bactra robustana
(p.294)

Endothenia gentianaeana
(p.294)

Endothenia oblongana
(p.295)

Endothenia marginana
(p.295)

Endothenia ustulana
(p.295)

Endothenia nigricostana
(p.295)

Endothenia ericetana
(p.296)

Endothenia quadrimaculana
(p.296)

Eudemis profundana
(p.296)

Eudemis porphyrana
(p.296)

Pseudosciaphila branderiana
(p.296)

Apotomis semifasciana
(p.297)

Apotomis lineana
(p.297)

Apotomis turbidana
(p.297)

Apotomis betuletana
(p.298)

Apotomis capreana
(p.298)

Apotomis sororculana
(p.298)

Apotomis sauciana
(p.298)

Orthotaenia undulana
(p.298)

x 2.5

Plum Tortrix
Hedya pruniana
(p.299)

Marbled Orchard Tortrix
Hedya nubiferana
(p.299)

Hedya ochroleucana
(p.299)

Hedya salicella
(p.300)

Metendothenia atropunctana
(p.300)

Celypha striana
(p.300)

Celypha rosaceana
(p.300)

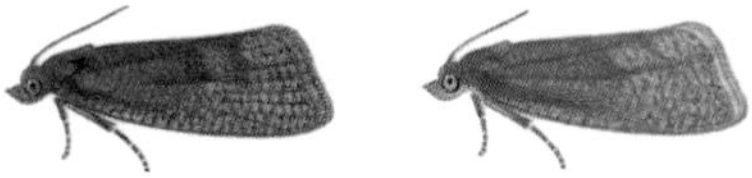

———— *Celypha rufana* ————
(p.300)

Celypha woodiana
(p.301)

Celypha cespitana
(p.301)

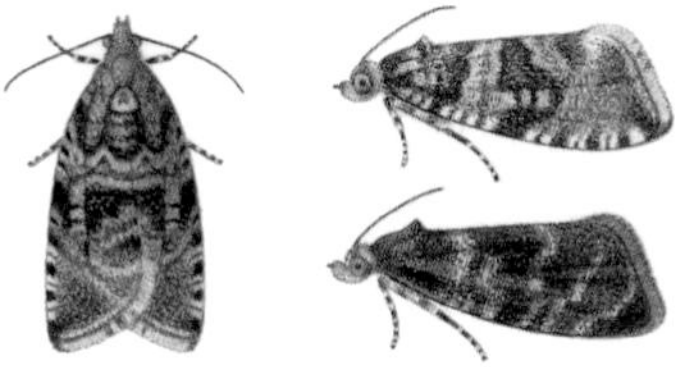

Celypha rivulana
(p.302)

Celypha aurofasciana
(p.302)

———— *Celypha lacunana* ————
(p.301)

Phiaris metallicana
(p.302)

Phiaris schulziana
(p.302)

Phiaris palustrana
(p.303)

Phiaris micana
(p.303)

———— *Phiaris obsoletana* ————
(p.303)

Argyroploce arbutella
(p.303)

Stictea mygindiana
(p.303)

x2.5

Olethreutes arcuella
(p.304)

Piniphila bifasciana
(p.304)

Lobesia occidentis
(p.304)

Lobesia reliquana
(p.304)

Lobesia abscisana
(p.304)

Lobesia littoralis
(p.305)

Eucosmomorpha albersana
(p.305)

Cherry Bark Tortrix
Enarmonia formosana
(p.305)

Ancylis achatana
(p.305)

Ancylis comptana
(p.306)

Ancylis unguicella
(p.306)

Ancylis uncella
(p.306)

Ancylis geminana
(p.306)

Ancylis diminutana
(p.306)

Ancylis subarcuana
(p.307)

Ancylis mitterbacheriana
(p.307)

Ancylis upupana
(p.307)

Ancylis obtusana
(p.307)

Ancylis laetana
(p.307)

Ancylis tineana
(p.307)

Ancylis unculana
(p.308)

Ancylis badiana
(p.308)

Ancylis paludana
(p.308)

Ancylis myrtillana
(p.309)

Ancylis apicella
(p.309)

Eriopsela quadrana
(p.309)

Thiodia citrana
(p.310)

Rhopobota ustomaculana
(p.310)

x 2.5

Holly Tortrix
Rhopobota naevana
(p.310)

———— *Rhopobota stagnana* ————
(p.310)

Rhopobota myrtillana
(p.311)

Bud Moth
Spilonota ocellana
(p.311)

Spilonota laricana
(p.311)

Acroclita subsequana
(p.311)

———— *Epinotia pygmaeana* ————
(p.311)

Epinotia subsequana
(p.312)

Epinotia subocellana
(p.312)

Epinotia bilunana
(p.312)

———— *Epinotia ramella* ————
(p. 312)

Epinotia demarniana
(p.312)

———— *Epinotia immundana* ————
(p.313)

Epinotia tetraquetrana
(p.313)

form *cinereana*

———————— *Epinotia nisella* ————————
(p.313)

———— Nut Bud Moth ————
Epinotia tenerana
(p.313)

Epinotia nemorivaga
(p.314)

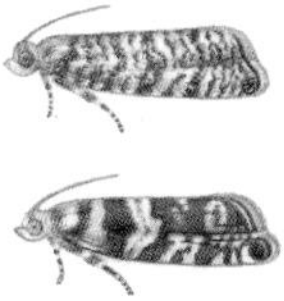

Epinotia tedella
(p.314)

x2.5

Epinotia fraternana
(p.314)

Epinotia signatana
(p.314)

Epinotia nanana
(p.315)

Epinotia rubiginosana
(p.315)

Willow Tortrix
Epinotia cruciana
(p.315)

Epinotia mercuriana
(p.316)

Epinotia crenana
(p.316)

Epinotia abbreviana
(p.316)

Epinotia abbreviana
(p.316)

Epinotia trigonella
(p.316)

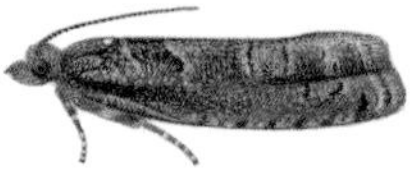

Epinotia maculana
(p.316)

Epinotia sordidana
(p.317)

Epinotia caprana
(p.317)

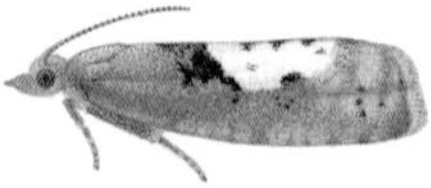

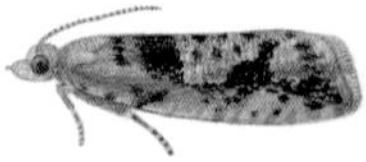

Epinotia brunnichana
(p.318)

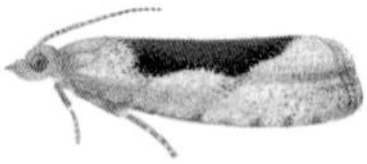

Epinotia solandriana
(p.318)

Spruce Bud Moth
Zeiraphera ratzeburgiana
(p.318)

Zeiraphera rufimitrana
(p.318)

Zeiraphera isertana
(p.319)

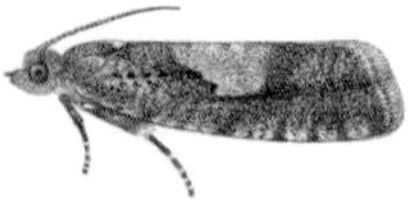

Larch Tortrix
Zeiraphera griseana
(p.319)

Crocidosema plebejana
(p.319)

Phaneta pauperana
(p.319)

Pelochrista caecimaculana
(p.320)

x 2.5

Eucosma aspidiscana
(p.320)

Eucosma conterminana
(p.320)

Eucosma tripoliana
(p.320)

Eucosma aemulana
(p.320)

Eucosma lacteana
(p.321)

Eucosma metzneriana
(p.321)

Eucosma campoliliana
(p.321)

Eucosma pupillana
(p.321)

Eucosma hohenwartiana
(p.321)

Eucosma cana
(p.322)

Eucosma obumbratana
(p.322)

Gypsonoma aceriana
(p.322)

Gypsonoma sociana
(p.323)

———— *Gypsonoma dealbana* ————
(p.323)

Gypsonoma oppressana
(p.323)

Gypsonoma minutana
(p.324)

Epiblema grandaevana
(p.324)

Epiblema turbidana
(p.324)

Epiblema foenella
(p.324)

Epiblema scutulana
(p.324)

Epiblema cirsiana
(p.325)

Epiblema cnicicolana
(p.325)

Epiblema sticticana
(p.325)

Epiblema costipunctana
(p.325)

Notocelia cynosbatella
(p.326)

Bramble Shoot Moth
Notocelia uddmanniana
(p.326)

x2.5

Notocelia trimaculana
(p.326)

Notocelia rosaecolana
(p.326)

Notocelia roborana
(p.327)

Notocelia incarnatana
(p.327)

Notocelia tetragonana
(p.327)

Pseudococcyx postica
(p.327)

Pine Bud Moth
Pseudococcyx turionella
(p.328)

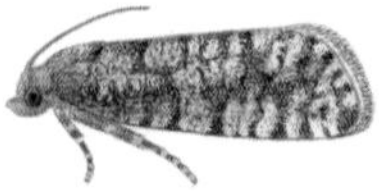

Pine Resin-gall Moth
Retinea resinella
(p.328)

Clavigesta sylvestrana
(p.328)

Pine Leaf-mining Moth
Clavigesta purdeyi
(p.328)

Pine Shoot Moth
Rhyacionia buoliana
(p.328)

Rhyacionia pinicolana
(p.329)

Spotted Shoot Moth
Rhyacionia pinivorana
(p.329)

Elgin Shoot Moth
Rhyacionia logaea
(p.329)

Dichrorampha petiverella
(p.331)

Dichrorampha alpinana
(p.331)

Dichrorampha plumbagana
(p.331)

Dichrorampha senectana
(p.332)

Dichrorampha sequana
(p.332)

Dichrorampha acuminatana
(p.332)

Dichrorampha consortana
(p.332)

Dichrorampha simpliciana
(p.333)

Dichrorampha sylvicolana
(p.333)

Dichrorampha montanana
(p.333)

Dichrorampha vancouverana
(p.333)

Dichrorampha plumbana
(p.333)

Spruce Seed Moth
Cydia strobilella
(p.334)

Cydia ulicetana
(p.334)

x 2.5

Cydia microgrammana
(p.335)

Cydia servillana
(p.335)

Pea Moth
Cydia nigricana
(p.335)

Cydia millenniana
(p.336)

Cydia fagiglandana
(p.336)

——————— *Cydia splendana* ———————
(p.336)

Codling Moth
Cydia pomonella
(p.336)

Cydia amplana
(p.337)

Cydia cognatana
(p.337)

Cydia pactolana
(p.337)

Cydia illutana
(p.337)

Cydia cosmophorana
(p.337)

Cydia coniferana
(p.338)

Cydia conicolana
(p.338)

Lathronympha strigana
(p.338)

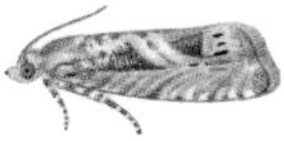

Selania leplastriana
(p.339)

Grapholita caecana
(p.339)

Grapholita compositella
(p.339)

Grapholita internana
(p.339)

Grapholita pallifrontana
(p.340)

Grapholita gemmiferana
(p.340)

Grapholita janthinana
(p.340)

*Grapholita
tenebrosana*
(p.341)

Plum Fruit Moth
Grapholita funebrana
(p.341)

Grapholita lobarzewskii
(p.341)

Grapholita lathyrana
(p.341)

Grapholita jungiella
(p.342)

x2.5

Grapholita lunulana
(p.342)

Grapholita orobana
(p.342)

Pammene splendidulana
(p.343)

Pammene giganteana
(p.344)

Pammene argyrana
(p.344)

Pammene albuginana
(p.344)

Pammene spiniana
(p.344)

Pammene populana
(p.345)

Pammene aurita
(p.345)

Pammene regiana
(p.345)

Pammene trauniana
(p.346)

Pammene fasciana
(p.346)

Pammene herrichiana
(p.346)

Pammene germmana
(p.346)

Pammene ochsenheimeriana
(p.346)

Fruitlet Mining Tortrix
Pammene rhediella
(p.347)

Pammene gallicana
(p.347)

Pammene aurana
(p.347)

Strophedra weirana
(p.347)

Strophedra nitidana
(p.347)

Pyralidae

x 1.5

Bee Moth
Aphomia sociella
(p.349)

Aphomia zelleri
(p.349)

Lesser Wax Moth
Achroia grisella
(p.349)

Wax Moth
Galleria mellonella
(p.350)

Synaphe punctalis
(p.350)

Meal Moth
Pyralis farinalis
(p.350)

Pyralis lienigialis
(p.350)

Large Tabby
Aglossa pinguinalis
(p.351)

Gold Triangle
Hypsopygia costalis
(p.351)

Hypsopygia glaucinalis
(p.351)

Endotricha flammealis
(p.351)

x 2
Cryptoblabes bistriga
(p.352)

x 2
Salebriopsis albicilla
(p.352)

x 2
Elegia similella
(p.352)

x 2
Ortholepis betulae
(p.352)

x 2
Matilella fusca
(p.352)

x 2
Moitrelia obductella
(p.353)

x 2
Pempeliella ornatella
(p.353)

x 2
Delplanqueia dilutella
(p.353)

x 2
Irish specimen

x 2
Sciota hostilis
(p.353)

x 2
Sciota adelphella
(p.353)

x 2
Pima boisduvaliella
(p.354)

x 2

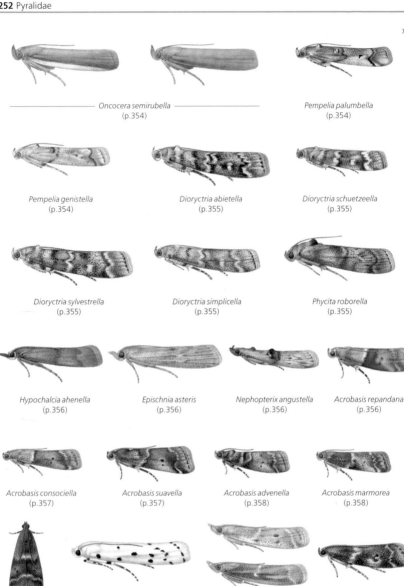

Oncocera semirubella
(p.354)

Pempelia palumbella
(p.354)

Pempelia genistella
(p.354)

Dioryctria abietella
(p.355)

Dioryctria schuetzeella
(p.355)

Dioryctria sylvestrella
(p.355)

Dioryctria simplicella
(p.355)

Phycita roborella
(p.355)

Hypochalcia ahenella
(p.356)

Epischnia asteris
(p.356)

Nephopterix angustella
(p.356)

Acrobasis repandana
(p.356)

Acrobasis consociella
(p.357)

Acrobasis suavella
(p.357)

Acrobasis advenella
(p.358)

Acrobasis marmorea
(p.358)

Apomyelois
bistriatella
(p.358)

Thistle Ermine
Myelois circumvoluta
(p.359)

Gymnancyla canella
(p.359)

Assara terebrella
(p.359)

Euzophera cinerosella
(p.359)

Euzophera pinguis
(p.360)

Nyctegretis lineana
(p.360)

Ancylosis oblitella
(p.360)

x 2

Homoeosoma nebulella (p.360)

Homoeosoma sinuella (p.360)

Phycitodes binaevella (p.361)

Phycitodes saxicola (p.361)

Phycitodes maritima (p.362)

Vitula biviella (p.362)

Indian Meal Moth *Plodia interpunctella* (p.362)

Cacao Moth *Ephestia elutella* (p.362)

Ephestia unicolorella ssp. *woodiella* (p.363)

Mediterranean Flour Moth *Ephestia kuehniella* (p.363)

Rhodophaea formosa (p.363)

Anerastia lotella (p.363)

Crambidae

x 2

Scoparia subfusca (p.365)

Scoparia pyralella (p.365)

Scoparia ambigualis (p.366)

Scoparia basistrigalis (p.366)

Scoparia ancipitella (p.366)

Eudonia lacustrata (p.367)

— *Eudonia pallida* — (p.367)

— *Eudonia alpina* — (p.367)

Eudonia murana (p.367)

Eudonia truncicolella (p.368)

Eudonia lineola (p.368)

Eudonia angustea (p.368)

Eudonia delunella (p.369)

Eudonia mercurella (p.369)

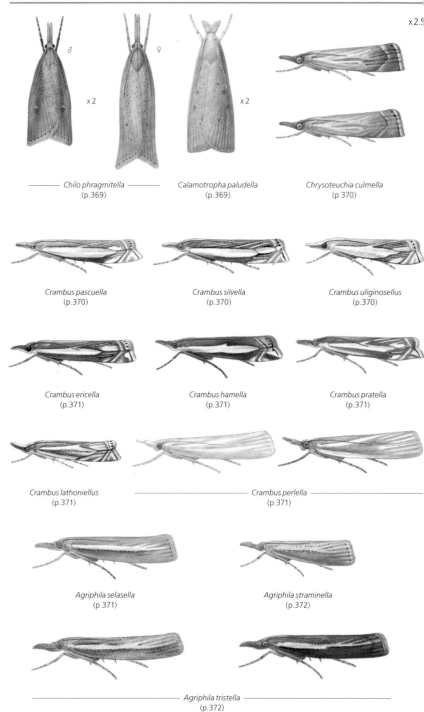

x 2.5

♂   ♀

x 2   x 2

Chilo phragmitella
(p.369)

Calamotropha paludella
(p.369)

Chrysoteuchia culmella
(p.370)

Crambus pascuella
(p.370)

Crambus silvella
(p.370)

Crambus uliginosellus
(p.370)

Crambus ericella
(p.371)

Crambus hamella
(p.371)

Crambus pratella
(p.371)

Crambus lathoniellus
(p.371)

Crambus perlella
(p.371)

Agriphila selasella
(p.371)

Agriphila straminella
(p.372)

Agriphila tristella
(p.372)

x 2.5

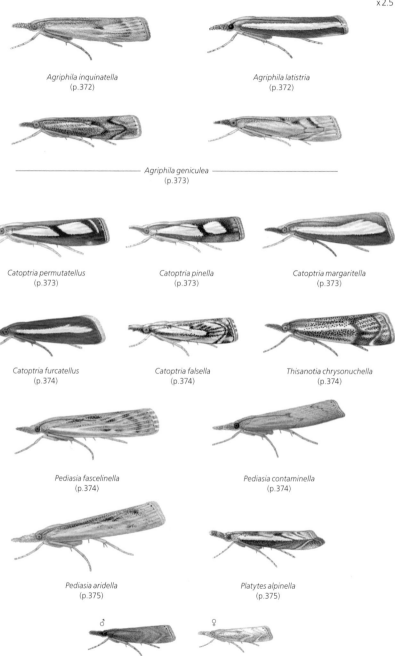

Agriphila inquinatella
(p.372)

Agriphila latistria
(p.372)

Agriphila geniculea
(p.373)

Catoptria permutatellus
(p.373)

Catoptria pinella
(p.373)

Catoptria margaritella
(p.373)

Catoptria furcatellus
(p.374)

Catoptria falsella
(p.374)

Thisanotia chrysonuchella
(p.374)

Pediasia fascelinella
(p.374)

Pediasia contaminella
(p.374)

Pediasia aridella
(p.375)

Platytes alpinella
(p.375)

Platytes cerussella
(p.375)

x 1.5

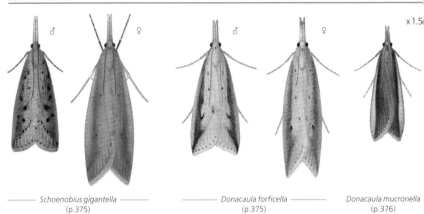

Schoenobius gigantella
(p.375)

Donacaula forficella
(p.375)

Donacaula mucronella
(p.376)

x 2

Brown China-mark
Elophila nymphaeata
(p.376)

Water Veneer
Acentria ephemerella
(p.376)

Small China-mark
Cataclysta lemnata
(p.377)

Ringed China-mark
Parapoynx stratiotata
(p.377)

Beautiful China-mark
Nymphula nitidulata
(p.377)

Cynaeda dentalis
(p.378)

Garden Pebble
Evergestis forficalis
(p.378)

Evergestis limbata
(p.378)

Evergestis extimalis
(p.378)

Evergestis pallidata
(p.378)

x 1.5

underside

——————— *Pyrausta aurata* ———————
(p.379)

——————— *Pyrausta purpuralis* ———————
(p.379)

underside

——————— *Pyrausta ostrinalis* ———————
(p.379)

Pyrausta sanguinalis
(p.379)

——————— *Pyrausta despicata* ———————
(p.380)

Pyrausta nigrata
(p.380)

Pyrausta cingulata
(p.380)

Nascia cilialis
(p.380)

Sitochroa palealis
(p.380)

Sitochroa verticalis
(p.381)

x 1.5

Small Magpie
Anania hortulata
(p.381)

Anania lancealis
(p.381)

Anania coronata
(p.381)

Anania perlucidalis
(p.381)

Anania stachydalis
(p.382)

Anania funebris
(p.382)

Anania verbascalis
(p.382)

Anania terrealis
(p.382)

Anania crocealis
(p.382)

Anania fuscalis
(p.383)

——————— European Corn-borer ———————
Ostrinia nubilalis
(p.383)

Bordered Pearl
Paratalanta pandalis
(p.383)

Paratalanta hyalinalis
(p.383)

Udea lutealis
(p.383)

Udea fulvalis
(p.384)

Udea prunalis
(p.384)

x 1.5

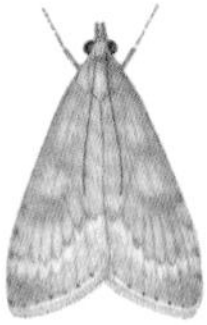

Udea decrepitalis
(p.384)

Udea olivalis
(p.384)

Udea uliginosalis
(p.384)

Rusty-dot Pearl
Udea ferrugalis
(p.385)

Mother of Pearl
Pleuroptya ruralis
(p.385)

Mecyna flavalis
(p.385)

Mecyna asinalis
(p.385)

Agrotera nemoralis
(p.386)

Palpita vitrealis
(p.386)

Dolicharthria punctalis
(p.386)

Rush Veneer
Nomophila noctuella
(p.386)

Tortricidae

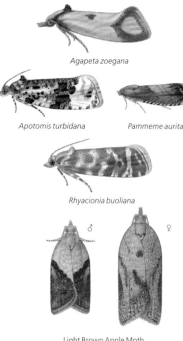

Agapeta zoegana

Apotomis turbidana

Pammeme aurita

Rhyacionia buoliana

♂ ♀

Light Brown Apple Moth
Epiphyas postvittana

This is a very diverse family of moths, with 399 species on the checklist. The resting positions of adults are variable and, although the body is always held horizontally, the wings may be flat and overlapping, tent-like or wrapped around the body, and the antennae can be laid back above the wings, hidden beneath them, or diverging and pointing upwards. They vary from small to large micro-moths, with the forewing length 3.5-15mm. **A key character of the family is the rather broad forewing**, with a distinct tornal angle. Many species have a wide base to the wing, giving a typical bell-shape to the adult at rest. Some species have raised scales (e.g. *Acleris*). The males of many species have a flap of scaled wing membrane from the base reaching a short way along the costa, known as the costal fold, and this can be helpful in species identification (e.g. *Capua vulgana*, *Dichrorampha* spp.). The hindwings are as broad or broader than the forewings. The wings are much reduced in female *Exapate*. The head has erect or raised scales on the crown, the face being smooth. The antennae are short, less than two-thirds the length of the forewing. A very few species have bipectinate antennae (e.g. *Philedone* and *Philedonides*). The labial palps are short, drooping, pointing forwards or slightly curved upwards, and the second segment is densely scaled or tufted, with the third segment short and blunt, except *Sparganothis* which has long palps pointing straight forward. The tongue is variably developed, but not scaled.

New species to the British list are regularly added, including in recent years *Aethes bilbaensis*, *A. fennicana*, *Cnephasia pumicana*, *Dichelia histrionana*, *Epinotia granitana*, *Cydia inquinatana*, *Pseudococcyx tessulatana* and *Gravitarmata margarotana*. Some are imported adventives, whilst others are presumed immigrant species, and *Cnephasia pumicana* may have been overlooked for years. Light Brown Apple Moth *Epiphyas postvittana* is perhaps one of our most readily encountered species these days, occurring over much of Britain. It is an adventive here, and native to Australia; it was first found breeding in Britain in Newquay, Cornwall, in 1936, and has now spread east and north. It is extremely common over much of its range. Some species have a very restricted distribution, such as *Grapholita lathyrana*, which has declined to near-extinction as a result of loss of unimproved neutral grasslands. A few species are considered extinct, including *Pristerognatha penthinana* and *Gibberifera simplana*.

Many species are distinctive and can be readily identified on external characters. There are, however, difficult groupings. A common trait across several genera in Tortricidae is to mimic bird droppings, and these species, numbering 60 at least, need careful examination. For some genera or species groups, such as *Cnephasia* species or the recently separated *Acleris emargana* and *A. effractana*, reference to the genitalia is often needed to confirm identification. One regular misidentification made by newcomers to micro-moths is to try to find the macro-moth Oak Nycteoline *Nycteola reveyana* (Nolidae) among the Tortricidae, especially *Acleris*. It isn't there!

Coverage here of the Tortricidae is not complete, although the majority of species are included below. The Checklist at the back of the guide provides a reference for those omitted, which are some of the rarest and most restricted species.

In such a large group it is not surprising that the life histories are diverse. Larvae of many species are internal feeders, boring into flowerheads, seeds, stems or roots, whereas others feed from spinnings on foliage of their foodplant. A number of species are polyphagous, with others restricted to one or a very few foodplants. The majority are single-brooded, but many have two or more generations a year. Most species pass the winter as a young, or fully fed larva in a cocoon, while a number of *Acleris* hibernate as adults. Some adults fly mainly by day, others from dusk and at night, and many are attracted to light, sometimes in abundance.

Further reading
British and Irish species: Bradley *et al.* (1973 & 1979); Clifton & Wheeler (2011)
European species: Razowski (2002 & 2003)

Phtheochroa inopiana page 233   921 (4187)

Local. Rare in the north. **FL** 8-10mm. Male forewing pale greyish brown, mottled brown in the middle, with wavy brown cross-lines in the outer half of the wing; a paler triangular area from the base to the costa at one-third, outwardly edged brown, and a black dot at two-thirds, beyond which the wing is concavely creased to the apex and tornus. The female is usually unicolorous brown or brownish orange, with the forewing narrower than in the male. **Similar species *Sparganothis pilleriana* and *Aphelia viburnana* rest with wings in a roof-like posture at a shallow, not steep, angle. Viewed from above, male *P. inopiana* has a brown arrowhead pointing into the thorax.** **FS** Single-brooded, May-August. Sits on the foodplant by day and comes to light. **Hab** Coastal landslips, damp neutral grasslands, ditches, woodland rides. **Fp** Fleabane, Field Wormwood; in the roots.

Phtheochroa sodaliana page 233   923 (4193)

Very local. **FL** 7-8mm. Head, thorax and palps white. Forewing white, dappled grey, white near the base with a black mark on the costa; a greyish-brown cross-band before one-half, incorporating at least one black mark, and interrupted towards the costa, a blackish band before the termen and an orange or red apical spot. Hindwing whitish and grey, with darker wavy cross-lines. **Similar species *Hysterophora maculosana* has a dark head and thorax; *Cochylis hybridella* is smaller, has a much smaller reddish mark at the apex, and a plain grey hindwing. FS** Single-brooded, June-July. Comes to light. **Hab** Hedgerows and scrub on chalk. **Fp** Buckthorn; in the berries, spinning a few together.

▶ Larvae of
*Phtheochroa
rugosana*. The
wrinkly pale
green larva
usually turns red
when fully fed
prior to pupation.

Phtheochroa rugosana page 233

925 (4214)

Local. **FL** 6-9mm. Head, thorax and palps white, tegulae mottled brownish and white. Forewing with raised scales, giving a rough texture, mottled greyish brown with darker brown forming an ill-defined cross-band at one-half, with reddish-brown marks in the middle and three short white marks reaching the costa. **FS** Single-brooded, late April-early August. Recorded in November. Comes to light. **Hab** Hedges, scrub and open woodland on dry soils. **Fp** White Bryony; in the flowers and unripe berries, later in the stem.

Hysterophora maculosana page 233

924 (4219)

Local. Very local in Scotland. **FL** 5-7mm. Head, thorax and palps mottled dark brown and blackish; the crest on the back of the thorax is mixed orangey or reddish brown and blackish. Forewing whitish, dappled grey, especially at the base and towards the termen; a more or less complete grey cross-band before one-half, the outer margin mixed with black and orangey brown marks and deeply indented before the costa; usually orangey brown along the termen. Hindwing whitish in the dorsal half in the male, grey throughout in the female. **Similar species** *Cochylis hybridella* **has the head and thorax whitish;** *Phtheochroa sodaliana*. **FS** Single-brooded, April-June. Can be numerous in woodland. Flies by day, when the white coloration of the male is conspicuous, and comes to light. **Hab** Woodlands, especially ancient woodland. **Fp** Native Bluebell, possibly Spanish Bluebell; in the seed-capsule.

Cochylimorpha alternana page 233

935 (4251)

Rare. **FL** 9-13mm. Forewing dull whitish yellow with darker yellow-brown or brownish markings and blackish scale-tufts along the cross-band. **Similar species** *C. straminea*. **FS** Single-brooded, July-August. Can be numerous. Flies at dusk and comes to light. **Hab** Chalk grassland, sand dunes. **Fp** Greater Knapweed; in spring in developing flower buds.

Cochylimorpha straminea page 233

936 (4247)

Common. Local and mostly coastal in Scotland. **FL** 7-10mm. Forewing dull whitish yellow with darker yellow-brown or brownish markings and an oblique cross-band before one-half from the dorsum to the middle of the wing, interrupted or faint towards the costa. **Similar species** *C. alternana* **is a much larger species, with paired blackish scale-tufts along the inner and outer margins of the cross-band;** *Aethes smeathmanniana* **has a richer yellow forewing with two interrupted cross-bands. FS** Double-brooded, May-July, August-September. Comes to light. **Hab** Grasslands wherever the foodplant is present, but not at high altitude. **Fp** Common Knapweed; in the stem below the shoots and flowers.

Phalonidia manniana page 233

926 (4255)

Local. Very local further north. **FL** 5-6mm. Forewing creamy white or pale brown suffused darker brown, often with hint of pale greenish brown; a pale greenish-brown cross-band from the dorsum before one-half, with blackish markings towards the middle, narrowing before the costa, and without an extension to the tornus; the cross-band before the apex is usually well defined, curving and narrowing towards the tornus. **Similar species**

Gynnidomorpha alismana has the cross-band before the apex pronounced only near the costa. **FS** Single-brooded, mid May-mid September. Flies over the foodplants in the evening and comes to light. **Hab** Fens, ditches, ponds, wet woodland rides, occasionally found elsewhere. **Fp** Water Mint, Gypsywort; in the stem.

Phalonidia affinitana page 233   932 (4256)

Local. Very scarce in Scotland. **FL** 5-7mm. Forewing dirty white, pale yellowish or pale brownish; an oblique orangey brown or brown cross-band from before one-half on the dorsum to near the middle only, often obscure, sometimes blackish edged, and a blackish-brown tornal dot. **This is a plain species, usually with ill-defined markings, especially in the female; the tornal dot is visible even in poorly marked examples. Similar species *Gynnidomorpha vectisana* is smaller and has the central cross-band reaching the costa.** **FS** Probably one extended brood, May-August. **Hab** Saltmarshes. **Fp** Sea Aster; in the flowers and seeds.

Phalonidia gilvicomana page 233   933 (4253)

Very local in the south, but probably overlooked. **FL** 5-6mm. Forewing deep yellowish brown from the base to the central cross-band and at the tornus; the cross-band and most of the outer part of the wing is blackish brown. **Similar species *P. curvistrigana* is larger and paler, with the yellowish colour more extensive beyond the cross-band. The arched costa of *P. gilvicomana* and *P. curvistrigana* separate these two species from others in related genera.** **FS** Single-brooded, June-July. Flies over the foodplants in the evening, but is rarely seen as an adult. **Hab** Open chalky soils in woodland, walls. **Fp** Nipplewort, Wall Lettuce; in the flowers.

Phalonidia curvistrigana page 233   934 (4254)

Very local. **FL** 6-7mm. Forewing yellowish, mixed orangey brown from the base to the central cross-band and beyond; the cross-band and area before the apex is mixed brown and blackish brown. **Similar species** *P. gilvicomana*. **FS** Single-brooded, July-August. Flies over the foodplant in the evening, but rarely comes to light. **Hab** Woodland, coastal cliffs. **Fp** Goldenrod; in the flowers and seeds.

Gynnidomorpha minimana page 233   927 (4264)

Very local. Best known from the fens of East Anglia. **FL** 6-7mm. Forewing creamy orangey brown, suffused with darker orangey brown or brown markings; an oblique dark cross-band from the dorsum at one-third to near the costa, slightly angled back to meet the costa at one-half, often with an irregular extension from the outer angle to the tornus. **Similar species *G. permixtana*, a smaller species, has greyish or brownish shading along the costa, especially in the basal half, and more suffused markings generally.** **FS** Double-brooded, May-June, late July-August. Flies over the foodplants on warm evenings. **Hab** Fen, boggy heathland. **Fp** Marsh Lousewort, Bogbean; in the seed-capsules and seed-pods.

Gynnidomorpha permixtana page 233   928 (4265)

Very local, but likely to be overlooked. **FL** 5-6mm. Forewing creamy orangey brown, suffused with darker brown or grey especially along the costa; an oblique dark cross-band from the dorsum at one-third to near the costa, angled back to meet the costa at one-half, often with an extension from the outer angle to the tornus. **Similar species** *G. minimana*. **FS** Double-brooded, May-June, late July-August. Can be disturbed by day, flies over the foodplant in the evening and comes to light. **Hab** Sparsely vegetated grasslands, sand dunes, stabilised shingle beaches. **Fp** Red Bartsia; in the flowers and seeds.

Gynnidomorpha vectisana page 233   929 (4263)

Local. **FL** 4-6mm. **A small and often obscurely marked species.** Variable. Forewing creamy brown, with darker brown or orangey brown markings, sometimes heavily reticulated in the outer half of the wing; the central cross-band is narrow and nearly perpendicular to the dorsum, usually darker in the middle and with an extension from the middle to the tornus. **Similar species** *Phalonidia affinitana*. **FS** Double-brooded, May-June, July-September. Can be abundant. Flies over the foodplants at dusk and comes to light. **Hab** Mainly coastal saltmarshes, also rocky shorelines, boggy heaths, fens. **Fp** Sea Arrowgrass, Marsh Arrowgrass; in the summer in the flowerheads, in the autumn in the shoots and top of the roots.

Gynnidomorpha alismana page 233   930 (4266)

Local. Very local in Scotland. **FL** 6-7mm. Forewing creamy brown, with darker brown markings; a slightly oblique cross-band from the dorsum at one-third to near the costa, angled back to meet the costa at one-half, often with a blackish-brown band near the middle, and with a cross-band before the apex, this pronounced only near the costa; there is no extension of the cross-band to the tornus. **Similar species** *Phalonidia manniana*. **FS** Single-brooded, late May-August. Flies at dusk and comes to light, sometimes well away from typical habitat. **Hab** Ponds, slow-flowing waterbodies. **Fp** Water-plantain; in the pith of the flowering stem, often several larvae to a stem. Small holes in dead stems in winter indicate the presence of larvae.

Gynnidomorpha luridana page 233   931 (4262)

Very local. **FL** 5-6mm. Forewing pale yellowish brown or pale brownish, with darker orangey brown markings; an oblique narrow cross-band from the dorsum at one-third to near the costa, angled back to meet the costa at one-half, with a blackish-brown smudge at the outer angle and with a pre-apical cross-band, this pronounced only near the costa; there is no extension of the cross-band to the tornus. **The pale colour and markings of this species are characteristic. FS** Double-brooded, May-August. Flies at dusk and comes to light. **Hab** Rough, dry ground on chalk or limestone. **Fp** Red Bartsia, also Scented Mayweed, burdocks; in the flowerheads.

Agapeta hamana page 233   937 (4268)

Common. **FL** 8-12mm. Forewing usually whitish yellow, occasionally bright or deep yellow, with a variable number and strength of orangey or brownish marks, but always including a distinct broad line from beyond the middle of the wing to the tornus. **Similar species *A. zoegana* has an almost circular brownish ring in the outer half of the wing. FS** Probably single-brooded, mid April-September. Occasionally seen in numbers. Can be disturbed by day, flies in the evening and comes to light. **Hab** Wherever thistles occur, except at high altitudes. **Fp** Reared from Musk Thistle, probably on other thistles; in the roots.

Agapeta zoegana page 233   938 (4271)

Common. Very local in Scotland. **FL** 8-12mm. Forewing usually bright yellow, occasionally brownish orange; a characteristic, almost circular brownish ring beyond the middle, incorporating the termen, enclosing a patch of the ground colour. Examples with brownish-orange ground colour often have the ring including leaden-grey scales. **Similar species** *A. hamana*. **FS** Single-brooded, May-September. Flies from sunset and occasionally comes to light. **Hab** Grasslands, roadside verges, quarries, mainly on chalk and limestone. **Fp** Common Knapweed, Small Scabious, probably Field Scabious; in the roots.

Eupoecilia angustana page 233   954 (4287)

Common. **FL** 5-8mm. Forewing creamy whitish with patches of orangey brown suffusion; a cross-band at one-half, narrow about the middle, and a subterminal cross-band mixed black and brown. Named variations occur: f. *fasciella*, on heaths and moors, is smaller, white and the central cross-band is without brown; ssp. *thuleana*, confined to Shetland, has a narrower, more pointed forewing, and is pale brown with indistinct darker brown or reddish-brown markings. **Similar species *Cochylis atricapitana*, *C. dubitana* and *C. hybridella* all have an interrupted or incomplete cross-band. FS** Single-brooded, May-September; flies late July-August on moorland. Flies on warm afternoons and comes to light. **Hab** Open woodland, meadows, downland, heathland, moorland. **Fp** Many plants, including plantains, Yarrow, Marjoram, Heather (on heaths and moors), Sitka Spruce (in Scotland); in spun flowers and seeds.

Vine Moth *Eupoecilia ambiguella* page 233   955 (4288)

Very local. An occasional immigrant. **FL** 6-8mm. Forewing whitish yellow, with whitish-orange suffusion in patches; a mixed blackish and brown cross-band at one-half, usually complete and wider on the costa, narrowing at or ending before the dorsum; a small dark mark is usually present on the tornus. **FS** Single-brooded, May-September. Flies at dusk and comes to light. **Hab** Scrub on wet heaths, wet woodland; immigrants usually in coastal areas. **Fp** Alder Buckthorn, less often on Ivy, Dogwood, Honeysuckle; feeds within a few berries spun together and pupates in a case cut from a leaf. A pest of grape vines in Europe.

Commophila aeneana page 233   952 (4291)

Very local. **FL** 6-8mm. Forewing bright orange with scattered shining white scales; broad black cross-bands at one-half and near the termen, more or less distict and scattered with blue metallic scales. **FS** Single-brooded, late May-early July. Can be disturbed on sunny days, and sits on the foodplant at dusk. **Hab** Rough ground on heavy clay soils. **Fp** Common Ragwort; in the roots.

Aethes tesserana page 234   939 (4310)

Common. Rare in northern England and Scotland. **FL** 5-8mm. Variable. Forewing yellowish to greenish yellow, with markings of orangey brown or brown, often appearing as four quadrate blotches, but sometimes as cross-lines, or cross-bands which may be edged darker. **FS** Single-brooded, May-August. Flies in sunshine by day and comes to light. **Hab** Sparsely vegetated grasslands, chalk and limestone grassland, sand dunes, stabilised shingle beaches. **Fp** Bristly Oxtongue, Hawkweed Oxtongue, hawkweeds, hawk's-beards, Ploughman's-spikenard; in the roots.

Aethes hartmanniana page 234   941 (4294)

Local. Very local in the north. **FL** 6-9mm. Forewing pale brownish yellow, sometimes suffused greyish, with markings orangey brown or reddish brown, edged with slightly raised white or silver metallic spots or lines; an oblique cross-band from the dorsum beyond one-third, interrupted near the costa, and angled back to meet the costa at one-half. **Similar species *A. piercei*, considered by some authorities to be the same species, is usually much bigger, brighter and more contrasting in colour, and occurs mostly in damp habitats and slightly earlier in the year. FS** Single-brooded, June-early August. Can be disturbed by day and flies at dusk. **Hab** Chalk and limestone grassland. **Fp** Probably Field Scabious, Small Scabious; possibly in the rootstock.

Aethes piercei page 234   942 (4295)

Very local. Records from northern England unconfirmed. Coastal in northern Scotland. **FL** 7-12mm. Forewing reddish brown, markings blackish brown, contrasting strongly with white or silvery edging. **Similar species** *A. hartmanniana*. **FS** Single-brooded, late May-early July. Can be disturbed by day, flies in the late afternoon and occasionally comes to light. **Hab** Damp and wet neutral grassland, fens, bogs; in The Burren, Ireland, dry limestone grassland. **Fp** Devil's-bit Scabious; in the roots.

Aethes williana page 234   944 (4296)

Very local. **FL** 5-8mm. The female is larger than the male. Forewing with a rounded apex and termen; pale yellow suffused grey, with shining white spots or lines; the cross-band at one-half greyish brown or blackish brown, interrupted below the costa, with raised scale-tufts about the middle. **FS** Single-brooded, May-early August. Flies by day and at dusk, and comes to light. **Hab** Sparsely vegetated grasslands, chalk and limestone grassland, coastal slope, sand dunes, stabilised shingle beaches. **Fp** Wild Carrot; in the lower part of the stem and top of the root.

Aethes cnicana page 234   945 (4326)

Common. **FL** 7-8mm. Forewing creamy whitish, more rarely yellowish, usually suffused with patches of brownish yellow and greyish; a reddish or reddish-brown oblique cross-band from the dorsum beyond one-third, sometimes interrupted near the costa, and angled back to meet the costa at one-half, the outer margin of the cross-band hardly indented near the dorsum; other reddish or reddish-brown markings may be present. **Similar species** *A. rubigana* **is, on average, slightly bigger and broader winged, with the cross-band broader and almost always interrupted, its outer margin deeply indented before the dorsum.** **FS** Single-brooded, late May-early August. Can be disturbed by day, flies at dusk and comes to light. **Hab** Wherever the foodplants occur. **Fp** Thistles; at first in the seeds, later in the pith of the stem.

Aethes rubigana page 234   946 (4327)

Common. Very local in Scotland. **FL** 7-10mm. Forewing creamy whitish suffused with patches of brownish yellow and greyish, with a reddish or reddish-brown cross-band. **Similar species** *A. cnicana*. **FS** Single-brooded, early June-August. Flies at dusk and comes to light. **Hab** Open woodland, scrub, waste ground, field margins. **Fp** Greater Burdock, possibly other burdocks; in the seeds.

Aethes smeathmanniana page 234   947 (4309)

Local. **FL** 6-9mm. Forewing pale yellowish, with some yellowish-brown suffusion; two orangey brown or reddish-brown oblique cross-bands from the dorsum, indistinct beyond the middle except for paler markings on the costa at one-half and before the apex. **Similar species** *A. dilucidana*, *A. francillana* **and** *A. beatricella*, **all of which have at least one complete cross-band**; also *Cochylimorpha straminea*. **FS** Single-brooded, late May-mid August; possibly double-brooded in northern England, early May-mid June, late July-mid September. Can be disturbed on warm days, flies at dusk and comes to light. **Hab** Grasslands, rough ground, coastal cliff and slope, quarries. **Fp** Yarrow, Common Knapweed, Corn Chamomile; in the seeds, the larva remaining in the seedhead over winter.

Aethes margaritana page 234 948 (4303)

Rare. **FL** 6-8mm. Forewing silvery white, contrasting strongly with several oblique yellow-brown cross-bands, these sometimes interrupted. **FS** Single-brooded, July-August. Readily disturbed on warm days, and flies at dusk. **Hab** Waste ground, chalk downland, shingle beaches. **Fp** Mainly Yarrow, also Scented Mayweed, Oxeye Daisy; in the flowers and seeds.

Aethes dilucidana page 234 949 (4314)

Local. Rare in northern England. **FL** 6-8mm. Forewing pale yellow, with two brown oblique narrow cross-bands from the dorsum, the outer reaching the costa, the inner obsolete beyond the middle except for a narrow mark on the costa. **Similar species *A. francillana* and *A. beatricella* both have two complete cross-bands**; also *A. smeathmanniana*. **FS** Single-brooded, June-August. Can be disturbed by day, flies weakly around the foodplant at dusk and comes to light. **Hab** Chalk and limestone grassland, quarries. **Fp** Wild Parsnip; in the seeds, spinning several together.

Aethes francillana page 234 950 (4321)

Local. Very local in the north and in Wales. Sometimes very common on the coast. **FL** 7-9mm. Forewing pale yellow, with two brown oblique narrow cross-bands from the dorsum reaching the costa, the costal end of the inner cross-band extending beyond the dorsal end of the outer cross-band. **Similar species *A. beatricella* has the cross-bands marginally broader and less oblique, so that the costal end of the inner cross-band does not extend beyond the dorsal end of the outer cross-band**; also *A. dilucidana*, *A. smeathmanniana*. **FS** Single-brooded, June-September. Can be disturbed by day, flies around the foodplants at dusk and comes to light. **Hab** Chalk and limestone grassland, coastal cliff and slope, quarries. **Fp** Wild Carrot, Rock Samphire, Hog's Fennel, Hemlock; in the flowers and seeds, later in the lower part of the stem.

Aethes beatricella page 234 951 (4317)

Local. Possibly spreading north and west. **FL** 7-9mm. Forewing pale yellow, with two narrow brown oblique cross-bands from the dorsum reaching the costa. **Similar species** *A. francillana*, *A. dilucidana*, *A. smeathmanniana*. **FS** Single-brooded, June-August. Flies at dusk and comes to light. **Hab** Rough ground, field margins, open woodland, roadside verges, coastal areas. **Fp** Hemlock, Alexanders; in the seeds, later in the upper part of the stem.

Cochylidia implicitana page 234 956 (4339)

Local. Very local in the north and west. **FL** 4-7mm. Forewing creamy whitish, sometimes greyish towards the base, slightly flushed with pinkish orange, often with indistinct fine darker lines before the apex; a reddish-brown or brown oblique cross-band from the dorsum at one-third, narrowed or interrupted and slightly angled before the costa, and often a brownish mark before the tornus; cilia concolorous with the ground colour, except at the tornus where they are grey and sometimes barred. **Similar species** *C. heydeniana* can be distinguished by its smaller size, lack of pinkish flush, more mottled appearance towards the apex, with a black tornal mark and distinct grey bars in the cilia along the termen; also *C. subroseana*, which is usually bigger with a broader central cross-band, and is creamy whitish suffused pale orangey brown, with orangey brown shading before the termen. **FS** Possibly single-brooded, May-early September. Can be disturbed during the afternoon, flies at dusk and comes to light. **Hab** Rough, open ground, coastal areas, open woodland. **Fp** Scentless Mayweed, Scented Mayweed, Stinking Chamomile, Goldenrod; in the flowers and seeds, and in the stem and shoot.

Cochylidia heydeniana page 234   957 (4338)

Very local. **FL** 4-6mm. Forewing creamy whitish, with a reddish-brown oblique cross-band from the dorsum, narrowing before the costa. **Similar species** *C. implicitana*, *C. subroseana*. **FS** Double-brooded, June, late July-August. Can be disturbed during the afternoon and flies at dusk. **Hab** Waste ground, walls, sand dunes. **Fp** Blue Fleabane, Canadian Fleabane; first brood in the flowers and seeds, second brood in the central shoot beneath the flowerhead.

Cochylidia subroseana page 234   958 (4335)

Very local. **FL** 5-7mm. Forewing creamy whitish suffused pale orangey brown, with a broad orangey brown oblique cross-band from the dorsum, narrowing before the costa, and orangey brown shading before the termen. **Similar species** *C. implicitana*, *C. heydeniana*. **FS** Single-brooded, June-early August. Can be numerous. Flies at dusk and comes to light. **Hab** Open woodland. **Fp** Goldenrod; in the flowers and seeds.

Cochylidia rupicola page 234   959 (4334)

Common in the south. Local or very local further north. **FL** 5-7mm. **A broad-winged dark species with a somewhat arched costa and rounded apex, separating this species from other *Cochylidia*.** Forewing ground colour varying from yellowish white to brownish white, paler examples with orangey brown markings, darker ones with dark brown markings, and with a slightly oblique broad median cross-band at one-half from the dorsum, narrowing before the costa, usually with a diffuse or narrow extension to the tornus, and a narrow dark subterminal cross-band. **FS** Single-brooded, June-July. Flies on warm days, at dusk, and comes to light. **Hab** Cliffs, fens, marshes, riverbanks, roadside verges. **Fp** Hemp-agrimony; in the flowers and seeds.

Cochylis roseana page 234   962 (4348)

Common. Local in northern England. **FL** 5-8mm. Forewing whitish, suffused orangey pink or pink mainly along the costa and in the outer half of the wing, with an orangey brown or brown oblique cross-band of variable width from the dorsum before one-half, extending to near the middle, then faint to the costa, and usually with a blackish mark at the tornus; cilia orangey brown, with grey bars. **Similar species C. flaviciliana has a clear, creamy white ground colour near the base and between the cross-band and the tornus, an orangey brown cross-band suffused pink, and a strong pink flush in the outer half of the wing, darker pink at the apex, and without grey bars in the cilia; also *Falseuncaria degreyana*, which has a narrower forewing, with a whitish-grey**

► Larva of *Cochylis roseana* exposed within a split seedhead of Wild Teasel. The adjacent hole was caused by the larva feeding on a seed nearby.

ground colour and rosy tint. **FS** Single-brooded, May-August. Flies in afternoon sunshine, at dusk, and comes to light. **Hab** Open woodland, open grassland, scrub, waste ground, roadside verges, field margins. **Fp** Wild Teasel; in the seeds.

Cochylis flaviciliana   page 234   963 (4349)

Very local. **FL** 5-8mm. Forewing creamy white near the base and between the cross-band and the tornus, with an orangey brown cross-band suffused pink and a strong pink flush in the outer half of the wing. **Similar species** *C. roseana, Falseuncaria degreyana*. **FS** Single-brooded, late June-early August. Rests by day under a flowerhead, flies in the evening and occasionally comes to light. **Hab** Dry grassland, roadside verges on chalk. **Fp** Field Scabious; in the flowers and seeds.

Cochylis dubitana   page 234   964 (4353)

Local. Very local in Scotland. **FL** 5-7mm. **Head white, thorax dark grey.** Forewing whitish with pale brown or grey suffusions, with a blackish basal cross-band extending a short way along the costa, a blackish broad quadrate mark on the dorsum at one-half, a narrow mark on the costa opposite, and blackish before the termen and apex. **Similar species C. atricapitana has a blackish head, broad wings, the ground colour often with a pinkish hue in the female, and a blackish and brown cross-band, irregularly margined and narrowly interrupted before the costa; C. hybridella has a whitish head and thorax**; also *Eupoecilia angustana*. **FS** Double-brooded, late May-June, mid July-August. Can be disturbed in the afternoon, flies in the evening and comes to light. **Hab** Grassland, rough ground. **Fp** Ragworts, hawk's-beards, hawkweeds, Goldenrod; in the flowers and seeds.

Cochylis molliculana   page 234   964a (4354)

Local. A recent colonist, well established along the south coast, and may be spreading inland. **FL** 6-8mm. Head and thorax yellowish brown, mottled darker brown. Forewing creamy with pale orangey brown and pale greyish-brown suffusions, especially before the termen; a broad oblique cross-band from the dorsum at one-half, narrower on the costa and somewhat interrupted beneath, this orangey brown with blackish flecks, and greyish brown at the costa. **Similar species C. hybridella has a whitish head and thorax, and a very similar wing pattern but with a more contrasting white ground colour and dark markings; C. pallidana is smaller and has a clear white head.** **FS** Probably double-brooded, mid May-early October. Can be abundant. Easily disturbed on warm afternoons, flies in the evening and comes to light. **Hab** Coastal cliff and slope, allotments, trackways, waste ground, field margins. **Fp** Bristly Oxtongue; in the flowers and seeds.

► Feeding signs of *Cochylis molliculana* in the seedheads of Bristly Oxtongue. The pappus of the seeds usually remains attached if a larva has been feeding within, and frass may be visible. A pupal exuvium is also visible above the right-hand seedhead.

Cochylis hybridella page 234 965 (4351)

Local. Very local in northern England. **FL** 6-8mm. **Head and thorax whitish.** Forewing whitish with brownish or grey suffusions, with a mixed blackish, dark grey and reddish-brown oblique cross-band from the dorsum at one-half, narrow or very narrow on the costa and usually widely interrupted beneath, and a very small reddish or reddish-brown mark at the apex. **Similar species** *C. dubitana*, *C. molliculana*, *C. atricapitana*, *Phtheochroa sodaliana*, *Eupoecilia angustana*. **FS** Probably single-brooded, mid June-September. Flies in the evening and comes to light. **Hab** Chalk, limestone and sandy grassland. **Fp** Bristly Oxtongue, Hawkweed Oxtongue, hawk's-beards; in the flowers and seeds.

Cochylis atricapitana page 234 966 (4355)

Common. Local in Scotland. **FL** 6-9mm. **Head and thorax blackish.** Forewing rather broad, whitish in the male, whitish with a pinkish hue in the female, with brownish or grey suffusions and pale lines, a broad irregularly margined blackish and grey cross-band from the dorsum at one-half, narrowly interrupted before the costa, and blackish before the termen and apex. **Similar species** *C. dubitana*, *C. hybridella*, *Eupoecilia angustana*. **FS** Two overlapping broods, mid April-early October. Can be disturbed by day, flies at sunset and comes to light. **Hab** Rough grassland, coastal cliff and slope, allotments, trackways, waste ground, field margins. **Fp** Ragwort; in the flowers and seeds.

Cochylis pallidana page 234 967 (4358)

Very local. Mainly a coastal species. **FL** 5-7mm. Head white, thorax dark greyish brown mixed whitish. Forewing whitish with pale greyish suffusions, with an oblique cross-band from the dorsum before one-half, narrower on the costa beyond one-half, and widely interrupted beneath, brown mixed with black, greyish black only on the costa, and with a blackish dot before the tornus. **Similar species** *C. molliculana*. **FS** Single-brooded, June. Flies at dusk. **Hab** Sand dunes, coastal cliffs, stone walls. **Fp** Sheep's-bit; in the flowers and seeds.

Cochylis nana page 234 968 (4347)

Common. Local in Scotland. **FL** 4-7.5mm. Head white. Forewing whitish, heavily suffused pale yellowish brown, grey in the basal area extending along the costa, with a greyish mixed black cross-band, very broad on the dorsum, narrowing to the costa, the outer margin indented in the middle. **FS** Single-brooded, mid May-early July. Sometimes common around birch trees. Can be disturbed by day, flies at dusk and comes to light. **Hab** Woodland, preferring mature birch woodland. **Fp** Birches; in the catkin.

Falseuncaria ruficiliana page 234 960 (4365)

Local. Very local in north-west England. **FL** 5-7mm. **The narrow forewing is characteristic of *Falseuncaria*.** Forewing ground colour whitish grey, sometimes pale orangey brown or reddish, with a broad brown or reddish-brown oblique cross-band from the dorsum at one-third extending to beyond the middle, edged whitish, and suffused and paler to the costa, and with the area before the termen reddish brown; the cilia are contrasting yellowish or orangey brown. **Similar species** *F. degreyana* has a rosy flush to the whitish-grey ground colour and a narrow cross-band. **FS** Double-brooded, late April-June, July-August. Can be disturbed during the day, flies at dusk and comes to light. **Hab** Chalk and limestone grassland, heathland, moorland, bogs. **Fp** Cowslip, Lousewort, Goldenrod, Yellow-rattle; in the seeds.

Falseuncaria degreyana page 234 961 (4364)

Rare. Only in Breckland in East Anglia, and rarely seen recently. **FL** 5-7mm. Forewing ground colour whitish grey with a rosy tint, with a narrow reddish-brown oblique cross-band from the dorsum at one-third extending to the middle, and suffused and paler to the costa at two-thirds. **Similar species** *F. ruficiliana, Cochylis roseana, C. flaviciliana*. **FS** Double-brooded, May-early June, July-early August. May be disturbed on warm afternoons and flies at dusk. **Hab** Roadside verges adjacent to sandy heaths. **Fp** Common Toadflax, Ribwort Plantain; in the flowers and seeds.

Spatalistis bifasciana page 235 1034 (4368)

Very local, but more widespread in the south in recent years. **FL** 6-7mm. Forewing blackish brown with metallic grey and bluish-grey scales, with an obscure yellowish-brown mark on the costa near one-half, sometimes extended towards the middle of the wing, a small yellowish-brown blotch at about one-quarter on the dorsum, and with the termen bordered yellowish brown, narrowing towards the tornus. **Similar species Dark examples of *Pseudargyrotoza conwagana* have yellowish cilia, not a yellowish terminal band. FS** Single-brooded, May-July. Flies at dusk and comes to light. **Hab** Open woodland, hedgerows. **Fp** Many trees, including Sweet Chestnut, oaks, Beech, Silver Birch, Hornbeam, Wild Cherry; in brown, withered leaves hanging above ground, the larva feeding from within folded or rolled leaf edges or a spun case of leaf fragments.

Green Oak Tortrix *Tortrix viridana* page 235 1033 (4370)

Common. Probably an occasional wanderer or immigrant. Numbers fluctuate annually, sometimes abundant in southern England. **FL** 9-12mm, occasionally smaller. Forewing light green, narrowly edged yellow along the costa; the cilia whitish. Rarely the forewing can be yellowish. **Similar species Unlike any other species of micro-moth; can be confused with Cream-bordered Green Pea *Earias clorana* (Nolidae), which has a white, not greyish, hindwing with a white-edged costa, and rests in a tent-like position. FS** Single-brooded, mid May-mid July, occasionally August. Easily disturbed by day and comes to light. **Hab** Woodland, scrub, hedgerow trees. **Fp**. Oak, sometimes Beech, Hornbeam, Sweet Chestnut, sallows, Bilberry and other plants; in a rolled or folded leaf.

Aleimma loeflingiana page 235 1032 (4372)

Common. Local in Scotland. **FL** 7-9mm. Variable. Forewing yellowish brown, with net-like brown markings, these varying in intensity from virtually absent to heavily marked but usually with a brown costal mark at one-third and one-half, sometimes joined, and the cilia typically with a broad brown basal line. **FS** Single-brooded, June-August, occasionally

May. Sometimes seen in numbers. Flies at dusk and comes to light. **Hab** Woodland, hedgerows. **Fp** Oaks, occasionally Hornbeam, maples; in a rolled leaf or pocket formed from a rolled leaf.

Acleris bergmanniana page 235 1035 (4376)

Local. Very local in Scotland. **FL** 6-7mm. Head and thorax yellow. Forewing yellow, speckled orange, particularly in the outer two-thirds, with bluish-grey markings edged dark brown, comprising a basal mark and narrow cross-bands from the costa at one-third to the dorsum at one-half, strongest on the costa, another on the costa just beyond one-half curving to the tornus, and one along the termen. Lightly marked and almost unicolorous examples occur. **FS** Single-brooded, June-early August. Flies from late afternoon and comes to light. **Hab** Open woodland, scrub, hedgerows, sand dunes, gardens. **Fp** Roses; in spun leaves and shoots.

Acleris forsskaleana page 235

Common. Local in southern Scotland. **FL** 6-8mm. Forewing yellow, with reddish-brown net-like markings, a dark greyish-brown cross-line at about one-half, angled near the middle of the wing, varying in intensity, sometimes with a dark greyish-brown dorsal blotch covering part of the cross-line and extending over much of the middle of the wing, and with a dark greyish-brown line bordering the termen. **FS** Single-brooded, June-mid September, with a partial second brood, October. Can be recorded in numbers. Comes to light. **Hab** Woodland, scrub, hedgerows, parks, orchards, gardens. **Fp** Field Maple, Sycamore; in spun leaves or flowers, later in a rolled leaf.

Acleris holmiana page 235

Common. Local in the north. **FL** 6-7mm. Forewing yellowish brown, shaded reddish brown, paler towards the base and apex, with a large triangular white mark on the costa just beyond one-half. **FS** Single-brooded, mid June-early September. Easily disturbed by day, flies at dusk and comes to light. **Hab** Open woodland, hedgerows, scrub, orchards, gardens. **Fp** Hawthorn, roses, apples, pears, Bramble, plums, Blackthorn; between two spun leaves.

Acleris laterana page 235

Common. **FL** 7-9mm. Variable. Forewing silvery grey, grey, pale yellowish brown or greyish brown, variably mixed with grey and black scales, markings reddish brown, with a large triangular costal blotch, sometimes obscure, sometimes extended diffusely towards the tornus, the inner edge usually well defined. The most typical colour combinations are yellowish brown with a dark brown blotch, and greyish with a reddish-brown blotch. **Similar species** *A. comariana* is typically smaller and less brightly marked, although there is size overlap, and *A. comariana* flies earlier and later in the year, but again there is overlap; normally genitalia examination will be required to confirm identity of *A. comariana*. *A. schalleriana* is more oblong-shaped with a larger costal blotch which extends further towards the apex. **FS** Single-brooded, mid July-early October. Readily disturbed by day, flies at dusk and comes to light. **Hab** Woodland, scrub, hedgerows, gardens. **Fp** Hawthorns, plums, Blackthorn, roses, Bramble, Rowan, sallows, Bilberry; between spun leaves and flowers.

Strawberry Tortrix *Acleris comariana* page 235

Common. Local in the south-west, Ireland and Scotland. **FL** 6-8mm. Variable. Forewing pale yellowish brown, whitish grey or light grey, with a bluish-black or light brown to reddish-brown triangular costal blotch. One form is dark grey or dark greyish brown with the markings obscure, but sometimes outlined with raised blackish scales. **Similar species** *A. caledoniana* has a narrower and more pointed forewing. Some forms of *A. laterana* and *A. schalleriana*. **FS** Double-brooded, June-early November. Flies in the afternoon, at dusk and comes to light. **Hab** Wet heathland, bogs, marshes, gardens. Can be common around cultivated strawberry beds. **Fp** Great Burnet, Marsh Cinquefoil, Water Avens, Wild Strawberry, Garden Strawberry cultivars; in folded or spun leaves. Has been reported as an occasional pest.

Acleris caledoniana page 235 1040 (4377)

Common in Scotland, local in northern England and Wales, very local in the south-west. **FL** 6-8mm. Forewing relatively narrow and pointed; mainly reddish brown more or less tinged grey, less often yellowish brown, with an almost triangular, slightly darker triangular costal blotch, the markings usually indistinct or obsolete. **Similar species** Some forms of *A. comariana*. **FS** Single-brooded, July-September. Flies by day in warm or sunny weather. **Hab** High moorland, mountain mosses and bogs, coastal heathland in Scotland. **Fp** Bilberry, also Cowberry, Bog-myrtle, Lady's-mantle, Cloudberry, cinquefoils; between spun leaves or in the upper shoots.

Acleris sparsana page 235 1041 (4383)

Common. **FL** 8-10mm. Variable. Forewing relatively broad; the typical form is grey with faintly darker oblique cross-bands from the costa and a few scattered raised black scales; one form has the forewing finely speckled whitish, sometimes with obscure reddish-brown streaks towards the costa; another has a large reddish-brown triangular blotch on the costa at about one-half, and the reddish may be suffused over a greater part of the wing; a further form has the forewing whitish grey with a greyish-black blotch. **FS** Single-brooded, August-January, but has been found in February and March; mostly recorded in late autumn. Flies from dusk and comes to light and Ivy blossom. **Hab** Woodland, parkland, gardens. **Fp** Beech, Sycamore; between two spun leaves.

Rhomboid Tortrix *Acleris rhombana* page 235 1042 (4384)

Common. **FL** 6-9mm. Variable. Forewing extended to a point at the apex, the costa shallowly concave beyond the middle; yellowish brown to reddish brown, veins and fine cross-lines darker, resulting in a net-like pattern especially prominent on pale examples, with a darker cross-band at about one-half, variable in extent, narrow or broad, sometimes Y-shaped, often narrow or obscure before the dorsum, sometimes strong on the dorsum and weak towards the costa; the cilia along the termen are pale yellow or whitish, brownish at the apex and around the tornus. One form has the forewing dark brown with the markings almost obsolete. **Similar species** *A. shepherdana* does not have the wing apex as strongly pointed, does not have a distinct cross-band, and the cilia are the same colour as the forewing. **FS** Single-brooded, August-early November, and has been recorded in late July, December and February. Readily disturbed by day, flies at dusk, and comes to light and Ivy blossom. **Hab** Hedgerows, scrub, orchards, occasionally gardens. **Fp** A range of trees and shrubs, including hawthorns, plums, Blackthorn, apples, pears, roses; within a spinning in the leaves.

Acleris aspersana page 235 1043 (4391)

Common. **FL** 5-8mm. Variable. Forewing relatively narrow, the apex slightly pointed; yellowish brown, the veins weakly orange-brown, with a reddish-brown broadly triangular costal blotch, extending to the apex, usually better defined in the male. One form has the veins strongly outlined, resulting in a net-like pattern. **FS** Single-brooded, July-August, and has been recorded in June and early September. Flies in the afternoon and evening, and comes to light. **Hab** Open situations, including chalk downland, coastal cliff and slope, heathland, marshes. **Fp** A range of plants, such as cinquefoils, Meadowsweet, Mountain Avens, Salad Burnet, Wild Strawberry, Common Rock-rose; in a folded leaf or in spun leaves.

Acleris ferrugana page 235   1044 (4402)

Common. Local in northern England, and scarce in Scotland where confined to oak woods. **FL** 6-8mm. Variable. Forewing yellowish brown to dark reddish brown, variably speckled with black scales, occasionally coarsely speckled black, with an almost triangular rather obscure, brown costal blotch at about one-half, the tip approaching the middle of the wing, this triangle sometimes comprising two or three separate marks; there is a black scale-tuft just above the dorsum at about one-third and a black, mixed white, scale-tuft at about one-half. Rarely, the forewing colour is creamy white. **Similar species Very difficult to distinguish from *A. notana* on external characters.** In fresh examples, *A. ferrugana* has a straighter termen and a small prominent black scale-tuft just above the dorsum at about one-third; *A. notana* has the apex slightly more pointed, and the black mark above the dorsum is very small or absent. In most cases, examination of the genitalia will be required to confirm identification (see p.391 for male and female genitalia). **FS** Double-brooded, July-mid August, September-May; the adult hibernates. Can be tapped from dense cover in winter. Comes to light. **Hab** Woodland, parkland, hedgerows, scrub. **Fp** Oaks, also sallows; between two spun leaves.

Acleris notana page 235   1045 (4403)

Common. **FL** 6-9mm. Variable. **Similar species** *A. ferrugana* (see p.391 for male and female genitalia). **FS** Double-brooded, mid June-mid August, October-late April, single-brooded in Scotland, June-July; the adult hibernates. Comes to light. **Hab** Woodland, parkland, scrubby heathland, grassland. **Fp** Birches, also Alder, Bog-myrtle; in spun leaves.

Acleris shepherdana page 235   1046 (4392)

Very local. Perhaps most frequent in East Anglia. **FL** 6-8mm. Forewing slightly pointed at the apex; reddish brown, the veins darker brown, resulting in a faint net-like pattern, with a darker, often obscure, almost triangular costal blotch, the outer margin variably suffused blackish leaden, sometimes with a weak greyish-black mark on the dorsum near one-quarter, this edged with a black scale-tuft just above the dorsum, and a yellowish, mixed black, scale-tuft in the middle of the wing at about one-half. **Similar species** *A. rhombana*. **FS** Single-brooded, July-September. Flies at dusk and comes to light. **Hab** Fens, marshes, water meadows. **Fp** Meadowsweet; in a spinning amongst the young shoots.

Acleris schalleriana page 235   1047 (4386)

Local. Very local in northern England. **FL** 6-9mm. Variable. Forewing whitish grey, yellowish brown, reddish brown or greyish brown, variably speckled brownish, with a broadly triangular darker costal blotch, this sometimes incomplete or obscure, nearly reaching the apex, and with a broken line of raised scales on its inner margin, a raised scale-tuft towards the middle of the wing near the point of the triangle, and another at about one-third above the dorsum. Darker forms with more indistinct markings occur. **Similar species** *A. comariana*, *A. laterana*. **FS** Double-brooded, late July-late August, mid October-May; the adult hibernates. Comes to light. **Hab** Open woodland, scrub, hedgerows, gardens; mainly on calcareous or damp soils. **Fp** Wayfaring-tree, Guelder-rose, cultivated Viburnums; at first in a small spinning between veins on the underside of a leaf, resembling an underside mine of *Phyllonorycter lantanella* (Gracillariidae), later in a spinning of twisted leaves.

▲ Larva of *Acleris schalleriana*.

► Spinning and feeding signs of *Acleris schalleriana* on Guelder-rose.

Garden Rose Tortrix *Acleris variegana* page 236   1048 (4390)

Common. **FL** 6-9mm. Extremely variable. Forewing white, occasionally yellowish brown, speckled greyish, the outer half pale reddish brown, sometimes brown, with raised scales along the junction, several raised scales towards the apex and in the tornal area, and with a reddish-brown patch on the dorsum near the base containing a large scale-tuft. One form has the basal half almost wholly white, the outer half blackish brown; another form has the basal half yellowish brown, sometimes with a darker mark towards the base on the dorsum and an almost triangular dark greyish-brown or blackish blotch on the costa; another form is whitish, speckled with blackish scales, with a dark greyish-brown triangular costal blotch. A melanic form occurs, with the forewing bluish black and blackish irregular marbling. **Similar species *A. permutana* has the forewing costa shallowly concave towards the apex, and is brighter yellowish brown and reddish yellowish brown, with a less variegated pattern. FS** Single-brooded, mid June-early November; has been recorded in May. Easily disturbed by day, flies at dusk and comes to light. **Hab** Gardens, orchards, hedgerows, scrub, woodland margins. **Fp** Roses, including garden varieties, also hawthorns, Bramble, Blackthorn, apples, pears, Salad Burnet and various other plants; in a folded leaf or in loosely spun leaves.

Acleris permutana page 236   1049 (4396)

Rare. **FL** 6-9mm. Forewing pale yellowish brown, speckled with yellowish-brown, black and grey scales, with a reddish-brown mark, suffused leaden grey, from the dorsum at one-quarter extending half way across the wing, supporting a large tuft of reddish-brown scales near its apex; the outer half of the wing is shaded darker yellowish brown, with a diffuse darker triangular mark on the costa just before the apex extending towards the tornus. **Similar species** One form of *A. variegana*. **FS** Single-brooded, August-September. Flies at dusk and comes to light. **Hab** Sandhills, vegetated shingle, downland. **Fp** Burnet Rose; in spun leaves and shoots.

Acleris kochiella page 236   1050 (4405)

Local. **FL** 7-9mm. Variable. Forewing costa is shallowly concave. Summer brood: forewing creamy white, sparsely speckled with small black scale-tufts, a triangular blotch more or less represented by three dark brownish blotches, two on the costa and one in the middle below. Autumn brood: forewing grey to silver grey, sometimes with yellowish and black scales, the surface roughened throughout with raised scales, and the blotches dark greyish or purplish brown, the middle blotch with raised black scales sometimes forming a streak. **Similar species *A. logiana* is on average larger, less strongly marked; in Scotland, only *A. logiana* occurs. The summer broods of the two species are harder to separate than the autumn broods**; where there is doubt, identity should be confirmed by examination of genitalia. **FS** Double-brooded, June-early August,

► Pattern left by *Acleris kochiella* when feeding between flat spun leaves of Wych Elm.

September-April; the adult hibernates. Can be disturbed from elm by day, flies at dusk and comes to light. **Hab** Woodland, hedgerows. **Fp** Wych Elm, English Elm; initially in a folded or curled leaf, later between two spun leaves.

Acleris logiana page 236 1051 (4407)

Very local. Formerly confined to Scotland, but in recent years has been found in southern England and is spreading there. **FL** 8-10mm. Variable. Forewing pale greyish white or dusky cream, variably speckled with small black scale-tufts arranged in curved lines, the markings typically reduced to small dark patches on the costa. One form has the triangular blotch reduced to two dark marks on the costa and one in the middle below. Another form has diffuse longitudinal orangey brown streaks or patches, these sometimes reduced to a patch on the dorsum. **Similar species** *A. kochiella*. **FS** Single brooded in Scotland, September-April; double-brooded in England, mid June-July, September-April; the adult hibernates. Can be found at rest on branches and tree trunks by day, and comes to light. **Hab** Birch woodland, scrub on heathland. **Fp** Silver Birch, Downy Birch; in spun leaves.

Acleris umbrana page 236 1052 (4388)

Rare. Has become more frequent in recent years. **FL** 8-10mm. Variable. Forewing yellowish brown to blackish brown, with a broad blackish streak from the base to the apex, varying in thickness and sometimes extending to the dorsum in the basal half, narrower towards and sometimes not reaching the apex; a tuft of ridge-like scales near the middle, with a broad scale-tuft above the dorsum towards the base and sometimes small tufts in the outer part of the wing. One form has the streak obscure, occasionally fractured into thin radiating lines. Another form has the forewing almost unicolorous dark purplish brown. **Similar species A. hastiana is on average larger and the costa is shallowly concave. FS** Double-brooded, July-early August, September-April; the adult hibernates. Comes to light and has been recorded at Ivy and sallow blossom. **Hab** Woodland rides, scrub. **Fp** Blackthorn; among spun leaves.

Acleris hastiana page 236 1053 (4394)

Common. **FL** 9-11mm. Extremely variable, with many named forms. Forewing costa strongly curved from the base, coarsely fringed with scales, and shallowly concave from before the middle towards the apex; the inner margin of the cross-band, if visible, is edged by roughened scales, with a scale-tuft

near the middle of the wing, usually with two small scale-tufts near the base, and one or two minute scale-tufts near the tornus, although these can become lost with wear. Pattern broadly conforms to one of three types: longitudinal markings, typically a stripe or stripes from the base to the apex and/or along the dorsum; or transverse markings, often in the form of a cross-band from the costa at one-fifth to the dorsum before one-half; or uniform in appearance. **Similar species** *A. cristana* **usually has a large raised tuft near the middle of the wing, the costa is more deeply concave, and the apex slightly less pointed;** *A. abietana* **has more obvious rough scales along the costa, more numerous small scale-tufts, and the pattern is often dark brown marbled with paler brown patches;** *A. maccana* **occurs only in Scotland, and has a smooth-scaled costa which is more evenly curved from the base to the costa at one-half**; also *A. umbrana*. **FS** Double-brooded, June-early May; mostly single-brooded in Scotland, appearing from September; the adult hibernates. Occasionally flies at dusk, and comes to light. **Hab** Woodland, heathland, sand dunes, scrub, fens, hedgerows. **Fp** Sallows, also White Poplar, Bog-myrtle; between spun or folded leaves.

Acleris cristana page 236 — 1054 (4389)

Common from the Midlands southwards. Local in Wales. **FL** 9-11mm. Extremely variable, with over 130 forms known. The forewing costa somewhat concave near one-half with a large tuft of raised scales in the middle of the wing, although this can be reduced, and with other scattered smaller tufts, including several in the tornal area, two approaching the apex and two towards the base, although these tufts can become lost with wear. Forms with prominent longitudinal markings on the dorsum are frequently encountered, some of which are contrastingly coloured. **Similar species** *A. hastiana*. **FS** Single-brooded, mid July-May; the adult hibernates. Can be tapped from scrub by day, flies at dusk, comes to light, and has been recorded at Ivy blossom. **Hab** Woodland, scrub, thick hedgerows. **Fp** Blackthorn, also hawthorns, apples; in spun leaves.

Acleris hyemana page 236 — 1055 (4397)

Common, perhaps more so further north. **FL** 6-9mm. Variable. **A narrow-winged species with a pointed apex.** Forewing silvery grey, with reddish-brown to purplish-brown markings, variably speckled blackish, usually including a mark on the costa at the base, an obscure narrow cross-band at about one-half, a large mark on the costa before the apex and diffuse markings on the dorsum. The reddish-brown shading is sometimes extensive, covering much of the wing, leaving a few silvery grey patches. Almost unicolorous reddish-brown examples occur occasionally. The female is smaller. **FS** Single-brooded, September-mid April; the adult hibernates. Flies in afternoon sunshine and comes to light. **Hab** Heathland, moorland. **Fp** Heather, heaths, bilberries, also recorded from Lodgepole Pine, Sitka Spruce; in a slight web on the shoots and among the flowers.

Acleris lipsiana page 236 — 1056 (4411)

Rare. **FL** 8-11mm. **A broad-winged species.** Forewing variable, reddish brown to greyish purple, variably speckled blackish, sometimes with two small yellowish scale-tufts near the middle of the wing. **Similar species** *A. rufana*, **greyish-purple form, which has a narrower forewing with a rather pointed apex, and usually has some whitish scaling along the costa.** **FS** Single-brooded, late August-May; the adult hibernates. **Hab** Bogs, moorland, mountains. **Fp** Bilberry, Bog-myrtle, Cowberry, also possibly birches, apples; in a tube of spun leaves.

Acleris rufana page 236 — 1057 (4412)

Very local. Local in northern England and Scotland. **FL** 8-10mm. Variable. Forewing pale yellowish brown, variably sparsely speckled black and reddish brown, sometimes with a darker mark on the dorsum at about one-third and a large pale grey triangular mark, suffused leaden grey, on the costa. Three other distinct forms: forewing almost unicolorous purplish brown; forewing with the costa shaded whitish basally, the rest of the wing pale yellowish brown; and forewing greyish white, with a longitudinal streak from near the base to the apex, sometimes with a similar streak along the dorsum. **Similar species** *A. lipsiana*. **FS** Single-brooded, August-April; the adult hibernates. Can be tapped from the foodplant by day, flies in the evening and comes to light. **Hab** Wet heathland, moorland, bogs, damp woodland. **Fp** Bog-myrtle, also reported on White Poplar, Goat Willow, Meadowsweet; in a spinning drawn towards the stem.

Acleris lorquiniana page 237   1058 (4387)

Rare. Most frequent in the fens and broads of East Anglia. **FL** 7-10mm. **Forewing with a pointed apex**, light yellowish brown, with scattered blackish scales and a prominent blackish spot in the middle of the wing, although this can be reduced. One form has a longitudinal reddish-brown streak from the base to the apex, this streak sometimes darkened blackish brown. Another form is almost unicolorous, with a few black scales in the middle. **Similar species Some *Agonopterix* (Elachistidae) are similarly patterned, but none has the apex pointed. FS** Double-brooded, June-July, September-October. Comes to light. **Hab** Marshes, fens. **Fp** Purple-loosestrife; first generation feed in the young shoots, the second generation in the flower spikes.

Acleris abietana page 237   1059 (4380)

Rare. May be spreading. Recorded new to Britain in 1965 from Perthshire. **FL** 9-11mm. Variable. **Forewing costa fringed with cilia-like scales**. Forewing dark brown with mid-brown or yellowish-brown patches, and fine short blackish-brown lines, with a concave ridge of raised scales at about one-quarter, extended in the middle as a short spur, another ridge bordering the inner margin of the cross-band at about one-half, sometimes including a prominent scale-tuft in the middle, with further scale-tufts towards the apex, above the dorsum near the base and a smaller tuft towards the tornus. One form is more unicolorous brown, with whitish scales along the dorsum and near the short spur. **Similar species** *A. hastiana* (some forms). **FS** Single-brooded, August-May; the adult hibernates. Comes to light. **Hab** Coniferous woodland. **Fp** Noble Fir, Continental authors give other firs and pines; in a slight web.

Acleris maccana page 237   1060 (4382)

Local. **FL** 8-10mm. Variable. Forewing dark grey with reddish-brown or purplish-brown markings, including an oblique cross-band from the costa at one-third to the tornus, the inner margin edged with roughened dark reddish-brown and/or white scales and the outer margin diffuse, with cross-bands near the base and the apex usually present. One form is whitish with the darker markings developed in the costal half of the wing. **Similar species** *A. hastiana* (some forms). **FS** Single-brooded, August-May; the adult hibernates. Comes to light. **Hab** Moorland, open woodland on high ground. **Fp** Bilberry, Cowberry, Bog-myrtle; in a tubular spinning amongst leaves.

Acleris literana page 237   1061 (4409)

Common. Very local in northern England and Scotland. **FL** 9-11mm. Very variable. Forewing light green, sometimes partly suffused pale yellowish brown or reddish brown, with variable black markings, ranging from much reduced to sharply defined shapes. Some forms have the forewing relatively smooth-scaled, with scattered groups of raised scales, these sometimes tipped whitish. Other forms are rough-scaled throughout, with the scale-tufts less prominent. In some forms the green can be darker or is replaced with light grey or, rarely, yellowish brown. **FS** Single-brooded, mid July-May; the adult hibernates. Abundant in oak woodland in some years. Can be tapped from tree trunks and branches by day, comes to light, and is occasionally recorded at sugar. **Hab** Oak woodland, sometimes gardens. **Fp** Oaks; in a spinning between the leaves.

Acleris emargana page 237   1062 (4385)

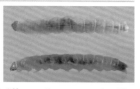

Common. Owing to confusion with *A. effractana*, the distribution is imperfectly known. **FL** 8-11mm. Variable. Forewing with the apex extended to a point, the costa scalloped, pale yellowish brown to reddish brown, veins variably lined brown or greyish brown, sometimes obscure, with a dark or greyish-brown cross-band at about one-half, occasionally obscure, this diffuse on the outer margin with the brownish colour sometimes extending towards the termen. **Similar species** *A. effractana* **has a shallower costal scallop**; most forms are difficult to identify, but dull grey or dull greyish-brown nearly unicolorous examples with groups of blackish raised scales and a weak costal scallop are likely to be *A. effractana*. Examples with a deep costal scallop that are well patterned, or pale yellowish brown, are likely to be *A. emargana*. Examination of genitalia is required to confirm identification of *A. effractana*. **FS** Single-brooded, July-October. Comes to light. **Hab** Woodland, scrub, hedgerows, marshes. **Fp** Goat Willow, occasionally birches; in folded leaves or between two leaves.

Acleris effractana page 237   1062a (not listed)

Local. Very recently raised to species level, having formerly been treated under *A. emargana*. Consequently, its distribution is imperfectly known and no map is included. Apparently this species is more widespread in the north than in the south, but it could be found throughout and careful searching is required. **FL** 8-11mm. **Similar species** *A. emargana*. **FS** Single-brooded, July-October. **Hab** Scrub, damp areas with the foodplants. **Fp** Purple Willow, Grey Willow; in spun leaves or tubed leaves.

Neosphaleroptera nubilana page 237   1027 (4420)

Local. **FL** 6-8mm. Male forewing dark greyish brown, sparsely and finely speckled whitish, with blackish-brown somewhat obscure markings, usually including a cross-band from the costa at one-half to the dorsum before the tornus. The female forewing is usually lighter, speckled greyish brown, with dark greyish-brown markings, including a cross-band from the costa at one-quarter not reaching the dorsum, a slightly oblique cross-band at one-half and a mark on the costa before the apex extended towards the tornus. Hindwing of both sexes greyish brown. **The brownish shading in the forewing and hindwing distinguishes** *N. nubilana* **from all** *Cnephasia* **species, which are grey.** **FS** Single-brooded, July-August. Flies in sunshine and towards dusk, and comes to light. It has been recorded at some clearwing moth (Sesiidae) pheromone lures. **Hab** Hedgerows, scrub, gardens. **Fp** Hawthorn, Blackthorn, occasionally apples; in spun shoots.

Exapate congelatella page 237   1026 (4436)

Common. Very local in the south. **FL** Male 10-11mm, female 5mm. Male forewing slate grey, suffused whitish in a narrow wedge from the base to near the apex, markings including a dark blotch at about one-third and a smaller blotch at about two-thirds, and with a short diagonal streak on the costa before the apex. Examples from Scotland have the forewing plain dark grey with obscure markings. Female with much reduced wings, whitish grey with a dark spot on the costa at about one-half. **FS** Single-brooded, October-December. Readily disturbed by day or found at rest on posts or vegetation. Flies on still afternoons, before dusk and comes to light. **Hab** Hedgerows, woodland edges, lowland heathland, open moorland, frequenting scrub and woodland in the south, and moorland and mosses further north. **Fp** Many shrubs and herbaceous plants; in spun shoots.

Tortricodes alternella page 237
1025 (4439)

Common. Local in Scotland. **FL** 9-11mm. Variable. Male forewing creamy white, silvery grey to pale yellowish brown, variably marked brown and blackish brown, with a yellowish-brown to brown basal patch, a broad cross-band at about one-half, sometimes rather obscure, and a more-or-less wedge-shaped mark on the costa just before the apex. Female forewing extensively shaded blackish brown, the markings less distinct. Obscurely marked grey forms occur. **FS** Single-brooded, late January-early April; has been recorded in May. Sometimes abundant. Can be found flying weakly in sunshine and at rest on tree trunks, and comes to light. **Hab** Deciduous woodland, scrub, occasionally gardens. **Fp** Many deciduous trees and shrubs; in spun leaves or under a turned-down leaf.

Eana osseana page 238
1029 (4443)

Common. **FL** 8-11mm. Forewing pale yellowish brown, variably shaded darker, sometimes tinged reddish brown, speckled with blackish-brown scales, with a few diffuse brown spots near the middle of the wing, although almost unicolorous examples occur. Male and female forewing pointed, more so in the female. **Similar species Male *Cnephasia longana* is comparatively blunt-winged, and does not have the diffuse brown spots near the middle of the wing. FS** Single-brooded, late June-August, into September and October in Scotland. Can be seen in numbers. May be disturbed by day and comes to light. **Hab** Downland, rough grassland, heathland, moorland. **Fp** Many herbaceous plants; in a silken tube among rootstocks, roots and under stones.

Eana incanana page 238
1030 (4450)

Local. Very local in Scotland. **FL** 8-11mm. Forewing light grey, lightly speckled darker grey, the markings dark greyish brown, variably edged and mixed black, including a well-defined curved cross-band at one-quarter, not reaching the dorsum, an oblique cross-band from the costa at about one-half, and a spot, which can be obscure, on the costa just before the apex. **Similar species** Sometimes very difficult to separate from *Cnephasia stephensiana* on wing pattern. **FS** Single-brooded, late May-early August. Can be found in numbers. Comes to light. **Hab** Woodland. **Fp** Bluebell, Oxeye Daisy; in a silken web over the blossom.

Eana penziana page 238
1031 (4462)

Local. Two variations occur in Britain: ssp. *colquhounana* is quite widespread on the coast; f. *bellana* is found inland, but it is now rare or possibly even extinct. **FL** 9-13mm. The *bellana* form has a whitish forewing, speckled blackish, with black markings, including a cross-band at one-quarter, disappearing towards the dorsum, an oblique cross-band from the costa at about one-half, sometimes broken near the middle, and a line or diffuse mark from just before the apex towards the termen. Ssp. *colquhounana* is extensively suffused grey, the markings less distinct. The intensity of the markings of both forms is variable. **Similar species** Some larger *Cnephasia*. **FS** Single-brooded, June-July (inland form), May-September (coastal ssp.). Flies at dusk and comes to light. **Hab** Inland hills, mountains, coastal cliffs. **Fp** Inland form on Sheep's-fescue; in a silken gallery at the roots. Coastal form on Sea Plantain, Thrift; in a silken tube feeding on the root-crowns, roots and leaves.

Cnephasia longana page 238   1016 (4493)

Common. Very local in northern England, rare in Scotland. **FL** 7-11mm. Male forewing unicolorous, whitish grey, pale yellowish brown to dark yellowish brown, rarely speckled with black scales. Female forewing pale yellowish brown, sometimes darker, with darker rather diffuse markings, including an oblique cross-band at one-quarter, an oblique cross-band from the costa at about one-half, the inner margin deeply indented below and above the middle of the wing, and often merged with a mark on the costa just before the apex. **Similar species** *Eana osseana* with male *C. longana* only. **FS** Single-brooded, June-early August. Readily disturbed by day, flies at dusk and comes to light. **Hab** Coastal and calcareous downland, vegetated shingle, waste ground. **Fp** Many herbaceous plants, including Wild Carrot, Thrift, Sea Aster, chamomiles; older larvae in a spinning on plant tips, particularly flowers.

Grey *Cnephasia* species group

The greyish *Cnephasia* are very difficult to determine as the forewing markings are so variable within and between species. The only secure way of identifying them to species is to examine the genitalia, and it is very likely that most records will be accepted only following dissection. The value of doing so is highlighted by the recent finding that *C. pumicana* is present in Britain, and it is possible there are other species awaiting discovery.

There are eight species currently resident, more or less separable into two size groups. The typical forewing of a *Cnephasia* can be described as light grey with darker grey markings, including a cross-band at about one-quarter, usually not reaching the dorsum, an oblique cross-band from the costa at one-half to the tornus, and a diffuse mark on the costa just before the apex. Markings can be pronounced or obscure, speckled or clearly defined, and unicolorous whitish and melanic forms can occur. The life histories of most species are known. They are normally single-brooded, flying in the summer, and the larvae feed in spring on a range of herbaceous plants.

The following notes may help to identify the species, or at least to determine which examples need further study. At the end are comments on four species, *Neosphaloptera nubilana*, *Eana incanana*, *Eana penziana* and *Isotrias rectifasciana*, which are very similar to *Cnephasia*.

Large species	(typically FL 8-11mm)	*C. communana*, *C. conspersana*, *C. stephensiana*
Small species	(typically FL 7-9mm)	*C. pasiuana*, *C. pumicana*, *C. incertana*
Can be large or small		*C. asseclana* (can be as large as *C. stephensiana* in Scotland), *C. genitalana*

Cnephasia communana page 238 1018 (4482) In England and Wales only; local in the south and east but can be numerous where it occurs, rarer or absent in the north and west. **FL** 9-11mm. Narrow, pointed forewing. **FS** Single-brooded, May-July. Usually the only large species flying in May. Easily disturbed by day, flies in the early morning, rests on tree trunks, and comes to light. **Hab** Rough grassland, chalk and limestone downland, fens. **Fp** Unknown in Britain, various herbs on the Continent.

Cnephasia conspersana page 238 1019 (4491) Local. Very local in Scotland, widespread in Ireland; often numerous in coastal areas, and found on chalk in southern England. **FL** 8-11mm. Narrow, pointed forewing. Usually whitish-grey ground colour; the cross-band at one quarter typically extends weakly to the dorsum, this not apparent in the other large species. A chalk-white form without markings occurs on downland. **FS** Single-brooded, June-mid August, possibly double-brooded in Ireland, May-September. Easily disturbed by day in sunny weather, rests on posts and comes to light. **Hab** Coastal cliffs and adjoining fields, chalk downland. **Fp** Oxeye Daisy and other Asteraceae; draws the petals together in a spinning. Also in the shoots of Sea Campion.

Grey Tortrix *Cnephasia stephensiana* page 238 1020 (4474) Common. Local in Scotland. Occasionally seen in numbers. **FL** 9-11mm. Broad, blunt forewing. Ground colour very variable, from pale grey to blackish brown, often mid grey; markings can be prominent or obscure. An almost unicolorous white form occurs on the Kent coast. Another form, frequent in northern England and Scotland, has a white forewing with markings well defined. **FS** Single-brooded, June-August; also recorded in April, May and October. Flies at dusk and regularly comes to light. **Hab** Open woodland, scrub, hedgerows, rough grassland, fens, marshes, gardens. **Fp** Daisies, buttercups, docks, plantains; usually in a spun tube or folded leaf.

Flax Tortrix *Cnephasia asseclana* page 238 1021 (4477) Common. The most widespread *Cnephasia* in Scotland. **FL** 7-10 mm. Usually has a more variegated or mottled pattern than other small species, with a narrower central cross-band and a costal mark near the apex often extending to the termen. A melanic form occurs which has most wing markings obscured. **FS** Single-brooded, late May-August; in Scotland emerges later than *C. incertana*. Readily disturbed by day, flies at dusk and regularly comes to light. **Hab** Open woodland, scrub, hedgerows, grasslands, gardens. **Fp** Oxeye Daisy, buttercups, docks, peas; usually in a spinning of the leaves and flowers.

Cnephasia pasiuana page 238 1022 (4479) Local in England from the Midlands southwards, but can be numerous where it occurs. Very local or absent elsewhere. **FL** 7-9mm. Usually greyish brown and obscurely patterned, but with a relatively broad central cross-band, its inner margin being more strongly defined. **FS** Single-brooded, June-July; also recorded in May and August. Flies in the evening and comes to light. **Hab** Rough grassland, fens, marshes. **Fp** Oxeye Daisy and other Asteraceae; draws the petals together in a spinning.

Cnephasia pumicana page 238 1022a (not listed) Very local; currently recorded south of a line between Hampshire and Suffolk. **FL** 7-9mm. A uniformly grey species most similar to *C. pasiuana*, with only a hint of typical *Cnephasia* markings. **FS** Probably single-brooded, late June-early August. Has so far been recorded only at light. **Hab** Recorded in open countryside and urban areas. **Fp** Unknown in Britain.

Cnephasia genitalana page 238 1023 (4480) Very local. Resident in coastal counties in central southern and south-east England, probably an immigrant or wanderer elsewhere. **FL** 8-11mm. Often slightly larger than others in the group of small species. Typically whitish grey with faint markings. **FS** Single-brooded, mid June-August. Often the latest of the plain species. Comes to light. **Hab** Chalk downland, rough grassland, open woodland, gardens, coastal cliff and slope. **Fp** Oxeye Daisy and other Asteraceae, buttercups and peas; in spun flowers and leaves.

Light Grey Tortrix *Cnephasia incertana* page 238 1024 (4471) Common. Local in Scotland. **FL** 6-9mm. Typically the forewing is cleanly marked, not mottled as in *C. asseclana*. The ovipositor is exceptionally long and can be seen protruding from the abdomen of the female, unlike the females of other *Cnephasia*. **FS** Single-brooded, late May-mid August, recorded in September; **usually the earliest of the smaller species on the wing**. Can be disturbed by day and regularly comes to light. **Hab** Hedgerows, verges, woodland, gardens. **Fp** Oxeye Daisy, plantains, docks, buttercups; twists the young leaves together in a spinning.

Species which are similar to the *Cnephasia* group

Neosphaloptera nubilana is most similar to *C. asseclana* but has the dark forewing distinctly mixed brownish, which is absent in *Cnephasia*. *Eana incanana* has an almost identical wing pattern to some forms of *C. stephensiana*, but *E. incanana* has a more arched costa, and the curved cross-band at one-quarter is clearly defined against the pale grey background. However, identification from photographs of adults at rest is very difficult as the arched costa is not always apparent. *E. penziana* is generally larger than most *Cnephasia*, although there is some overlap. *E. penziana* has a long, narrow forewing, with a noticeably rounded apex and curved termen. *Isotrias rectifasciana* (p.293) can resemble a small species of *Cnephasia*, although it has a distinct basal patch, lacks the short, incomplete cross-band at one-quarter, and the cross-band at one-half is almost directly transverse, not oblique.

Sparganothis pilleriana page 239   1012 (4517)

Very local. **FL** 7-10mm. Variable. **Labial palps long.** Male forewing yellowish brown, sometimes tinged greyish, occasionally paler, with brown markings, including an oblique narrow cross-band from the costa at one-third to the dorsum at one-half, and a dark quadrate mark on the costa at about two-thirds, often with a narrow extension towards the tornus; markings can be obscure. Female forewing almost uniform reddish brown, sometimes yellowish brown. Darker forms predominate on saltmarshes, paler forms being more frequent on heathland. **Similar species** *Phtheochroa inopiana*. **FS** Single-brooded, July-August. Flies from dusk and comes to light. **Hab** Fens, wet heaths, bogs, saltmarshes. **Fp** Many herbaceous plants, including Common Sea-lavender, Bog Asphodel, woundworts, knapweeds, plantains; in folded or spun leaves.

Eulia ministrana page 239   1015 (4520)

Common. **FL** 8-10mm. Variable. Forewing pale yellow, overlaid orangey brown to reddish brown in the basal area, middle and towards the termen, the suffusion in the basal and central areas connected, usually with a conspicuous white spot or mark at about three-quarters; any markings ill defined. In extreme forms the wing is almost entirely reddish brown. **FS** Single-brooded, late April-mid July. Readily disturbed by day, flies at dusk and comes to light. **Hab** Woodland, scrub, hedgerows, isolated trees in mosses and bogs. **Fp** Many trees and shrubs, including Hazel, Ash, Bilberry, but mainly birch in Scotland; older larvae in a tubular spinning between two leaves.

Pseudargyrotoza conwagana page 239   1011 (4522)

Common. **FL** 5-7mm. Forewing yellowish, variably suffused pale reddish brown, with rather ill-defined dark greyish-brown markings and leaden metallic streaks or spots, and a small, pale yellow dorsal blotch at about one-half. In strongly marked examples the small yellowish dorsal blotch contrasts distinctly with the blackish suffusion. A form occurs rarely with a deep orange forewing and bright silvery spots. **Similar species** *Spatalistis bifasciana*. **FS** Double-brooded, May-August; recorded in September and November. Can be numerous where found. Flies around tree-tops on warm sunny days and at dusk, and comes to light. **Hab** Woodland, scrub, hedgerows, gardens. **Fp** Ash, privets; in the seeds or berries.

Red-barred Tortrix *Ditula angustiorana* page 239   1010 (4525)

Common. Very local in Scotland. **FL** 6-8mm. Male smaller than the female. Male forewing brown, with a dull yellowish-brown semicircular patch on the dorsum before one-half, a narrow reddish-brown cross-band at about one-half, obscure towards the costa, and an ill-defined reddish-brown mark on the costa just before the apex, extended towards the tornus. Female forewing orangey brown, with a small yellowish-brown patch on the costa at one-half and less black in the outer half of the wing. **FS** Single-brooded, mid May-August, with an occasional partial second brood, September-October; has been recorded in April. Flies in sunshine, at dusk and comes to light. **Hab** Gardens, hedgerows, orchards, scrub, woodland. **Fp** Many species of tree, shrubs and occasionally herbaceous plants, including apples, Ivy, Larch, junipers, oaks, Mistletoe; usually among spun leaves and the developing fruit buds.

Epagoge grotiana page 239   1006 (4531)

Common. Very local in northern England, rare in Scotland. **FL** 6-8mm. Forewing yellowish brown, coarsely speckled reddish brown, with dark reddish-brown markings, including an ill-defined basal patch, an oblique cross-band from just before one-half on the costa, broadest towards the dorsum, and a large quadrate mark on the costa before the apex. **FS** Single-brooded, June-early August; occasionally recorded in late May. Flies at dawn and dusk, and comes to light. **Hab** Woodland, hedgerows, sandhills. **Fp** Oaks, hawthorns, Bramble; in a spinning in the leaves. Also feeds on dead oak leaves on the ground.

Periclepsis cinctana page 239   1005 (4539)

Rare. Recently found on Tiree in the Inner Hebrides. Formerly known from the Dover and Canterbury areas of Kent. **FL** 7-8mm. Forewing white, speckled leaden, with reddish-brown markings, speckled black and leaden, comprising a basal patch, a cross-band near one-half and a costal spot just before the apex. **FS** Single-brooded, late June-July. Flies in sunshine. **Hab** Chalk grassland, machair grassland. **Fp** Continental authors give Broom, greenweeds, Kidney Vetch, Common Bird's-foot-trefoil, Field Wormwood; in a tubular silken gallery.

Philedone gerningana page 239 1008 (4541)

Local. More local in parts of its range, very local in East Anglia. **FL** 6-9mm. **Male has bipectinate antennae.** Male forewing pale yellowish brown, suffused reddish brown near the base, the outer half reddish brown, sometimes with yellowish brown bordering the outer margin of an oblique reddish-brown cross-band at about one-half. Female forewing is narrower and extended to a point at the apex, similarly marked but suffused dull purplish brown. **Similar species Male *Epiphyas postvittana* is typically larger, with a strongly arched costa and thread-like antenna. FS** Single-brooded, July-September. Flies by day in sunshine from sunrise, in the evening and comes to light. **Hab** Heathland, moorland, bogs, limestone grassland. **Fp** Bilberries, rock-roses, Thrift, cinquefoils, occasionally Sitka Spruce; in spun leaves and flowerheads.

Capua vulgana page 239 1007 (4547)

Common. More frequent in the south. **FL** 7-9mm. Forewing with an evenly arched costa, ground colour yellowish brown, variably suffused greyish; **the male has a costal fold to near one-half and a pale yellowish-brown patch near the base obliquely edged to the costa at one-third,** beyond which are brownish markings, often indistinct; female is unicolorous. **FS** Single-brooded, May-mid July; recorded in late April. Flies by day around trees and over Bilberry, also at dusk and comes to light. **Hab** Woodland, scrub, heathland, moorland, occasionally gardens. **Fp** Several species of trees and shrubs, including Hornbeam, alders, whitebeams, Bilberry.

Philedonides lunana page 239 1009 (4549)

Local. **FL** 6-8mm. Male larger than the female and the forewing costa of both sexes is shallowly concave. **Male has bipectinate antennae.** Male forewing pale greyish, with dark brown or reddish-brown markings, sparsely speckled black, including a diffuse basal patch, a well-defined cross-band from the costa just before one-half to near the tornus, broadening towards the dorsum, and a diffuse spot on the costa before the apex. Female forewing similar, the ground colour whitish. **FS** Single-brooded, March-May. Males fly by day from sunrise to sunset. **Hab** Heathland, moorland, bogs, acid grassland. **Fp** Many herbaceous plants, including cinquefoils, Alexanders, Bog-myrtle, Bilberry, Heather; in spun or folded leaves and flowers.

◄ Larva of *Archips oporana*.

▲ Spinning of *Archips oporana* on Giant Fir. Note the cluster of dead needles in which the larva hides, and the individual needle in the silk spinning; this needle was severed at the base from near the shoot tip, and like others, has been pulled back by the larva into the spinning to be eaten. This is a rare example of a moth larva which brings its food back 'home' before it eats.

Archips oporana page 239   976 (4555)

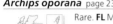

Rare. **FL** Male 9-10mm, female 11-14mm. Male forewing purplish brown, with reddish-brown markings, thinly edged white, including an oblique cross-band at one-half, a mark on the costa before the apex and a mark along the termen; the apex is slightly pointed. Female forewing pale purplish brown, markings reddish brown, with a coarse net-like pattern over much of the wing; the apex is strongly pointed. Hindwing greyish brown, at least the outer half with a coppery suffusion in the male, orangey in the female. **FS** Single-brooded, late June-July. Flies from mid-afternoon and comes to light. **Hab** Coniferous woodland. **Fp** Scots Pine, European Silver Fir, White Spruce, Giant Fir, junipers; in a silken tube amongst the needles, severing a needle and drawing it into the spinning to feed.

Large Fruit-tree Tortrix *Archips podana* page 239   977 (4557)

Common. Rare in Scotland. **FL** Male 9-11mm, female 9-14mm. Male forewing yellowish brown, suffused pale purplish in the basal two-thirds, the veins finely lined in the outer one-third, with dark reddish-brown markings, including a short bar from the dorsum near the base, and an obscure cross-band at about one-half; the apex is extended to a point. Hindwing grey, tinged orange at the apex. Female forewing with an extensive net-like pattern, the darker markings weak, and the apex is strongly extended to a point. Hindwing brownish grey, the outer half bright orange. A dark purplish-brown form occurs with the markings rather obscure and the orange on the hindwing reduced or absent. **FS** Single-brooded, late May-July, with a partial second brood from mid August-mid October. Flies from late afternoon and comes to light. **Hab** Woodland, orchards, hedgerows, gardens, occasionally more open habitats. **Fp** Many deciduous trees, shrubs and occasionally conifers; in spun leaves or flowers.

Brown Oak Tortrix *Archips crataegana* page 239   979 (4558)

Common. More frequent in the south. **FL** Male 9-10mm, female 11-13mm. Male forewing light greyish brown, with dark purplish-brown markings, variably suffused darker, including a short bar from the dorsum near the base, a broad oblique cross-band from the dorsum beyond the middle not reaching the costa, the inner margin edged paler, with a near triangular mark on the costa before the apex extended towards the termen; the apex is slightly extended to a point. Female forewing usually darker than the male, often with a net-like pattern, the cross-band narrowing before reaching the costa, and the apex is extended to a point. **Similar species Variegated Golden Tortrix *A. xylosteana* is slightly narrower winged and is smaller, more yellowish brown and variegated, and in the male the cross-band reaches the costa.** **FS** Single-brooded, June-August. Can be disturbed by day, flies at dusk and comes to light. **Hab** Woodland, orchards, hedgerows, occasionally scrubby situations. **Fp** Many deciduous trees and shrubs, including oaks, Ash, elms, limes, sallows; older larvae roll a leaf into a tube.

Variegated Golden Tortrix *Archips xylosteana* page 239   980 (4559)

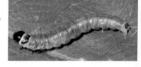

Common. Local in northern England, very local in Scotland. **FL** 7-11mm. Forewing pale yellowish brown, sometimes darker, suffused greyish brown, with reddish-brown markings, variably suffused darker, including a short bar from the dorsum near the base, and an oblique cross-band at one-half, usually joined to a mark on the costa before the apex, this sometimes extended towards the tornus. Female longer winged than the male, with a more pointed apex. **Similar species** *A. crataegana*. **FS** Single-brooded, late May-mid August; recorded in mid September. Occasionally recorded in numbers. Readily disturbed by day, flies at dusk and comes to light. **Hab** Woodland, hedgerows, gardens. **Fp** Many deciduous trees and shrubs, including oaks, elms, Hazel, Ash, fruit trees, also Bramble, Honeysuckle, St John's-wort; in a rolled leaf.

Rose Tortrix *Archips rosana* page 239 981 (4560)

Common. Local in northern England, very local in Scotland. Appears to be declining. **FL** Male 7-8mm, female 8-11mm. Male forewing very broad, light brown to purplish brown, with darker brown markings, including a basal patch, an oblique cross-band at about one-half and a mark on the costa just before the apex, this with a dark line extending towards the tornus. Female forewing is longer, the apex extended to a point, with a slight net-like pattern and the markings weaker; the hindwing is usually tinged orangey yellow towards the apex. **FS** Single-brooded, late June-early September; recorded in early October. Readily disturbed by day, flies at dusk and comes to light. **Hab** Woodland, orchards, hedgerows, scrub, gardens. **Fp** Many deciduous trees and shrubs, especially fruit trees, raspberry, Black Currant, Hop, and occasionally conifers; in a rolled leaf, spun leaves or on flowers and young fruit.

Choristoneura diversana page 239 982 (4562)

Very local. **FL** 7-10mm. Forewing yellowish brown, sometimes greyish brown, with darker brown markings, sometimes purplish brown, including a basal patch, and an oblique cross-band at about one-half with the inner margin fairly straight, the outer margin irregular and sometimes diffuse, especially above the tornus, the band often interrupted above the middle and narrowing to the costa. **Similar species** *Syndemis musculana* **has ground colour greyish white; Barred Fruit-tree Tortrix** *Pandemis cerasana* **and Dark Fruit-tree Tortrix** *P. heparana* **both have the outer margin of the cross-band relatively straight or gently curved and well defined. FS** Single-brooded, late June-July. Flies in the evening, at dusk and comes to light. **Hab** Woodland, orchards, scrub. **Fp** Many deciduous trees, shrubs and herbaceous plants, including Field Maple, birches, apples, oaks, sallows, Honeysuckle, clovers; in spun leaves.

Choristoneura hebenstreitella page 240 983 (4564)

Common. Rare in northern England. Possibly declining. **FL** Male 9-12mm, female 11-15mm. Forewing yellowish brown, tinged greyish, weakly lined brown, resulting in a net-like pattern, with brownish markings, including a basal patch, an oblique cross-band at about one-half with the inner margin fairly straight, the outer margin irregular and expanded above the tornus, the band sometimes interrupted above the middle and narrowing to the costa, and with an almost semi-circular mark on the costa at about two-thirds; in the female the apex is pointed. **Similar species** *Lozotaenia forsterana* **has a broader forewing and always has the cross-band widely interrupted in the middle, and does not have a basal patch. FS** Single-brooded, late May-July; recorded in August. Occasionally recorded in numbers. Flies at dusk and comes to light. **Hab** Woodland, moorland, gardens. **Fp** On a variety of deciduous trees and shrubs, including oaks, birches, Bog-myrtle, Ivy, Bilberrry; in spun or rolled leaves.

Argyrotaenia ljungiana page 240 974 (4568)

Common. **FL** 6-8mm. Forewing silvery white, suffused grey, with reddish-brown markings, speckled black, including a basal patch, a broad oblique cross-band at about one-half and a mark on the costa just before the apex, sometimes extended to the tornus. The population in urban areas in south-east England has a brighter silver ground colour than the heathland and moorland populations elsewhere. **FS** Double-brooded in the south, mid April-May, late June-August; also recorded in February, March and October. Single-brooded in the north, late April-June. Flies in late afternoon and comes to light.

Hab Heathland, moorland, marshes, gardens. **Fp** Many deciduous trees, shrubs and herbaceous plants, including Heather, heath, Bog-myrtle, Bilberry, birches, pines, Larch, knapweeds, Wood Sage; in spun leaves.

Ptycholomoides aeriferanus page 240   987 (4572)

Common, spreading slowly. First recorded in Britain 1951 in Kent. **FL** 8-10mm. Forewing pale golden yellow with blackish-brown markings, comprising a basal patch, a broad oblique cross-band at about one-half, and a mark on the costa just before the apex, often with weak greyish iridescent mottling. Hindwing dark chocolate brown. **FS** Single-brooded, late June-mid August; recorded in May. Comes to light. **Hab** Larch woodland, gardens, occasionally elsewhere. **Fp** Larch; among spun needles.

Ptycholoma lecheana page 240   1000 (4574)

Common. More local in northern England and Scotland. **FL** 7-11mm. Female usually larger than the male. Forewing dark brown, variably suffused greenish yellow, the outer half shaded yellowish brown to pale reddish brown, with an obscure, slightly darker, cross-band at one-half partially edged metallic leaden, this occasionally absent, the outer line sometimes extended as a line towards the apex. Hindwing blackish brown. **FS** Single-brooded, mid May-mid August. Flies in the afternoon and evening, and comes to light. **Hab** Woodland, parks, orchards, hedgerows, gardens. **Fp** On a range of trees and shrubs, including apples, poplars, oaks, sallows, Larch; in rolled or spun leaves.

Chequered Fruit-tree Tortrix *Pandemis corylana* page 240   969 (4578)

Common. Local in Scotland. **FL** 9-12mm. Female usually larger than the male. Forewing pale yellowish brown, sometimes reddish brown, lined brown to dark brown, resulting in a net-like pattern, with a basal patch, the outer edge of which is almost parallel to the oblique cross-band at about one-half, these markings outlined darker; the cilia are often reddish brown. The intensity of the forewing markings can vary. Hindwing is grey in the basal half, pale yellowish in the outer half, with a few short darker cross-lines. **Similar species *P. cerasana* has a weaker net-like pattern, and the margins of the basal patch and the cross-band are usually more strongly divergent towards the dorsum; the hindwing is more uniformly coloured grey. *P. corylana* is rarely on the wing before mid July, at least a month after *P. cerasana* starts flying.** **FS** Single-brooded, July-September, occasionally to mid October. Readily disturbed by day and comes to light. **Hab** Deciduous woodland, scrub, hedgerows, gardens. **Fp** On a range of trees and shrubs, including Hazel, Ash, oaks, Bramble, Honeysuckle; in spun leaves or a folded leaf.

Barred Fruit-tree Tortrix *Pandemis cerasana* page 240   970 (4579)

Common. **FL** 8-11mm. Forewing yellowish brown to pale greyish brown with brown markings, including a weak net-like pattern, a basal patch and an oblique cross-band, these edged reddish brown. The intensity of the markings can vary and some examples have a dark greyish-brown suffusion in the dorsal half of the wing. **Similar species *P. heparana* is typically reddish brown and has an angular projection on the inner margin of the cross-band**; also *P. corylana*, *Clepsis rurinana*, *Choristoneura diversana*. **FS** Single-brooded, late May-September. Flies at dusk and comes to light. **Hab** Open woodland, parkland, scrub, orchards, hedgerows, gardens. **Fp** Many deciduous trees, shrubs and herbaceous plants, including oaks, alders, sallows, birches, Bilberry, loosestrifes; in a rolled or folded leaf.

Pandemis cinnamomeana page 240 971 (4577)

Local. Possibly expanding its range further north. **FL** 8-11mm. The face and part of the labial palps are white in the male. Forewing reddish brown with darker brown markings, including a basal patch, an oblique cross-band, the inner margin of which is nearly straight or gently sinuous, and a mark on the costa just before the apex. Hindwing of the female tinged reddish brown towards the apex. **Similar species Male *P. heparana* does not have any white on the head, and the female is without the reddish-brown suffusion towards the hindwing apex; both sexes of *P. heparana* have an angular projection on the inner margin of the cross-band.** **FS** Single-brooded, June-early August, sometimes with a partial second brood from late August-mid October. Readily disturbed by day. Comes to light. **Hab** Woodland, gardens. **Fp** Many deciduous trees and shrubs, including birches, plums, Larch, Rowan, Bilberry; among spun leaves.

Dark Fruit-tree Tortrix *Pandemis heparana* page 240 972 (4580)

Common. **FL** 8-12mm. Forewing yellowish brown to reddish brown with darker markings, including a weak net-like pattern, a basal patch, an oblique cross-band at about one-half, this typically with an angled projection on the inner margin near the middle of the wing, and a mark on the costa towards the apex. **Similar species** *P. cerasana*, *P. cinnamomeana*, *Choristoneura diversana*. **FS** Single-brooded, late May-September; in Scotland, August-September. Can be found in numbers. Readily disturbed by day, flies at dusk and comes to light. **Hab** Woodland, hedgerows, scrub, gardens. **Fp** Many deciduous trees, shrubs and herbaceous plants, including oaks, apples, pears, limes, sallows, Blackthorn, Honeysuckle, birches, Bog-myrtle, Hop; usually in a rolled leaf, occasionally on flowers.

Pandemis dumetana page 240 973 (4581)

Very local. **FL** 8-10mm. **Forewing rather broader than in other *Pandemis*.** Forewing dull brown, the veins lined dark greyish brown resulting in a net-like pattern, with dark greyish-brown markings, including a basal patch, a cross-band at about one-half and a mark on the costa just before the apex, with a dark line extending from this mark towards the tornus. **Hindwing light grey in the basal half, whitish grey with a slight net-like pattern in the outer half. FS** Single-brooded, July-August. Flies from sunset and comes to light. **Hab** Chalk downland, fens, damp woodland, gardens. **Fp** Mainly herbaceous plants, also deciduous trees and shrubs, including loosestrifes, knapweeds, mints, Bramble, oaks, Ivy; in rolled leaves, occasionally on flowerheads.

Syndemis muscularia page 240 986 (4584)

Common. Represented by ssp. *musculinana* on Orkney, Shetland and the Outer Hebrides. **FL** 8-10mm. Forewing greyish white, silvery grey to greyish brown, sparsely speckled black, with dark brown markings comprising an indistinct basal patch, a broad oblique

cross-band at about one-half and a spot on the costa before the apex. The intensity of the markings varies. **Similar species** *Choristoneura diversana*. **FS** Single-brooded, late April-mid July. Easily disturbed by day, flies freely from late afternoon, and comes to light. **Hab** Open woodland, parkland, hedgerows, scrub, moorland, gardens. **Fp** Many deciduous trees and shrubs, including Bog-myrtle, Bramble, oaks, birches; in a tube of spun leaves or a folded leaf.

Lozotaenia forsterana page 240

1002 (4590)

Common. **FL** 10-14mm. A broad-winged species. Female often larger than the male. Forewing light greyish brown, sparsely marked darker greyish brown, resulting in a faint net-like pattern, with an oblique cross-band from the costa at about one-half to the tornus, usually widely interrupted in the middle, but sometimes faintly discernible, with the outer margin diffuse near the tornus, and with a mark just before the apex. **Similar species** *Choristoneura hebenstreitella.* **FS** Single-brooded, late May-August; has been recorded in December. Comes to light. **Hab** Woodland, parks, orchards, scrub, hedgerows, moorland, gardens. **Fp** On a range of plants, especially Ivy, also including Honeysuckle, Bilberry, Larch, pines; between spun leaves.

Carnation Tortrix *Cacoecimorpha pronubana* page 240

985 (4592)

Common. First recorded at Bognor, Sussex, in 1905. May have declined after a period of abundance. **FL** Male 7-9mm, female 8-12mm. Male forewing dark yellowish brown, with a weak dark brown net-like pattern on the outer part of the wing, markings dark brown to purplish brown, including an oblique cross-band from just before the costa at one-half, the termen broadly shaded darker; hindwing orange, with a black suffusion at least along the termen, often more extensive. Female forewing long, extended to a point at the apex, similar to the male, with a clearer net-like pattern and without darker shading on the termen; hindwing orange, brownish towards the base. **The characteristic orange hindwing is easily seen when the adult flies in sunshine. FS** At least double-brooded, recorded all year round. Comes to light. **Hab** Gardens, hedgerows, sometimes in woodland, more coastal in some areas. **Fp** Many shrubs and herbaceous plants, including Sea-buckthorn, Evergreen Spindle, fuschias, carnations, Tamarisk; in a spinning.

Bilberry Tortrix *Aphelia viburnana* page 240

988 (4604)

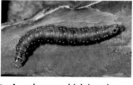

Common. Local in the south. **FL** 8-11mm. The female has a pointed apex. Forewing greyish brown to reddish brown, sometimes paler, unmarked or occasionally with dark brown or reddish-brown markings, including an oblique cross-band from just before one-half on the costa, approaching the tornus. **Similar species** Pale unicolorous forms are similar to *A. paleana*, which is paler, without a trace of brown; the two species can be readily separated by examining the genitalia; also *Phtheochroa inopiana.* **FS** Single-brooded, July-September; recorded in June. Can be found in numbers. Readily disturbed by day, flying in sunny conditions, and occasionally comes to light. **Hab** Heathland, moorland, mosses, bogs, saltmarshes, coastal chalk grasslands. **Fp** Many shrubs and herbaceous plants, including Bilberry, Bog-myrtle, sallows, cinquefoils, knapweeds, Meadowsweet, Sea Aster, also pines, firs; in spun leaves.

Timothy Tortrix *Aphelia paleana* page 241   989 (4596)

Common. More coastal in Scotland. **FL** 8-12mm. Female has a pointed apex. Forewing whitish grey to pale yellowish brown; in the male, the base of the wing, thorax and head are suffused orange yellow. The underside of the forewing is pale greyish brown. **Similar species** *A. unitana* is difficult to distinguish from *A. paleana*, and identification is often only possible by examination of genitalia (see p.389 for male genitalia only); male *A. unitana* is silver grey with whitish cilia and without an orange-yellow suffusion at the wing base and on the thorax, and the underside of the forewing is greyish; the female usually has a slightly broader forewing; also *A. viburnana*. **FS** Single-brooded, June-August; recorded in May. Readily disturbed by day, flies in the evening and comes to light. **Hab** Rough grassland, limestone grassland, scrub, waste ground, sandhills. **Fp** On various grasses, including Common Couch, and herbaceous plants such as knapweeds, plantains, meadowsweets, occasionally oaks and Beech; in spun leaves.

Aphelia unitana page 241   990 (4597)

Rare. Distribution poorly understood. **FL** 9-12mm. Female has a pointed apex. Male forewing silvery grey, sometimes with a slight yellow suffusion, the cilia whitish. Female forewing whitish yellow to pale yellow. **Similar species** *A. paleana* (see p.389 for male genitalia only). **FS** Single-brooded, June-early August. Readily disturbed by day. **Hab** High moorland, bogs, limestone dales. **Fp** A range of herbaceous plants, including Hogweed, angelicas, Bramble; in spun leaves.

Clepsis senecionana page 241   991 (4616)

Local. **FL** 6-8mm. Female has a narrower and more pointed forewing than the male. Male forewing grey, brownish towards the base, weakly speckled greyish brown or with weak short cross-lines towards the apex. Female forewing pale yellowish brown, sometimes darker towards the base, and weakly speckled greyish brown. **FS** Single-brooded, May-June. Flies in afternoon sunshine and in the evening. **Hab** Moorland, mosses, fens, marshes, bogs, damp grassland in woods, rough pastures. **Fp** Bilberry, Cowberry, Bog-myrtle, also Larch, spruces, pines; in a rolled leaf or spun terminal leaves.

Clepsis rurinana page 241   992 (4618)

Rare. Not recorded recently. **FL** 8-10mm. Forewing pale yellowish brown, markings slightly darker, including an oblique cross-band, well defined only along the inner margin, from the costa just before one-half, and with a mark on the costa just before the apex with a line extended towards the tornus. Hindwing pale grey, whitish in a wedge-shape widening towards the apex. **Similar species** Pale forms of *Pandemis cerasana* have a defined outer margin to the central cross-band, and the hindwing is uniform pale grey. **FS** Single-brooded, late June-July. **Hab** Open woodland, hedgerows. **Fp** On a range of deciduous trees and bushes, including Beech, oaks, roses, Honeysuckle; in rolled leaves.

Cyclamen Tortrix *Clepsis spectrana* page 241 993 (4623)

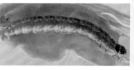

Common. Local in Scotland. **FL** 8-11mm. Variable. Forewing pale yellowish brown, sometimes reddish brown, variably speckled brown, with a brown oblique cross-band from before one-half on the costa, well defined on the costa, becoming diffuse towards the dorsum, and with a brown wedge-shaped mark just before the apex. Markings vary in strength, sometimes reduced, occasionally almost obsolete. Dark brown and unmarked orangey examples occur rarely. **FS** Double-brooded, May-September. Occasionally recorded in numbers. Flies at dusk and comes to light. **Hab** Bogs, fens, marshes, grazing levels, damp woodland, gardens. **Fp** Many plants, including willowherbs, cinquefoils, spiraeas, Hop, Sea Aster, Common Sea-lavender, Saltmarsh Rush; in spun leaves and flowers.

Clepsis consimilana page 241 994 (4629)

Common. **FL** 7-8mm. Male forewing yellowish brown, typically with weak markings, including an oblique cross-band on the costa at about one-half. Female forewing darker, speckled reddish brown and without markings except for a few small dark spots along the dorsum. Rarely, the male forewing can be yellowish with a reddish cross-band. **FS** Single-brooded over much of the range, June-September, double-brooded in the south, late May-mid November. Can be disturbed by day, flies in the afternoon and evening, and comes to light. **Hab** Hedgerows, scrub, waste ground, gardens. **Fp** Many trees and shrubs, including privets, Ivy, Lilac, Honeysuckle, apples; feeds on dead or withered leaves in a dense, untidy spinning.

Light Brown Apple Moth *Epiphyas postvittana* page 241 998 (4632)

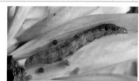

Common. Spreading. An adventive from Australia, first recorded in Cornwall in 1936. Populations are depressed after a cold winter. **FL** 7-12mm. Male smaller than the female. Variable. Male forewing with the basal half pale yellowish, sharply contrasting with the rest of the wing, this reddish brown, sometimes darker. Female forewing yellowish brown to orange-brown, obscurely speckled darker, with variable darker markings, including an oblique mark on the costa at about one-third and a mark near one-half, with one on the dorsum at about one-quarter and at the tornus. **Similar species** *Philedone gerningana* (male only). **FS** Continuously brooded, but mainly two overlapping generations, all year round. Can be abundant, especially in the autumn. Readily disturbed by day, flies from dusk and comes to light. **Hab** Gardens and a wide range of other habitats. **Fp** Many trees, shrubs and herbaceous plants; in a spinning among living and dead plant tissues, including flowers, leaves, fruits and seeds.

Lozotaeniodes formosana page 241 1001 (4635)

Common. Expanding its range. First recorded in 1945, in Surrey. **FL** 10-12mm. Forewing pale yellowish brown crossed with numerous interconnecting orange-brown cross-lines, extensively edged dark reddish brown, forming a strong net-like pattern. **FS** Single-brooded, late May-August, with an occasional partial second brood in October. Flies at dusk and comes to light. **Hab** Woodland, gardens, occasional wanderers elsewhere. **Fp** Scots Pine; in a silk tube along a twig.

Summer Fruit Tortrix *Adoxophyes orana* page 241   999 (4637)

Very local. First recorded in 1950, in Kent. **FL** 7-10mm. Male usually smaller than the female. Male forewing yellowish brown with darker brown markings, including a basal patch, an oblique, irregularly margined, rather narrow cross-band from the costa just before one-half to the dorsum before the tornus, a cross-line from a mark on the costa at about two-thirds to the tornus, with another thin line parallel to the termen. Female forewing slightly darker, markings similar to the male, often rather fainter. **FS** Double-brooded, late May-October. Flies at dusk and comes to light. **Hab** Gardens, orchards, parks. **Fp** Many trees, shrubs and other plants, including apples, pears, plums, sallows, Honeysuckle, roses; in a spinning on the leaves and fruits.

Olindia schumacherana page 241   1013 (4642)

Local. Most frequent in southern England. **FL** 5-8mm. Male usually smaller than the female. Forewing blackish brown, speckled black, mottled leaden near the base and towards the apex; the male has a narrow white cross-band before one-half, often interrupted below the costa, sometimes reduced to two spots; in the female, the cross-band is usually wide and complete, sometimes narrow. **FS** Single-brooded, late May-July, with an occasional partial second brood, September. Flies in afternoon sunshine and comes to light. **Hab** Woodland, downland, fens, marshes. **Fp** Lesser Celandine, columbines, mercuries, Bilberry, Hedge Woundwort and other herbaceous plants; in a spun or folded leaf.

Isotrias rectifasciana page 241   1014 (4646)

Common. Local in parts of northern England, very local in Scotland. **FL** 5-8mm. Forewing creamy white, speckled yellowish brown, the markings yellowish brown to dark brown, overlaid greyish, variably speckled black, including a basal patch, a cross-band at one-half, with the inner margin indented near the middle of the wing, although this indent is not always clear in the female. **Similar species** Small *Cnephasia*. **FS** Single-brooded, May-July; recorded in August. Flies from sunset and comes to light. **Hab** Open woodland, scrub, hedgerows, gardens. **Fp** Unknown, possibly hawthorns.

Bactra furfurana page 242   1110 (4656)

Local. **FL** 6-9mm. **A long-winged species.** Forewing yellowish brown or orangey brown, with small patches of dirty yellowish white forming markings across the wing, including an irregular curved cross-band at about one-third, usually linked by a narrow curved line through the middle of the wing to a mark below the costa at two-thirds, sometimes faintly extended into a curved cross-band from the costa to the tornus; a diffuse darker brown apical streak is usually discernible. **Similar species Unicolorous forms of *B. furfurana* and *B. lancealana* can be separated only by examination of genitalia** (see p.392 for male genitalia only). **FS** Single-brooded, June-mid August. Flies in the afternoon, at dusk and comes to light. **Hab** Marshes, bogs, meres, margins of freshwater ponds, soft-rock cliffs. **Fp** Common Club-rush, spike-rushes, Compact Rush; in the stems.

Bactra lancealana page 242   1111 (4655)

Common. **FL** 5-10mm. Very variable. Forewing grey to light brown, with markings varying from almost uniform to a darker brown in a fine net-like pattern, with a dark brown crescent-shaped mark in the middle of the wing at about two-thirds; a short brown apical streak is usually discernible, sometimes diffusely joined to the crescent. One form has the dorsal two-thirds dark brown, the costal third paler. In upland areas, very small pale examples occur regularly. **Similar species *B. lacteana* is very similar, slightly smaller and usually with a distinct brownish streak from the crescent to the apex; along this streak are two long fine dark parallel lines.** In *B. lancealana* these lines are rarely long or parallel. However, if *B. lacteana* is suspected examination of the genitalia is

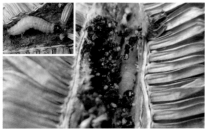
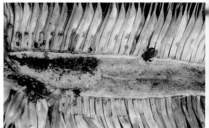

▲ Larva and larval workings of *Endothenia gentianaeana* and pupal exuvium within the centre of an old flowerhead of Wild Tea

necessary to confirm identification. *B. robustana* is typically larger, although there is overlap, and usually has darker hindwings; large and distinctly marked examples found in saltmarshes are likely to be *B. robustana*. For more obscurely marked or uniform examples of *Bactra* reference to genitalia may be needed to confirm identification (see p.392 for male genitalia only). **FS** Double-brooded, May-October. Can be abundant in rushy fields. Easily disturbed by day and comes to light. **Hab** Marshes, bogs, rush pastures, wet heathland, open wet woodland, moorland, mountains. **Fp** Rushes, including Compact Rush, club-rushes, Galingale; in the stems.

Bactra lacteana page 242   1111a (4657)

Very local, but likely to be overlooked. **FL** 5-7mm. Forewing very similar to well-marked *B. lancealana*. A uniform reddish brown form occurs. **FS** Double-brooded, May-August. Can be common very locally. Flies low to the ground on warm afternoons and very occasionally comes to light. **Hab** Grazed sedge lawns, moorland. **Fp** Carnation Sedge; mines the centre of the shoot just below ground level, causing the youngest leaves to wilt and die.

Bactra robustana page 242   1112 (4659)

Very local. **FL** 8-11mm. Variable, with a similar range of forms to *B. lancealana*. **FS** Single-brooded, late May-early September; has been recorded in October. Flies in the evening and comes to light. **Hab** Saltmarshes, coastal ditches, coastal gravel pits. **Fp** Sea Club-rush; in the stems.

Endothenia gentianaeana page 242   1097 (4665)

Common. Very local in the north and west. **FL** 6-9mm. Forewing basal two-thirds blackish and reddish brown, mixed leaden with a slight bluish or purplish reflection, with scattered black spots; often there is a pale blotch on the dorsum just before one-half, this in-filled darker, an obscure pale mark on the costa at about one-third, and a somewhat obscure whitish spot near the middle of the wing at about two-thirds; the apex is dark, inwardly edged with a short series of blackish dots. **Similar species** *E. oblongana* and *E. marginana*. *E. oblongana* is typically the smallest of the three species, with the narrowest whitish band in the outer part of the forewing. *E. marginana* is usually sized between the other two species and also has a whitish band, slightly broader; the male has a white hindwing, the female has the basal part of the hindwing pale. *E. gentianaeana* is typically the largest, with the pale outer band whitish, tinged pale yellowish brown and greyish; the hindwing is dark greyish brown. If there is doubt, confirmation of these *Endothenia* may require examination of genitalia. **FS** Single-brooded, May-mid August. Comes to light. **Hab** Waste ground, embankments, quarries, open woodland. **Fp** Wild Teasel; in the pith in the centre of the seedhead.

Endothenia oblongana page 242   1098 (4666)

Local. Scarce in the north. **FL** 5-7mm. **Similar species** Forewing pattern very similar to *E. gentianaeana* and *E. marginana* but hindwing pale brownish grey, darker towards the outer margin. **FS** Single-brooded, May-September. Readily disturbed by day. Flies in the evening and occasionally comes to light. **Hab** Chalk and limestone grassland, embankments, quarries, soft-rock cliffs. **Fp** Ribwort Plantain; in the rootstock.

Endothenia marginana page 242   1099 (4667)

Common. Local or very local in the north. **FL** 5-7mm. **Similar species** Forewing pattern very similar to *E. gentianaeana* and *E. oblongana* but hindwing in the male white, in the female pale towards the base. **FS** Double-brooded, May-mid September. Readily disturbed by day. Flies in the evening and occasionally comes to light. **Hab** Rough grassland, embankments, damp woods, boggy heaths, fens. **Fp** Betony, hemp-nettles, Lousewort, Marsh Lousewort, Yellow-rattle, Wild Teasel, Marsh Gentian; in the flowers, seedheads or seed capsules.

Endothenia ustulana page 242   1101 (4669)

Very local. **FL** 4-6mm. Forewing heavily mixed leaden grey and black, with an obscure paler blotch on the dorsum just before one-half, this in-filled darker, and with an indistinct narrow paler cross-band, leaden mixed whitish, from the costa at about three-quarters to the tornus; orangey brown along the termen and around the apex. **Similar species** *E. pullana* (not illustrated), a scarce species in southern England on Marsh Woundwort, is typically slightly paler, with the markings less well defined; reference to the genitalia is likely to be needed to confirm identification. **FS** Single-brooded, June-July; has been recorded in early September. Flies in the afternoon and occasionally comes to light. **Hab** Damp open woodland, hedge banks, waste ground, gardens. **Fp** Bugle, including garden cultivars; at first in the roots, later in the central stem causing stunting of growth and sometimes wilting.

Endothenia nigricostana page 242   1102 (4671)

Local. Very local in northern England. **FL** 5-7mm. Forewing blackish grey, finely speckled white, particularly in the apical half, with a diffuse, occasionally obscure, large pale yellowish-brown or yellowish blotch on the dorsum at about one-half. **FS** Single-brooded, late May-early July. Sometimes flies in the afternoon, is active at dusk and comes to light. **Hab** Open woodland, woodland margins, hedgerows. **Fp** Hedge Woundwort; in the flower stalk and roots. The larva ascends to near the top of an old stem to pupate in spring.

► Larva of *Endothenia nigricostana* within an old stem of Hedge Woundwort. The larva hibernates in the root and ascends to the top of a dead stem in early spring to pupate.

Endothenia ericetana page 242

Local. Very local in northern England and Scotland. **FL** 7-9mm. Forewing orangey brown or brown, sometimes tinged reddish brown, suffused with whitish or greyish brown; markings are rather darker brown and include a triangle in the middle of the wing with the angle nearest to the dorsum a right-angle or slightly acute. **Similar species E. quadrimaculana has a larger, longer, paler forewing, and the dark triangle in the middle of the wing has the angle nearest to the dorsum obtuse. FS** Single-brooded, late June-mid September, occasionally to early October. Flies in the evening and comes to light. **Hab** Field margins, rough grassland, open woodland. **Fp** Marsh Woundwort, possibly Hedge Woundwort, Corn Mint; in the roots.

Endothenia quadrimaculana page 242

Common. Local in northern England and Scotland. **FL** 9-11mm. Forewing whitish brown, pale yellowish brown or dark yellowish brown, usually with distinctly darker brown markings, sometimes mixed black, with a dark brown triangle in the middle of the wing. Darker, obscurely marked forms occur. **Similar species** *E. ericetana*. **FS** Single-brooded, June-mid August, occasionally to September. Flies from late afternoon and comes to light. **Hab** Fields, damp meadows, fens, marshes, hedgerows, woodland rides. **Fp** Marsh Woundwort, sometimes on Spear Mint and possibly Hedge Woundwort; in the roots.

Eudemis profundana page 242

Common. Very local in northern England but may be expanding its range there. **FL** 6-9mm. Variable. Forewing blackish brown, with the basal patch brown mixed whitish grey except along the dorsum where it is blackish brown, and with a whitish, whitish-brown or greyish-brown cross-band from the costa at one-third to the dorsum at one-half, often joined to a whitish or white oblique mark beyond, the inner margin of the join shallowly concave. Obscurely marked dark brown or blackish-brown forms occur. **Similar species** *E. porphyrana* **has less white, more greyish or leaden marking and is less variable, and has the base more uniformly brown suffused whitish grey; the dorsal mark and cross-band are less often fused, and if fused, then the inner margin is deeply concave. FS** Single-brooded, July-early September; recorded in mid June. Occasionally seen in numbers. Can be found resting on oak trunks, is readily disturbed by day, and comes to light. **Hab** Woodland, parkland, scrub. **Fp** Oaks; in a rolled leaf.

Eudemis porphyrana page 242

Very local, in the south only. **FL** 8-10mm. Forewing dark brown, with a uniformly coloured basal patch and a whitish-grey or leaden cross-band infrequently fused to a dorsal mark. **Similar species** *E. profundana*. **FS** Single-brooded, late June-August. Flies in the evening and comes to light. **Hab** Woodland, hedgerows, orchards. **Fp** Apples, particularly Crab Apple, possibly oaks; in a rolled leaf, or two or more leaves rolled together along a twig.

Pseudosciaphila branderiana page 242

Local. Scarce or extinct in the north of its range. **FL** 10-13mm. Forewing greyish brown with greyish-white to light brownish-grey cross-bands from the costa at one-third and in the apical third. A unicolorous form has the forewing light brownish grey, weakly tinged pale greyish, with markings virtually indistinguishable. **FS** Single-brooded, June-August, sometimes early September. Comes to light. **Hab** Open woodland, scrub, parks, gardens. **Fp** Aspen; in a folded or rolled leaf or between two spun leaves.

The spinning
f *Apotomis*
etuletana in a
oung shoot of
ilver Birch.

Apotomis semifasciana page 242   1089 (4692)

Local. Very local in Scotland. **FL** 8-9.5mm. Forewing whitish grey to grey, sparsely
speckled black, with a darker basal patch, this sometimes obscure, and a large, almost
triangular, greyish-brown to blackish-brown blotch on the costa, the inner margin nearly
perpendicular to the costa. The intensity of the markings can vary. **Similar species** Paler
forms are similar to *A. lineana*, which is narrower winged and has a dark grey suffusion
along the dorsum; *A. infida* (not illustrated), a very rare Scottish species, is similar but
typically lighter in colour, with the inner edge of the costal mark much more oblique to the
costa. If *A. infida* is suspected, examination of genitalia will be necessary to confirm
identification. **FS** Single-brooded, late June-August. Flies at dusk and comes to light.
Hab Damp woodland, margins of osier beds, fens, banks of ponds and streams, slumping
coastal slopes, hedgerows. **Fp** Sallows; on catkins and later in spun leaves.

Apotomis lineana page 242   1091 (4697)

Very local. **FL** 8-10mm. Forewing dark grey, especially in a broad band along the dorsum
to the tornus, and dark grey mixed brownish and black in the middle of the wing, a broad
greyish-white cross-band from the costa at one-third, ending before the dorsum in an
incomplete oval, the outer margin of the cross-band with a projection which almost links
to the greyish-white apical third. **Similar species A. turbidana has whitish, not
greyish-white, markings and has no oval**; also *A. semifasciana*. **FS** Single-brooded,
June-August. Readily disturbed by day. **Hab** Damp meadows, banks of rivers and streams,
bogs, marshes, gardens. **Fp** White Willow, Crack Willow; between spun leaves.

Apotomis turbidana page 242   1092 (4700)

Common. **FL** 9-10.5mm. Forewing dark grey, especially in a broad band along the dorsum
to the tornus, and in a central cross-band where mixed with brownish and black, with
distinct whitish markings near the base extending in a broad streak into the middle of the
wing at about one-half, and in a blotch on the costa at one-third; the apical third is
whitish, sometimes tinged yellowish. **Similar species** *A. lineana*. **FS** Single-brooded, late
May-July, occasionally to late August. Locally numerous in areas of birch, rarely found
elsewhere. Flies towards dusk and comes to light. **Hab** Woodland, parkland, heathland,
mosses, scrub. **Fp** Birches; in spun leaves.

Apotomis betuletana page 242   1093 (4701)

Common. **FL** 8-10.5mm. Forewing blackish brown in the basal two-thirds, mixed leaden and blackish, with the apical third white, sometimes with patches of yellowish brown, and with the outer edge of the basal area oblique, almost straight, with a small whitish indentation near the middle, this sometimes hook-like; the cilia are white, tipped grey. **Similar species *A. capreana* is broader winged, has a more strongly arched costa, a greyish-white blotch on the costa at one-third, a distinct hook-like white indentation in the outer margin of the dark basal patch, the margin being concave; *A. sororculana* is slightly smaller and narrower winged, has the outer edge of the dark basal patch more oblique, and has the apex blackish and the cilia darker. FS** Single-brooded, mid June-September. Can be disturbed by day, flies towards dusk and comes to light. **Hab** Open woodland, woodland margins, scrub, heathland, waste ground. **Fp** Birches; between spun leaves.

Apotomis capreana page 242   1094 (4703)

Local. Rare in northern England and Scotland. **FL** 8-10.5mm. Forewing dark brown in the basal two-thirds, variably marked leaden and black, with a rather indistinct greyish-white blotch from the costa at one-third to the middle of the wing, the outer edge of the basal area somewhat concave, with a hook-like indentation extending into the dark area near the middle, and with the apical third white with a short series of black dots towards the apex; the cilia are pale brownish yellow, tipped greyish. **Similar species** *A. betuletana*, *A. sororculana*. **FS** Single-brooded, June-August. Flies towards dusk and comes to light. **Hab** Open woodland, woodland margins, heathland, gravel pits, marshes, streambanks. **Fp** Goat Willow, probably Grey Willow; in tightly spun leaves on young terminal shoots.

Apotomis sororculana page 242   1095 (4706)

Local. **FL** 8-9.5mm. Forewing dark brown in the basal two-thirds, mixed leaden and spotted black, in part weakly shaded greyish white, with a small whitish blotch on the costa at one-third, sometimes extended towards the middle of the wing, the outer edge of the basal area oblique, almost straight, the apical third white, somewhat suffused grey and with several black spots, and dark brown towards the apex; the cilia are dark grey. **Similar species** *A. betuletana*, *A. capreana*. **FS** Single-brooded, May-July. Flies towards dusk and comes to light. **Hab** Woodland, heathland, mosses. **Fp** Birches; between flatly spun leaves.

Apotomis sauciana page 242   1096 (4708)

Local. Very local in southern England. **FL** 6.5-8mm. Forewing with the basal two-thirds blackish brown, mixed black and grey, the outer edge oblique, with a hook-like indentation extending into the dark area, the apical third creamy white, more or less extensively shaded blackish brown towards the apex and bordering the termen. The Scottish subspecies, *grevillana*, is typically smaller, with the basal two-thirds lighter. **FS** Single-brooded, late May-August. Flies in the afternoon and evening, and comes to light. **Hab** Open woodland, moorland, including high moorland of 650-950m. **Fp** Bilberry, Bearberry, also bred from Mountain Willow; in a loose spinning in the terminal leaves of a shoot.

Orthotaenia undulana page 242   1087 (4711)

Common. Local in Scotland. **FL** 7-9.5mm. Forewing dark brown, variably overlaid with yellowish brown or yellowish grey, with two broad pale brownish-white or dirty cream cross-bands from the costa at one-third and beyond one-half, the outer cross-band curving to

the tornus, a yellowish-brown bar beyond, sometimes interrupted, from below the costa to below the mid point of the termen. **Similar species** *Celypha lacunana* **has the outer edge of the inner cross-band with a finger-like projection into the middle of the wing.** *O. undulana* **is typically paler in the apical third, the dark markings more clearly defined. FS** Single-brooded, May-July, occasionally to mid August. Readily disturbed by day, flies towards sunset and comes to light. **Hab** Open woodland, woodland margins, hedgerows, scrub, damp heathland, gardens. **Fp** A range of trees, shrubs and herbaceous plants, including birches, Honeysuckle, Bilberry, Bog-myrtle; in loosely spun leaves.

Plum Tortrix *Hedya pruniana* page 243 1082 (4715)

Common. **FL** 7-8.5mm. Forewing with the basal two-thirds blackish and leaden bluish grey, sparsely speckled whitish, the outer edge of the dark basal patch angled outwards near the middle, the apical third whitish, mottled silvery greyish and dark brown; **there are two blackish dots placed beyond the**  **angle of the outer edge of the dark basal patch, and a dark brown quarter-circle mark at the apex.** A light-coloured form in which the dark markings are almost absent occurs rarely. **Similar species** *H. nubiferana* **is typically longer winged, with two black dots just above the angle of the outer edge of the basal patch towards the costa, and with pale greyish or greyish-brown patches in the outer third.** **FS** Single-brooded, May-August. Readily disturbed by day and comes to light. **Hab** Open woodland, hedgerows, scrub, vegetated shingle, orchards, gardens. **Fp** Blackthorn, Wild Plum, Wild Cherry, less often apples, pears, hawthorns, Hazel; at first on the young shoots, later in a folded leaf or between two spun leaves.

Marbled Orchard Tortrix *Hedya nubiferana* page 243 1083 (4714)

Common. Local in Scotland. **FL** 7-10mm. Forewing with the basal two-thirds dark brown or dark yellowish brown, mixed bluish grey and blackish, with the apical third white with pale greyish or greyish-brown patches. A rare form occurs which is largely pale whitish grey, with the markings almost absent. **Similar species** *H. pruniana*. **FS** Single-brooded, May-August. Readily disturbed by day and comes to light. **Hab** Woodland, parkland, scrub, hedgerows, orchard, moorland, gardens. **Fp** Hawthorns, Crab Apple, plums, less often other trees and shrubs; in spun flowers, leaves and shoots.

Hedya ochroleucana page 243 1084 (4717)

Common. More local in the west and very local in northern England. **FL** 7.5-10mm. Forewing with the basal two-thirds brown, mixed leaden and black, sparsely speckled pale yellowish brown, the outer edge of the basal patch slightly convex, **the apical third pale yellowish brown**. The pale yellowish brown fades over time and may become very faint, in which case the two small pale indentations in the outer margin of the dark basal patch will usually help determine this species. **FS** Single-brooded, June-August. Readily disturbed by day and comes to light. **Hab** Chalk downland, hedgerows, scrub, waste ground, orchards, woodland, gardens. **Fp** Roses, occasionally apples; in a spinning among leaves.

Hedya salicella page 243

Local. Very local in northern England. **FL** 9-11mm. Thorax white, marked darker. Forewing basal area and dorsal third whitish, thinly speckled black and shaded greyish near the dorsum, the remaining part of the wing blackish brown, overlaid black and mottled pale blue-grey. **FS** Single-brooded, June-August. Comes to light. **Hab** Open woodland, scrub on heathland, parkland, hedgerows, marshes, streambanks, sometimes parks and gardens. **Fp** White Willow, Grey Willow, Aspen, Black Poplar; in spun shoots or a folded leaf.

Metendothenia atropunctana page 243

Common. Mainly a northern species, more local in the south. **FL** 6-7.5mm. Forewing with the basal two-thirds dark brown, mixed blackish, leaden and pale yellowish brown, with a pale yellowish-brown mark usually present on the costa at one-third, sometimes indistinct, with the outer edge of the dark basal patch slightly convex, **and the apical third creamy, mixed yellowish brown in patches, often with a pinkish tinge, and a distinct black spot just above the middle beyond the outer edge of the dark basal patch**. **FS** Single-brooded, with an occasional second brood in the south, May-August. Flies in the evening. **Hab** Moorland and mosses, wet heathland, damp woodland, occasionally hedgerows. **Fp** Bog-myrtle, sallows, birches, occasionally other trees; in a spinning, the leaf spun into a pod when feeding on sallows and birches.

Celypha striana page 243

Common. Local in parts of northern England and may be spreading. **FL** 7.5-10.5mm. Forewing pale brown to reddish brown, with an irregularly edged, brown or greyish-brown cross-band at about one-half and another, angled, often less distinct, from the apex along the termen. **FS** Single-brooded, May-September. Male flies by day in sunshine. Comes to light. **Hab** Waste ground, chalk and limestone grassland, vegetated shingle, fields, embankments, woodland, hedgerows, gardens. **Fp** Dandelion; initially under a silken web on the surface of the tap root, later in the roots.

Celypha rosaceana page 243

Local. Records from the north of the range are considered doubtful owing to confusion with *C. rufana*. **FL** 7-9mm. Forewing pale yellowish brown, tinged pinkish brown or pinkish, often lightly dusted darker in the outer part of the wing; the cilia are yellowish brown to pink. The intensity of the pink coloration varies and can be bright purplish pink, or faint, with the forewing more yellowish. **Similar species C. rufana is relatively broader winged, much darker and without a pink hue when fresh, although it may look rather pinkish when worn. FS** Single-brooded, late May-September. Comes to light. **Hab** Rough fields, embankments, roadsides, vegetated shingle, sand dunes, gardens. **Fp** Perennial Sow-thistle, Prickly Sow-thistle, Dandelion; on the roots, in a silken tube.

Celypha rufana page 243

Very local in north-west England and Wales, wanderers elsewhere. **FL** 7.5-9mm. Forewing dull yellowish brown to dark greyish brown, variably tinged reddish brown, often resulting in a fine net-like pattern, especially in the outer half of the wing, and a small dark patch on or above the dorsum before the tornus; the cilia are yellowish brown or reddish brown. **Similar species** *C. rosaceana*. **FS** Single-brooded, May-early August. Flies in midday sunshine, in the late afternoon and evening, later coming to light. **Hab** Waste ground, gardens. **Fp** Tansy, Mugwort; on the roots, living beneath silken webbing.

▲ The leaf mine of *Celypha woodiana* on Mistletoe. The mines can be searched for in spring and summer in old orchard trees and other trees bearing Mistletoe, using close-focus binoculars.

Celypha woodiana page 243

1066 (4729)

Rare. **FL** 7.5-8.5mm. Forewing whitish, lightly mottled blackish, grey, pale greyish brown and dorsally bluish grey, with a prominent bluish-grey blotch on the costa at about one-half and a black mark below. **FS** Single-brooded, July-August. Occasionally comes to light. **Hab** Orchards, gardens, grazing levels in Somerset. **Fp** Mistletoe; in a mine on a leaf.

Celypha cespitana page 243

1067 (4728)

Local. Most frequent on the coast. **FL** 5.5-7.5mm. Variable. Forewing greyish to dull yellowish brown, mixed blackish and reddish brown, with two white or silvery white cross-bands from the costa at one-third and beyond one-half, these sometimes shaded yellowish brown; the inner edge of the outer cross-band has a deep, narrow indentation near the middle and sometimes another towards the costa and the blotch before the tornus is variable but often triangular and oblique. Forewing markings can be rather diffuse. **Similar species C. rivulana is generally broader and larger, with markings well defined, the dark mark near the tornus being triangular or arch-shaped.** **FS** Single-brooded, June-August. Flies in sunshine and occasionally comes to light. **Hab** Chalk and limestone downland, cliffs, waste ground, vegetated shingle, sand dunes. **Fp** Thyme, Thrift, sea-lavenders, clovers, Common Bird's-foot-trefoil; in a silken tube, usually along the ground.

Celypha lacunana page 243

1076 (4731)

Common. **FL** 6.5-8.5mm. Variable. Forewing predominantly blackish brown, variably overlaid with yellowish brown or greyish brown, often extensive in the outer half of the wing, with two creamy or silvery white cross-bands from the costa at one-third and beyond one-half, the outer edge of the inner cross- 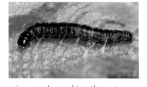 band with a finger-like projection into the middle of the wing, rarely reaching the outer cross-band, the projection sometimes obscure, but rarely obliterated. An almost unicolorous form occurs, with the forewing blackish, speckled metallic leaden grey. **Similar species C. rurestrana (not illustrated) is a rare species found in the south-west and south Wales. It is very similar but the cross-bands are noticeably whiter, and the outer edge to the inner cross-band is straighter, with a weakly defined projection.** If *C. rurestrana* is suspected, confirmation by examination of genitalia is likely to be required; also *Orthotaenia undulana*. **FS** Double-brooded, May-early November. Can be numerous where found. Readily disturbed by day and comes to light. **Hab** Open woodland, woodland margins, marshes, meadows, hedgerows, roadside verges, gardens. **Fp** Many herbaceous plants, occasionally trees or shrubs; in spun leaves, shoots and flowers.

Celypha rivulana page 243   1068 (4733)

Local. Perhaps most frequent in southern England. **FL** 7-9mm. Forewing predominantly reddish brown or chocolate brown, with silvery white cross-bands from the costa at one-third and beyond one-half, these containing yellowish-brown cross-lines, and with a distinct and usually large brown blotch, almost triangular or arch-shaped, just before the tornus. **Similar species** *C. cespitana*. **FS** Single-brooded, mid June-August. Flies in sunshine and occasionally comes to light. **Hab** Damp meadows, marshes, boggy heathland, open woodland, chalk and limestone grassland, scrub, sand dunes. **Fp** On a variety of plants, including Meadowsweet, Dyer's Greenweed, hawkweeds, plantains, orchids; in the flowers and on terminal shoots.

Celypha aurofasciana page 243   1069 (4735)

Very local. Possibly expanding its range. **FL** 5.5-6.5mm. **A small species.** Labial palps and face white, the crown yellow. **Forewing rather narrow, with the costa smoothly arched and the termen rounded**; blackish brown, sometimes shaded greyish, with a pair of narrow yellowish cross-bands from the costa at about one-third, each containing a metallic leaden grey cross-line, and a similar single cross-band, outwardly curved, beyond one-third; the cilia are mainly yellowish. A form occurs with orange scales scattered over the blackish brown, the yellowish much reduced and the leaden colour expanded on the cross-bands. **FS** Single-brooded, June-July. Flies in late afternoon and early evening, and comes to light. **Hab** Open woodland. **Fp** Under moss or liverworts growing on old or rotten bark of trees; in silken galleries.

Phiaris metallicana page 243   1072 (4743)

Local. A northern species. **FL** 7.5-9mm. Variable. Forewing dark brown, heavily suffused pale yellowish grey, with poorly defined pale yellowish-grey cross-bands from the costa at one-third and two-thirds, these edged with sparse metallic silver scales. An almost unicolorous form occurs, and in well-marked examples the silvery metallic spots can be more conspicuous. **Similar species** *P. obsoletana* has a narrower forewing with a more acute apex; it is a mountain species, found higher up than *P. metallicana*. **FS** Single-brooded, June-July. Flies from late evening and at dusk. **Hab** Hillsides, highland woodland and can occur at lower elevations of about 200-300m. Can be especially numerous where Caledonian pinewoods have a Bilberry ground layer. **Fp** Bilberry; among spun leaves.

Phiaris schulziana page 243   1073 (4744)

Local. Predominantly northern, rare in southern England. **FL** 6.5-11mm. Forewing reddish brown to orangey brown, variably speckled black, with narrow white cross-bands or spots, these overlaid shining metallic leaden grey, and a white spot just above the middle of the wing beyond one-half and the cilia white, chequered blackish brown. **FS** Single-brooded, June-August. Easily disturbed by day and flies in evening sunshine. **Hab** Heathland, mosses, mountain moorland, occurring up to 1,000m or more. **Fp** Crowberry, Heather, and has been bred from Trailing Azalea; in a silken tube.

Phiaris palustrana page 243   1074 (4750)

Local in Scotland. Very local or scarce elsewhere. **FL** 6-7.5mm. Forewing reddish brown or orangey brown, variably mixed blackish brown and black, with a white cross-band from the costa at one-third speckled with leaden scales, a whitish spot, and irregular whitish cross-lines in the outer half of the wing sometimes forming a cross-band. **Similar species _P. micana_ has a broader forewing, more rounded at the apex, with markings less well defined, and with an outer cross-band usually discernible. FS** Single-brooded, late May-August. Can be numerous very locally at lower elevations in the Scottish Highlands. Flies on sunny afternoons, in the evening and at dusk. **Hab** Heathland with mixed birch and pine, preferring sheltered situations. **Fp** Mosses; in a silken spinning.

Phiaris micana page 243   1075 (4749)

Very local. Predominantly a Scottish species, rare in southern England and Wales. **FL** 6-8.5mm. Forewing dark brown suffused with yellowish grey or orangey brown, or with scattered paler patches, and two white cross-bands from the costa at one-third and at two-thirds, with a white spot adjacent to the inner margin of the outer cross-band near the middle, these markings usually with ill-defined margins; the paler markings may be speckled with silvery or grey metallic scales. A dull greyish-brown form occurs. **Similar species** _P. palustrana_. **FS** Single-brooded, June-August. Easily disturbed in sunshine and flies towards sunset. **Hab** Boggy places. **Fp** Not known in Britain and Ireland.

Phiaris obsoletana page 243   1077 (4741)

Very local. A northern and upland species. **FL** 7-8.5mm. Forewing dark brown, variably suffused with yellowish-brown scales, two creamy white or pale brownish-white cross-bands from the costa at one-third and at two-thirds, sometimes scattered with silvery scales, the inner margin of the outer cross-band with a narrow projection into the middle of the wing, sometimes linked to the inner cross-band, and often creamy white along the termen. A unicolorous pale form occurs. **Similar species** Some forms of _P. metallicana_. **FS** Single-brooded, June-July. Readily disturbed in sunny weather from Bearberry and low-growing vegetation, and flies from dawn until after sunrise. **Hab** Montane, from 650-1,000m. **Fp** Not known.

Argyroploce arbutella page 243   1071 (4765)

Local. A northern species. Reported once from Northumberland in the 19th century and once from Essex as a vagrant. **FL** 6-7mm. Forewing reddish purple, suffused reddish brown in the basal half, with irregular whitish-edged leaden cross-lines. **Similar species _Stictea mygindiana_ is generally larger and has a broader forewing, with a duller reddish-purple colour. FS** Single-brooded, May-June; recorded once in August. Flies in afternoon sunshine over the foodplant. **Hab** Sheltered hollows on the higher parts of mountains. **Fp** Bearberry; in spun terminal leaves.

Stictea mygindiana page 243   1070 (4774)

Local in Scotland. Very local in the Midlands and northern England. **FL** 8-9.5mm. Forewing purplish brown, in a net-like pattern, speckled leaden grey and darker grey, with a rather obscure dark purplish-brown cross-band at one-half, although this can be more developed in some examples. The wing colour varies and sometimes any pattern can be obscured with greyish colouring. **Similar species** _Argyroploce arbutella_. **FS** Single-brooded, May-June. Flies from late afternoon, sometimes earlier in sunshine. **Hab** Heathland, mosses, moorland. **Fp** Crowberry, occasionally Bearberry, Bog-myrtle; in a spinning.

Olethreutes arcuella page 244 1080 (4776)

Very local. **FL** 6.5-8.5mm. Forewing deep orangey brown with several silver marks, including three short streaks near the base, a cross-line at one-third and three short transverse marks towards the apex, and beyond the cross-line an ocellus-like marking comprising a pale yellowish area and a short thick black streak containing four silver spots; all silver markings reflect blue in bright light. **FS** Single-brooded, May-August. Flies by day in sunshine, may be found at rest on leaves, and occasionally comes to light. **Hab** Open woodland, woodland rides, recently coppiced areas in woodland, wooded heathland. **Fp** Decaying leaves on the ground.

Piniphila bifasciana page 244 1079 (4778)

Common. Local in northern England and rare in Scotland. Expanding its range. **FL** 5.5-7.5mm. Forewing brownish grey, sometimes darker, with two broad cross-bands from the costa at one-third and occupying the apical third, greyish white, suffused pale yellowish brown, orangey or salmon pink, particularly towards the apex. **FS** Single-brooded, June-August. Comes to light. **Hab** Conifer woods, heathland with conifers, gardens. **Fp** Scots Pine, Maritime Pine; in a spinning in the shoots and among the male flowers.

Lobesia occidentis page 244 1105 (4789)

Very local. Possibly declining. **FW** 4-6mm. Forewing dark brown with a broad cross-band from the costa at one-third curving gently to the dorsum before one-half, pale yellowish in the dorsal half, darker and mixed leaden in the costal half, a narrow yellowish cross-band from the tornus to the costa at two-thirds and pale yellowish along the termen. **FS** Double-brooded, June-August; occasionally a third generation occurs. Flies in afternoon sunshine and at dusk. **Hab** Open or coppiced woodland, coastal cliffs, sand dunes. **Fp** Wood Spurge, Sea Spurge; in spun shoots.

Lobesia reliquana page 244 1106 (4794)

Local. Scarcer in northern England and Scotland. **FL** 5-6mm. Forewing relatively narrow and pointed, orangey brown, with a cross-band formed of a brown, dark grey or black triangular mark on the dorsum just before one-half, and extended by pale yellowish to the costa at one-third, and another cross-band, rather irregular, formed of a blackish triangular mark at the tornus merging with a diffuse broad dark grey, brown or blackish triangular patch on the costa just beyond one-half. The male has a whitish hindwing, shaded darker towards the apex. **FS** Single-brooded, May-June. Flies by day and comes to light. **Hab** Open woodland, parkland, scrub, hedgerows, gardens. **Fp** Oaks, birches, Blackthorn; in the terminal shoots or spun leaves.

Lobesia abscisana page 244 1108 (4793)

Common across most of England. Appears to be spreading northwards. Recorded new to Ireland in 2010. **FL** 5-6mm. Forewing dark brown, sometimes patchily suffused yellowish grey, with two whitish cross-bands speckled dark brown, one on the dorsum just before one-half to the costa at one-third, with another, slightly more obscure, from the tornus to the costa at about two-thirds, this angled on the outer margin, with **a small rounded brown spot along the inner margin of the outer cross-band**; the apex is dark brown. **FS** Double-brooded, May-early September. Occasionally flies in the afternoon, flies from dusk and comes to light. **Hab** Grassland, open woodland, vegetated shingle, waste ground. **Fp** Creeping Thistle; in the shoot, spinning leaves together and burrowing into the heart.

Lobesia littoralis page 244   1109 (4806)

Local. **FL** 5-8mm. **Forewing rather narrow and pointed**, whitish, speckled darker, the markings yellowish brown to reddish brown, comprising a diffuse basal patch with a darker patch on the dorsum at about one-third to the middle of the wing, a cross-band from the dorsum at about two-thirds to the costa near one-half, and a large, usually obscure patch before the termen. Flies in afternoon sunshine and comes to light. **FS** Double-brooded, late May-October. **Hab** Coastal cliffs, where it can be numerous on rocky coastlines, saltmarshes, vegetated shingle, gardens. **Fp** Thrift, also Common Bird's-foot-trefoil; first generation in the young shoots or flowerheads, second generation in the flowerheads and on the seeds.

Eucosmomorpha albersana page 244   1217 (5053)

Local. Becoming increasingly scarce further north. **FL** 6-7mm. Forewing brownish or reddish brown in the basal two-thirds, pale orangey brown in the apical third, with metallic leaden streaks on the costa; there is a weak ocellus laterally edged metallic leaden, containing two or three black dashes. **FS** Single-brooded, May-June. Flies in afternoon sunshine and occasionally comes to light. **Hab** Woodland rides and clearings, particularly oak woods. **Fp** Honeysuckle; in a pod made from a folded leaf or two spun leaves.

Cherry Bark Tortrix *Enarmonia formosana* page 244   1216 (5055)

Common. Very local in Scotland. May be spreading. **FL** 6-9mm. Forewing orangey brown, patterned dark brownish and metallic leaden, with a few small white streaks on the costa, these more distinctive towards the apex; the ocellus is thickly edged leaden laterally, containing three or four thick black dashes and the cilia are dark grey with three or four distinct orangey patches. **Similar species** *Grapholita lobarzewskii* **is typically smaller and narrower winged, lacks the metallic and contrasting white markings, and the cilia are brown with a creamy basal line. FS** Single-brooded, late May-September. Flies in afternoon and evening sunshine, and occasionally comes to light. **Hab** Parkland, gardens, orchards, hedgerows, suburban roadside trees, scrub woodland near moorland. **Fp** Various rosaceous trees, especially mature apple trees, ornamental and wild cherries, Rowan; often in a wound in the bark, exuding reddish frass, sometimes under loose bark.

▶ Reddish frass of Cherry Bark Tortrix *Enarmonia formosana* on the tree bark.

Ancylis achatana page 244   1115 (5074)

Local. Possibly expanding its range slowly northwards. **FL** 6-9mm. Forewing reddish brown to dark brown with two greyish-white to silvery grey cross-bands, one from the dorsum before one-half to the costa at one third, angled in the middle, linked obliquely from the outer margin by a thin band to the second cross-band, from the tornus to the costa at two-thirds, **with a large brown near-triangular blotch on the dorsum before the tornus. FS** Single-brooded, late May-mid August. Flies from dusk and comes to light. **Hab** Woodland, scrub, hedgerows. **Fp** Hawthorn, Blackthorn, cotoneasters; spinning one or two leaves into a tube, feeding on adjacent leaves.

Ancylis comptana page 244   1116 (5062)

Local. Very local in the west and northern England. **FL** 5-6mm. **A small, narrow-winged species.** Forewing extended into a distinct lobe at the apex, brownish, with a narrow greyish cross-band from the costa at one-third to the dorsum at three-quarters, extended beyond and looping over a brown patch near the tornus, and with small whitish streaks along the outer half of the costa. **FS** Double-brooded, April-September. Can be abundant. Readily disturbed by day and comes to light. **Hab** Chalk and limestone grassland, coastal cliffs. **Fp** Salad Burnet, cinquefoils, strawberries, thymes; in a folded leaf.

Ancylis unguicella page 244   1117 (5057)

Local. More frequent in the north of its range. **FL** 6-9mm. Forewing extended into a small lobe at the apex, dull orangey brown or pale reddish brown, with whitish or whitish-grey markings, sometimes obscure, including two fairly straight oblique cross-bands from the dorsum beyond one-half to the costa at one-third, and from the tornus to the costa at two-thirds, sometimes two or three dark brown dashes in the middle of the wing , and with whitish paired streaks on the costa. **FS** Single-brooded, late April-July. Flies late afternoon, at dusk, in the evening and occasionally comes to light. **Hab** Heathland, moorland. **Fp** Bell Heather, occasionally other heathers; in a spinning among the leaves. Also reared from Bearberry, the larva in short tunnels between the soil and leaves.

Ancylis uncella page 244   1118 (5058)

Local. **FL** 7-9mm. Forewing extended into a small lobe at the apex, reddish brown, variably shaded paler and darker, with a large pale greyish patch on the dorsum at one-half and another, irregularly shaped, at the tornus, extending just over halfway across the wing, and with short paired whitish streaks on the costa. **FS** Single-brooded, sometimes double-brooded in the south, May-early August; recorded in April. Can be disturbed by day and comes to light. **Hab** Damp heathland, wet moorland, mosses, open woodland, scrub. **Fp** Heathers, birches; spins leaves together.

Ancylis geminana page 244   1119 (5064)

Local. **FL** 6-8mm. Forewing extended into a small lobe at the apex, pale greyish brown, sometimes tinged reddish brown, the dorsum broadly bordered pale greyish by an undulating streak from the base to the termen, although this may be obscure towards the base, this streak broadened at about one-half into the middle of the wing and arched above the tornus, extending to beyond halfway across the wing. **Similar species *A. geminana*, *A. diminutana* and *A. subarcuana* are very similar and all have been treated as *A. geminana* in the past. *A diminutana* is more reddish brown, with the dorsal streak speckled dark brown and nearly straight, not arched, above the tornus; *A. subarcuana* is pale grey to whitish grey, with brown markings weak, restricted to the middle of the wing, and the dorsal streak is more obscure.** **FS** Single-brooded, sometimes with a second brood, May-August. Flies at dusk, occasionally coming to light. **Hab** Damp woodland, bogs, fens, marshes, wet moorland. **Fp** Sallows and willows; spins a leaf into a pod.

Ancylis diminutana page 244   1119a (5066)

Local in England. Probably not recorded from Scotland. **FL** 6-8mm. **Similar species** *A. geminana*, *A. subarcuana*. **FS** Single-brooded, May-August. Flies at dusk and occasionally comes to light. **Hab** Open woodland, heathland. **Fp** Sallows; spins a leaf into a pod.

Ancylis subarcuana page 244 1119b (5065)

Local. **FL** 6-8mm. **Similar species** *A. diminutana, A. geminana.* **FS** Single-brooded, possibly double-brooded, late April-August. Flies in the afternoon and occasionally comes to light. **Hab** Heathland, dry grassland, sand dunes; can be numerous locally along Scottish coasts. **Fp** Creeping Willow; spins a leaf into a pod.

Ancylis mitterbacheriana page 244 1120 (5076)

Common. Very local in northern England and Scotland. **FL** 6-8mm. Forewing slightly extended into a small lobe at the apex, yellowish brown over most of the wing and along the costa towards the base, with a dark reddish-brown area from the base along the dorsum, edged with a fine white line just before one-half, and with small whitish streaks on the costa in the apical half. **FS** Single-brooded, late April-mid July. Readily disturbed by day, flies in the evening and comes to light. **Hab** Deciduous woodland. **Fp** Oaks, Beech; spins a leaf into a pod.

Ancylis upupana page 244 1121 (5063)

Rare. **FL** 5-8mm. Forewing slightly extended into a small lobe at the apex, dark brown, the apical area and termen orangey brown, with small white costal streaks in the apical half. **FS** Single-brooded, May-June; has been recorded in July. Flies from early afternoon until just before sunset, and occasionally comes to light. **Hab** Woodland, heathland, hedgerows. **Fp** Birches, elms; in a folded leaf or between two spun leaves.

Ancylis obtusana page 244 1122 (5061)

Very local. **FL** 5-6mm. **A small, broad-winged species.** Forewing hardly extended at the apex, orangey brown shaded reddish brown, with a narrow whitish cross-band from the costa at one-third curved to the dorsum beyond one-half, extended beyond and looping around a narrow, almost triangular, brown mark near the tornus, and with small whitish costal streaks. **FS** Single-brooded, late April-July. Flies in afternoon sunshine and occasionally comes to light. **Hab** Woodland, heathland. **Fp** Buckthorn, Alder Buckthorn; between spun leaves.

Ancylis laetana page 244 1123 (5059)

Local. Very local in western and northern England, and in Scotland. **FL** 6-7mm. Forewing extended into a small lobe at the apex, white at the base and broadly along the costa, grey to dark grey or black dorsally, with an obscure bluish-grey ocellus, edged black along the costal margin, and shaded reddish brown towards the apex. **FS** Single-brooded, May-June. Flies from dusk and comes to light. **Hab** Open woodland, parkland, hedgerows, gardens. **Fp** Aspen, rarely Black Poplar; within a folded leaf or between two spun leaves.

Ancylis tineana page 244 1124 (5077)

Rare. Found only in the Scottish Highlands and at one site in England. **FL** 5-7mm. Forewing extended into a small lobe at the apex, greyish, variably speckled dark brown, with a distinct large whitish ocellus, extending to near the apex, and with greyish-white costal streaks in the apical half, interspersed darker. **FS** Single-brooded, June. Flies by day. **Hab** Scrub on heathland and grassland, boggy moorland, mosses. **Fp** Birches; in a rough silken tube among spun leaves on trees stunted by grazing.

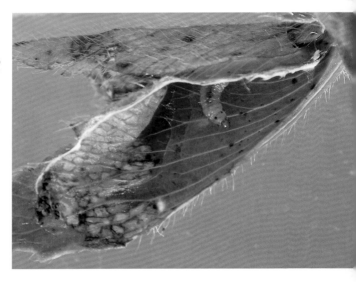

► Larva of
Ancylis badiana
exposed within
its spinning on
clover. The larva
spins a leaflet into
a pod, sealing the
edges with silk.

Ancylis unculana page 244 — 1125 (5068)

Local. Much scarcer away from southern and eastern counties. **FL** 6-8mm. Forewing extended into a small lobe at the apex, reddish brown, with a narrow whitish speckled brown or grey cross-band from the dorsum at three-quarters to the costa before one-half, with a creamy extension along the costa to the base, and fused with a large irregular greyish patch above the tornus; there are short paired whitish streaks along the apical half of the costa. **FS** Single-brooded, possibly double-brooded in some areas, May-August. Flies from early evening and occasionally comes to light. **Hab** Chalk downland, scrub, damp woodland, hedgerows. **Fp** Buckthorn, Alder Buckthorn; spins a leaf edge to form a pod.

Ancylis badiana page 244 — 1126 (5073)

Common. **FL** 5-8mm. Forewing extended into a broad lobe at the apex, appearing distinctly hooked, dark brown, suffused orangey towards the apex, with a broad creamy whitish or greyish-white cross-band, rather indistinct on the dorsum at three-quarters, broadening and distinct towards the middle of the wing and extended along the costa to the base, and with short whitish streaks along the apical half of the costa, and long creamy cilia along the termen before the apex. **Similar species** *A. paludana* is lighter, with a broader creamy band along the costa to the base, and with three or four black dots at the base of the cilia before the apex, where there is never more than one in *A. badiana*. **FS** Double-brooded, late April-October. Flies at sunrise, in the afternoon and evening, and occasionally comes to light. **Hab** Grasslands, woodland rides, chalk-pits, waste ground, vegetated shingle. **Fp** Peas, vetches, clovers; spins leaves together to form a pod.

Ancylis paludana page 244 — 1127 (5071)

Rare. East Anglia, formerly Northumberland. **FL** 6-7mm. **Similar species** *A. badiana*. **FS** Double-brooded, May-August. Flies at dusk. **Hab** Fens. **Fp** Marsh Pea; spins two leaves together to form a pod.

Larval working of *Ancylis myrtillana* on Bilberry. The larva has spun a leaf downwards to form a pod in which it feeds.

Ancylis myrtillana page 244

1128 (5069)

Local. Scarce in the south and east. Can be locally numerous in Scotland. **FL** 6-8mm. Forewing extended into a small lobe at the apex, brown to dark brown dorsally in a patch from the base to beyond one-half, a broad pale brownish-white or greyish-white cross-band from the dorsum at three-quarters to the costa before one-half, and extended along the costa to the base, with the outer third of the wing pale brownish white, mixed with a few darker markings. The pale markings can be suffused with short wavy brown lines. **FS** Single-brooded, May-July. Readily disturbed by day, flies in sunshine and towards sunset. **Hab** Open woodland, heathland, moorland, mosses. **Fp** Bilberry, Cowberry, Bog Bilberry; spinning the edges of a leaf together to form a pod.

Ancylis apicella page 244

1129 (5070)

Very local. **FL** 6-9mm. Forewing extended into a lobe at the apex, appearing distinctly hooked; pale yellowish brown along the costa, dark brown or mixed dark brown and pale yellowish or greyish brown along the dorsum, with a creamy white suffused pale brownish longitudinal streak from the base to about

three-quarters, a narrow extension from the middle of the wing to the dorsum at three-quarters joined to a pale ocellus containing at least one dark brown mark, and with short pale yellowish-brown streaks on the costa. **FS** Double-brooded, May-August. Readily disturbed by day, flies from dusk and comes to light. **Hab** Damp open woodland, heathland, fens, bogs, marshes. **Fp** Alder Buckthorn, Buckthorn; in a folded leaf, later between two leaves flatly spun together.

Eriopsela quadrana page 244

1189 (4810)

Very local. **FL** 5.5-7.5mm. Forewing whitish or greyish white, coarsely mixed dark brown, with a dark brown dorsal blotch at about one-third, this sometimes curved outwards, and another large brown blotch just before the tornus, the blotches edged orangey brown in the middle of the wing and in some examples the orangey brown is more extensive in the outer third of the wing, and with a short black streak in the middle of the wing at four-fifths. **FS** Single-brooded, late April-early June. Flies in late afternoon and comes to light. **Hab** Open woodland, chalk and limestone grassland, grassy escarpments, embankments. **Fp** Goldenrod; in a spinning on the lower leaves.

Thiodia citrana page 244 1204 (4817)

Local. Most widely found in the south-east. **FL** 7.5-10mm. Forewing pale yellow, variably overlaid with yellowish brown or brown markings, including an oblique cross-band towards the base, an irregular cross-band from the tornus to the costa at about one-half, sometimes narrow or absent at the costa, with an extension from the middle to the apex. **FS** Single-brooded, July-August; has been recorded in May. Occasionally sits on the flowers of the foodplant on sunny afternoons. Flies at dusk and comes to light. **Hab** Open ground, especially near the coast, including vegetated shingle, sand dunes, waste ground, Breckland. **Fp** Yarrow, Field Wormwood, Stinking Chamomile; in a spinning in the flowerheads.

Rhopobota ustomaculana page 244 1158 (4828)

Local. A northern species. **FL** 5.5-7mm. Forewing dark brown with silvery white markings, including a broad, somewhat triangular dorsal blotch, and a bar from the tornus sometimes forming an angled cross-band to the costa at two-thirds, and with a dark dot at the apex. **FS** Single-brooded, June-July. Flies in afternoon sunshine and in the evening. **Hab** Moorland. **Fp** Cowberry, possibly also Bilberry; between spun leaves.

Holly Tortrix *Rhopobota naevana* page 245 1159 (4829)

Common. **FL** 5.5-7.5mm. **Although the wing pattern is extremely variable, this moth rests with a characteristic crease in the wing from the apex to the middle at two-thirds, and the forewing is extended to a lobe at the apex with a distinct notch below on the termen.** Forewing typically brown and blackish brown, variably suffused greyish, with a greyish-white cross-band from the dorsum beyond one-half, obtusely angled in the middle, reaching the costa at one-third, the inner margin clearly defined, the outer margin rather more rounded and less distinct, and with a black dot or short streak just beyond the outer angle and a large whitish ocellus. In lighter forms the ground colour is shining white with the markings contrasting strongly. Uniform dull yellowish-brown, reddish-brown or blackish examples occur, but the black streak or dot is usually visible. **FS** Single-brooded, late May-mid September. Can occur in numbers. Flies at dusk and comes to light. **Hab** Found in a wide variety of habitats, including hedgerows, gardens, orchards, woodland, moorland. **Fp** Many trees and shrubs, including Holly, Blackthorn and, on moorland, Bilberry; in a spinning, sometimes in spun flowers.

Rhopobota stagnana page 245 1161 (4827)

Local. Very local from the Midlands northwards. **FL** 5-9mm. The female is usually much smaller than the male, with a narrower forewing and more contrasting whitish and dark reddish-brown markings. Forewing dark reddish brown, suffused grey, with a broad greyish cross-band at one-half, obtusely angled in the middle, reaching the costa at one-third, with another cross-band, typically more obscure, from the tornus usually reaching the costa, with the inner margin deeply indented and a black dot in the indentation. **FS** Double-brooded, April-September. Can be disturbed on warm days, flies in early evening and comes to light. **Hab** Chalk downland, limestone pavement, damp woodland, scrub, boggy areas. **Fp** Small Scabious, Devil's-bit Scabious; in the flowerhead and later boring into the stems, late-summer brood on young leaves at the base of the plant.

Rhopobota myrtillana page 245 1162 (4826)

Local. Mainly a northern species, but occurs very locally in southern England and Wales. **FL** 4.5-6mm. Forewing dark reddish brown, variably suffused greyish, especially towards the base, with a silvery greyish-white cross-band at one-half, somewhat angled in the middle or towards the costa, reaching the costa at one-third, with another angled cross-band from just before the tornus reaching the costa at two-thirds, and usually with a blackish-brown apical mark. **FS** Single-brooded, May-June. Flies in late afternoon and evening sunshine. Occasionally abundant. **Hab** Heathland, moorland and montane habitats. **Fp** Bilberry; between two leaves flatly spun together.

Bud Moth *Spilonota ocellana* page 245 1205 (4831)

Common. Local in Scotland. **FL** 5.5-7.5mm. Variable. Forewing typically whitish, sometimes tinged pale yellowish brown, variably suffused grey, with the basal third blackish brown and a usually prominent blackish triangular mark just before the tornus; the ocellus is elongate, nearly reaching the apex, and is partially edged leaden metallic, containing several black dashes towards the apex. A unicolorous dark form, with a black triangle near the tornus, is frequent, as are intermediates with greyish-brown or greyish-white central areas. **Similar species** *S. laricana* is very similar to the intermediate forms. Although *S. laricana* has a narrower forewing than most *S. ocellana*, this feature is not reliable, and confirmation of identity is likely to require examination of genitalia, although the differences in these are subtle and there is variability. Also dark forms of *Gypsonoma dealbana*, which have a small black streak in the middle of the wing. **FS** Single-brooded, late May-early September. Comes to light. **Hab** Open woodland, parkland, scrub, hedgerows, orchards, gardens. **Fp** A range of trees and shrubs, including oaks, apples, hawthorns, sallows, Sea-buckthorn, Bog-myrtle, Sea Spurge; larger larvae spin leaves together in an untidy spinning.

Spilonota laricana page 245 1205a (4832)

Local, but distribution unclear owing to confusion with *S. ocellana*. **FL** 5-7.5mm. Formerly treated as a form of *S. ocellana*. The forewing is narrower and more greyish with black markings, but otherwise is very similar. **FS** Single-brooded, June-August. Comes to light. **Hab** Larch plantations, woods. **Fp** Larch, sometimes Sitka Spruce; at first mines the leaves, later in a spinning.

Acroclita subsequana page 245 1160 (4834)

Very local. **FL** 5-8.5mm. **Forewing yellowish brown, with small patches of somewhat raised scales in the middle of the wing between one-third and the termen**, with thinly distributed black speckling and a slightly darker basal patch angled and marked blackish along the outer edge. A form occurs with more extensive blackish marking at one-third and at the tornus. **FS** Double-brooded, April-October. Flies from evening onwards and occasionally comes to light. **Hab** Limestone cliffs, vegetated shingle, sand dunes. **Fp** Sea Spurge, Portland Spurge; in a silken gallery in the shoot, or spins leaves to the stem.

Epinotia pygmaeana page 245 1130 (4866)

Local. Very local in northern England and Scotland. **FL** 5.5-6.5mm. **Antenna of the male with very long, diverging cilia.** Forewing brownish, variably suffused whitish grey and blackish brown, with a pale dorsal blotch, sometimes obscure, and an obscure ocellus, this often formed of just a few black dots or short dashes. The strength of the markings can vary, with reddish-brown examples occurring. Hindwing white, darker in the apical third.

Similar species *E. subsequana* has a stronger ocellus, with leaden edges and several short black streaks, the hindwing has darker scales streaked across the white area, and the antenna of the male has very short cilia. **FS** Single-brooded, April-early June. Can be tapped from host trees and flies in midday sunshine around tree-tops. **Hab** Plantations, sometimes isolated trees. **Fp** Spruces, Silver Fir; older larvae feed from a spinning among the leaves.

Epinotia subsequana page 245 1131 (4867)

Rare. **FL** 5-6mm. **Similar species** *E. pygmaeana*. **FS** Single-brooded, April-May. Can be very locally numerous. Flies in afternoon sunshine until late evening, high up round the host tree. **Hab** Plantations. **Fp** Giant Fir, Silver Fir, Noble Fir, Norway Spruce; older larvae feed from spun terminal shoots.

Epinotia subocellana page 245 1132 (4863)

Common. **FL** 4.5-6.5mm. Forewing whitish, sprinkled blackish brown, with a weak darker basal patch bordered by a broad whitish cross-band and darker in the apical half, with a series of paired whitish costal streaks and a large ocellar patch with metallic bluish-grey streaks mixed with black spots and dashes; the apical area is tinged reddish or orangey brown. **FS** Single-brooded, May-early August; has been recorded in mid April. Rarely seen in numbers. Flies from dusk and comes to light. **Hab** Woodland, scrub, fens, wet heaths, bogs. **Fp** Sallows; between two leaves spun flatly together.

Epinotia bilunana page 245 1133 (4877)

Common. More local in Scotland. **FL** 6-8mm. Forewing creamy white to greyish white, sparsely speckled blackish brown, with a blackish mark on the dorsum at about one-third, sharply defined on the outer edge, and another prominent blackish mark on the dorsum just before the tornus. There is variation in the strength of the dark markings. **Similar species** Some forms approach *E. ramella*, which flies later in the year, and has a more solid triangular patch on the dorsum at about one-third. The male of *E. bilunana* has a costal fold. **FS** Single-brooded, May-July. **Hab** Open woodland, parkland, scrub, heathland, mosses, gardens. **Fp** Birches; in a catkin.

Epinotia ramella page 245 1134 (4870)

Common. **FL** 6-7.5mm. Forewing whitish to greyish white, variably suffused light grey and blackish brown, with a large, sharply defined, blackish triangular dorsal mark at about one-third and a dark mark towards the tornus; the ocellus is weak, containing a series of black streaks. The extent of the darker suffusion varies. One common form has a broad blackish band from near the base to the apex, with only a narrow white area along the costa and the dorsal triangle just discernible. **Similar species** *E. bilunana*. **FS** Single-brooded, July-October, occasionally to November. Flies from dusk and comes to light. **Hab** Woodland, scrub, heathland, mosses. **Fp** Birches; in a twig or catkin.

Epinotia demarniana page 245 1135 (4861)

Very local. **FL** 6-7.5mm. Head and labial palps yellowish white. Forewing blackish brown, with a large white dorsal blotch at about one-half, marked near the middle with a small vertical blackish-brown line from the dorsum, and another smaller, more irregular, whitish patch above the tornus, and with a series of paired white streaks along the outer half of the costa, and the apical area tinged reddish brown. **Similar species** *E. trigonella* is

larger, with an evenly arched costa and a dark head, and generally flies later in the year. FS Single-brooded, late May-August. Flies high above the trees in the evening, later comes to light. **Hab** Open woodland, parkland, riverbanks, fens, boggy moorland, chalk pits. **Fp** Birches, Alder, Goat Willow; in a catkin.

Epinotia immundana page 245 — 1136 (4853)

Common. **FL** 5.5-7mm. A narrow-winged species with a shallow, evenly arched costa. Variable. Forewing blackish brown and greyish brown, mixed orangey brown, with a usually conspicuous, almost triangular, whitish, greyish or orangey brown dorsal blotch distinctly marked blackish brown along the inner margin; the ocellus is weakly leaden. Sometimes the dorsal blotch is obscure. **Similar species *E. tetraquetrana* is typically slightly larger, although there is overlap, is usually lighter, with the paler dorsal mark nearly perpendicular to the dorsum along the inner margin, and there is usually a conspicuous blackish mark on the inner margin of the ocellus.** Also similar to some *Pammene*, most of which have a shorter forewing and more prominent ocellus. **FS** Double-brooded, April-September, single-brooded in Scotland. Flies from dusk and comes to light. **Hab** Open woodland, heathland, banks of canals and streams, also recorded from chalk pits and gardens. **Fp** Alders, birches, and recorded on roses; the first brood in catkins or buds, those of the second in rolled leaves.

Epinotia tetraquetrana page 245 — 1137 (4864)

Common. **FL** 5.5-7.5mm. A narrow-winged species with a rather straight costa. Variable. Forewing yellowish brown with a few blackish-brown markings, a rather obscure greyish-white quadrate blotch on the dorsum, sometimes extended into a cross-band to the costa; the ocellus is greyish white with a blackish mark on the inner margin. Dark examples can occur, the wing appearing almost entirely blackish. **Similar species** *E. immundana*. **FS** Single-brooded, April-June. Occasionally numerous. Can be tapped from trees by day, flies in the evening and comes to light. **Hab** Woodland, heathland, moorland. **Fp** Birches, Alder; at first in a twig, later in a turned-down edge of a leaf.

Epinotia nisella page 245 — 1138 (4878)

Common. **FL** 5.5-8mm. Very variable, with a number of different forms. Forewing typically with a blackish-brown basal patch, its outer edge acutely angled near the middle, and with a broad area to the termen whitish mixed greyish and blackish, or blackish brown mixed whitish, or pale orangey brown; typically, the ocellus is blackish brown mixed whitish, with some leaden markings. One frequent form has a triangular or oblong-shaped black or reddish-brown blotch on the dorsum, sometimes extended into the basal patch. In another form the reddish brown extends over much of the wing. Form *cinereana*, which may be a separate species and is very local, is like the typical form but mixed blackish, greyish and whitish. **FS** Single-brooded, July-early October; sometimes recorded in late June and November. Flies from dusk and comes to light. **Hab** Woodland, parkland, boggy heathland, riverbanks, hedgerows, gardens. **Fp** Sallows, poplars, f. *cinereana* on Aspen; in the catkins, between two leaves spun flatly together, or in the shoot.

Nut Bud Moth *Epinotia tenerana* page 245 — 1139 (4869)

Common. **FL** 5.5-7.5mm. Variable. Forewing orangey brown or dark brown, suffused yellowish brown, with a whitish quadrate blotch containing fine brownish lines on the dorsum at about one-half, the inner edge sharply defined and nearly perpendicular to the dorsum, the blotch sometimes linked diagonally to a suffused whitish blotch on the costa to form a cross-band, constricted near the middle; the ocellus is weak, containing three or four black dashes, these often reduced or absent. One form is obscurely marked, with the whitish replaced with pale orangey brown. **FS** Single-brooded, mid June-early October.

Flies from dusk and comes to light. **Hab** Woodland, hedgerows, gardens. **Fp** Hazel, Alder; at first in the catkins, later in the buds.

Epinotia nemorivaga page 245
1141 (4847)

Local. A northern species, also found in western Ireland. **FL** 4.5-5.5mm. Head and labial palps greyish brown. Forewing greyish brown to blackish brown, with silvery white markings, including a triangular dorsal blotch at one-half, often obscurely extended in a cross-band to the costa, and another cross-band from the tornus to about two-thirds on the costa; the silvery white markings are speckled or contain fine lines of greyish brown or blackish brown. **Similar species** *E. tedella* **has the head and labial palps creamy white, and is brownish compared with the greyish-brown appearance of** *E. nemorivaga*. **FS** Single-brooded, late May-July. Flies in afternoon sunshine. **Hab** Mountain moorland, sometimes at lower elevations. **Fp** Bearberry; in the spring spins young leaves of a shoot, forming bladder-like mines.

Epinotia tedella page 245
1142 (4875)

Common. **FL** 4.5-6.5mm. Head and labial palps creamy white. Forewing brown to blackish brown, overlaid with whitish cross-bands or cross-lines. There is considerable variation in the strength of the paler markings, in some these appear net-like, in other forms the whitish is suffused greyish, with the markings blurred, and in extreme cases the whole wing is almost uniformly blackish brown. **Similar species** *E. fraternana* **is often slightly larger, has the forewing suffused with orangey brown and usually a more clearly contrasting narrow cross-band at one-third**; also *E. nemorivaga*. **FS** Single-brooded, May-early July; also recorded in August. Can be tapped from the branches of the host trees by day, flies in afternoon sunshine and comes to light. **Hab** Conifer plantations, gardens. **Fp** Norway Spruce, Sitka Spruce; in an untidy spinning among the needles.

Epinotia fraternana page 246
1143 (4876)

Local. Very local in some areas. **FL** 4.5-6.5mm. Head and labial palps creamy white. Forewing dark brown, more or less suffused with orangey brown, two narrow whitish cross-bands, one at one-third, the other from the tornus to the costa at two-thirds, both outwardly angled in the middle. **Similar species** *E. tedella*. **FS** Single-brooded, May-mid August. Can be tapped from branches of the host trees by day, and flies high around them in afternoon sunshine. **Hab** Plantations, mixed woodland, parkland, gardens. **Fp** Silver Fir, Giant Fir, occasionally other fir species; mines the leaves, later instar larvae constructing a spinning.

Epinotia signatana page 246
1144 (4849)

Local. Very local or scarce from the Midlands northwards. **FL** 6.5-7.5mm. Forewing dark brown, heavily shaded greyish brown, speckled pale yellowish and flecked blackish brown, with two rather diffuse whitish cross-bands often discernible, the inner margin of the inner band indented about the middle **and a prominent, sometimes interrupted, longitudinal black streak from near the middle of the wing through the outer cross-band into the apical area. FS** Single-brooded, June-July. Flies at dusk and occasionally comes to light. **Hab** Open woodland, scrub, hedgerows. **Fp** Blackthorn, Crab Apple, possibly hawthorns; in a folded leaf or spun shoot.

▲ Larva of *Epinotia rubiginosana*.

◀ Larval working of *Epinotia rubiginosana* on Scots Pine. The larva spins the needles together, usually twisting them slightly.

Epinotia nanana page 246 1145 (4858)

Local. Scarce in northern England and very local in Scotland. **FL** 4-5mm. **A very small species.** Forewing blackish brown, somewhat suffused greyish or with poorly defined whitish speckled cross-bands, narrow near the base, at one-third, and from the tornus to the costa at two-thirds. **Similar species** *E. nigricana* (not illustrated), a rare species on Silver Fir, is slightly larger (FL 5.5-6.5mm), has a costal fold in the male and a slender protruding ovipositor in the female, neither character is present in *E. nanana*. **FS** Single-brooded, June-August. Flies towards dusk and comes to light. Can be tapped from host trees by day. **Hab** Plantations, mixed woodland, parkland, gardens. **Fp** Norway Spruce, Sitka Spruce; mines the needles, spinning a silken tube between them.

Epinotia rubiginosana page 246 1146 (4872)

Local. **FL** 6-7mm. Forewing blackish brown, shaded grey, speckled pale orangey brown or reddish brown, this pronounced in the outer part of the wing, with a whitish dorsal blotch obscurely extended greyish to the costa in a cross-band and with a diffuse dark brown mark before the weak ocellus. **FS** Single-brooded, late May-July. Can be disturbed from the branches of the host trees by day, flies around the tops of trees in afternoon sunshine until sunset, and comes to light. **Hab** Conifer plantations, heathland. In Scotland particularly associated with stunted Scots Pine on wet moorland. **Fp** Scots Pine, Stone Pine; in a tubular spinning, slightly twisting the needles.

Willow Tortrix *Epinotia cruciana* page 246 1147 (4850)

Common. In north-west England numerous only where Creeping Willow grows. **FL** 5.5-7mm. Forewing orangey brown or reddish brown, suffused yellowish brown at the base, with a creamy white or yellowish-grey cross-band, sometimes with silvery scales, from the dorsum at two-thirds to the costa at one-third, sometimes broad along the dorsum and extended towards the base along the costa, and a pale blotch on the costa at three-quarters. Unicolorous orangey brown or dark reddish-brown forms occur rarely. **FS** Single-brooded, mid May-early August. Flies in late afternoon and comes to light. **Hab** Open woodland, scrub, boggy heathland, moorland, streambanks, coastal slopes, sandhills, gardens. **Fp** Sallows, willows, especially Grey Willow, Creeping Willow, Eared Sallow; in spun leaves.

Epinotia mercuriana page 246
1148 (4852)

Local. Scarce in north-west England. **FL** 5-6.5mm. **A narrow-winged species.** Forewing reddish brown, with pale yellowish-brown to yellowish-brown markings, these thickly edged silver and thinly speckled blackish. **FS** Single-brooded, July-September. Flies in afternoon sunshine and at dusk. **Hab** Moorland, sometimes found near sea level. **Fp** Heather, sallows, willows, Bilberry, Mountain Avens, possibly other herbaceous plants; in spun leaves.

Epinotia crenana page 246
1149 (4857)

Rare, with a thinly scattered distribution in Scotland. Recorded new to Ireland in 2004. **FL** 6-7mm. Variable. Forewing with the dorsal third white or suffused grey, sometimes broken into two or three blotches, and the costal two-thirds blackish brown. There are forms with the whole wing suffused grey or blackish brown, whilst in another the dorsal area is chestnut brown. **FS** Single-brooded, August-April; the adult hibernates. Flies in sunshine around the foodplant. **Hab** Boggy heathland, streambanks, moorland. **Fp** Grey Willow, Eared Sallow, possibly other sallows and willows; in spun terminal shoots.

Epinotia abbreviana page 246
1150 (4845)

Common. Local in Scotland. **FL** 5.5-7.5mm. Variable. Forewing orangey brown to dark reddish brown, with a whitish, creamy white or yellowish-brown oblique cross-band from the dorsum at one-third, sharply angled in the middle, then inwardly oblique to the costa at one-third, and a similar narrow cross-band from the tornus to the costa beyond one-half, overlaid with silvery leaden scales; the costa has a series of paired whitish streaks, more prominent in the apical half. Almost unicolorous dark reddish-brown examples occur. **FS** Single-brooded, May-August. Flies at dusk and comes to light. **Hab** Woodland, scrub, hedgerows, gardens. **Fp** English Elm, Wych Elm, less often Field Maple, and has been recorded on Common Nettle and Greater Stitchwort; in a developing bud, later in a spun shoot.

Epinotia trigonella page 246
1151 (4840)

Common. Local in north-west England. **FL** 7.5-10mm. Forewing with the costa evenly arched, blackish brown, mixed paler brown in the outer half, speckled black, leaden and whitish, with an almost square, sometimes triangular, white dorsal blotch just before one-half, often containing yellowish-brown or blackish-brown markings from the dorsum, and a similar, almost round, blotch at the tornus. **Similar species** *E. demarniana*. **FS** Single-brooded, late July-September. Flies at dusk and occasionally comes to light. **Hab** Woodland, scrub, hedgerows, open heathland, mosses, bogs. **Fp** Birches, favouring small trees; in spun leaves, usually at the end of a shoot.

Epinotia maculana page 246
1152 (4843)

Local. Very local in northern England and Scotland. **FL** 8-11mm. Forewing dark brown, purplish brown or greyish brown, more or less overlaid with whitish, with a black mark of varying size, sometimes outwardly curved and triangular, on the dorsum at one-third, and an ill-defined blackish blotch just before the tornus; the paler ocellus is dotted with a series of black marks. **FS** Single-brooded, August-October. May be tapped from trees by day, flies at dusk and comes to light. **Hab** Mixed woodland. **Fp** Aspen, sometimes poplars; in a rolled leaf.

▲ Larval working of *Epinotia sordidana* on Alder. The spinnings are usually found near the growing tip of coppice regrowth in open situations.

Epinotia sordidana page 246

1153 (4838)

Local. Recorded new to Ireland in 2002. **FL** 8.5-11mm. Forewing creamy brownish to pale greyish brown, suffused darker brown, with poorly defined markings, including a slightly paler dorsal blotch often edged darker, especially along the inner margin; the cilia are coloured as the forewing. The hindwing is uniformly grey. **Similar species Near-unicolorous forms of *E. caprana* and *E. solandriana*; *E. caprana* is usually slightly smaller and has a greyish-white hindwing, grey near the apex; *E. solandriana* often has contrasting darker and lighter patches in the cilia.** If there is doubt, examination of genitalia will be required to confirm identity. **FS** Single-brooded, August-October. Flies at dusk and comes to light. **Hab** Fens, marshes, bogs, the margins of lakes and streams. **Fp** Alder, preferring young coppice regrowth in open situations; in a rolled or puckered leaf, later spinning a leaf into a pod.

Epinotia caprana page 246

1154 (4839)

Common. **FL** 8-10.5mm. Variable. Forewing ranging from dark yellowish brown through brownish to reddish brown, occasionally darker, sometimes with greyish suffusion, especially along the costa; markings are generally indistinct, in most forms with an obscure, large and slightly paler triangular dorsal blotch, edged somewhat darker, especially along the inner margin. One form has the costal half blackish brown, the dorsal half whitish, with an irregular margin between the two. Another form has the costal third dark brown, much of the rest of the wing reddish brown extending towards the apex, and a dark brown triangular dorsal blotch which extends towards the base. **Similar species** *E. sordidana*, *E. solandriana*. **FS** Single-brooded, end July-October. Flies at dusk and comes to light. **Hab** Fens, marshes, gravel pits, wet moorland, mosses. **Fp** Sallows, willows, Bog Myrtle, and has been recorded on Lodgepole Pine; in spun terminal leaves.

▲ Larva of *Epinotia brunnichana*.

◀ Transverse roll in a leaf of birch made by the larva of *Epinotia brunnichana*.

Epinotia brunnichana page 246   1155 (4842)

Common. **FL** 8.5-10.5mm. Variable. Forewing fairly constant pale greyish brown, with a large, usually four-sided dorsal blotch, whitish grey-brown or white, edged black, sometimes strongly marked black on the inner edge, and with black patches elsewhere on the wing. One form has a semicircular black dorsal blotch. In another form the general colour is greyish, mixed blackish, with a paler dorsal blotch visible and a paler area in the tornus, this often containing a series of blackish dots. **Similar species** An obscurely marked form of *E. solandriana* is similar, but has a distinctly triangular shaped, not four-sided, dorsal blotch. **FS** Single-brooded, July-September, sometimes later. Flies at dusk and comes to light. **Hab** Woodland, parkland, scrub, gardens. **Fp** Birches, Hazel, sallows, willows; in a tranversely rolled leaf.

Epinotia solandriana page 246   1156 (4844)

Common. **FL** 7.5-10mm. Extremely variable. Typical forms: forewing yellowish brown, reddish brown, dark brown or grey, with a large dorsal blotch, usually triangular, either slightly paler than the ground colour, or white. Another range of forms has the forewing white, creamy, orangey white or brownish white, with a brown or dark brown elongate dorsal blotch with a concave margin in the middle of the wing. A scarce form has the forewing dark brown with a yellowish-brown longitudinal streak through the middle of the wing. **Similar species** *E. sordidana*, *E. caprana*, *E. brunnichana*. **FS** Single-brooded, July-August; also recorded in mid June. Flies at dusk and comes to light. **Hab** Open woodland, scrub. **Fp** Birches, Hazel, Goat Willow, occasionally roses; in a rolled leaf.

Spruce Bud Moth *Zeiraphera ratzeburgiana* page 246   1163 (4884)

Common. **FL** 5.5-7mm. Forewing blackish brown, usually heavily suffused yellowish brown or orangey brown, especially in the outer half, with a whitish almost triangular pale blotch on the dorsum faintly extended to the costa in a cross-band, and whitish marks elsewhere, all often suffused yellowish brown. **Similar species** *Z. rufimitrana* is slightly larger, with a more obvious and broad cross-band, constricted near the middle, and also has darker hindwings. **FS** Single-brooded, late June-August. Comes to light. **Hab** Conifer plantations, mixed woodland, gardens. **Fp** Norway Spruce, Sitka Spruce, Morinda Spruce, occasionally Grecian Fir, Scots Pine, Stone Pine; in a shoot.

Zeiraphera rufimitrana page 246   1164 (4883)

Rare. **FL** 5.5-7.5mm. Forewing blackish brown, well defined in the basal area, with a pale yellowish-brown cross-band, outwardly angled or constricted in the middle, and usually more distinct towards the dorsum, and a similarly coloured, more or less defined cross-band from the tornus to the costa at three-quarters. **Similar species** *Z. ratzeburgiana*. **FS** Single-brooded, July. Flies at dusk and comes to light. **Hab** Conifer plantations, mixed woodland. **Fp** Silver Fir, Grecian Fir, Stone Pine; spins a terminal shoot.

Zeiraphera isertana page 246 1165 (4885)

Common. Local in Scotland, where it is apparently absent in the east. **FL** 7-9.5mm. Variable. **Costa smoothly arched.** Forewing blackish or blackish brown with whitish or greyish markings, including an oblique dorsal blotch not extending beyond the middle of the wing; the forewing may be heavily suffused yellowish brown, brown or yellowish grey. The markings can be obscure and an almost unicolorous blackish-brown form occurs. **Similar species** *Z. griseana* is typically larger and longer winged, with a larger **pale dorsal blotch extended towards the costa. FS** Single-brooded, late May-mid September. Can be found in numbers. Rests on tree trunks, from which it readily flies if disturbed, flies at dusk and comes to light. **Hab** Woodland, parkland, hedgerows, gardens. **Fp** Oaks; between spun leaves or in a folded leaf. Has been recorded feeding in the soft galls of cynipid wasps.

Larch Tortrix *Zeiraphera griseana* page 246 1166 (4882)

Common. **FL** 7.5-10.5mm. **A long-winged species.** Forewing greyish brown, heavily speckled pale yellowish brown, with a large, almost quadrate or triangular, speckled whitish dorsal blotch to beyond the middle of the wing, often obscurely extended and inwardly angled to the costa.

Almost unicolorous examples occur. **Similar species** *Z. isertana.* **FS** Single-brooded, June-July; has been recorded in mid August. Flies high around host trees in the late afternoon and comes to light. **Hab** Larch plantations, conifer woodland, parks, gardens. **Fp** Sitka Spruce, larches, pines, occasionally Silver Fir; in spun needles.

Crocidosema plebejana page 246 1157 (4887)

Very local. Breeds in coastal areas but fairly regularly recorded inland, where it may be an immigrant or wanderer. **FL** 5.5-8mm. Male blackish brown, especially along the dorsum towards the base and in the outer half along the costa, with a creamy white to pale yellowish-brown triangular dorsal mark with a finger-like projection towards the apex, meeting the silver-edged whitish ocellus, containing two or three black dots or dashes. Female forewing similar, with extensive yellowish-brown suffusion. **FS** Overlapping broods, February-December, mainly seen in the autumn. Flies from late afternoon and comes to light. **Hab** Cliffs, rocky places, vegetated shingle, gardens. **Fp** Tree-mallow; on the seeds, and burrows into the shoots and soft stems, stunting growth.

Phaneta pauperana page 246 1198 (4889)

Rare. **FL** 6-8mm. **A narrow-winged species.** Forewing greyish white, suffused creamy white towards the apex, the basal third grey, mixed dark grey, black and reddish brown, the outer edge of this basal patch being oblique from the costa to just below the middle of the wing, with a short series of raised black scales near the apex; the ocellus is obscure, suffused grey. **FS** Single-brooded, April-early June. Can be tapped from vegetation on cool days, flies on warm evenings and later comes to light. **Hab** Coastal cliffs, chalk grassland, gravel pits, gardens. **Fp** Dog-rose; in spun flowers and hips.

Pelochrista caecimaculana page 246   1188 (4896)

Very local. **FL** 6-9.5mm. Forewing pale greyish brown, tinged yellowish brown, speckled darker brown, with paired pale costal streaks and a dirty whitish ocellus, sometimes containing a few black marks. **Similar species Separated from *Eucosma* species by the obscure ocellus, which lacks metallic edging. FS** Single-brooded, June-July; has been found in August. Readily disturbed by day and occasionally comes to light. **Hab** Chalk and limestone grassland, waste ground, quarries, coastal cliffs. **Fp** Common Knapweed, possibly other knapweeds and Saw-wort; in the roots.

Eucosma aspidiscana page 247   1190 (4966)

Very local. **FL** 6-9mm. Forewing light golden brown mixed dark brown, the basal area dark dorsally, paler towards the costa, with a pale grey dorsal blotch sometimes extended and inwardly angled to the costa, and whitish costal streaks prominent in the apical half; the ocellus is well developed, laterally edged metallic leaden, containing three black dashes. **FS** Single-brooded, late April-July. Easily disturbed in sunshine, flies at sunset and comes to light. **Hab** Coastal cliffs, limestone pavement, quarries, open woodland. **Fp** Goldenrod, possibly also Michaelmas Daisy, Goldilocks Aster; in the flowerheads, later burrowing into the stem.

Eucosma conterminana page 247   1192 (4963)

Very local. Possibly expanding its range, and sometimes a wanderer or immigrant. **FL** 7-9mm. Forewing brown suffused pale yellowish brown, especially along the costa, with a large and conspicuous, almost triangular, dorsal blotch, dirty yellowish white, sharply defined along the inner margin, somewhat more diffuse along the outer margin; the ocellus is conspicuous, laterally edged pale golden or silvery metallic, containing two or three broken black dashes. **FS** Single-brooded, mid June-September; has been recorded in early October. Readily disturbed by day, flies at sunset and comes to light. **Hab** Chalk grassland, quarries, gardens, waste ground, roadside verges. **Fp** Great Lettuce, Prickly Lettuce; initially in the flowerheads, then burrowing in the seedheads.

Eucosma tripoliana page 247   1193 (4948)

Local. Coastal, but occasionally wanders inland. Recently recorded new to Ireland. **FL** 6-8mm. Variable. Forewing pale yellowish brown, variably marked brown, the basal patch obscure and diffuse, with a dirty creamy dorsal blotch, this often rather diffuse and sometimes marked with fine brown lines, and the costa with a series of creamy streaks; the ocellus is well developed, laterally edged silvery metallic, containing two or three black dashes. The markings are occasionally indistinct and darker examples can occur. **Similar species *E. aemulana* has dark brown mottling near the base and along the costa, with the cilia at the apex dark brown; the yellowish brown is tinged greyish.** *E. rubescana* (not illustrated), is a rare saltmarsh species on Sea Aster in the east and south-east of England. It is somewhat larger and plainer, pale yellowish brown overlaid reddish or greyish brown, and appears almost unicolorous, and starts flying in June, earlier than *E. tripoliana*. If there is any doubt, identification should be confirmed by examination of genitalia. **FS** Single-brooded, July-August. Flies in late afternoon and comes to light. **Hab** Saltmarshes. **Fp** Sea Aster; in a loose spinning in the flowerheads.

Eucosma aemulana page 247   1194 (4945)

Very local. Possibly declining. Predominantly south-eastern. **FL** 5-7.5mm. Forewing whitish to pale yellowish brown, tinged greyish and mottled darker brown; the markings in the central part of the wing are often rather diffuse or obscure, but a dirty creamy dorsal blotch and paler costal streaks are usually discernible; the ocellus is laterally edged silvery metallic, containing two to four black dashes. **Similar species** *E. tripoliana*. **FS** Single-brooded, July-August. Flies in late afternoon and comes to light. **Hab** Open woodland, particularly recently coppiced woodland, chalk downs, coastal cliffs. **Fp** Goldenrod; in the flowerheads, feeding on unripe seeds.

Eucosma lacteana page 247   1195 (4949)

Rare. Coastal. **FL** 6.5-8mm. Forewing white variably suffused pale greyish brown, with whitish costal streaks, these interspersed darker, with other markings faint, although in more strongly marked examples the inner edge of the paler dorsal blotch is partly edged dark brown or black; the ocellus is weakly developed, laterally edged silvery metallic, containing two or three broken black dashes. **FS** Single-brooded, July-August. Flies from late afternoon onwards. **Hab** Drier margins of saltmarshes. **Fp** Sea Wormwood; in a tubular spinning on a flower spike.

Eucosma campoliliana page 247   1197 (4943)

Common. More local in northern England and Scotland. Most frequent near the coast. **FL** 6.5-8.5mm. Forewing white, sometimes weakly tinged grey, with a distinct black dorsal mark at about one-third and another smaller blackish mixed orangey brown dorsal mark at three-quarters; the ocellus is edged silvery leaden and contains two or three black dashes, with a black mark towards the apex; the costa has several blackish or reddish-brown marks, with a reddish-brown apical spot. **FS** Single-brooded, June-September. Flies from late evening and comes to light. **Hab** Chalk grassland, sand dunes, vegetated shingle, waste ground, open woodland. **Fp** Common Ragwort; in the flowerheads, feeds on the seeds, occasionally burrowing into the stem.

Eucosma pupillana page 247   1199 (4968)

Very local. Possibly declining. **FL** 7-9mm. Forewing whitish, suffused pale brownish grey with a narrow inwardly angled brownish-grey cross-band from the dorsum at about one-half, blackish brown near the middle, not quite reaching the costa, and another from the costa at one-half merging with a triangular mark near the tornus; the ocellus is distinct, white with a longitudinal silvery metallic bar marked with a series of black dots. **Similar species *E. metzneriana* (page 247), a rare species resident in south-east England on wormwoods, is much larger (FL 9-11.5mm), with more diffuse greyish-brown markings on the dorsum and an ill-defined ocellus. FS** Single-brooded, June-August. Flies at dusk and has been recorded at light. **Hab** Field margins, quarries, gravel pits, waste ground, particularly in coastal districts. **Fp** Wormwood; bores into the stems beneath the lower shoots, and into the rootstock.

Eucosma hohenwartiana page 247   1200 (4935)

Common. **FL** 7-10.5mm. Forewing greyish brown to reddish brown, somewhat suffused pale brownish white, often along the costa, with a greyish-white or pale brownish-white

dorsal blotch; the ocellus is pale brownish white, laterally edged silvery or pale golden metallic, containing three fragmented black dashes. **Similar species *E. cana* is paler, rather whitish and brownish, with the dorsal blotch appearing to have whitish longitudinal or slightly oblique streaks through it, whereas *E. hohenwartiana* is darker and greyish brown, without white, and the dorsal blotch is pale grey or brown and unstreaked.** Also '*E. hohenwartiana*' group. **FS** Single-brooded, June-August. Flies at dusk and comes to light. **Hab** Rough grassland, meadows, marshes, chalk and limestone grassland, embankments, woodland rides. **Fp** Common Knapweed; in the flowerheads and seedheads.

▲ Larva of *Eucosma hohenwartiana* feeding within the developing seedhead and the top of the stem of Common Knapweed.

Similar species in the *'E. hohenwartiana'* group

Based on constant differences in the female genitalia, *E. fulvana* and *E. parvulana* have recently been restored to species level, having been treated as forms of *E. hohenwartiana*. No character in the male genitalia can be used to separate these species. All three species are superficially very similar on external characters. *E. parvulana* (not illustrated) is the smallest of the three, and tends to be fairly uniformly greyish brown; it is rare in central southern England and is associated with Saw-wort, although it is occasionally found elsewhere, suggesting it may have an alternative foodplant. *E. fulvana* (not illustrated) is the largest and palest, whitish brown or pale yellowish brown, sometimes tinged orangey; it is widespread in southern England, rare elsewhere, and is associated with Greater Knapweed. *E. hohenwartiana* is between these two in size, although there is overlap, is greyish brown, and is associated with Common Knapweed. Where there is overlap in geographical range, confirmation of identification normally requires examination of the female genitalia.

Eucosma cana page 247   1201 (4932)

Common. **FL** 7-11mm. Forewing greyish brown or dark brown, heavily suffused whitish brown or pale yellowish brown, the dark coloration forming diffuse longitudinal or slightly oblique streaks in the dorsal half of the wing and through the dorsal blotch; the dorsal blotch is pale greyish or whitish brown, sometimes obscure, and the pale costal streaks are sometimes fused to form a long whitish mark on the costa before the apex; the ocellus is edged silvery metallic, containing up to four broken dashes. **Similar species** *E. hohenwartiana* and closely related species. **FS** Single-brooded, late May-August. Flies from late afternoon and comes to light. **Hab** Chalk and limestone grassland, meadows, waste ground, coastal cliffs, coastal slopes, vegetated shingle. **Fp** Thistles, Common Knapweed; in the flowerheads and seedheads.

Eucosma obumbratana page 247   1202 (4926)

Local. Very local in Wales, northern England and Scotland. **FL** 7-9.5mm. Forewing yellowish brown in the dorsal third, brown to reddish brown over the rest of the wing; the ocellus is weak, obscurely edged laterally with silvery metallic, usually containing two or three long, often broken, black dashes. **FS** Single-brooded, mid June-August. Flies in the evening and comes to light. **Hab** Arable field margins, embankments, quarries, waste ground, chalk grassland, parkland, vegetated shingle, sand dunes. **Fp** Perennial Sowthistle; in the flowerheads.

Gypsonoma aceriana page 247   1167 (4989)

Local. Very local in Wales and northern England. **FL** 5-7mm. Labial palps and face brownish. Forewing creamy white, with a well-defined dark brown or blackish basal patch, paler at the base on the costa, the outer margin weakly angled, with a small blackish mark near the tornus and suffused greyish along the termen, with a black spot at the apex. A form occurs which has the wing heavily suffused pale greyish brown. **Similar species G. sociana has the face white and the forewing with the apical spot reddish brown, mixed blackish; the typical form of G. dealbana has more greyish in the outer half of the wing, a short black dash in the middle at two-thirds, and an orangey brown apical spot. FS** Single-brooded, late June-August. Can be found sitting on poplar trunks, flies from dusk, and occasionally comes to light. **Hab** Open woodland, parkland, hedgerows, suburban habitats. **Fp** Various poplars, including Black-poplar, Lombardy-poplar, White Poplar, Eastern Balsam-poplar; in the shoots and leaf stalks, later in the stem.

◄ A bud infected by the larva of *Gypsonoma dealbana*, showing frass at the base of the bud.

► The characteristic feeding pattern and frass-covered silk tube of *Gypsonoma dealbana* on the underside of a leaf.

Gypsonoma sociana page 247   1168 (4987)

Local. Very local in Scotland. **FL** 5-7mm. Face white. Forewing white, occasionally tinged pale yellowish brown, with a well-defined dark brown or blackish basal patch, suffused with whitish scales, the outer margin sometimes angled, with a diffuse blackish mark on the costa at about one-half, an often obscure blackish triangular mark near the tornus, and a reddish-brown mixed blackish apical spot. **Similar species** *G. aceriana*, *G. dealbana*. **FS** Single-brooded, June-early August. Can be found sitting on poplar trunks, flies in the evening and comes to light. **Hab** Open woodland, parkland, gardens, banks of rivers, streams, canals and ponds. **Fp** Aspen, Black-poplar, sallows; feeds in leaf buds, and in sallow catkins.

Gypsonoma dealbana page 247   1169 (4985)

Common. Scarce in Scotland. **FL** 5-6mm. Variable. Head brown, sometimes mixed greyish white. The typical form has the forewing white, with a well-defined dark brown basal patch, the outer margin angled, with a broad area of greyish suffusion from the tornus to the costa incorporating **a small black streak near the middle of the wing at about two-thirds**, the apex orangey brown. Darker forms can occur in which the whitish markings are partially or totally obscured greyish or purplish grey, although the black streak usually remains a prominent diagnostic feature. **Similar species** *G. aceriana*, *G. sociana*, dark forms of *Spilonota ocellana*. **FS** Single-brooded, late May-August; has been recorded in late September and mid October. Flies from dusk and comes to light. **Hab** Open woodland, parkland, scrub, hedgerows, gardens. **Fp** Hawthorn, oaks, poplars, sallows, Hazel; lives within a frass-covered tube on the underside of a leaf in the autumn, hibernates at the base of a bud then eats into buds, shoots and sallow and poplar catkins in spring.

Gypsonoma oppressana page 247   1170 (4986)

Very local. Scarce in north-west England. **FL** 6-7mm. **Forewing with a shallow arched costa and rather rounded apex**, dark brown, suffused with whitish and greyish, sometimes mixed with yellowish brown, especially in the outer third, with a whitish speckled brownish and greyish cross-band from the dorsum before one-half to the costa at one-third, with short blackish streaks or marks towards the termen, and with a diffuse dark greyish apical spot. **FS** Single-brooded, June-mid August. Can be found sitting on poplar trunks, flies at dusk and comes to light. **Hab** Woodland rides, parkland, suburban habitats. **Fp** Black-poplar, White Poplar; in spring on the buds, later in a tube lying flat against a twig.

Gypsonoma minutana page 247 1171 (4983)

Very local. **FL** 5-7mm. Forewing pale reddish brown or orangey brown, overlaid with many rather diffuse darker brown cross-lines, with an obscure basal patch, its outer edge angled; the ocellus is distinct, and includes up to seven horizontal parallel black dashes. Flies on warm sunny afternoons and occasionally comes to light. **Hab** Woodland rides, parkland, suburban habitats. **FS** Single-brooded, July-early August. **Fp** Poplars, including White Polar, Black-poplar, Aspen; between two leaves spun flatly together.

Epiblema grandaevana page 247 1181 (5003)

Rare. **FL** 10-15mm. **A large, pale and obscurely marked species.** The male is usually smaller than the female. Forewing light brownish, obscurely speckled darker brown, greyish white and white. **FS** Single-brooded, June-July. Comes to light. **Hab** Sand pits, old brickyards, spoil tips, ballast tips. **Fp** Colt's-foot, Butterbur; in the roots, affected plants usually wilting.

Epiblema turbidana page 247 1182 (5002)

Very local. **FL** 8-11.5mm. The male is usually smaller than the female. Forewing brown, suffused pale greyish, with the dorsal area sometimes darker brown either side of a pale greyish-brown dorsal blotch; the ocellus is weakly developed, containing a series of black dashes. **FS** Single-brooded, end June-July. Flies from late afternoon and sunset on warm evenings, and occasionally comes to light. **Hab** Damp meadows, copses, gardens, streambanks. **Fp** Butterbur, possibly Winter Heliotrope; in the roots, silk tunnels in the crown indicating the presence of a larva.

Epiblema foenella page 247 1183 (4998)

Common. Local in Wales and north-west England, and scarce in Scotland. May be expanding its range. **FL** 8-12mm. Forewing purplish brown to reddish brown, with a conspicuous large white angled mark from the dorsum and a pale ocellus, usually containing three or four black or brown dots, just above the tornus. The white dorsal mark can vary and is sometimes extended, sometimes suffused grey or occasionally obliterated. **FS** Single-brooded, late June-September. Flies from late afternoon and comes to light. **Hab** Waste ground, marshes, riverbanks, coastal cliffs, disturbed ground in open woodland. **Fp** Mugwort, Southernwood; in the roots and stems.

Epiblema scutulana page 247 1184 (4994)

Common. **FL** 8.5-11mm. The male is usually larger than the female. Male forewing mottled pale to dark grey, leaden, black, brown, sometimes mixed yellowish brown and speckled whitish, and often tinged reddish brown towards the apex; there is a prominent, almost square, creamy white dorsal blotch at one-half, and a large ocellus, variably shaded creamy white, light yellowish brown and leaden, containing a series of black dots. Male hindwing is pale greyish brown. Female forewing is mottled blackish and leaden, sometimes with scattered yellowish-brown scales, with a contrasting white dorsal blotch and the ocellus mainly leaden, with white along the termen. Female hindwing is blackish brown. **Similar species** *E. cirsiana* and *E. scutulana* are very similar and both could be confused with *E. sticticana* and *E. cnicicolana*. The male of *E. cirsiana* is smaller, with the forewing darker, similar in colour and size to that of the female *E. scutulana*, but the male hindwing is brown. The females of both species are almost inseparable and the differences in the genitalia are also subtle. *E. sticticana* is typically paler, heavily suffused yellowish brown or yellowish grey. *E. cnicicolana* is usually smaller than *E. scutulana*, and the whitish dorsal blotch tends to be narrower and extends to nearer the costa than in *E. scutulana* and *E. cirsiana*. **FS** Single-brooded over a long period, or possibly with a second brood, May-mid August. Comes to light. **Hab** Waste ground, rough grassland, coastal cliffs, scrub, open woodland, frequenting damp or marshy ground. **Fp** Musk Thistle, Spear Thistle; in the roots and stems.

Epiblema cirsiana page 247 — 1184a (4995)

Common. Until recently considered a form of the preceding species, and old records could refer to either species. **FL** 5.5-9.5mm. Both sexes are of similar size. **Similar species** *E. scutulana*, *E. sticticana*, *E. cnicicolana*. **FS** Single-brooded, May-June. Comes to light. **Hab** Grassland, marshes, woodland. **Fp** Marsh Thistle, Common Knapweed; in the roots and stems.

Epiblema cnicicolana page 247 — 1185 (4997)

Rare. **FL** 6-8.5mm. Forewing blackish brown and leaden grey, sometimes mixed yellowish brown and reddish brown, with a narrow, almost oblong, whitish dorsal blotch at about one-half reaching beyond the middle of the wing; the ocellus is ill-defined, whitish or greyish, edged laterally with two thick leaden lines. **Similar species** *E. costipunctana* is broadly similar in markings, but the forewing appears more variegated and has stronger suffusion of yellowish brown or reddish brown; the dorsal blotch contains a series of weak brown lines, and the blotch may be weakly extended greyish to the costa. *Notocelia tetragonana* is slightly broader winged and darker, and has metallic bluish-grey reflections, especially in the outer half of the wing, and the dorsal blotch is arch-shaped, not oblong. Also *E. scutulana*, *E. cirsiana*. **FS** Single-brooded, mid May-June. Sits on the leaves of the foodplant in the afternoon, flies freely in the evening and occasionally comes to light. **Hab** Ditches, wet meadows, coastal landslips. **Fp** Common Fleabane; in the lower part of the stem.

Epiblema sticticana page 247 — 1186 (4993)

Common. Local in northern Scotland. **FL** 7-9mm. Forewing dark brown, often only visible near the base or along the dorsum, heavily suffused yellowish brown or yellowish grey, with fine silver leaden lines, a large creamy white dorsal blotch at about one-half; the ocellus is laterally edged metallic leaden and contains a series of black dots. Occasionally darker forms occur, with the dorsal blotch suffused grey and appearing less pronounced. **Similar species** *E. scutulana*, *E. cirsiana*. **FS** Single-brooded, May-July. Flies in afternoon sunshine until dusk, and occasionally comes to light. **Hab** Waste ground, sand pits, embankments, coastal landslips, shingle, sandhills, open woodland. **Fp** Colt's-foot, occasionally Winter Heliotrope; in the roots and flowering stems.

Epiblema costipunctana page 247 — 1187 (4999)

Common. Local in north-west England. **FL** 6-8.5mm. Forewing dark brown, variably suffused yellowish brown or reddish brown, with a rather diffuse paler basal patch, a somewhat square whitish dorsal blotch at about one-half, this weakly lined darker and occasionally weakly extended greyish towards the costa; the ocellus is rather obscure, paler, laterally edged leaden, containing a series of black dashes. **Similar species** *E. cnicicolana*. **FS** Double-brooded, although single-brooded in the north, May-August. Flies in late afternoon sunshine and occasionally comes to light. **Hab** Waste ground, rough grassland, disturbed ground in open woodland, embankments, sandhills. **Fp** Common Ragwort; in the rootstock.

▲ Larva of *Notocelia cynosbatella* concealed within a shoot of rose.

Notocelia cynosbatella page 247 1174 (5019)

Common. **FL** 7.5-10.5mm. **Labial palps yellowish.** Forewing basal third blackish brown, slightly extended along the costa, much of the rest of the wing creamy white, suffused pale yellowish brown or pale greyish brown, with a pale greyish-brown blotch near the tornus, the apex greyish brown. **FS** Single-brooded, May-July; has been recorded in early August. Flies from late afternoon and comes to light. **Hab** Hedgerows, scrub, orchards, gardens. **Fp** Roses, including cultivars, Bramble; in flower buds, young shoots or between two leaves.

Bramble Shoot Moth *Notocelia uddmanniana* page 247 1175 (5021)

Common. Very local in Scotland. **FL** 7-10mm. Forewing grey, with a large arch-shaped chocolate-brown blotch near the tornus, broadly bordered pale grey either side by a cross-band, each of which often extends to the costa. **FS** Single-brooded, with an occasional partial second brood in some years, mid May-September; has been recorded in October.

Comes to light. **Hab** Bramble patches, hedgerows, scrub, open woodland, gardens. **Fp** Bramble, Raspberry, Loganberry; in an untidy spinning.

Notocelia trimaculana page 248 1176 (5026)

Common. Local in Scotland. **FL** 7-8.5mm. **A narrow-winged species.** Forewing ground colour white, partly suffused leaden grey, with a near oblong whitish blotch on the dorsum bordering a dark brown basal patch and with a pale brownish area marked with a few black dots above the dorsum at three-quarters; the ocellus is white, broadly edged silvery, the silvery marks sometimes more extensive in the outer third of the wing. **Similar species** *N. rosaecolana* is usually larger and has a broader forewing, with the costa evenly arched, the forewing colour is often paler and the costal streaks are finer and more oblique. *N. roborana* is usually larger than both *N. trimaculana* and *N. rosaecolana*; the male of *N. roborana* has a long dark brown costal fold extending the colour of the basal patch along the costa to about one-half; in both sexes of *N. roborana* the outer two-thirds of the forewing is typically more uniform than those of the other two species.
N. incarnatana has a similar forewing pattern but has a pronounced pink suffusion and a more pointed apex than *N. trimaculana*, *N. rosaecolana* and *N. roborana*; *N. incarnatana* can also be distinguished by the small blackish triangular mark projecting into the pale ocellus, which lacks silvery marks, and the apex which is more heavily marked black.
FS Single-brooded, May-August. Comes to light. **Hab** Hedgerows, open woodland, chalk downland. **Fp** Hawthorn; in spun shoots.

Notocelia rosaecolana page 248 1177 (5025)

Common. Scarce in Scotland. **FL** 7.5-9.5mm. **Similar species** *N. trimaculana*, *N. roborana*, *N. incarnatana*. **FS** Single-brooded, late May-mid August. Comes to light. **Hab** Chalk and limestone grassland, parkland, open woodland, hedgerows, gardens. **Fp** Roses, especially Sweet-briar, including cultivars; in a spinning on young shoots.

◄ The resinous gall of Pine Resin-gall Moth *Retinia resinella* on a twig of Scots Pine.

Notocelia roborana page 248 | 1178 (5022)

Common. Local in Scotland. **FL** 7.5-11mm. **Similar species** *N. trimaculana*, *N. rosaecolana*, *N. incarnatana*. **FS** Single-brooded, mid June-August. Comes to light. **Hab** Open woodland, hedgerows, gardens, chalk and limestone grassland, coastal cliffs, vegetated shingle, sandhills. **Fp** Roses, including cultivars; in spun shoots and leaves.

Notocelia incarnatana page 248 | 1179 (5024)

Very local. Predominantly coastal. **FL** 6.5-8mm. **Similar species** *N. trimaculana*, *N. rosaecolana*, *N. roborana*. **FS** Single-brooded, June-early September. Flies from late afternoon and comes to light. **Hab** Coastal limestone grassland, limestone pavement, chalk grassland, sandhills, occasionally open woodland. **Fp** Roses, Burnet Rose; in spun leaves.

Notocelia tetragonana page 248 | 1180 (5020)

Very local. Scarce in Wales, northern England and Scotland. **FL** 6-7.5mm. Forewing relatively broad, blackish brown, with metallic bluish-grey reflections and a scattering of yellowish-brown scales, especially in the outer half of the wing, and with a small white arch-shaped dorsal blotch at about one-half. **Similar species** *E. cnicicolana*. **FS** Single-brooded, June-July. Flies from late afternoon and occasionally comes to light. **Hab** Open woodland, woodland rides, hedgerows, scrub. **Fp** Roses; in spun leaves.

Pseudococcyx posticana page 248 | 1208 (5028)

Local. More frequent in the east. **FL** 6-7mm. Head and labial palps dark orangey brown. Forewing blackish brown, mottled pale greyish, suffused orangey brown towards the apex. The hindwing is dark greyish brown. **Similar species** **Clavigesta sylvestrana is slightly smaller, has a grey head and labial palps, a paler forewing and hindwing, and flies later in the year; P. turionella is larger and paler, with a whitish hindwing in the male, pale grey in the female. FS** Single-brooded, late April-June. Flies in the evening and occasionally comes to light. **Hab** Pine woods, heathland with scattered trees, suburban gardens. **Fp** Scots Pine, seemingly preferring smaller trees; in the stem of a lateral shoot, aborting the bud above.

Pine Bud Moth *Pseudococcyx turionella* page 248 1209 (5029)

Local. Scarce in north-west England and Scotland. Perhaps most frequent in the south-east and East Anglia. **FL** 7-10mm. Labial palps, head and much of the thorax pale orangey brown. Forewing dark greyish brown near the base and along the costa to about one-half, the rest of the wing suffused orangey brown, especially in the apical third; the markings are mottled greyish white, giving the effect of paler cross-bands, with an obscure pale mark above the tornus. The hindwing is whitish in the male, pale grey in the female. **Similar species** *P. posticana*. **FS** Single-brooded, late April-June. Flies from late afternoon and later comes to light. **Hab** Pine woods, heathland with young pines. **Fp** Scots Pine, sometimes Lodgepole Pine or Corsican Pine; in a large central bud or a whorl of buds, including the terminal bud of the tree, tunnelling a short distance into the stem.

Pine Resin-gall Moth *Retinea resinella* page 248 1214 (5033)

Very local. Predominantly northern, with a few scattered records in England. **FL** 7.5-10.5mm. Forewing blackish brown, silvery metallic and leaden, and speckled whitish, resulting in a pattern of irregular incomplete bands across the wing. The hindwing cilia are white. **FS** Single-brooded, June-early July. Has a two-year life cycle, with peak emergences in odd-numbered years. Can be locally numerous. Occasionally comes to light. **Hab** Conifer plantations, pine woods. **Fp** Scots Pine, Lodgepole Pine, Monterey Pine; usually on young trees, forming in a resinous gall on a twig (see p.327).

Clavigesta sylvestrana page 248 1206 (5041)

Very local. **FL** 5.5-7mm. Labial palps, head and thorax grey. Forewing brownish grey mixed silvery grey, forming obscure paler irregular bands across the wing, with obscure greyish streaks on the costa; in some examples the outer third of the wing is speckled or suffused yellowish brown. **Similar species** *Pseudococcyx posticana*. **FS** Single-brooded, late June-early September. Flies in the late afternoon and occasionally comes to light. **Hab** Conifer woods, heathland, parkland. **Fp** Maritime Pine, Stone Pine, possibly Scots Pine, Weymouth Pine; in a silken gallery on the shoots, buds and the male flowers.

Pine Leaf-mining Moth *Clavigesta purdeyi* page 248 1207 (5042)

Common. Local in northern England. Recently recorded new to Ireland and expanding its range in England. **FL** 4.5-6mm. Forewing narrow and rather pointed, greyish brown mixed whitish grey, forming obscure paler irregular lines across the wing, and orangey brown in at least the apical third, sometimes with a brownish blotch at the tornus. **FS** Single-brooded, end June-mid September. Flies in the evening and comes to light. **Hab** Coniferous woodland, suburban habitats. **Fp** Scots Pine, Corsican Pine, Lodgepole Pine; initially mining a needle, in spring on the bases of the new growth of needles.

Pine Shoot Moth *Rhyacionia buoliana* page 248 1210 (5044)

Common. Local in northern England. **FL** 8-11.5mm. Thorax with a distinct posterior crest. Forewing bright orange, somewhat suffused yellowish brown or reddish brown, with irregular silvery cross-lines; the orange mark before the tornus is tall and triangular, though the sides of the triangle are not always joined near the costa. The hindwing is dark grey. **Similar species** *R. pinicolana* **has an insignificant thoracic crest, is duller orange, more strongly marked along the costa, and the mark before the tornus is arch-shaped, not extending beyond the middle of the wing; the hindwing is grey.** **FS** Single-brooded, mid June-August. Flies in the late evening and later comes to light. **Hab** Conifer plantations. **Fp** Scots Pine, other pines; in the shoots. If the leading shoot is attacked this results in a subsequent characteristic bend in the main stem.

Rhyacionia pinicolana page 248   1211 (5045)

Common. Local in the west and northern England, rare in Scotland. **FL** 8-11.5mm. **Similar species** *R. buoliana*. **FS** Single-brooded, June-early September. Flies from dusk, later coming to light. **Hab** Coniferous plantations, gardens. **Fp** Scots Pine; in the shoots, causing them to distort as they grow.

Spotted Shoot Moth *Rhyacionia pinivorana* page 248   1212 (5048)

Common. More local in Scotland. **FL** 7-9mm. Forewing dull orangey brown or reddish brown, somewhat mixed blackish brown, with silvery grey markings resulting in a pattern of ill-defined bands across the wing. **FS** Single-brooded, mid May-July; has been recorded in late April and early August. Occasionally flies in bright sunshine, flies in the evening and comes to light. **Hab** Conifer plantations, mixed woodland, parkland, heathland with scattered pines. **Fp** Scots Pine; in young side shoots.

Elgin Shoot Moth *Rhyacionia logaea* page 248   1213 (5050)

Very local. Scotland only. **FL** 6.5-9mm. Forewing narrow and pointed, dark brown with obscure greyish cross-lines, with at least the apical third suffused dull reddish brown, reflecting yellowish brown in bright light. **FS** Single-brooded, April-early May. Flies freely during the day when mild. **Hab** Pine forests, perhaps with a preference for younger trees. **Fp** Scots Pine, Lodgepole Pine, Sitka Spruce; in the buds and mining young shoots, later boring into the stem.

'Dichrorampha' group

At first sight most of the species in this group appear dull, and they tend to be dismissed as 'too difficult to identify'. Under magnification, however, all species are revealed to be much more attractive, and most are not especially hard to determine to species if examples are in good condition. The group divides into two, based on whether or not the male has an overlapping flap of wing membrane along the costa, the 'costal fold'. Scales on the forewing have pale-coloured tips, usually yellowish brown or orange, mostly in the outer third of the wing, although sometimes more extensive; this is described below as 'speckling', and is often dense. The group of three species without a costal fold in the male is tricky to sort, particularly if examples are not reared from the early stages, and examination of genitalia is normally required to identify these species. Worn examples of any *Dichrorampha* may require examination of genitalia to confirm identity.

The table below summarises some of the key features of the males given in the species descriptions, and may be helpful in separating out *Dichrorampha*. Only one species, *D. acuminatana*, is seen regularly in moth traps, others just occasionally, most often on very warm nights in the summer. Most species can be found by day on warm afternoons, when adults fly close to or rest on the foodplants.

D. alpinana and *D. flavidorsana* are somewhat similar to two *Pammene* species, *P. regiana* and *P. trauniana*; all have conspicuous yellow or orange marks on the dorsum, although the shapes of these marks separate the genera fairly easily, those in the *Pammene* species being more substantial.

Male of Dichrorampha	Costal fold	Dorsal blotch	Other features
D. petiverella	Yes, to ⅓	Contrasting pale yellow and curved	
D. alpinana	Yes, to ½	Contrasting yellow or yellowish orange and slightly curved	Hindwing dark brown, paler near base
D. flavidorsana	Yes, to ⅓	Contrasting yellow or yellowish orange and slightly curved	Hindwing uniform dark brown
D. plumbagana	Yes, to ½	Obscure pale grey and triangular	Silvery scales on pale markings
D. senectana	Yes, to ½	Obscure, hardly visible	Dense speckling across whole forewing
D. sequana	Yes, to ½	Contrasting creamy whitish and quadrate	
D. acuminatana	Yes, to ⅓	Somewhat contrasting brownish white and triangular	Pointed forewing, underside of which greyish brown
D. consortana	Yes, to ⅓	Obscure whitish and narrow triangular	Small species, mixed dark and light
D. simpliciana	Yes, to ½	Obscure pale greyish brown and triangular	Large species, obscurely marked
D. sylvicolana	Yes, to ½	Obscure yellowish orange and triangular	Short, blunt forewings
D. montanana	Yes, to ⅓	Obscure whitish grey and triangular	Underside of forewing dark greyish brown with greyish-white scales between dark veins towards termen
D. vancouverana	Yes, to ½	Obscure pale orangey brown or yellowish brown and very oblique	Narrow forewing, dull orangey brown or yellowish brown
D. plumbana	No	Obscure greyish brown and triangular	See p.390 for male genitalia only
D. sedatana	No	Obscure greyish brown and triangular	Dense speckling across whole forewing; see p.390 for male genitalia only
D. aeratana	No	Obscure greyish brown and triangular	See p.390 for male genitalia only

Dichrorampha petiverella page 248   1273 (5249)

Common. Local in Scotland. **FL** 5-6.5mm. Male with costal fold to one-third. Forewing dark brown with a curved pale yellow blotch from the dorsum to the middle of the wing, dense yellowish-brown speckling, and with a few short creamy marks on the costa before the apex and a series of black dots or short streaks along the termen. **Similar species *D. sequana* has a broad, creamy whitish blotch perpendicular to the dorsum, reaching beyond the middle of the wing.** **FS** Single-brooded, late May-August. Flies around the foodplant in sunshine and in the evening, and occasionally comes to light. **Hab** Grasslands, rough ground. **Fp** Yarrow, Sneezewort, Tansy; in the roots.

Dichrorampha alpinana   page 248   1274 (5248)

Common. Local in northern England, scarce in Scotland. **FL** 6.5-7.5mm. Male with costal fold to one-half. Forewing dark brown, with dense dull or bright orange speckling on the outer third, a curved yellow or yellowish-orange blotch from the dorsum to near the middle of the wing and black dots along the termen. The hindwing is dark brown, paler near the base. **Similar species *D. flavidorsana* (not illustrated) has a costal fold in the male only to one-third, a slightly narrower dorsal cross-band, and a uniformly dark brown hindwing. Females can be separated reliably only by examination of genitalia.** Finding examples sitting on flowers of their respective foodplants is a good indicator of species. **FS** Single-brooded, late May-early August. Flies in sunshine and sits on the flowers of the foodplant, flies at dusk and comes to light. **Hab** Chalk and limestone grassland, meadows, rough ground, roadside verges, gardens. **Fp** Oxeye Daisy; in the roots.

Dichrorampha flavidorsana (not illustrated)   1275 (5247)

Local. Very local in northern England and scarce in Scotland. **FL** 6.5-8mm. Male with costal fold to one-third. **Similar species** *D. alpinana*. **FS** Single-brooded, July-August. Occasionally flies in sunshine, more active at dusk and comes to light. **Hab** Rough ground, roadside verges, gardens. **Fp** Tansy; in the roots.

Dichrorampha plumbagana page 248   1276 (5251)

Common. Very local in parts of northern England, local and mainly coastal in Scotland. **FL** 5-7.5mm. Male with costal fold to near one-half. Forewing dark brown, densely speckled yellowish grey, with pale greyish markings comprising an oblique, almost triangular, blotch from the dorsum to the middle of the wing or beyond, and **a few streaks in the outer third, these contrasting somewhat with the ground colour, and marked with silvery scales**; there are three black dots along the termen. The underside of the forewing is dark greyish brown, more or less speckled greyish white toward the termen, with greenish bronzy reflections. **Similar species *D. senectana* is larger and more evenly and densely speckled, with any markings hardly discernible, and the black terminal dots minute or absent.** **FS** Single-brooded, May-June. Sits on flowers of the foodplant, flies in sunshine and occasionally comes to light. **Hab** Chalk and limestone grassland, neutral grassland, sand dunes, roadside verges, rough ground. **Fp** Yarrow; in the roots.

Dichrorampha senectana page 248   1277 (5243)

Very local. **FL** 6-8mm. Male with costal fold to near one-half. **Similar species** *D. plumbagana*. **FS** Single-brooded, June. Flies in the afternoon and at dusk. **Hab** Coastal cliff and undercliff, quarries. **Fp** Oxeye Daisy; in the roots.

Dichrorampha sequana page 248   1278 (5240)

Local from the Midlands southwards. **FL** 4.5-6.5mm. Male with costal fold to near one-half. Forewing dark brown with dense yellowish-brown speckling, especially on the outer half of the wing, and with a broad creamy whitish blotch perpendicular from the dorsum to beyond the middle of the wing; there are three or four black dots on the termen. **Similar species** *D. petiverella*. **FS** Single-brooded, May-June. Sits on the foodplant in the afternoon, flies at sunset, and occasionally comes to light. **Hab** Chalk and limestone grassland, neutral grassland, roadside verges, rough ground, open woodland. **Fp** Yarrow; in the roots.

Dichrorampha acuminatana page 248   1279 (5232)

Common. Local in northern England, very local in Scotland. **FL** 5-7.5mm. Male with costal fold to one-third. Forewing with a pointed apex, dark brown, sometimes with a faint purplish or pinkish hue, more or less speckled with dull yellowish brown, and a broad brownish-white almost triangular blotch from the dorsum to beyond the middle of the wing; there are three to six black dots on the termen, sometimes obscure, and the cilia are shining dark grey, with a central white band. Underside of forewing greyish brown, with greyish-white and slightly bronzy reflections. **Similar species** The pointed forewing and somewhat contrasting dorsal blotch help to distinguish this species from other obscurely marked *Dichrorampha*, as well as the contrasting shining grey-and-white-banded cilia, but only if the example is in good condition. Most similar to *D. montanana*, which has a blunter forewing and a rather more obscure dorsal blotch, and the underside of the forewing is dark greyish brown, with greyish-white scales between the dark veins towards the termen, and with greyish, greenish and coppery reflections. **FS** Double-brooded, mid April-September. Flies in the afternoon and evening, and regularly comes to light. **Hab** Chalk and limestone grassland, rough ground, roadside verges, gardens. **Fp** Yarrow, Tansy; in the roots.

Dichrorampha consortana page 248   1280 (5226)

Very local. Scarce in northern England, rare in Scotland. **FL** 4.5-6mm. Male with costal fold to one-third. Forewing blackish brown with dull yellowish-brown speckling and a narrow, whitish, almost triangular blotch from the dorsum to beyond the middle of the wing, and with leaden blue-metallic oblique streaks from the costa in the outer third, a leaden ocellus sometimes white-edged, and three to five black dots on the termen. **The small size and contrasting dark and light markings distinguish this species from other *Dichrorampha*. FS** Single-brooded, July-August. Flies in the afternoon and occasionally comes to light. **Hab** Chalk and limestone grassland, rough ground. **Fp** Oxeye Daisy; in the lower part of the stem, stunting growth and causing lateral shoots to grow taller than the tenanted shoot.

Dichrorampha simpliciana page 248   1281 (5239)

Local. Scarce in Scotland. **FL** 6-8mm. Male with costal fold to near one-half. Forewing greyish brown, speckled dull pale grey or yellowish grey, with a broad pale greyish-brown almost triangular blotch from the dorsum to the middle of the wing, and with three to five black dots on the termen. **The large size, broad forewing and obscure markings distinguish this species from** other *Dichrorampha*. **FS** Single-brooded, July-September. Can be disturbed by day, flies in the evening and comes to light. **Hab** Rough ground, roadside verges. **Fp** Mugwort; in the roots.

► Larva of *Dichrorampha simpliciana* within a root of Mugwort.

Dichrorampha sylvicolana page 248   1282 (5238)

Scarce. **FL** 5-6mm. Male with costal fold to near one-half. Forewing short and blunt, dark brown, densely speckled dull orange on the outer half of the wing, with a broad dull yellowish-orange almost triangular blotch from the dorsum to the middle of the wing, hardly contrasting with the ground colour, and with three black dots along the termen. **FS** Single-brooded, June-July. Can occasionally be disturbed by day, but prefers to hide among vegetation. **Hab** Woodland rides, wet meadows and pastures, riverbanks, ditches. **Fp** Sneezewort; in the roots.

Dichrorampha montanana page 248   1283 (5255)

Common from Wales and the Midlands northwards. The most widespread of the genus in Scotland. **FL** 5.5-7mm. Male with costal fold to one-third. Forewing pale greyish brown, sometimes darker in the female, densely speckled dull yellowish brown, with a broad dull whitish-grey almost triangular blotch from the dorsum to beyond the middle of the wing, hardly contrasting with the ground colour, and with small pale greyish marks on the costa and up to five black dots on the termen; the cilia are dark greyish brown with a central whitish band. **Similar species** *D. acuminatana*. **FS** Single-brooded, June-August. May be disturbed by day and flies in the evening. **Hab** Limestone, neutral and sandy grasslands, rough ground. **Fp** Yarrow, Tansy; in the roots.

Dichrorampha vancouverana page 248   1284 (5246)

Common. Local in northern England, scarce in Scotland. **FL** 6-7.5mm. Male with costal fold to near one-half. Forewing narrow, dark brown, densely speckled dull orangey brown or dull yellowish brown, with a sharply oblique dull pale orangey brown or yellowish-brown blotch from the dorsum to the middle of the wing, hardly contrasting with the ground colour, and with none to five black dots on the termen. **FS** Single-brooded, June-July. Flies by day in sunshine, in the evening and comes to light. **Hab** Chalk and limestone grassland, rough ground, coastal areas, gardens. **Fp** Yarrow, Tansy; in the roots.

Dichrorampha plumbana page 248   1285 (5214)

Common. Local in Scotland. Distribution uncertain owing to confusion with *D. aeratana*. **FL** 6-7mm. **Male without costal fold.** Forewing dark brown, densely speckled dull yellowish brown on the outer two-thirds only, speckling very weak on the inner third, with a greyish-brown dorsal blotch, sometimes extending to near the costa, hardly contrasting with the ground colour, and with several black dots on the termen. **Similar species** *D. sedatana* (not illustrated) is slightly larger, with dense speckling across all the forewing, and flies later in summer; *D. aeratana* (not illustrated) is, on average, slightly bigger, but otherwise is very similar to *D. plumbana*. Examination of genitalia is required to distinguish these three species (see p.390 for male genitalia only), unless examples have been reared.

FS Mainly single-brooded, May-June, occasionally September. Flies in warm or sunny weather, in the evening and occasionally comes to light. **Hab** Grassland, rough ground, coastal areas. **Fp** Oxeye Daisy, Yarrow; in the roots.

Dichrorampha sedatana (not illustrated)   1286 (5215)

Local. Very local in northern England. **FL** 7-8mm. **Male without costal fold. Similar species** *D. plumbana*, *D. aeratana* (see p.390 for male genitalia only). **FS** Single-brooded, mid May-July. Can be seen in numbers. Flies in sunshine, in the evening and occasionally comes to light. **Hab** Grassland, rough ground, coastal areas. **Fp** Tansy; in the roots.

Dichrorampha aeratana (not illustrated)   1287 (5218)

Local. **FL** 6-8mm. **Male without costal fold. Similar species** *D. plumbana*, *D. sedatana* (see p.390 for male genitalia only). **FS** Single-brooded, May-June. Can be seen in numbers. Flies in sunshine and sits on the flowers of the foodplant, flies in the evening and occasionally comes to light. **Hab** Grassland, rough ground. **Fp** Oxeye Daisy; in the roots.

Spruce Seed Moth *Cydia strobilella* page 248   1254 (5139)

Local. Very local away from southern England. **FL** 5-7mm. Forewing dark brown, speckled golden yellowish brown over much of the outer half, densest towards the termen, the costa with silvery whitish or yellowish streaks, some extended as weak leaden metallic cross-lines, with a pair of cross-lines at one-half, often joined in the middle; the ocellus is obscure, laterally edged metallic leaden, containing up to four black dots. The hindwing is brown. **Similar species C. conicolana is dark greyish brown, has two narrow leaden cross-bands from the dorsum at one-half and the tornus, and a more strongly marked ocellus; C. cosmophorana is dark greyish brown, has two narrow silvery whitish cross-bands, the one at one-half narrowed or interrupted near the middle, with distinct white marks on the costa and on the termen below the apex. FS** Single-brooded, late April-May. Occasionally seen in numbers. Flies in afternoon sunshine around spruces and occasionally comes to light. **Hab** Coniferous woodland, plantations, gardens, parks. **Fp** Norway Spruce, Serbian Spruce; in a cone.

Cydia ulicetana page 248   1255 (5116)

Common. **FL** 5-7mm. Variable. Forewing pale grey to dark grey, variably speckled pale yellowish brown or yellowish grey, the costa with several small whitish streaks, darker between; the ocellus is somewhat obscure and laterally edged silvery leaden metallic, typically containing three black dashes, and there is a black dash or a few dots in the wing near the ocellus. Strongly marked examples have the central part of the forewing almost white, blackish beyond, with the ocellus well developed. In another form the markings are virtually obsolete, except for the costal streaks and ocellus. **FS** Double-brooded, late March-September, single-brooded in the north. Can be abundant flying around gorse bushes in sunshine, is easily disturbed from the foodplant in dull weather and comes to light. **Hab** Scrub, woodland margins, heathland, grassland, wherever gorse is abundant. **Fp** Gorses, Broom, bird's-foot-trefoils, greenweeds; on unripe seeds in the pods.

Cydia microgrammana page 249   1220 (5128)

Very local. Mainly coastal. **FL** 4.5-6mm. Head grey or whitish grey, labial palps whitish with the terminal segment grey. Forewing short and blunt, grey, slightly tinged brownish, the outer third densely speckled pale yellowish grey, the rest of the wing speckled greyish white, somewhat arranged into wavy cross-lines, though this speckling may be reduced, the costa with many short whitish streaks, interspersed darker grey, with a few extended into oblique leaden lines in the outer half; the ocellus is obscure and laterally edged leaden metallic, containing two or three black dashes. **Similar species** The Alfalfa Moth *C. medicaginis* (not illustrated) has a slightly narrower, more pointed forewing (similar in shape to *C. ulicetana*), the labial palps are whitish, the terminal segment yellowish white. It is a rare species associated with medick species, and has been recorded from Hampshire, Kent, Surrey, Essex and Suffolk. *Grapholita caecana* is a long-winged species, usually larger, with several narrow dark brown streaks in the middle of the wing and oblique from the costa. *G. gemmiferana* has an orangey brown suffusion towards the termen, and a more developed ocellus. **FS** Single-brooded, late May-early August. Flies in the afternoon, at dusk and comes to light. **Hab** Chalk and limestone grassland, waste ground, coastal cliff and slope. **Fp** Common Restharrow; in the green seed-pods.

Cydia servillana page 249   1256 (5146)

Local from the Midlands southwards. **FL** 5.5-7mm. Forewing whitish, variably suffused grey or creamy white, the basal area and apical third variably shaded blackish grey; the ocellus is somewhat obscure, laterally edged bluish metallic, containing three or four black dashes. The hindwing of the male is white, darker towards the apex. **FS** Single-brooded, May-June. Males fly in afternoon sunshine, both sexes fly in the evening. Occasionally comes to light. **Hab** Open woodland, woodland edges, parkland, scrub, margins of ponds and rivers, coastal cliffs. **Fp** Goat Willow, Grey Willow; in a slender gall on a twig.

▶ Gall of *Cydia servillana* in a one-year-old twig of Grey Willow. Note the very slight swelling in the twig and the silk-capped emergence hole just above the dead bud.

Pea Moth *Cydia nigricana* page 249   1257 (5111)

Common. Local in northern England, scarce in Scotland. **FL** 5.5-7.5mm. Labial palps mottled pale brownish or dark greyish above, whitish below. Forewing rather uniform in appearance, greyish brown or dark brown, sometimes slightly tinged yellowish grey, with a series of short whitish streaks along the costa, these interspersed blackish; the ocellus is rather obscure, containing three or four black dots or dashes. Hindwing dark brown with contrasting white cilia. **Similar species** *Grapholita funebrana* and *G. tenebrosana* have weak costal streaks and pale greyish-brown hindwings; *G. funebrana* has dark greyish-brown palps, a more rounded apex and termen, with an obscure whitish ocellus and dorsal blotch; *G. tenebrosana* has whitish palps, the third segment tinged brownish. **FS** Single-brooded, possibly with a small second brood, May-August. Flies in afternoon sunshine and comes to light.

▶ A discoloured pea pod indicating the presence of the larva of Pea Moth *Cydia nigricana* within.

Hab Grasslands, gardens, waste ground, open woodland, coastal slopes. **Fp** Various wild and cultivated species of pea, Garden Pea; in the pods.

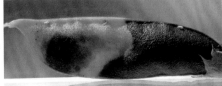

Cydia millenniana page 249   1258 (5142)

Very local. **FL** 6.5-8.5mm. Forewing blackish grey, with a blackish blotch in the middle of the wing at about two-thirds, the costa with several short white streaks; the ocellus is laterally edged leaden, containing four, often wedge-shaped, blackish dashes. **FS** Single-brooded, May-June. Flies in afternoon sunshine. **Hab** Larch plantations. **Fp** Larch; in a resinous gall at a junction of twigs, taking two years to develop.

Cydia fagiglandana page 249   1259 (5153)

Common. Local in northern England, where slowly expanding its range; rare in Scotland. **FL** 6-9mm. Forewing blackish, variably speckled whitish grey or greyish, with a whitish-grey cross-band from the dorsum before one-half, angled in the middle of the wing, to the costa at one-third, the inner margin usually distinctly edged blackish, the rest often obscured by dense speckling, and the costa with whitish streaks of variable length and intensity; the ocellus is large, laterally edged metallic leaden, containing three or four black dashes or dots, these occasionally obsolete. **Similar species C. splendana has a large purplish ocellus, edged on the inner margin by a curved triangular black streak. FS** Single-brooded, mid May-August. Occasionally seen in numbers. Flies in the evening and comes to light. **Hab** Woodland, hedgerow trees, isolated trees on downland. **Fp** Beech; in the nuts.

Cydia splendana page 249   1260 (5152)

Common. Very local in Scotland. Numbers may be boosted by immigration. **FL** 7-10mm. Forewing with the base greyish, speckled paler, and a whitish-grey cross-band from the dorsum at about one-half, narrowing about the middle of the wing, then  broadening to the costa between one-third and two-thirds, the costa with slender, often paired, greyish streaks, interspersed darker; the ocellus is large, laterally edged metallic leaden, containing four or five blackish dashes, with a triangular blackish streak from the tornus wrapping round the inner margin of the ocellus. A form occurs which is blackish brown with a slight purplish hue, with most of the markings obscure except for the ocellus and black streak. **Similar species C. pomonella has a longer forewing, a large coppery ocellus and fine transverse brownish lines**; also *C. fagiglandana*. **FS** Single-brooded, July-early October. Flies at dusk and comes to light. **Hab** Gardens, hedgerows, woodland. **Fp** Oaks, Sweet Chestnut, Walnut; in an acorn or nut.

Codling Moth *Cydia pomonella* page 249   1261 (5144)

Common. Very local in Scotland. **FL** 6.5-10.5mm. Forewing grey, with fine brownish transverse lines and a large dark purplish-brown ocellus, this thickly edged coppery or golden, with a triangular blackish streak from the tornus wrapping round the inner margin of the ocellus. **Similar species** Dark examples of *C. splendana*. **FS** Single-brooded, with a possible second brood, end May-October. Flies at dusk and comes to light.
Hab Woodland, hedgerows, gardens, orchards, parkland. **Fp** Mainly apples, less often pears, Quince, Walnut, plums, Sweet Chestnut, Fig, Common Whitebeam; in the fruit or spun berries.

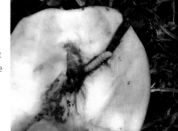

▶ Exposed larva of *Cydia pomonella* in an apple.

Cydia amplana page 249 1262 (5154)

Scarce immigrant. Has become more frequent in recent years. **FL** 7.5-10mm. Forewing orangey brown, dark brown over much of the dorsal half, with a cream coloured dorsal blotch at one-half, this containing a few narrow and vertical brownish lines. **FS** Single-brooded, mid July-early October; mainly seen in August. Comes to light, occasionally in numbers. **Hab** Recorded from a range of habitats, though predominantly coastal. **Fp** Not confirmed to have bred in Britain; on the Continent on oaks, Sweet Chestnut, Walnut, possibly Beech and Hazel; in an acorn or a nut.

Cydia cognatana page 249 1265 (5132)

Very local. Scotland only. **FL** 5.5-7mm. Forewing dark greyish brown, tinged greyish yellow, with a slender curved white dorsal blotch, this bisected blackish brown vertically, and with a series of short paired whitish streaks along the costa; the ocellus is laterally edged bluish leaden metallic, containing four or five, often fragmented, black dashes. **Similar species** *C. coniferana, C. illutana, C. indivisa*. **FS** Single-brooded, June. Flies around the tops of pines in afternoon sunshine. **Hab** Old pine forests, isolated old pine trees. **Fp** Scots Pine; in the bark.

Cydia pactolana page 249 1266 (5140)

Very local. First recorded in Britain in 1965. **FL** 5.5-8mm. **A broad-winged species without a distinct dorsal blotch.** Forewing dark brown, with the outer third densely speckled yellowish brown and often suffused yellowish grey or pale grey towards the base, with an S-shaped cross-band from the costa at one-third, formed of two pale greyish cross-lines with dark brown between, diverging before the dorsum, and with silvery whitish or yellowish costal streaks, some interspersed violet metallic; the ocellus is laterally edged silver metallic leaden, sometimes with violet reflections, containing three or four black dashes or a series of dots. **FS** Single-brooded, May-July. May be disturbed by day and comes to light. **Hab** Coniferous woodland, plantations. **Fp** Norway Spruce, Larch; in the bark of a twig.

Cydia illutana page 249 1266a (5133)

Very local. First recorded in Britain in 1975. **FL** 5.5-6.5mm. Forewing blackish brown, usually with ill-defined paler markings, including a narrow whitish dorsal blotch, curving just before the middle and sometimes containing dark brown marks, and small white marks on the costa, some interspersed weakly leaden metallic or bluish; the ocellus is laterally edged weakly leaden metallic or bluish, with four or five black dashes. **Similar species** *C. coniferana, C. cognatana, C. indivisa*. **FS** Single-brooded, late May-June. Comes to light. **Hab** Coniferous or mixed woodland, also gardens. **Fp** European Larch, Norway Spruce, Silver Fir, probably other conifers; in the green cones.

Cydia cosmophorana page 249 1267 (5138)

Local. Very local in northern England. **FL** 4-6.5mm. Forewing dark greyish brown, densely speckled yellowish-brown or yellowish-grey over the ocellus, with two conspicuous silvery whitish cross-bands, the inner one at one-half narrowed or interrupted near the middle, and with short whitish marks on the costa and one on the termen below the apex, those on the costa interspersed with metallic bluish or leaden purplish streaks; the ocellus is distinct and laterally edged metallic bluish leaden, containing four or five black dashes. **Similar species** *C. strobilella, C. conicolana*. **FS** Single-brooded, May-July. Flies in afternoon sunshine around the tops of pines and occasionally comes to light. **Hab** Coniferous woodland, plantations, heathland. **Fp** Scots Pine, Corsican Pine; in a resinous nodule on the bark, in the one-year-old gall of *Retinia resinella*, or associated with the old feeding damage of a *Dioryctria* (Pyralidae) species.

Cydia coniferana page 249 1268 (5136)

Local. **FL** 4.5-7mm. Forewing dark greyish brown, weakly speckled yellowish grey in the outer half, with a narrow whitish dorsal blotch containing a short dark brown vertical line, curved at about the middle and often weakly joined to form a silvery white cross-band to the costa at one-third, and a weakly defined silvery cross-band from the tornus to the costa at two-thirds, and with almost paired small white streaks along the costa and a small white mark on the termen below the apex; the ocellus is laterally edged metallic leaden, containing three or four long black dashes. **Similar species** C. *cognatana* has a rather larger and more curved dorsal blotch and the costal streaks are more evenly spaced apart; *C. illutana* is blackish brown, usually with ill-defined markings; *C. indivisa* (not illustrated), first found in Britain in 2010, in Buckinghamshire, has a narrow, evenly curved dorsal blotch, with a rather large and distinctive ocellus; *C. conicolana* does not have a dorsal blotch and the black dashes in the ocellus are less distinct. **FS** Single-brooded, but probably double-brooded in the south, mid May-August. Flies around pines in afternoon sunshine and at dusk, and occasionally comes to light. **Hab** Coniferous forests, plantations. **Fp** Scots Pine, Corsican Pine, Maritime Pine, Serbian Spruce, Sitka Spruce; in a silk-lined tunnel under the bark, exuding reddish frass.

Cydia conicolana page 249 1269 (5134)

Very local. **FL** 4.5-6mm. Forewing dark greyish brown, finely speckled pale yellowish brown across the outer two-thirds, with two narrow leaden metallic cross-bands from the dorsum just beyond one-half and from the tornus, and short white streaks along the costa; the ocellus is laterally edged bluish leaden metallic, containing three or four long black dashes. **Similar species** C. *strobilella*, C. *coniferana*, C. *cosmophorana*. **FS** Single-brooded, May-June. Flies around the tops of pines by day and occasionally comes to light. **Hab** Coniferous woodland, small copses and isolated trees on heathland and moorland. **Fp** Scots Pine, Corsican Pine; in a cone, emergence holes in the wings of fallen cones indicating the presence of the moth on that tree.

Lathronympha strigana page 249 1219 (5163)

Common. Local in Scotland.

FL 5.5-8mm. Forewing orangey brown, sometimes suffused brown, the costa with a few obscure short brown streaks, these becoming more leaden in the apical half; the ocellus is laterally edged pale golden metallic, containing two or three fine black dashes. **FS** Double-brooded, end May-September. Occasionally found in numbers. Flies at sunrise and in the evening, is readily disturbed by day and comes to light. **Hab** Chalk and limestone grassland, embankments, open woodland. **Fp** St John's-worts, particularly Perforate St John's-

▶ The larval spinning of *Lathronympha strigana* on St John's-wort. The infected shoot tends to grow at right-angles to the stem.

wort, Common Bird's-foot-trefoil; in an untidy spinning in a shoot, causing the shoot to bend at right-angles as it continues to grow.

Selania leplastriana page 249   1218 (5159)

Rare. **FL** 5.5-6.5mm. Forewing dark brown, suffused yellowish grey and yellowish brown towards the costa and termen, with a dorsal blotch at about one-half to beyond the middle, comprising a curving triangular whitish mark containing a thin brown line, several short pale yellowish-white transverse lines between the dorsal blotch and the base reaching the middle of the wing, and a series of distinct pale streaks, some leaden, on the costa; the ocellus is laterally edged silvery, containing two or three black dashes. The hindwing is greyish brown. **Similar species** *Grapholita jungiella* **is smaller, has no short transverse lines between the dorsal blotch and the base, the pale markings are silvery, and the male has white hindwings. FS** Single-brooded, July; occasionally recorded in October, indicating a partial second brood. Flies in afternoon sunshine and occasionally comes to light. **Hab** Coastal cliffs and slopes. **Fp** Wild Cabbage; in the stem, side shoots, flowers and seed-pods.

Grapholita caecana page 249   1240 (5093)

Rare. An occasional wanderer or immigrant. **FL** 5.5-8mm. **A long-winged species.** Forewing grey, speckled yellowish grey, with several very thin dark brown streaks in the middle of the wing, creamy whitish along the costa with a series of short, dark brown streaks, two or three of which are obliquely extended into the wing towards the termen, with leaden metallic lines between; the silvery ocellus is faint. **Similar species** *Cydia micrigrammana*, Alfalfa Moth *C. medicaginis* (not illustrated), *G. gemmiferana*. **FS** Single-brooded, late May-July. Flies in afternoon and evening sunshine. **Hab** Chalk downland and cliffs. **Fp** Sainfoin (the native species, not fodder cultivars); in the shoots and stems.

Grapholita compositella page 249   1241 (5084)

Common. Local in upland parts of northern England, very local in Scotland. **FL** 4-4.5mm. Forewing blackish brown with a dorsal blotch beyond one-half, represented by four short whitish lines reaching the middle of the wing, the costa with distinct short whitish streaks. The hindwing of the male is white, with the apex and termen darker. **FS** Double-brooded, May-August. Often seen in small numbers. Males fly in afternoon sunshine, females towards sunset. Very occasionally recorded at light. **Hab** Rough, grassy places, chalk and limestone grassland, meadows, sand dunes, old railway cuttings, open woodland. **Fp** Clover, Common Bird's-foot-trefoil; first brood in the stems, those of the second in spun leaves and flowerheads, sometimes in the pods.

Grapholita internana page 249   1242 (5091)

Local. **FL** 4-5.5mm. Forewing dark brown, with the dorsal blotch at about one-half represented by two distinct short curved creamy white lines reaching the middle of the wing, and with a series of short mixed silvery and creamy white streaks along the costa. The hindwing of the male is white, the apex darker. **Similar species** *G. pallifrontana* **is blackish brown, without speckling, and the paler markings are pale yellowish; the male has dark brown hindwings. FS** Single-brooded, April-June; has also been recorded in August. Often found in numbers flying in sunshine around Gorse bushes, when the white hindwings are obvious; flies with *Cydia ulicetana*. Very occasionally seen at light. **Hab** Heathland, moorland, coastal scrub on grassland, quarries. **Fp** Gorse; in the pods.

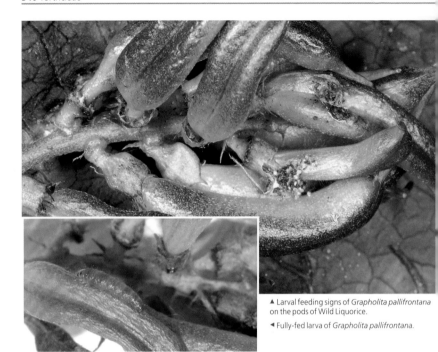

▲ Larval feeding signs of *Grapholita pallifrontana* on the pods of Wild Liquorice.

◀ Fully-fed larva of *Grapholita pallifrontana*.

Grapholita pallifrontana page 249   1243 (5087)

Very local. Declining. **FL** 4.5-5.5mm. Forewing blackish brown, the dorsal blotch at about one-half represented by two distinct short slightly curved pale yellowish lines reaching the middle of the wing, and with a series of shorter pale yellowish streaks along the costa. **Similar species** *G. internana*. **FS** Single-brooded, late May-July. Males fly over the foodplant in afternoon sunshine. **Hab** Rough grassland, scrub margins, particularly on chalk and limestone grassland. **Fp** Wild Liquorice; in the pods.

Grapholita gemmiferana page 249   1244 (5095)

Rare. **FL** 5-6mm. Forewing greyish brown, densely speckled pale orangey brown in the outer third, with a series of short whitish or pale yellowish and greyish-black streaks along the costa; the ocellus is laterally edged silvery, containing four or five black dots. **Similar species** *Cydia microgrammana*, Alfalfa Moth *C. medicaginis* (not illustrated), *G. caecana*. **FS** Single-brooded, late May-June. Flies in sunshine. **Hab** Coastal cliffs and landslips. **Fp** Narrow-leaved Everlasting-pea; in a spun leaf or between two spun leaves.

Grapholita janthinana page 249   1245 (5104)

Common. Very local in southern Scotland. **FL** 5-6mm. **A short-winged species with a rounded termen.** Head and thorax orangey brown. Forewing dark brown, suffused purplish or sometimes pinkish, mixed reddish brown, extensively mixed with dark brown, the termen and apex yellowish brown, with a rather obscure, slightly curved dorsal blotch formed from four short greyish lines; the ocellus is poorly developed, containing three or four black dashes. **Similar species** *G. lobarzewskii* **is longer winged, without any purplish suffusion, and the thorax and forewing are of similar colour, whereas in *G. janthinana* the contrast is obvious. FS** Single-brooded, late June-August. Flies in afternoon sunshine until dusk and occasionally comes to light. **Hab** Hedgerows, gardens, woodland edges, scrub. **Fp** Hawthorn; in the berries.

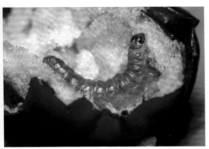

▲ Larva of *Grapholita janthinana* feeding in a hawthorn berry.

▲ Two hawthorn berries spun together by the larva of *Grapholita janthinana.*

Grapholita tenebrosana page 249   1246 (5103)

Local. Scarce in northern England and Scotland. **FL** 5-6.5mm. Labial palps whitish, the third segment tinged brownish. Forewing dark purplish brown, sometimes tinged reddish, with weak bluish leaden streaks and weakly speckled greyish white along the costa; the ocellus is obscure, laterally edged bluish leaden, usually containing four irregular black spots. **Similar species** *Cydia nigricana*, *G. funebrana*. **FS** Single-brooded, late May-July. Flies in sunshine and occasionally comes to light. **Hab** Woodland, moorland, scrub on chalk and limestone grassland, hedgerows, urban areas, gardens. **Fp** Roses, especially Dog-rose, Rowan; in the fruit.

Plum Fruit Moth *Grapholita funebrana* page 249   1247 (5102)

Common. Local in northern England. Recorded new to Scotland in 2006. **FL** 5.5-7mm. Forewing with a rather rounded apex and termen, blackish brown, more or less suffused with greyish scales forming an indistinct greyish dorsal blotch and an ocellus, and with weak pale streaks on the costa; the ocellus usually contains four black dots. **Similar species The Oriental Fruit Moth *G. molesta* (not illustrated), an occasional accidental import on fruits, is often larger with better defined markings and an obscure whitish spot near the middle of the wing at about two-thirds**; also *Cydia nigricana*, *G. tenebrosana*. **FS** Single-brooded, May-October. Flies in the evening and comes to light. **Hab** Hedgerows, woodland margins, scrub, gardens, orchards. **Fp** Blackthorn, wild plums and domestic cultivars; in the fruit.

Grapholita lobarzewskii page 249   1249 (5106)

Very local. **FL** 6-6.5mm. Thorax orangey brown. Forewing suffused orangey, pinkish or reddish brown, mixed brownish, with a few obscure white costal streaks and an obscure dorsal blotch, this divided into four lines; the ocellus is poorly developed, laterally edged bluish leaden, containing three or four black dashes. **Similar species** *Enarmonia formosana*, *G. janthinana*. **FS** Single-brooded, May-early July. Comes to light. **Hab** Gardens, roadside verges. **Fp** Apple in Britain; in the fruit.

Grapholita lathyrana page 249   1250 (5099)

Rare. Now restricted to very few sites. **FL** 4-5mm. Forewing dark brown, suffused yellowish brown and yellowish grey in the apical third, with two silvery whitish cross-lines from the dorsum at about one-half to the costa at one-third, often interrupted in the middle, and another from the tornus to the costa at two-thirds, again sometimes interrupted; the ocellus is laterally edged silvery or golden, usually containing four black dashes. **FS** Single-brooded, late March-May. Flies in sunshine. **Hab** Grassland. **Fp** Dyer's Greenweed; in a spinning on unexpanded flower buds, flowers or young shoots.

▲ Larva of *Grapholita lathyrana* on Dyer's Greenweed.

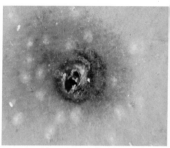

▲ Larval workings of *Grapholita lobarzewskii* on an apple. Note the two holes and the faint reddish mine under the skin made by the larva. Codling Moth *Cydia pomonella* makes a single hole in the fruit, without a mine.

Grapholita jungiella page 249   1251 (5098)

Common. **FL** 4.5-6mm. Forewing greyish brown to yellowish brown along the costa and in the outer third, dark brown in much of the dorsal area, with a distinct dorsal blotch at about one-half, represented by two white curved lines reaching the middle of the wing, and with white costal streaks, a few extended as silvery leaden metallic lines into the wing; the pale ocellus is laterally edged silvery leaden metallic, usually containing three or four black dashes, these sometimes broken. The hindwing of the male is whitish, shaded darker at the apex. **Similar species** *Selania leplastriana*. **FS** Single-brooded, sometimes with a small second brood in the south, April-June and August. Flies in the early morning, in afternoon sunshine and comes to light. **Hab** Woodland, chalk and limestone grassland, wet grassland, scrub. **Fp** Bitter-vetch, Bush Vetch; in the pods or between tightly spun leaves.

Grapholita lunulana page 250   1252 (5100)

Very local. Local in northern England, where in the recent past it was spreading and quite common for a while. **FL** 5-8mm. **A narrow-winged species.** Forewing blackish brown, tinged greyish or yellowish grey in the inner third, with a narrow curved whitish dorsal blotch at one-half, this approaching the middle of the wing, and with a series of fine short whitish markings along the costa; the ocellus contains three or four small black dots or dashes. **Similar species G. orobana is broader winged, with the dorsal blotch more strongly curved and often broader, and with a more well-developed ocellus**; also *Glyphipterix haworthana* (Glyphipterigidae). **FS** Single-brooded, May-June. Can be found in numbers. Flies in sunshine. **Hab** Woodland margins, hedgerows, rough hillsides, pastures, embankments, sand dunes. **Fp** Bitter-vetch, probably Meadow Vetchling or other vetches, also Tufted Vetch, Garden Pea; in spun leaves or mining the stems, later in the pods.

Grapholita orobana page 250   1253 (5101)

Rare. **FL** 5-7mm. **Similar species** *G. lunulana*; also *Glyphipterix haworthana* (Glyphipterigidae). **FS** Single-brooded, July-August. Flies in late afternoon sunshine and comes to light. **Hab** Woodland, marshes, cliffs. **Fp** Vetches, especially Wood Vetch, Tufted Vetch, Meadow Vetchling, Marsh Pea; in the pods.

'*Pammene*' group

This is a group of 21 species that are quite diverse in their superficial wing markings. A few species are easy to identify, such as *Pammene aurana*, which has striking orangey yellow spots. However, most species are difficult to determine, look rather similar to others in the genus and can be variable within a species. Some could be confused with a few species in the genus *Epinotia*, such as *E. immundana*. Only one species is reasonably numerous, *P. fasciana*, and this is regularly seen in moth traps.

P. regiana and *P. trauniana* are very similar to each other, with conspicuous yellow or orange marks on the dorsum, and are somewhat similar to two *Dichrorampha* species, *D. alpinana* and *D. flavidorsana*, although the shapes of these marks should enable separation of the genera.

Most tricky is the '*P. albuginana*' group, which contains five species that all look rather similar. Only *P. albuginana* is covered below. The other four are infrequently recorded. *P. suspectana* is a rare species found in southern England, associated with the bark of Ash. *P. agnotana* is also a rare southern England species; it is associated with old hawthorn trees and most examples have been found by tapping branches onto a beating tray. *P. obscurana* is somewhat more widespread, and is occasionally found in various parts of England, Wales and Scotland, associated with birch. *P. ignorata* has been recorded fewer than ten times in southern England, but most of these records have been in moth traps in the past few years; its foodplant is unknown.

Pammene splendidulana page 250   1223 (5175)

Common. Scarce in northern England and Scotland. **FL** 4.5-5.5mm. Forewing dark brown, speckled greyish yellow, densely in the outer third, with two silvery white or metallic greyish cross-bands, broad on the dorsum at one-half to the costa beyond one-third, narrowing to the middle and usually interrupted, the other from the tornus to the costa, incorporating a distinct black spot near the middle. The hindwing of the male is whitish basally, broadly darker towards the apex and along the margin. **Similar species** **P. leudersiana** (not illustrated), a rare species known only from a few Scottish examples, has a more strongly angled basal patch, and does not have a black spot in the middle of the wing. If *P. leudersiana* is suspected, confirmation by examination of genitalia is required. **FS** Single-brooded, April-early June. Flies in afternoon and evening sunshine, and occasionally comes to light. **Hab** Woodland, parkland, heathland. **Fp** Oaks; in a silken gallery between two leaves flatly spun together.

Pammene giganteana page 250

Local. Scarce in Wales and Scotland. **FL** 6.5-7.5mm. **A long- and narrow-winged species.** Forewing blackish brown, faintly speckled whitish grey, with an obscure oblique whitish-grey dorsal blotch at about one-half; the ocellus is laterally edged silvery grey, containing indistinct black dashes. Hindwing creamy white, broadly darker towards the apex and along the margin. **Similar species P. argyrana is smaller and shorter winged, with a broader dorsal blotch. FS** Single-brooded, mid March-May. Flies in afternoon sunshine, visits sallow blossom and comes to light. **Hab** Woodland, parkland, heathland. **Fp** Oaks; in a young oak apple or spongy gall.

Pammene argyrana page 250

Common. Local in Wales and Scotland. **FL** 4.5-6mm. Forewing blackish brown, mottled whitish near the base and towards the termen and variably speckled yellowish brown and yellowish grey throughout, with a conspicuous whitish dorsal blotch, variably marked blackish, sometimes slightly extended beyond the middle towards the costa; the ocellus is laterally edged leaden metallic, containing indistinct black dashes or scales. The hindwing of the male is whitish or pale greyish brown, in a broad wedge-shape from the base, broadly margined dark brownish. **Similar species P. albuginana has much less pale mottling, with a creamy or dirty whitish dorsal blotch and a speckling of orangey brown scales in the outer half; the male hindwing has a less distinct wedge-shaped patch, which is pale brownish.** Also *P. giganteana*. **FS** Single-brooded, April-early June. Can be found at rest on oak trunks by day, flies in the afternoon and comes to light. **Hab** Woodland, parkland, scrub. **Fp** Oaks; in a spongy gall or other wasp gall.

Pammene albuginana page 250

Very local. Scarce from the Midlands northwards. Recently recorded new to Wales. **FL** 5-6mm. Forewing blackish brown, mottled brownish near the base, distinctly speckled orangey brown usually in small patches in the outer half, with a large creamy or dirty whitish dorsal patch, containing up to three fine blackish lines; the dorsal blotch may be much reduced and obscure; the ocellus is laterally edged leaden, containing several fine black dashes. **Similar species** *P. argyrana*. There are a few rare *Pammene* (none illustrated) which look superficially similar to *P. argyrana* and *P. albuginana*: *P. suspectana* is speckled pale yellowish brown in the apical third, including the ocellus; *P. agnotana* and *P. obscurana* are both obscurely marked, the former having an obscure ocellus, the latter typically slightly larger and with a more evenly arched costa than in *P. agnotana*; *P. ignorata* is most similar to *P. albuginana* but has a broader forewing and broader dorsal blotch. *P. suspectana*, *P. agnotana* and *P. ignorata* have been found in southern England, *P. obscurana* has been recorded In England, Wales and Scotland. If any of these rare species is suspected, identification must be confirmed by examination of genitalia. **FS** Single-brooded, May-July. Flies in late afternoon and evening, and comes to light. **Hab** Oak woodland. **Fp** Oaks; in a spongy gall or other wasp gall.

Pammene spiniana page 250

Local. Very local in Wales and northern England. **FL** 4.5-5.5mm. Forewing blackish brown, the outer third of the wing variably speckled yellowish brown, with a small triangular white dorsal blotch at about one-half, the apex of the triangle pointing towards the costa at or before one-half, and sometimes divided by a short dark brown streak at the dorsum; the ocellus is laterally edged violet metallic, containing two or three black dashes. **Similar species The dorsal blotch of *P. populana* is larger, whitish to pale yellow, almost always containing a blackish-brown line, and the apex of the triangle points towards the outer half of the costa. FS** Single-brooded, August-September. Very

▶ Pupal cocoons
and larva of
Pammene regiana
exposed behind
the bark of
Sycamore. Larvae
descend from the
seeds when fully
fed and often
pupate under the
bark.

occasionally found in numbers. Flies in sunshine. **Hab** Woodland margins, hedgerows, scrub. **Fp** Blackthorn, hawthorns; in spun flowers.

Pammene populana page 250   1232 (5192)

Local. Very local in northern England and scarce in Scotland. **FL** 4.5-6.5mm. Forewing brownish or brownish grey, variably speckled orangey brown in the outer two-thirds, with a white to pale yellow dorsal patch at about one-half, this divided dark brown at the dorsum, and with an oblique leaden metallic cross-band from the tornus to the costa at two-thirds and other shorter leaden streaks near the costa; the ocellus is laterally edged leaden metallic, containing three or four obscure black dashes. **Similar species** *P. spiniana*. **FS** Single-brooded, July-mid September. Comes to light. **Hab** Woodland, fens, marshes, hedgerows, scrub, sand dunes. **Fp** Willows, especially Goat Willow, Osier, Creeping Willow; in a spinning in the buds and young shoots.

Pammene aurita page 250   1233 (5197)

Common. Expanding its range in the north. **FL** 6.5-7mm. Forewing brownish, variably shaded orangey brown, with a broad orangey yellow dorsal patch and a series of short whitish costal streaks. **FS** Single-brooded, June-August. Flies in the afternoon and comes to light. **Hab** Woodland, parkland, gardens, hedgerows, scrub. **Fp** Sycamore; in the seeds.

Pammene regiana page 250   1234 (5196)

Common. More local in Ireland. **FL** 6-7.5mm. Forewing dark brown, with a large pale yellow to orangey yellow dorsal blotch, sometimes tinged slightly orange, its margin typically with a distinct angle, and the outer edge perpendicular or slightly oblique to the dorsum, meeting the dorsum well before the tornus, and with a series of short whitish or creamy white costal streaks. **Similar species** *P. trauniana* typically has a pale yellow dorsal blotch, more evenly curved without a distinct angle along its margin, but sometimes with one or two angles, meeting the dorsum near the tornus, nearer than in *P. regiana*, and with thicker, more pronounced, pale yellowish costal streaks. Some examples may require examination of genitalia to confirm identity. **FS** Single-brooded, May-mid August, occasionally late August. Occasionally comes to light. **Hab** Woodland, hedgerows, suburban gardens. **Fp** Sycamore, Norway Maple, possibly Field Maple; in the seeds.

Pammene trauniana page 250
1235 (5194)

Rare. Perhaps most frequent in the south-east. **FL** 4.5-6mm. Forewing blackish brown, with a large pale yellow, rarely orangey yellow, dorsal blotch and series of broad pale yellowish costal streaks. **Similar species** *P. regiana*. **FS** Single-brooded, May-July. Flies in sunshine high around Field Maple trees and occasionally comes to light. **Hab** Woodland, hedgerows, scrub. **Fp** Field Maple; in the seeds.

Pammene fasciana page 250
1236 (5173)

► Evidence of *Pammene fasciana* in two acorns pulled apart to reveal the feeding signs.

Common. Very local in Scotland. **FL** 5-8mm. Forewing variably mottled greyish, blackish and white near the base, with a large whitish dorsal blotch extended and curved towards the apex, and with a series of short, paired costal streaks; the large ocellus is laterally edged bluish metallic, usually containing six or seven black dashes, these often broken, and the area around the ocellus is variably shaded orangey brown; there are a few irregular black spots between the ocellus and the dorsal blotch. **Similar species P. herrichiana is darker, grey to greyish black, with the dorsal blotch much reduced or obscure, the orangey brown shading reduced and the ocellus weakly marked, and starts flying about a month earlier. FS** Single-brooded, late June-August. Males fly at sunrise and in the afternoon sunshine. Comes to light. **Hab** Woodland, suburban habitats. **Fp** Oaks, Sweet Chestnut; in an acorn or nut.

Pammene herrichiana page 250
1236a (Not listed)

Very local. **FL** 5-8mm. **Similar species** *P. fasciana*, some authors considering *P. herrichiana* to be a form of that species. **FS** Single-brooded, May-June. Comes to light. **Hab** Woodland. **Fp** Beech; in a nut.

Pammene germmana page 250
1237 (5205)

Very local, with older records only in northern England. **FL** 5-6.5mm. Forewing pointed, with the termen obtusely angled to the dorsum, dark brown with a satin sheen and a series of short creamy costal streaks, some faintly extended leaden metallic a short way into the wing, and with the cilia greyish brown. **Similar species Strophedra weirana is usually smaller, the forewing slightly shorter and rounded at the apex, the cilia at the apex with a whitish central band. FS** Single-brooded, May-early July. Flies early in the morning, in the afternoon and evening, and occasionally comes to light. **Hab** Woodland, hedgerows. **Fp** Oaks, Beech.

Pammene ochsenheimeriana page 250
1238 (5200)

Very local. **FL** 3.5-5mm. Forewing blackish grey, evenly but very sparsely speckled whitish across the middle of the wing and in the outer third, with two narrow black cross-bands, the inner at one-third obscure, the outer, between the tornus and the costa beyond one-half, edged with silver scales, and with a series of short whitish costal streaks. **FS** Single-brooded, May-June. Can be tapped from branches by day, flies in the late afternoon and comes to light. **Hab** Coniferous woodland. **Fp** Giant Fir, Norway Spruce, Sitka Spruce, possibly other conifers; has been reared from aborted buds and from an old aphid gall on spruce.

Fruitlet Mining Tortrix *Pammene rhediella* page 250 1239 (5190)

Common. Local in Scotland. **FL** 4-5.5mm. Forewing dark purplish brown, the outer third reddish brown, paler towards the termen, with an angled leaden cross-line from the costa to the tornus. **FS** Single-brooded, late April-June. Flies in sunshine, typically high along hawthorn hedges. Occasionally comes to light. **Hab** Woodland margins, parkland, hedgerows, orchards, scrub, suburban habitats. **Fp** Hawthorns, sometimes Dogwood, apples, pears, Wild Plum, Dwarf Cherry; in spun flowers or on the surface of the fruit.

Pammene gallicana page 250 1271 (5168)

Local. Very local in northern England and scarce in Scotland. **FL** 5-6mm. Forewing pale brown with a violet hue, variably suffused dark brown, with scattered violet leaden markings, an obscure dorsal blotch, and a series of obscure whitish costal streaks; the ocellus is obscure, narrow, laterally edged metallic violet leaden, containing three to five thick black dashes, these sometimes joined. **FS** Single-brooded, July-August. Flies in afternoon sunshine, at dusk and occasionally comes to light. **Hab** Dry grassland, open downland, waste ground, marshes, sand dunes. **Fp** Wild Carrot, Wild Angelica, Hogweed, other umbellifers; in a loose spinning amongst the seeds.

Pammene aurana page 250 1272 (5167)

Common. Local in Scotland. Often absent from areas where the foodplant occurs. **FL** 4-6mm. Forewing dark brown with two rounded orangey yellow blotches, one on the dorsum before one-half, the other towards the apex. **Similar species** *Phaulernis fulviguttella* (Epermeniidae). **FS** Single-brooded, June-July; has been recorded in mid May. Rests on flowerheads of the foodplant, flies in sunshine and occasionally comes to light. **Hab** Open woodland, hedgerows, grassland, suburban habitats, waste ground. **Fp** Hogweed; in a spinning on the seeds.

Strophedra weirana page 250 1221 (5207)

Local. **FL** 4.5-5.5mm. Forewing relatively slender, rounded at the apex, dark brown, sometimes with a very faint slightly paler cross-band, and with a few small whitish marks on the costa. **Similar species** *Pammene germmana*. **FS** Single-brooded, occasionally with a second brood in some areas, May-August. Flies at sunrise, in afternoon and evening sunshine, and comes to light. **Hab** Woodland, particularly on calcareous soils. **Fp** Beech; between two spun leaves.

Strophedra nitidana page 250 1222 (5208)

Very local. Scarcer from the Midlands northwards. **FL** 4-4.5mm. Forewing rounded at the apex, dark brown, with a faint curved cross-band at about one-half, pale greyish brown containing a central dark brown line, and with a few short white costal streaks. **FS** Single-brooded, May-July. Flies in sunshine in the afternoon and early evening, and occasionally comes to light. **Hab** Oak woods. **Fp** Oaks; in a spinning between two leaves.

▶ Larva of *Pammene aurana* emerging from a seed of Hogweed.

Pyralidae

Pyralis farinalis

Hypsopygia costalis

Oncocera semirubella

Myelois circumvoluta

There are 89 species in this family, which includes familiar and regularly encountered species such as the Bee Moth *Aphomia sociella*. The group also contains the so-called 'tabbies' and 'knot-horns'. The forewing length is variable within the family, 5-22mm. Several resting postures are exhibited, and a few species sit with the lower part of the abdomen curved into the air, rather like some species of macro-moths in the Geometridae. Most, however, sit with the wings held roof-like, more or less appressed to or around the abdomen. The antennae are typically laid back over the thorax and abdomen. The forewings range from elongate to broadly triangular, are very varied in markings, and have a distinct tornal angle. The hindwings are broad, typically broader than the forewings, and with the dorsal cilia short. The head has raised scales on the crown, the face smooth or with raised or erect scales. The antennae are long, from one-half to about three-fifths the length of the forewing. The males of some species have a scale-tuft at the base of the antennae, the so-called 'knot-horn'. The labial palps are short to long, forward pointing or slightly upwardly curved, and the tongue is typically covered with scales at the base. A characteristic of the Pyraloidea (Crambidae and Pyralidae) are the tympanal organs on the second abdominal segment, although these are difficult to discern.

Many species in the family are more prevalent in the southern half of Britain, such as the Gold Triangle *Hypsopygia costalis* and *Endotricha flammealis*. Thistle Ermine *Myelois circumvoluta* was also a southern species but is spreading north and west and has now been found on Shetland and in Ireland. A few species do not appear to be as frequent as formerly, including the Large Tabby *Aglossa pinguinalis* and its scarcer relative, the Small Tabby *A. caprealis*; these have possibly declined through changing farming practices and the tidying up of outbuildings and barns. *Eurhodope cirrigerella* is now considered extinct. Several others are scarce adventives, for example *Cryptoblabes gnidiella*, which is sometimes found on imported Pomegranates. Others are rare immigrants, such as *Acrobasis tumidana*, a species which may have been temporarily established in parts of the south-east in the latter half of the 19th century. A few are considered to be minor pests of stored food products, such as the Indian Meal Moth *Plodia interpunctella* and Mediterranean

▼ Larva of *Cryptoblabes gnidiella* feeding in the calyx of a Pomegranate.

Flour Moth *Ephestia kuehniella*. Both *Trachonitis cristella* and *Catastia marginea* are considered not to be British and were either recorded erroneously or dubiously.

The larvae of many species feed on dried vegetable matter, several being associated with human habitations, and a few are associated with the nests of bees and wasps. Others are associated with the foliage of herbaceous plants, including various trees and shrubs, feeding within stems, cones, berries or fungi. Many feed from within silken tubes, galleries or webs, and a few feed gregariously. Adults fly from dusk or at night, and many come to light. Some fly naturally by day or can be easily disturbed by day from their resting place. A few species can be found at rest in shady places on walls in buildings. Many are distinctive and can be easily identified, although there are some groups of species, such as *Cadra/Ephestia* and *Phycitodes/Homoeosoma*, that require care to ensure accurate identification, including examination of genitalia.

Further reading
British and Irish species: Goater (1986)
European species: Palm (1986); Slamka (1997) (covers most British and Irish species); Slamka (2006) (Pyralidae part)

Bee Moth *Aphomia sociella* page 251   1428 (5568)

Common. **FL** 12-17mm. Male forewing creamy buff in the basal third, variably shaded greyish brown beyond, with darker dentate cross-lines at one-third and about three-quarters. Female forewing browner, usually greenish along the costa and termen, more extensively shaded greyish brown than the male, the cross-lines being more conspicuous and with a prominent round black spot between the cross-lines. **FS** Single-brooded, May-mid September; has been recorded from early April and into October. Larvae can be abundant where found. **Hab** Woodland, scrub, parkland, hedgerows, gardens. **Fp** Nests of bumblebees and wasps (nests above ground are preferred); larvae feed on old cells and debris, as well as the brood itself. The larva has been found feeding on dead insects. Masses of pupal cocoons made of dense, tough silk can be found in dry, dark places, such as among stacked wooden slats.

Aphomia zelleri page 251   1429 (5574)

Rare. Resident on only a few coastal sites, rarely seen elsewhere. Perhaps formerly resident on the Isle of Wight. **FL** 9-17mm. Forewing pale sandy brown, often reddish towards the costa, variably shaded darker and can appear almost blackish; the cross-line at about three-quarters is elbowed, usually weak, and typically there are two dark spots either side of one-half. **FS** Single-brooded, June-August. On still, warm evenings the females are said to make short flights, with the males found on sand or low vegetation; also flies in early morning sunshine. Comes to light. **Hab** Coastal sandhills, sandy shingle. **Fp** The moss *Brachythecium albicans* and Common Restharrow; in a vertical silk tube in sand beneath the plant.

Lesser Wax Moth *Achroia grisella* page 251   1426 (5587)

Local. Very local in Scotland. Possibly declining. **FL** 7-11mm. **Oval-shaped at rest.** Female larger than the male. Head golden yellow. Forewing pale shining greyish brown, unmarked. **FS** Single-brooded, with a possible partial second brood, June-August, occasionally September-October; has been recorded in January and April. **Hab** Beehives. Found in a range of habitats, including gardens. **Fp** Wax in beehives, preferring old wax. The larva has also been found in a wasps' nest.

► Larva of Wax
Moth *Galleria
mellonella* in an
old honeycomb
cell of a
honeybee.

Wax Moth *Galleria mellonella* page 251   1425 (5589)

Common. More local in northern England and infrequently recorded in Scotland.
FL 14-19mm. **Rests in a characteristic fashion resembling a noctuid moth.** Female usually larger and darker than the male. Forewing brownish with the cross-line at about two-thirds comprising a series of short dark streaks; a dark basal streak is usually present, with a number of dark marks along the dorsum. Markings can be obscure. **FS** Probably with two overlapping broods, June-October. Comes to light and recorded at sugar. **Hab** Beehives and probably the nests of other Hymenoptera. Found in a range of habitats, including gardens. **Fp** Honeycomb, preferring old combs.

Synaphe punctalis page 251   1414 (5620)

Very local in the south. Predominantly coastal, but found inland. **FL** 10-13mm. Forewing brownish, the male variably shaded darker, the first cross-line brown and the second, at two-thirds, yellowish brown. The female is a paler yellowish brown, smaller and narrower-winged, and is heavier-bodied than the male. **FS** Single-brooded, late May-mid September. The male frequently flies by day and comes to light, the female is seldom seen. Can be abundant where found. **Hab** Vegetated shingle, dune-slacks, saltmarshes, sheltered hollows on chalk downland, grazed acid grassland, lowland heathland. **Fp** Mosses, such as *Hypnum cupressiforme*; in a silk tube in the moss.

Meal Moth *Pyralis farinalis* page 251   1417 (5627)

Common. In Scotland, recently recorded only in the south. **FL** 11-14mm. Forewing yellowish brown between two white cross-lines, with the area to the first cross-line and beyond the second darker brown, the latter clouded lilac. Often rests with its abdomen upturned. **Similar species** *P. lienigialis* (page 251), a scarce species with similar habits to *P. farinalis*, resident in southern central England and recently recorded new to Wales, is slightly smaller and darker, lacking the contrast between the pale median area and the dark basal and outer areas of *P. farinalis*. **FS** Single-brooded, sometimes with a partial second brood, late April-October. Flies at dusk and comes to light. Occasionally found at sugar and also inside buildings, resting on walls. **Hab** Stables, barns, mills, farms and animal shelters, sometimes in houses and gardens. Less frequently encountered in the wider countryside. **Fp** Stored cereals, cereal refuse. Possibly garden refuse.

arge Tabby *Aglossa pinguinalis* page 251   1421 (5633)

Local. Declining and apparently not recently recorded from Scotland. **FL** 12-18mm. Forewing drab yellowish brown, heavily shaded blackish brown, with two yellowish-brown cross-lines, the serrate cross-line at about two-thirds strongly curving around a blackish dot. **Similar species** The much scarcer Small Tabby *A. caprealis* (not illustrated) is smaller, with a contrastingly blackish marked forewing and paler hindwing. **FS** Single-brooded, June-September. Rests by day in dark corners of outhouses and barns and, if disturbed, runs rather than flies. Comes to light. **Hab** Barns, stables, animal shelters, neglected farm buildings. Occasionally in the wider countryside and gardens. **Fp** Refuse of cereals, seeds, hay, animal feed, sheep dung; in a thick silk tube.

Gold Triangle *Hypsopygia costalis* page 251   1413 (5652)

Common. Very local in southern Scotland. Recently recorded new to Ireland. **FL** 8-10mm. Forewing rosy purple with two fine yellow cross-lines, these enlarged at the costa to form conspicuous yellow spots; the termen and cilia are yellow. Darker examples occur rarely, with the cilia reddish or purplish. **FS** Double-brooded in parts of its range, April-mid November. Occasionally recorded at sugar and comes to light. **Hab** A range of habitats, including gardens, farmland, woodland. **Fp** Stored clover and hay, probably also thatch; on dead vegetation. Reported from squirrels' dreys.

Hypsopygia glaucinalis page 251   1415 (5658)

Common. Recently recorded new to Scotland and the Isle of Man. **FL** 10-14mm. Forewing brownish grey, tinged reddish brown, with two narrow yellow cross-lines. **FS** Single-brooded, sometimes with a partial second brood in parts of its range, June-November. Comes to light and sugar. **Hab** A range of habitats, including gardens, farmland, scrubby grassland, woodland. **Fp** On dead and decaying vegetable matter, including haystacks, thatch, birds' nests, piles of leaves and mowings; in a silken gallery.

Endotricha flammealis page 251   1424 (5661)

Local. Scarce in the north. Expanding its range. **FL** 8-10mm. Forewing orange-yellow, variably shaded reddish brown in the basal and outer areas, with whitish cross-lines at about one-third and three-quarters. Darker forms occur rarely, varying from purplish red to almost black. **FS** Single-brooded, with an occasional partial second brood, June-mid October, but usually June-August. Can be numerous in the south. Easily disturbed by day, flies from dusk and visits flowers, such as ragwort, and sugar. Comes to light. **Hab** Woodland, lowland heathland, waste ground, gardens, wetland sites, scrubby situations near the sea. **Fp** Mainly decaying leaves on the ground, usually starting on living vegetation on deciduous trees, shrubs and low plants; the larva rests, curved or coiled, in a light silk web in leaf litter.

Cryptoblabes bistriga page 251
page 251

1433 (5668)

Local. More local in the north, with scattered records in Scotland. **FL** 8-9mm. Forewing reddish brown to purplish, darker between two whitish cross-lines, sometimes greyer or whitish in the basal area and towards the termen. **FS** Single-brooded, late April-mid September in the south, June-July in the north. Can be disturbed by day from tall bushes and trees. Comes to light. **Hab** Oak woodland. **Fp** Oaks, sometimes alders and other deciduous trees; in a folded leaf.

Salebriopsis albicilla page 251

1446 (5676)

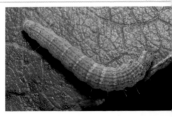

Rare. **FL** 8-10mm. Male with the head and thickly scaled base of the antennae white, the head of the female pale greyish brown, lacking the white and thickened antennal scales. Forewing grey, speckled darker, appearing sooty black. **FS** Single-brooded, mid May-July. **Hab** Woodland. **Fp** Small-leaved Lime; in a rolled leaf.

Elegia similella page 251

1449 (5679)

Very local. Possibly an occasional immigrant. **FL** 9-10mm. Forewing with a straight costa, sooty black, slightly tinged purplish when fresh, with a conspicuous white cross-band almost reaching the costa at one-third. **Similar species** The forewing of *Ortholepis betulae* has a broadly arched costa and a distinct black scale-tuft before a whitish cross-line; *Matilella fusca* does not have a conspicuous white cross-band extending to near the costa, and is also narrower-winged, with a straight costa, and very obscurely marked; *Apomyelois bistriatella* has the forewing with an arched costa, and a narrow, usually curved whitish cross-line which is obscure or absent beyond the middle towards the costa. **FS** Single-brooded, mid May-August. Comes to light. **Hab** Mature oak woods, parkland. **Fp** Oaks; in a silk web.

Ortholepis betulae page 251

1450 (5681)

Local, possibly declining. **FL** 9-13mm. Forewing broadly arched from the base to the apex, sooty grey, finely speckled paler, with a tuft of raised black scales before a rather obscure whitish cross-line just beyond one-third, this line most conspicuous near the dorsum. The tuft can be reduced or absent in worn examples. **Similar species** *Elegia similella*, *Matilella fusca*, *Apomyelois bistriatella*. **FS** Single-brooded, mid June-August. Comes to light. **Hab** Heathland with scattered birches, woodland, parkland. **Fp** Birches; in a web on the upperside of a leaf.

Matilella fusca page 251

1451 (5684)

Local. **FL** 11-13mm. Forewing narrow and elongate with a straight costa, sooty black, finely speckled paler, with an oblique rather obscure cross-line at one-third, this being most prominent near the dorsum, and with another obscure cross-line at about four-fifths. Between these lines are two darker dots. **Similar species** *Elegia similella*, *Ortholepis betulae*, *Apomyelois bistriatella*. **FS** Single-brooded, possibly two broods in parts of its range, May-September. Can be disturbed from heather and burnt twigs by day, sometimes congregating in burnt areas. Found at flowers, including Common Ragwort, and comes to light. **Hab** Heathland, moorland, woodland, also gardens. **Fp** Heathers, possibly including garden varieties; in a silken web.

Moitrelia obductella page 251   1444 (5770)

Rare. **FL** 10-12mm. Forewing deep purple to orangey brown, paler and finely speckled whitish along the costa, and paler brown or light orange along the dorsum; the cross-lines are obscure, indicated by white scales. **FS** Single-brooded, July-August. Can be disturbed from vegetation by day. Flies above the foodplant before dusk and comes to light.
Hab Chalk downland. **Fp** Wild Marjoram; older larvae in a web mixed with dead leaves in a terminal shoot.

Pempeliella ornatella page 251   1463 (5686)

Rare. Only recently from Kent and Ireland. **FL** 11-13mm. Forewing yellowish brown, mixed reddish brown along the costa and veins, especially towards the termen; the cross-line at about one-third is indistinct, consisting of whitish dots or arrowhead markings on the veins, and the whitish cross-line at four-fifths is oblique and nearly straight. **Similar species *Delplanqueia dilutella* tends to be smaller and darker, with a more defined cross-line at one-third, and an angled cross-line at four-fifths.** Examination of genitalia may be required to identify weakly marked examples of either species.
FS Single-brooded, late June-August. Can be disturbed from grassy vegetation by day, and comes to light. **Hab** Chalk downland, particularly steeper slopes with broken ground, limestone pavement in Ireland. **Fp** Wild Thyme; has been found in a web among roots.

Delplanqueia dilutella page 251   1462 (5690)

Local. Rather more frequent in the west and on the coast, but scarce in north-west England. **FL** 8-11mm. Forewing yellowish brown, mixed crimson and/or blackish brown, often in part shaded paler towards the costa, with a whitish cross-line at one-third and another at four-fifths, the latter with a near right-angled bend before the tornus, and between these two lines are two black spots, more conspicuous when the surrounding area is suffused white. Examples from the west of Ireland tend to be larger and more brightly marked. **Similar species** *Pempeliella ornatella*. **FS** Single-brooded, mid June-mid September. Easily disturbed by day from short calcareous turf, flies in the evening and comes to light. **Hab** Chalk and limestone grassland and rocky situations, quarries, sand dunes. **Fp** Wild Thyme growing on ants' nests; in a silk tube amongst the plant.

Sciota hostilis page 251   1447 (5725)

Rare. Perhaps resident only in Warwickshire and possibly in Kent. An infrequent immigrant to south-east England. **FL** 10-12mm. Forewing grey, speckled darker, usually with a dull reddish patch at the base and a cross-line at one-third, obscure towards the costa, within a prominent broad blackish cross-band, and another pale cross-line at four-fifths sometimes obscure. **Similar species *S. adelphella* is typically a brighter moth, with salmon pink or orange at the base and usually along part of the dorsum. Also similar to the very scarce immigrant *S. rhenella* (not illustrated), which is usually slightly paler with a more contrasting blackish bar near the base and an orangey basal patch.** **FS** Single-brooded, late May-July. Comes to light. Occurs at low density. **Hab** Woodland. **Fp** Aspen; in a silken tube between two leaves, one of which is often dead.

Sciota adelphella page 251   1447a (5727)

Very local. Probably now resident in East Sussex, Kent and Suffolk. Elsewhere likely to be a scarce immigrant or wanderer. **FL** 10-12mm. Forewing grey with a prominent salmon pink or orange basal patch, and usually similarly coloured along part of the dorsum to the pale cross-line at four-fifths, with the pale cross-line at one-third broadly edged blackish towards the base, giving a defined convex outer edge to the basal patch. **Similar species** *S. hostilis*, *S. rhenella*. **FS** Single-brooded, late May-mid August. Comes to light.
Hab Open woodland, gardens. **Fp** White Willow. On the Continent, on sallows, willows, occasionally poplars; within a spinning.

Pima boisduvaliella page 251   1453 (5740)

Rare. Resident on parts of the Kent, Suffolk and Norfolk coasts, and possibly north Yorkshire with wanderers or immigrants recorded elsewhere. **FL** 10-12mm. Forewing brown, the costa with a conspicuous white streak, this edged darker each side, and with a weakly defined yellowish-brown dorsal streak. **Similar species** *Anerastia lotella* **has a less distinctive and duller costal streak, with the forewing broader. FS** Possibly double-brooded, at least in part of its range, late May-August. Flies close to the ground on warm sunny evenings, and comes to light. **Hab** Shingle beaches, sandhills. **Fp** Sea Pea; also reported from Common Bird's-foot-trefoil, Kidney Vetch, Spiny Restharrow; in the seed-pod.

Oncocera semirubella page 252   1441 (5751)

Very local. Predominantly southern but recently recorded in Northumberland, with a single old record in southern Ireland. **FL** 11-14mm. Forewing light crimson or pink, occasionally tinged greyish to greyish brown, with the dorsum yellow, the costa usually having a white or grey stripe. **FS** Single-brooded, June-August, but occasionally to early October. Easily disturbed from vegetation in sunny weather, comes to light and has been found at sugar. **Hab** Chalk and limestone downland and cliffs, vegetated shingle, sparse open grassland near heathland. **Fp** Common Bird's-foot-trefoil, White Clover, possibly Black Medick, Horseshoe Vetch, restharrows; in a dense web amongst the foodplant, sometimes several larvae to a web.

Pempelia palumbella page 252   1442 (5767)

Local. Apparently not recently recorded in Scotland. **FL** 11-13mm. Forewing leaden brown, shaded greyish white, particularly towards the costa, with a pale reddish-brown cross-line at one-third, inwardly angled near the costa, and a second cross-line at three-quarters, reddish brown and distinctly sinuate. **FS** Single-brooded, late May-mid September. Readily disturbed from heather on sunny days, more sluggish in dull weather. Flies from dusk and comes to light. **Hab** Heathland, heathy cliff-edges; occasionally deciduous woodland, downland. **Fp** Heather, heaths; in a silk tube spun from below ground into the foodplant.

Pempelia genistella page 252   1443 (5761)

Very local. Primarily coastal. Recorded new to Wales in 2001. **FL** 12-13mm. Forewing light brown with thinly scattered black marks; the cross-line at one-third, elbowed inwards towards the costa, is slightly paler than the ground colour and partly edged black, with the cross-line at four-fifths sinuate near the middle and black-edged. **FS** Single-brooded, late June-September, with a few possible second-brood individuals recorded in October. Can be beaten from gorse by day. Comes to light. **Hab** Heathland, vegetated shingle, chalk and limestone downland. **Fp** Gorse, Dwarf Gorse; in a silken web.

Similar species in the '*Dioryctria*' group

Dioryctria species are difficult to separate and care is needed to ensure accurate identification. While most examples can be identified from superficial characters, examination of genitalia may be necessary, although differences in the genitalia of females are not obvious. *D. sylvestrella* is often the largest of the group, but overlaps with *D. abietella* in size. *D. sylvestrella* has the angled cross-line at four-fifths composed of short relatively straight lines, and also appears more chequered when viewed from above, with the orangey brown or brick-red patch inside the first cross-line being more obvious, and the white mark at two-thirds quadrate. *D. simplicella* has a rather straighter forewing costa than the other three species, and is more uniform grey, with the kidney-shaped spot hardly paler than the ground colour. *D. schuetzeella* is the smallest of the group, appearing brighter, with the broadest and whitest cross-lines relative to forewing size, and the cross-line at four-fifths is the most dentate. Photographs and a key are given in Parsons & Clancy (2002).

Dioryctria abietella page 252   1454 (5784)

Common. An occasional immigrant. **FL** 12-15mm. Forewing silvery grey, mottled darker, with a silvery grey cross-line at one-third, outwardly edged black, particularly towards the dorsum, and adjacent to a light brown or light reddish patch, with a strongly dentate cross-line at four-fifths; between these lines is a typically conspicuous kidney-shaped silvery grey or whitish spot. **FS** Single-brooded, June-mid October; has also been recorded in late April. Comes to light. **Hab** Conifer plantations, heathland with conifers, gardens. **Fp** Scots Pine, Monterey Pine, Norway Spruce, Noble Fir, Douglas Fir, Larch; in the shoots and cones.

Dioryctria schuetzeella page 252   1454a (5782)

Very local. Most regularly found in the south-east. **FL** 10-13mm. **Similar species** *D. abietella, D. sylvestrella, D. simplicella*. **FS** Single-brooded, late June-August. Comes to light. **Hab** Spruce plantations, mixed woodland. **Fp** Norway Spruce, White Spruce, Silver Fir; within a spinning.

Dioryctria sylvestrella page 252   1454b (5781)

Local. Recently established. Also an occasional immigrant. **FL** 13-17mm. **Similar species** *D. abietella* and, to a lesser extent, *D. simplicella*. **FS** Single-brooded, June-early October. Comes to light. **Hab** Pine forests, mixed woodland. **Fp** Scots Pine, Weymouth Pine, Corsican Pine, Maritime Pine, also White Spruce; on the living bark.

Dioryctria simplicella page 252   1455 (5783)

Local. Very local in Scotland. **FL** 11-14mm. Occasional melanic examples can occur. **Similar species** *D. abietella*. **FS** Single-brooded, late April-September. Comes to light. **Hab** Conifer plantations, heathland with conifers, gardens. **Fp** Scots Pine, Corsican Pine; in a cavity in wood, adjacent to a wound, feeding from within on the living bark.

Phycita roborella page 252   1452 (5796)

Common. Local in the north and just reaches Scotland. **FL** 11-14mm. Forewing dark reddish brown, variably shaded pale pinkish grey or blackish brown, with an obscure triangular dark patch extending from the dorsum at one-third towards the costa; the first cross-line at about two-fifths is obscure and oblique, and the cross-line at four-fifths is indented below the costa. **FS** Single-brooded, with occasional second-brood examples, June-October. Comes to light, sugar and flowers. **Hab** Oak woodland, hedgerows, scrub, also gardens. **Fp** Oaks, also recorded from Crab Apple, pears, Hazel; in spun leaves.

▲ Two spindle berries spun together by the larva of *Nephopterix angustella*.

◄ Larval workings of *Epischnia asteris* on Golden-samphire. The larval web is conspicuous but larvae are hidden deep within the spinning.

Hypochalcia ahenella page 252   1457 (5811)

Local. More local in the north. **FL** 9-14mm. Forewing greyish brown, brownish or reddish brown, with two faint darker cross-lines converging towards the dorsum. **FS** Single-brooded, May-August. Easily disturbed by day, flies from dusk and comes to light. Usually only seen singly or in small numbers. **Hab** Dry grasslands, including chalk downland, sand dunes, railway banks, quarries and other sparsely vegetated habitats. **Fp** Not known, but possibly including Common Rock-rose.

Epischnia asteris page 252   1456 (5834)

Very local. Coastal. Formerly recorded in Hampshire and Pembrokeshire. **FL** 11-14mm. Forewing light grey, veins variably highlighted darker, with the dorsum, particularly near the base, finely marked yellowish brown. **FS** Single-brooded, June-August, with a possible second brood in the Channel Islands where it has been recorded in early September. Occasionally disturbed from the larval foodplant by day, and comes to light. **Hab** Rocky coasts and shores, including limestone cliffs. **Fp** Golden-samphire; often in a conspicuous web on the shoots.

Nephopterix angustella page 252   1465 (5848)

Very local, probably spreading. Also an occasional immigrant. **FL** 9-11mm. Forewing narrow, greyish, variably speckled reddish brown and whitish, the veins beyond one-half variably highlighted blackish; the cross-line at one-third is pale, edged reddish and blackish, and preceded by a bar of raised black scales from the dorsum to just beyond the middle. **FS** Probably two overlapping broods, late April-November. Comes to light. **Hab** Hedgerows, scrub. **Fp** Spindle; in the berries, spinning two together.

Acrobasis repandana page 252   1436 (5854)

Common. Local in the north and the south-west. **FL** 9-11mm. Forewing grey, shaded dark pinkish or orangey red in the basal area to an oblique whitish cross-line at one-third, this outwardly edged blackish. **Similar species The scarce immigrant *A. tumidana* (not illustrated) has raised scales towards the wing base, most visible in profile, although these can be reduced or lost if worn. FS** Single-brooded, mid June-early September. Can be disturbed by day during warm weather. Comes to light and has been recorded at sugar. **Hab** Oak woodland, pasture woodland, scrubby habitats. **Fp** Oaks; in spun leaves, usually high up in a tree.

▲ Larva of *Acrobasis consociella*, dorsal and lateral views.

▶ Frass-covered silken tubes of the young larvae of *Acrobasis consociella* on oak.

Acrobasis consociella page 252   1437 (5869)

Common. Local in the north. No recent records from Scotland. **FL** 9-11mm. Forewing greyish with a weak coppery sheen or mauve shade, pale greyish to the first cross-line, this oblique and whitish, outwardly edged black. **FS** Single-brooded, late May-August; has been recorded in October. Hides by day among clusters of oak leaves and falls to the ground if disturbed. Comes to light and occasionally to sugar. **Hab** Oak woodland, parkland, hedgerows, scrubby habitats, gardens. **Fp** Oaks, also reared from Hornbeam; feeds gregariously in a mass of spun leaves, preferring saplings or young coppice.

Acrobasis suavella page 252   1438 (5857)

Local in southern England and south Wales, with a scattering of records elsewhere. Recorded new to Scotland from Shetland in 2003. **FL** 10-12mm. Forewing silvery grey, shaded coppery red or reddish brown, particularly towards the dorsum, with a slightly arched, oblique whitish cross-line at one-third, edged by an obscure blackish triangle on the costa, and with a fine whitish cross-line at four-fifths; the costa has a distinct silvery greyish patch between the two cross-lines and within this patch are two dots aligned perpendicular to the dorsum, and there is a blackish mark at the apex. **Similar species** *A. advenella* and *A. marmorea*, which are smaller than *A. suavella*. *A. advenella* does not have a distinct blackish apical patch, is without a distinct silvery costal patch, and the dots in the middle of the wing are obliquely aligned with the tornus. *A. advenella* appears broader-winged than *A. marmorea*, and also has the head and collar reddish brown. *A. marmorea* is typically the smallest of the three; it has a whitish patch on the dorsum, the cross-line at about one third reaches the dorsum at a near right angle, and the dots in the middle of the wing are usually joined. **FS** Single-brooded, June-August. Comes to light. **Hab** Blackthorn thickets and stunted bushes in open country, such as downland and coastal locations. Recorded from gardens. **Fp** Blackthorn, Cotoneaster and, less often, hawthorns and whitebeams; in a dense silk gallery on the stem, covered with frass and leaf fragments.

◄▲ Larvae of *Acrobasis advenella* on hawthorn.

Acrobasis advenella page 252   1439 (5856)

Common. More local in southern and central Scotland, with possible wanderers found further north. **FL** 9-10mm. Forewing silvery grey, shaded reddish brown or blackish brown, the basal area usually darker than the rest of the wing, with an oblique cross-line at about one-third, this more defined towards the dorsum and preceded by a fine silvery grey line, both enclosing a wedge of coppery red and black; the dots in the middle of the wing are obliquely aligned with the tornus. **Similar species** *A. suavella, A. marmorea.* **FS** Single-brooded, June-September. Flies from dusk and comes to light and sugar. **Hab** Hawthorn hedges, preferring old uncut hedges, woodland, gardens. **Fp** Hawthorns, less often Rowan; in a spinning.

Acrobasis marmorea page 252   1440 (5860)

Local. Very local in the north. **FL** 8-10mm. Forewing silvery grey, shaded dark crimson or reddish brown, particularly in the basal area, with a cross-line at about one-third, angled near the middle and reaching the dorsum almost at right angles, this line forming an oblong, sometimes triangular, whitish patch in the dorsal half, and with a silvery grey patch at about one-half extending obliquely across the wing, enclosing two slightly obliquely placed dots which are usually joined together; at four-fifths is a whitish cross-line, this usually edged crimson or reddish brown. **Similar species** *A. suavella, A. advenella.* **FS** Single-brooded, late May-mid September. **Hab** Hedgerows, scrub, woodland, particularly near the coast. **Fp** Blackthorn, also hawthorns and *Sorbus*, preferring young or stunted bushes; in a web among twigs, resembling sheep's wool.

Apomyelois bistriatella page 252   1486 (5873)

Very local. Late-season examples may be immigrants. Recently found in Ireland. **FL** 7-12mm. Forewing with the costa arched, especially towards the apex, dark grey, the first cross-line at one-third forming a conspicuous short whitish bar from the dorsum, usually curved or somewhat sinuate, with the cross-line at four-fifths being somewhat obscure. **Similar species** *Elegia similella, Ortholepis betulae, Matilella fusca.* **FS** Single-brooded, mid May-September. Comes to light. Usually found singly or in small numbers. **Hab** Heathland, downland, occasionally elsewhere. **Fp** Within the fungus *Daldinia concentrica* and probably other *Daldinia* species, growing on burnt, sickly or dead birches or on charred gorse; in a silk tube within the wood or fungus.

▲ Larva of Thistle Ermine *Myelois circumvoluta* in the dead stem of a thistle.

▲ Larva of *Gymnancyla canella* on Prickly Saltwort.

Thistle Ermine *Myelois circumvoluta* page 252   1458 (5898)

Common in England and Wales, but hardly reaches Scotland. Reported from Shetland and recently discovered in Ireland; appears to be extending its range. **FL** 13-17mm. Forewing glossy silvery white, sparsely covered with black dots, those nearer the base usually larger, with a row of fine black dots along the termen. **Similar species White species in the genus *Yponomeuta* (Yponomeutidae), which are always smaller. FS** Single-brooded, May-mid September. Can sometimes be found on thistles by day and after dark, and comes to light. **Hab** A wide range of habitats, including waste ground, field borders, chalk downland, grassland on light sandy soils, gardens. **Fp** Various thistles, including Woolly, Creeping, Cotton and Spear Thistle, also Greater Burdock, preferring larger plants; at first feeds on the seeds, then burrows into the stems.

Gymnancyla canella   page 252   1464 (5962)

Very local. Coastal. Can be common in the larval stage. **FL** 10-12mm. Variable. Forewing ranges from sandy white to pale brick red, through shades of pale grey and salmon pink, with the costa usually paler, and between the two cross-lines are two blackish dots at about two-thirds; some examples are heavily shaded darker brown or thinly but extensively speckled blackish. **FS** Single-brooded, June-September. Comes to light. **Hab** Sandy coasts, just above the high-tide mark. **Fp** Prickly Saltwort; in a web into which sand grains become trapped, initially burrowing into a side stem, later feeding on the seeds or within the main stem.

Assara terebrella   page 252   1461 (5986)

Very local. Probably an occasional immigrant. **FL** 8-12mm. Forewing blackish brown with an oblique whitish cross-line at one-third and another narrower line at four-fifths, with a whitish patch extending from the middle of the costa to the middle of the wing, enclosing two transversely placed dark dots. **FS** Single-brooded, June-August; has been recorded in September. Comes to light. **Hab** Mature spruce woods, occasionally gardens. **Fp** Norway Spruce; in a developing cone, stunting the growth and causing it to fall prematurely.

Euzophera cinerosella   page 252   1469 (5997)

Very local. **FL** 9-12mm. Obscurely marked. Forewing pale greyish brown, shaded paler in the costal half and towards the termen, with a darker zigzag-shaped cross-line at one-third, this strongly oblique and often obscure, and a whitish cross-line at four-fifths; between these lines is an outline of a narrow oval, this dark-edged and pale-centred. **FS** Single-brooded,

May-mid August. Comes to light. **Hab** Coastal habitats, waste ground. **Fp** Wormwood; in the crown of the root.

Euzophera pinguis page 252

Local. Extending its range and recently recorded in Northumberland. **FL** 11-13mm. Forewing light yellowish brown, sometimes pinkish tinged, the basal half blackish with an indistinct yellowish-brown basal patch and a usually pronounced cross-line at one-third, the cross-line at about two-thirds surrounded by blackish clouding. **FS** Single-brooded, June-September. Comes to light. **Hab** Woodland, hedgerows. **Fp** Ash; in a gallery in the wood, feeding from within on the living bark.

Nyctegretis lineana page 252

Rare. A probable vagrant in Lincolnshire and East Sussex. Formerly in Essex. **FL** 8-9mm. Forewing light brown or reddish brown, darker basally, with an oblique white cross-line at one-third and a white cross-line at four-fifths; often between the two lines is a white spot, this black-edged towards the costa. **FS** Single-brooded, July-August. Flies over the foodplant at dusk and comes to light. Can be found visiting flowers of Viper's-bugloss and other plants after dark. **Hab** Flat sandy ground behind coastal sandhills. **Fp** Common Restharrow, possibly also clovers and other low-growing plants; in a loose silken tube under the host plant.

Ancylosis oblitella page 252

Very local. Resident, immigrant and temporary colonist. Probably only resident in parts of south-east England. **FL** 7-10mm. Forewing light greyish brown, variably shaded darker, with a pale cross-line at one-third, often not reaching the costa and sometimes preceded on the dorsum by an irregular darker brownish blotch, and another cross-line at four-fifths, this dentate; between the two lines are two, sometimes obscure, black dots. **FS** Double-brooded, May-mid October. Readily disturbed by day, flies at night and comes to light. Has been found on the flowers of Common Fleabane. **Hab** Waste ground, coastal shingle, saltmarshes, sand dunes. Occasionally elsewhere. **Fp** Goosefoot.

Homoeosoma nebulella page 253

Very local. Probably an occasional immigrant. **FL** 9-13mm. Forewing pale brownish grey, speckled lighter, with a pale streak along the costa tapering towards the apex, the outer half finely edged dark grey along the leading edge of the costa, and with two faint spots representing a cross-line at nearly one-half and two faint slightly elongate spots at about two-thirds. **Similar species Weakly marked or worn examples of *Phycitodes binaevella*, which have three, not two, spots near one-half, and are without the fine dark costal edge. FS** Single-brooded, June-September. Flies from dusk, comes to light and visits flowers, such as ragworts and thistles. **Hab** Rough ground on chalky or sandy soils, Breckland. **Fp** Common Ragwort, Spear Thistle, possibly also Tansy, Oxeye Daisy; on the flowers and developing seeds.

Homoeosoma sinuella page 253

Local. Extending its range. Recently recorded in Northumberland and found new to Ireland. **FL** 8-11mm. Forewing pale yellowish brown, sometimes reddish tinged, variably darker along the costa, with two broad and irregular brownish or reddish cross-lines, broadest on the costa, occasionally broken. **FS** Single-brooded, May-mid September. Easily disturbed by day. Flies from dusk and comes to light. Can be numerous where found. **Hab** Dry situations where vegetation is sparse, such as chalk downland, heathland, Breckland, sand dunes, vegetated shingle, cliffs and grassy banks by the sea, waste ground, railway embankments. Also scrubby habitats. **Fp** Ribwort Plantain, other plantains; in the rootstock.

Homoeosoma nimbella (not illustrated) 1482 (6082)

Scarce. **FL** 8-11mm. Forewing narrow with the costa straight-edged, pale yellowish brown with a whitish costal streak. **Similar species** *Phycitodes saxicola*, *P. maritima* (see p.393 for male and female genitalia). **FS** Single-brooded, May-August. Has been found at rest on flowers by day, flies at dusk and occasionally comes to light. **Hab** Sand dunes, sea cliffs, stone walls on the coast. **Fp** Sheep's-bit; in flowers and developing seeds.

Phycitodes binaevella page 253 1483 (6087)

Common. More local in the north and scarce in Scotland. **FL** 10-13mm. Forewing chalky white, shaded light brown in the dorsal half and towards the termen, with a cross-line at about one-third comprising three elongate black spots arranged in a straight line, the middle and costal spots being close together or fused, and with two prominent blackish spots at about two-thirds, these sometimes close together. **The largest, most robust-looking and strongly marked of the group. Similar species** *Homoeosoma nebulella*. **FS** Single-brooded, late May-mid September. Comes to light and visits flowers at night, such as thistles and Common Ragwort. **Hab** A wide range of sites, including coastal habitats, woodland, waste ground, but particularly associated with sandy or chalk soils. **Fp** Spear Thistle, also Plymouth Thistle; on the seeds in a flowerhead.

Phycitodes saxicola page 253 1484 (6090)

Common. Predominantly coastal. **FL** 7-10mm. Forewing narrow with the costa straight-edged, light greyish brown with a whitish-grey costal streak, the cross-line at about one-third represented by three round dots arranged in an angled line, these often small or even obsolete, and usually with one or two small two black marks at about two-thirds, and a further faint, pale and straight cross-line at four-fifths. **Similar species *P. saxicola*, *P. maritima* and *Homoeosoma nimbella* are very similar, and they can be separated with confidence only by rearing from larvae or by genitalia examination (see p.393 for male and female genitalia); little reliance can be placed on identification from photographs.** However, well-marked examples of *P. maritima* are typically slightly larger and more broad-winged than *H. nimbella* and *P. saxicola*, with the cross-line at four-fifths slightly inwardly angled before the costa. **FS** Possibly two broods, May-October. Comes to light and visits flowers, such as Red Valerian, at night. **Hab** Coastal habitats, including chalk and limestone grassland, vegetated shingle, sand dunes, slumping cliff, waste ground, gardens; found inland occasionally, including on lowland heathland. **Fp** Composites, including Scentless Mayweed, Fleabane, Jersey Cudweed; in the flowerheads.

▼ Larva of *Phycitodes saxicola.*

▶ A mass of ragwort seedheads
spun together by the larvae of
Phycitodes maritima.

Phycitodes maritima page 253   1485 (6086)

Local. Possibly extending its range. Predominantly coastal. **FL** 8-10mm. Forewing with the costa arched, light yellowish brown mixed grey, with a broad whitish costal streak, the cross-line at about one-third represented by three dots arranged in an angled line, these occasionally rather obscure, and with two black spots at about two-thirds. **Similar species** *Homoeosoma nimbella*, *P. saxicola* (see p.393 for male and female genitalia). **FS** Possibly two broods, May-October. Comes to light and visits flowers, such as Common Ragwort, at night. **Hab** Coastal habitats, including vegetated shingle, waste ground, occasionally gardens; found inland increasingly frequently in various habitats. **Fp** Composites, including Yarrow, Common Ragwort, possibly also Tansy, Golden Samphire; initially in the leaf-axils and upper parts of the stems, later on the flowerheads.

Vitula biviella page 253   1478b (6094)

Very local. A recent colonist. **FL** 6-9mm. Forewing relatively broad, grey to greyish brown, with whitish cross-lines at one-third and at about four-fifths; between the lines are two dark dots, joined to form a transverse dash or weak crescent. **Similar species** *Ephestia elutella*, *E. unicolorella*, other *Ephestia* and *Cadra* species. **FS** Single-brooded, mid June-mid August. Comes to light. **Hab** Pine trees growing in gardens, open spaces and parkland. **Fp** Probably associated with pines, including Maritime Pine and Corsican Pine; probably in a spinning among needles.

Indian Meal Moth *Plodia interpunctella* page 253   1479 (6102)

Local. Synanthropic and an adventive. An occasional minor pest. **FL** 7-10mm. Forewing reddish brown with a contrasting creamy white to pale yellow basal patch inside the first cross-line. **FS** Probably a number of overlapping broods, February-November. Can be found indoors by day. Comes to light. Sometimes numerous where found. **Hab** Usually found in warehouses, pet shops, foodstores, barns and houses, occasionally recorded outdoors. **Fp** A wide range of stored foodstuffs, including grain, rice, flour, cereals, nuts, dried fruit; in a silken web among the foodstuff.

Cacao Moth *Ephestia elutella* page 253   1473 (6112)

Local. Synanthropic and an adventive, but possibly able to sustain populations outside. **FL** 6-10mm. Forewing grey to greyish brown with pale cross-lines at one-third and about four-fifths, these weakly convergent approaching the dorsum, which can be tinged reddish; between these lines are two dark dots, these sometimes weakly joined. **Similar species** *E. unicolorella* and *E. elutella* are very similar and little reliance can be

placed on identification from photographs. However, *E. unicolorella* is more likely to be encountered outdoors and in moth traps than *E. elutella*. *E. unicolorella* is usually larger and darker, with a rich reddish-brown tone, the cross-lines tending to converge towards the dorsum more strongly than those of *E. elutella*, which lacks the rich tone apart from a reddish tinge on the dorsum in some examples. *Vitula biviella* is similar, but broader-winged, always without the reddish tone and is usually smaller. In most cases, species in the genera *Ephestia*, *Cadra* and *Vitula* should be identified by genitalia examination. **FS** Two or three overlapping broods, depending on conditions, January-November. Comes to light and has been found on flowers. **Hab** Usually found in buildings, breeding in warehouses, barns, outhouses, animal shelters. Occasionally outdoors. **Fp** Associated with a wide range of stored produce, including cereals, coffee, tobacco, birdseed, chocolate, nuts, dried fruit; in a silken web among the foodstuff. Has been bred from dead insects found in an insect-o-cutor.

Ephestia unicolorella ssp. *woodiella* page 253   1474 (6113)

Local. **FL** 6-10mm. Forewing brownish grey, variably shaded darker, with the dorsum and, to a lesser extent, the termen of the male suffused reddish or reddish brown, and with pale cross-lines at one-third and about four-fifths which converge towards the dorsum; between these lines are two small black dots. **Similar species** *Vitula biviella*, *E. elutella*, other *Ephestia* and *Cadra* species. **FS** Single-brooded, May-September. Comes to light. **Hab** Gardens, woodland, parkland, coastal localities. **Fp** Not known; probably dried plant material, such as old berries, dead stems of Ivy.

Mediterranean Flour Moth *Ephestia kuehniella* page 253   1475 (6105)

Local. Synanthropic, and possibly a scarce immigrant. An occasional minor pest. **FL** 8-13mm. Forewing grey to greyish brown, with paler ill-defined cross-lines at one-third and four-fifths, both dark-edged, the first strongly angled above the dorsum. **FS** Probably continuously brooded in well-heated situations. Comes to light. **Hab** Found in warehouses, bakeries, flour-mills, kitchens. **Fp** Stored products, such as flour, oatmeal, rice, walnuts, cereal, dog biscuits, chicken feed, small animal pet food, dried currants; in a silken gallery, sometimes massed together forming larger webs.

Rhodophaea formosa page 253   1445 (5766)

Local. Possibly spreading and recently recorded as far north as Yorkshire and Northumberland. **FL** 9-11mm. Forewing pinkish red or brownish red, with a whitish cross-line beyond one-half almost obscured by a broad blackish cross-band, broadest towards the costa and often incorporating a black dot, and with a whitish cross-line at three-quarters, faint and sinuate, edged blackish and reddish. **FS** Single-brooded, late May-September. **Hab** Hedgerows. **Fp** Elm, preferring hedges and bushes; in a very slight web on the upperside of the leaves.

Anerastia lotella page 253   1432 (6123)

Local. Very local in Scotland. Mostly coastal, but inland in Breckland in East Anglia, and in Worcestershire and Yorkshire. **FL** 9-12mm. Forewing sandy brown, often tinted reddish, variably speckled darker, sometimes with darker longitudinal streaks, and usually with a pale creamy or whitish costal streak. Some inland examples are smaller and darker. **Similar species** *Pima boisduvaliella*. **FS** Single-brooded, late May-August, occasionally early September. May be abundant where it occurs. Easily disturbed by day. After dark can be found sitting on Marram and other maritime grasses, and comes to light. **Hab** Coastal sandhills; also vegetated shingle, sandy heathland, Breckland. **Fp** Marram, Sheep's Fescue, Grey Hair-grass, possibly other sandhill grasses; feeding from a silken tube at the stem base and among the rootstock.

Crambidae

Scoparia pyralella

Beautiful China-mark
Nymphula nitidulata

A very diverse group containing 145 species, including the familiar grass moths, and the china-marks, which have aquatic or sub-aquatic larvae. Species vary between small and large, with forewing length 4.5-22mm, although the female Water Veneer *Acentria ephemerella* has a wingless form. Most grass moths, *Crambus*, *Agriphila* and related genera, rest with the wings almost roof-like or rolled slightly around the body, with the antennae laid back over the thorax and abdomen. However, there are other resting postures, including flat (e.g. *Udea*) and extended (e.g.

Catoptria pinella

Pyrausta purpuralis

Small Magpie
Anania hortulata

Pyrausta nigrata). The forewings range from very elongate to broadly triangular, with very varied markings and a distinct tornal angle. The hindwings are broad, often broader than the forewings, sometimes with markings resembling those of the forewings and with the dorsal cilia short. The head has raised scales on the crown, the face is smooth or with raised or erect scales. The antennae are long, from one-half to almost as long as the length of the forewing. The labial palps are short to long, forward pointing or slightly downwardly or upwardly curved, and the tongue is typically covered with scales at the base. A characteristic of the superfamily Pyraloidea (Crambidae and Pyralidae) is the presence of tympanal organs on the second abdominal segment, although these are difficult to discern. The Crambidae were formerly included within the Pyralidae.

Within the family are two frequently encountered immigrants, the Rusty-dot Pearl *Udea ferrugalis* and the Rush Veneer *Nomophila noctuella*. It also includes several scarcer immigrants, including long-distance travellers from the far south of Europe or North Africa, such as *Euchromius ocellea*, *Antigastra catalaunalis* and *Uresiphita gilvata*, all of which have become more frequent in occurrence in the last 20 years or so. There are several adventives, with the recently noted Boxworm Moth *Cydalima perspectalis*, associated with various species of box, showing signs of becoming established. *Loxostege sticticalis* is considered extinct as a resident species, although immigrants of this moth still occur regularly. The yellow and pink *Pyrausta sanguinalis* is extinct in Britain, but can still be found very locally in parts of Ireland, including Northern Ireland, and occurs on the Isle of Man. Also on the British list are a very few species which are doubtfully British, such as *Herpetogramma centrostrigalis* which is known to science from only a single specimen.

A number of species are associated with grasses, the life history of several being imperfectly known. The larvae of others feed on mosses or lichens, various water plants or other herbaceous plants, and in a few cases shrubs and trees. A few species bore into stems of the foodplant. Many of the grass moths feed from within silken tubes or galleries, while some other species feed from a slight web under leaves or loosely spin or roll leaves together. Adults of some species can be readily disturbed by day, sometimes in abundance, and fly a short distance before settling again. Some members of *Scoparia/Eudonia* are easily encountered by day resting on tree trunks and fly readily when approached. Others fly naturally by day in sunshine, or fly naturally from dusk and after dark. Many are attracted to light, and can occur some distance from their breeding site.

Many species are distinctive and can be readily identified, although within *Scoparia/ Eudonia*, the grass moths, *Udea* and *Anania*, there are small groups or species pairs that look superficially similar, particularly when worn. The most daunting, and frequently seen in moth traps, is the *Scoparia/Eudonia* group. However, wing shape and pattern separate them into subgroups: *E. alpina*, *E. angustea* and *E. lineola* are relatively narrow-winged species; *S. pyralella* and *E. delunella* are typically characteristically marked; and *S. subfusca* and *E. pallida* are plainly marked. This leaves seven species which require particular attention to be sure of identification: *S. ambigualis*, *S. basistrigalis*, *S. ancipitella*, *E. murana* and *E. truncicolella* all look similar, as does the pair *E. lacustrata* and *E. mercurella*.

Further reading
British and Irish species: Goater (1986)
European species: Goater *et al*. (2005) (Crambidae part); Palm (1986); Slamka (1997) (covers most British and Irish species); Slamka (2008) (Crambidae part)

Scoparia subfusca page 253   1332 (6165)

Common. More local away from the coast in northern England and Scotland. **FL** 10-13mm. Variable, markings sometimes obscure. Forewing whitish grey, speckled light brownish grey, generally with two whitish cross-lines, one just before one-third, the other just beyond two-thirds, with an X-shaped, sometimes 8-shaped, darker marking present at two-thirds. A darker, brownish-grey form almost devoid of markings can occur, with smaller and darker examples usually found in the north, west and in montane districts. Those on Orkney and Shetland are small and dark, with the pale cross-lines fairly distinct. **FS** Single-brooded, May-August. Readily disturbed by day and comes to light. **Hab** Open habitats, including stony hillsides, quarries, vegetated shingle, marshes, waste ground. Also gardens, heathland, woodland. **Fp** Bristly Oxtongue, Colt's-foot; in a silk tube at the base of the plant, probably feeding on wilted leaves and the root.

Scoparia pyralella page 253   1333 (6172)

Common. More local in the north. **FL** 8-10mm. Forewing relatively short (FL 2.25-2.5 x width), whitish, variably mottled brown or greyish brown, with generally conspicuous white cross-lines at one-third and two-thirds, an 8-shaped or kidney-shaped mark just beyond one-third and another at about two-thirds, these shaded orange-brown, brown or greyish brown, the outer mark incompletely ringed darker. Paler forms occur, where the darker mottling is reduced or absent and the brownish markings are enlarged, forming two short cross-bands; another form is almost pure white with much-reduced markings. **Similar species** *S. ambigualis* has a form with the forewing tinged yellowish brown, especially when freshly emerged, and has a relatively long forewing, whereas in *S. pyralella* it is relatively short. **FS** Single-brooded, May-mid August. Readily disturbed by day, flies in warm dry weather from late afternoon to dusk, and comes to light. Also found at night on the flowers of Common Ragwort. **Hab** Open country on light soils, including chalk and limestone grassland, sand dunes, vegetated shingle, disused quarries, waste ground; occasionally gardens, woodland clearings. **Fp** Dead leaves at the base of Ribwort Plantain; in a slight web. Probably other plants; possibly also on roots.

Scoparia ambigualis   page 253   1334 (6168)

Common. **FL** 8-11mm. Forewing relatively long (FL 2.5-2.75 x width), greyish white, sometimes faintly tinged yellowish brown, speckled darker grey, with whitish cross-lines at one-third and two-thirds, between which is a yellowish-brown or greyish 8-shape with blackish edging forming a small X; a whitish line near the termen forms a rough X with the outer cross-line, and there is broad dark grey tornal blotch beneath this X. Smaller, darker and almost unmarked forms occur. **Similar species** *S. ambigualis* is variable in size and markings, and can be confused with several other species. *S. basistrigalis* is a large species which has a long and broad forewing with a blunt apex, and is blacker with the cilia more strongly chequered. *S. ancipitella* has a relatively short forewing, with a plain greyish-white base, and a wide irregular whitish or greyish-white cloud between the outer cross-line and the termen. *Eudonia murana* and *E. truncicolella* both have a more pointed forewing. Also *S. pyralella*. **FS** Probably single-brooded, late April-August; also recorded in October. Easily disturbed by day from tree trunks and rocks, and comes to light. **Hab** Woodland, scrub, heathland, boggy moorland, parkland, gardens. **Fp** On a wide range of mosses, including *Polytrichum commune* and *Mnium hornum*, and possibly dead plant material; in a silk tube deep in the moss.

Scoparia basistrigalis   page 253   1334a (6166)

Local. There is uncertainty over the authenticity of many of the northern and Irish records. **FL** 10-11mm. Forewing long and broad, with a blunt apex, greyish white, often with a faint yellowish-brown hue, heavily speckled darker grey and blackish, with whitish cross-lines at one-third and two-thirds and a partly obscure whitish line near the termen forming a broad X with the second cross-line; between the first two cross-lines is a blackish blotch or an obscure X, sometimes with an obscure 8-shape touching the first cross-line; the cilia are whitish, usually distinctly chequered. **Similar species** *S. ambigualis*. **FS** Single-brooded, June-mid August; also recorded in early September. Rests by day on tree trunks, flies at night and comes to light. **Hab** Deciduous and mixed woodland; also marshes, gardens. **Fp** Has been found on the moss *Mnium hornum* growing on the ground away from the base of a tree; probably on other mosses; in a slight silken web.

Scoparia ancipitella   page 253   1335 (6169)

Very local. Mainly northern and western. **FL** 9-11mm. Forewing relatively short (FL 2.5 x width), greyish white, sometimes faintly tinged yellowish brown, speckled darker, plain greyish white near the base with whitish cross-lines, edged darker, at one-third and two-thirds, the latter being notched near the costa, an obscure line near the termen merging with the second cross-line, resulting in a wide irregular whitish or greyish-white cloud; between the first two cross-bands is an X- or 8-shaped greyish-brown mark, with a yellowish-brown 8-shape touching the first cross-line. **Similar species** *S. ambigualis*. **FS** Single-brooded, late June-August. Flies at night and comes to light. **Hab** Deciduous woodland. **Fp** Continental authors list the mosses *Polytrichum commune* and *Mnium hornum*; possibly in a web.

Eudonia lacustrata page 253

1338 (6180)

Common. **FL** 8-9mm. Forewing whitish to yellowish brown, speckled darker, with cross-lines of ground colour just before one-third and just beyond two-thirds, a black X at two-thirds, and with a broad pale line near the termen interrupted at the middle and forming a broad X with the second cross-line; the central third of the wing is usually pale, with a hint of yellowish brown, or sometimes distinctly whitish. **Similar species** *E. mercurella* has a slightly narrower and more pointed forewing, the broad blackish shading along the outer margin of the inner cross-line extends to the dorsum, and the pale line near the termen is usually entire; *E. mercurella* never has any yellowish brown on the forewing. **FS** Single-brooded, mid May-August; also recorded in early October. Rests by day on tree trunks, stone walls and fences, flies at night and comes to light. **Hab** Open habitats, woodland, parkland, gardens. **Fp** Mosses, including *Hypnum cupressiforme*; in a silk tube among the moss.

Eudonia pallida page 253

1336 (6199)

Local. More local in Scotland. A possible occasional immigrant. **FL** 8-9mm. Forewing whitish to pale yellowish brown, with a whitish cross-line just beyond two-thirds, and at two-thirds a dot-like, star-shaped or elongate spot, this dark brown or blackish, which is preceded by two, often elongate spots, these sometimes obscure. **One of the smallest species of the group, generally distinguished by the weak markings. FS** Possibly double-brooded, late May-early October, occasionally November. Flies freely on warm evenings and comes to light. **Hab** Grasslands, fens, marshes, bogs, but occurs away from these habitats. **Fp** Has been found on the moss *Calliergonella cuspidata*; in a silken tube.

Eudonia alpina page 253

1337 (6186)

Very local. **FL** 10-12mm. Forewing long, narrow, grey, lightly speckled darker, with greyish-white somewhat indistinct cross-lines at one-third and two-thirds; between the cross-lines is a black X-shaped mark, preceded by a black-ringed obscure spot. A form occurs in Shetland with the ground colour clear whitish and grey. **Usually the largest of the narrow-winged species of the group. FS** Single-brooded, late May-July. Flies in afternoon sun, is readily disturbed from montane vegetation and comes to light. **Hab** High mountains (above 700m), and occurring locally at lower elevations on Speyside and Shetland. **Fp** Not known.

Eudonia murana page 253

1339 (6182)

Local. Very local in the southern part of its range. Records from south-west and lowland north-west England are considered doubtful. **FL** 9-12mm. Forewing elongate and pointed, white, speckled blackish, partly obscuring the outlines of the markings, with two whitish cross-lines at one-third and just beyond two-thirds, curving around an obscure blackish 8- or X-shape, the first with two sometimes obscure dark spots adjacent to its outer edge; an obscure whitish line near the termen forms a broad X-shape with the second cross-line. Some examples have the markings between the cross-lines more prominent and may include a round black smudge, an elongate black mark with a pale centre, and a distinct 8-shape. **Similar species** *E. truncicolella* has the forewing less elongate and more pointed, with the dark markings thinner and usually slightly more distinct; the distance between the two cross-lines at the dorsum is shorter in *E. murana*. *E. murana* also appears more speckled or peppered. Where these species occur together, *E. murana* usually starts flying much earlier in the season than *E. truncicolella*. The genitalia of both species are very similar. **FS** Single-brooded, late May-mid September. Sometimes flies freely before dusk and comes to light. **Hab** Moorland, mountains, often in rocky situations. **Fp** Probably mosses such as *Hypnum cupressiforme*, *Dicranum scoparium*, *Bryum capillare* and *Grimmia pulvinata*.

Eudonia truncicolella page 253

page 253

1340 (6193)

Common. **FL** 9-11mm. Forewing white, unevenly speckled blackish resulting in a mottled appearance, with two whitish cross-lines, one before one-third, the other just beyond two-thirds, with an obscure line near the termen meeting the second cross-line near the middle, giving rise to a broad X-shape, the dark tornal mark below the whitish X usually narrow; a blackish-edged mark, forming an X-shape, is usually visible at two-thirds, this preceded by two small oval or slightly elongate blackish marks adjacent to the first cross-line. **Similar species** *Scoparia ambigualis*, *E. murana*. **FS** Single-brooded, mid June-October. Easily disturbed by day and comes to light. **Hab** Woodland, scrub, heathland, moorland, parkland, occasionally gardens. **Fp** Mosses, including *Dicranum scoparium*, *Hypnum cupressiforme* and *Campylopus* species; in a silk tube among the moss.

Eudonia lineola page 253

1341 (6187)

Very local. Primarily coastal. **FL** 9-10mm. Forewing pointed, whitish, tinted pale yellowish grey, speckled darker, with a whitish cross-line just before one-third, and another just beyond two-thirds, the second cross-line curving around a black X-shaped mark, **this X-shaped mark often with a white dot above it, visible even in worn examples**; the termen has a series of black dots or short dashes along its length. **FS** Single-brooded, late June-August; also recorded in November. Can be disturbed by day, flies from dusk and comes to light. **Hab** Coastal scrub, rocks, isolated trees. **Fp** The lichen *Xanthoria parietina*; in a slight web under the lichen lobes, creating bare patches where the larva has been eating.

Eudonia angustea page 253

1342 (6184)

Common. Local in Scotland. Probably more frequent in coastal counties. **FL** 8-10mm. Forewing narrow, sharply pointed at the apex, whitish, tinted brownish, and variably lightly speckled blackish, with generally distinctive whitish cross-lines about one-quarter and just beyond two-thirds, the latter strongly angled around an X-shaped mark, and with a narrow oval mark, sometimes heavily edged black, at one-third.

▼ Silken tube of *Eudonia angustea* incorporating moss and frass.

FS Possibly continually brooded, January-December. Flies from dusk, comes to light and can be found on flowers such as Ivy blossom and Buddleia. **Hab** Coastal sandhills, vegetated shingle, gardens, woodland, parkland. **Fp** The mosses *Syntrichia ruralis* var. *ruraliformis*, *Tortula muralis*, *Pseudocrossidium revolutum*, *Homalothecium sericeum*; in a silk tube adorned with moss fragments.

Eudonia delunella page 253 1343 (6189)

Very local. **FL** 8-9mm. Forewing white or whitish, lightly speckled blackish, with two somewhat obscure white cross-lines, just before one-third and just beyond two-thirds, the first thickly edged black on the outer margin and with two black marks beyond, and a prominent black X before the second, the black marks usually expanded to the costa, forming quadrate dark brown or black blotches; there is a large dark brown or blackish mark at the tornus and another, smaller, in the middle of the termen. **The contrasting black and white pattern helps to distinguish this species.** **FS** Single-brooded, late May-mid September. Found at rest on trees, comes to sugar and to light. **Hab** Woodland, usually ancient woodland. Also gardens with apple trees. **Fp** Has been reared from the moss *Leucodon sciuroides* in Europe; maybe on lichens and other mosses on trees.

Eudonia mercurella page 253 1344 (6195)

Common. More local in northern Scotland. **FL** 7-9mm. Forewing whitish, variably speckled darker, with a whitish cross-line at one-third and another at two-thirds, and with an entire whitish line towards the termen forming a broad X-shape with the second cross-line, this more distinct in darker examples; at about two-thirds is a variably distinct blackish X. A melanic form occurs

where only the terminal white X is visible. Another form has a contrasting dark central third, with a white base and terminal area; this form occurs rarely on the south coast of Britain. **Similar species** *E. lacustrata*. **FS** Single-brooded, June-September, occasionally from late April. Readily disturbed by day and comes to light. **Hab** Woodland, scrub, moorland, grasslands. **Fp** Mosses, such as *Hypnum cupressiforme* and *Homalothecium sericeum*; in a silk tube among the moss.

Chilo phragmitella page 254 1290 (6222)

Local. Few records in Scotland. **FL** 12-19mm. Sexually dimorphic, the female usually larger than the male. Male forewing light, dark or reddish brown, sometimes suffused darker, the apex slightly pointed. Female forewing straw yellow, longer and narrower than the male, slightly concave near apex and produced to a narrow point. **Similar species The female is similar to the female of *Donacaula forficella*, which has the costa convex near the apex.** **FS** Single-brooded, May-mid September; has been recorded in October. Comes to light. **Hab** Reedbeds, fens, gravel pits, grazing levels, riverbanks, ditches. Occasionally wanders. **Fp** Common Reed, Reed Sweet-grass; within the stem and rootstock.

Calamotropha paludella page 254 1292 (6235)

Local. Expanding its range. **FL** 12-15mm. Forewing whitish brown to pale sandy brown, with an obscure row of dark dots at two-thirds, some or all of which can be absent. **Similar species The plain form of Silky Wainscot *Chilodes maritima* (Noctuidae), which has short, not long, palps, with the forewing narrower and pointed; also *Orthotelia sparganella* (Glyphipterigidae), which rests with wings roof-like, not flat.** **FS** Single-brooded, June-mid September. Comes to light. **Hab** Fens, marshes, broads, the margins of flooded gravel pits, wandering occasionally and appearing in gardens. **Fp** Bulrush, occasionally Lesser Bulrush.

Chrysoteuchia culmella page 254
1293 (6241)

Common. **FL** 9-12mm. Forewing sandy brown, variably suffused leaden grey, with a strongly elbowed cross-line towards the termen and a short series of blackish dots near the base of the termen, and with the cilia golden metallic. In darker examples, the veins are paler except towards the costa. **FS** Single-brooded, May-September, sometimes into October in the south, but flight season ends early August in the north. Readily disturbed by day and comes to light. Can be abundant. **Hab** Grasslands, such as meadows, chalk and limestone grassland, waste ground. **Fp** Grasses; at the base of the culms.

Crambus pascuella page 254
1294 (6243)

Common. **FL** 11-13mm. Forewing with a triangularly produced apex, yellowish brown, sometimes paler, with a broad shining white longitudinal streak extending from the base to about four-fifths, narrowing from about one-half, and with the cilia, in part, silvery metallic. A more uniform dark brown form occurs regularly in the New Forest and occasionally in Kent. **Similar species** *C. silvella* has the white longitudinal streak barely touching the costa and extending to the base of the cilia. *C. uliginosellus* is a paler species and usually smaller, having a relatively broader forewing, the longitudinal streak touching the costa to about one-half and with a tooth-mark on the dorsal edge near one-half, this sometimes obscure, and appears particularly whitish in flight. **FS** Single-brooded, May-August, sometimes into September. Readily disturbed by day and comes to light. **Hab** Grasslands, including pastures, damp woodland rides, moorland, marshes. **Fp** Sheep's-fescue, Deergrass, possibly other grasses and sedges; in a slight silken tube at the base.

Crambus silvella page 254
1296 (6244)

Rare. Restricted to the New Forest, east Dorset and a site in Cardiganshire, with occasional examples found elsewhere. **FL** 10-13mm. Forewing brown to light brown, with a white longitudinal streak approaching the costa at the base and extending to the termen, usually broken by an oblique line at about two-thirds and an elbowed cross-line near the termen; the cilia are partly metallic. **Similar species** *C. pascuella*. **FS** Single-brooded, July-mid September. Flies in late afternoon and comes to light. **Hab** Boggy heathland, mires. **Fp** Sedges.

Crambus uliginosellus page 254
1297 (6245)

Very local. **FL** 9-11mm. Forewing whitish to yellowish brown, with a broad white longitudinal streak touching the costa from the base to about one-half, extending to, and sometimes beyond, an elbowed cross-line near the termen, and interrupted by an oblique brownish line at about two-thirds, and with the cilia partly metallic. **Similar species** *C. pascuella*. **FS** Single-

brooded, late May-August. Readily disturbed by day, flies from dusk and comes to light. **Hab** Damp grassland, bogs, fens. **Fp** Carnation Sedge, Common Cottongrass, probably other grasses and sedges; in a silken tube.

Crambus ericella page 254
1298 (6246)

Very local. Recently found in Northern Ireland. **FL** 10-14mm. Forewing yellowish brown, with a narrow white longitudinal streak from the base to an elbowed cross-line near the termen, broken by an oblique line at about two-thirds, a tapering whitish streak along the dorsum, and the cilia partly metallic. **FS** Single-brooded, late June-August. Readily disturbed by day, flies freely in the late afternoon and comes to light. **Hab** Heathland, moorland on mountains, limestone crags. **Fp** Possibly Sheep's-fescue or hair-grasses.

Crambus hamella page 254
1299 (6252)

Very local. A very few old records in Scotland. **FL** 12-13mm. Forewing greyish brown, with a broad white longitudinal streak not touching the costa and a tooth-mark on the dorsal margin at just over one-half, the streak ending in a point just before an elbowed cross-line near the termen. **FS** Single-brooded, August-mid September. Occasionally flies by day but more usually from dusk, and comes to light. **Hab** Dry heathland. **Fp** Not known, but possibly fine-leaved grasses.

Crambus pratella page 254
1300 (6250)

Very local, but somewhat commoner in the north on coasts. Predominantly a northern species and probably resident in Wales, but an occasional immigrant or wanderer elsewhere. **FL** 10-12mm. Forewing light reddish brown, the apex produced, with a white longitudinal streak broadening from the base, interrupted at about two-thirds by an oblique line, the dorsal margin with a tooth-mark at about one-half, the leading edge of the streak tapered towards an elbowed cross-line near the termen; the cilia are whitish metallic. **Similar species C. lathoniellus is generally slightly smaller and does not have the costal margin of the longitudinal streak angled before the interruption. FS** Single-brooded, June-early September. Readily disturbed by day, has been found flying in evening sunshine and comes to light. **Hab** Dry, short-turfed grassland, including coastal sandhills, heaths. **Fp** Grasses, possibly preferring hair-grasses; at the roots and stem bases.

Crambus lathoniellus page 254
1301 (6251)

Common. **FL** 9-11mm. Forewing whitish brown to reddish brown with greyish or whitish streaks, with a narrow white longitudinal streak widening from the base, interrupted at about two-thirds by an oblique line, with the costal margin straight and the dorsal margin with a tooth-like mark at about one-half; the cilia are partly metallic. **Similar species** *C. pratella*. **FS** Single-brooded, May-August. Easily disturbed by day and comes to light. Sometimes abundant where found. **Hab** A wide range of grasslands. **Fp** Various grasses; at the stem bases.

Crambus perlella page 254
1302 (6253)

Common. **FL** 11-14mm. Forewing bright shining white or pale yellowish white. A streaked form occurs in most populations, sometimes predominantly, with the veins and ground colour of the forewing variably shaded dusky grey. **FS** Single-brooded, late May-September. Easily disturbed by day and comes to light. Can be abundant where found. **Hab** Grasslands, such as bogs, downland, vegetated shingle, dunes, woodland glades. **Fp** Various grasses, possibly with a preference for coarser grasses; at the base of the stems.

Agriphila selasella page 254
1303 (6266)

Local. More local in the north and in Scotland. **FL** 11-15mm. Forewing sandy yellow or yellowish brown, with a whitish longitudinal streak, this splitting into three or four fine branches extending along the veins, more or less bordered darker brown along the costal margin. **Similar species A. tristella has a narrower, yellowish or creamy, never white, longitudinal streak. A. selasella has a smoother texture to the forewing,**

with a more square wing tip, and the face is rounded, not slightly pointed, although this can be difficult to see. **FS** Single-brooded, June-September. Can be disturbed by day and comes to light. **Hab** Saltmarshes, fens, marshes, wet grassland and occasionally drier situations elsewhere, such as chalk and limestone grassland. **Fp** Various grasses, including Common Saltmarsh-grass, Small Cord-grass, Sheep's-fescue; on the ground among stems.

Agriphila straminella page 254   1304 (6267)

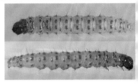

Common. **FL** 8-10mm. Forewing pale sandy brown, darker towards the costa, with a series of dark brown or blackish dots along the termen, and with the cilia metallic. **FS** Single-brooded, June-September. Flies by day and comes to light. Often abundant where found. **Hab** Grasslands, including chalk downland, heathland, marshes. **Fp** Various grasses, such as Sheep's-fescue; on the lower part of the stem.

Agriphila tristella page 254   1305 (6258)

Common. **FL** 12-14mm. Forewing pale sandy brown to dark brown, variably shaded darker towards the costa, with a pale creamy or yellowish longitudinal streak, although rarely this can be obsolete, branched towards and often reaching the termen, and sometimes with a rather indistinct weakly elbowed cross-line at just over five-sixths. **Similar species** *A. selasella*. **FS** Single-brooded, June-September. Readily disturbed by day and comes to light. **Hab** Grasslands, including chalk downland, heathland, waste ground. **Fp** Various grasses, such as Wavy Hair-grass and meadow-grass; on the lower part of the stem.

Agriphila inquinatella page 255   1306 (6260)

Common. More local in the north. **FL** 11-14mm. Forewing pale sandy brown to greyish brown, variably darker along the veins, with a fine pale longitudinal streak, this usually more prominent in darker examples, and with an oblique cross-line, frequently obsolete in part, passing through a short dark longitudinal dash near the middle of the wing, with another cross-line at about five-sixths, this curved and often thickened to form a chevron-like mark near the dorsum. **Similar species** *A. geniculea* has the cross-lines more strongly elbowed, the forewing appearing generally smoother and glossier. *Pediasia contaminella* typically rests head down in a declining posture, and is more uniform in appearance, with the cross-lines less clear and a dot in the middle of the wing which, if visible, is small and round, not a short bar. **FS** Single-brooded, June-October. Readily disturbed by day and comes to light. Occasionally numerous where it occurs. **Hab** Grasslands, particularly on light, dry soils, including grassy shingle, sandy heathland, dunes, downland, waste ground. **Fp** Grasses, such as Sheep's-fescue and Bristle Bent; in a slight silken tube among roots and stem bases.

Agriphila latistria page 255   1307 (6264)

Local. Scarce in Scotland, Ireland and the Channel Islands. **FL** 11-13mm. Forewing reddish brown, tinged greyish, with a broad white longitudinal streak extending into the cilia. **FS** Single-brooded, July-September. Comes to light. **Hab** Drier habitats, including coastal sandhills, dry heaths, downland. **Fp** Grasses, especially brome; on the soil surface among the roots.

Agriphila geniculea page 255 1309 (6275)

Common. More local in Scotland. **FL** 10-13mm. Forewing pale sandy brown, appearing slightly glossy, variably speckled brownish or greyish brown, with an indistinct pale longitudinal streak, and with an oblique elbowed cross-line at about one-half, thickened and darkened to form a short dash near the middle, with another cross-line at about three-quarters, strongly elbowed near the apex, less so near the tornus, and sometimes thickened near the middle; the cilia are silvery metallic. **Similar species** *A. inquinatella*. **FS** Single-brooded, July-October. Readily disturbed by day and comes to light. Occasionally abundant. **Hab** Dry grasslands, including sandhills, vegetated shingle, chalk grassland, grassy heaths, gardens. **Fp** Grasses; in a slight silken tube at the stem base.

Catoptria permutatellus page 255 1310 (6280)

Very local. Formerly recorded in Cumbria. **FL** 11-14mm. Forewing golden brown, darker bordering the white longitudinal streak, this bisected by oblique dark brown stripes at about one-half and five-sixths, resulting in a short white line running nearly parallel to the termen; the cilia are whitish, banded brown and weakly chequered. **Similar species** Similar to the more widespread *C. pinella*, which does not have the distinct narrow white line at the end of the longitudinal streak. Also similar are *C. osthelderi* (not illustrated), a very scarce immigrant, and *C. speculalis* (not illustrated), a species with a single 19th-century record; if either is suspected, genitalia examination will be necessary to confirm identification. **FS** Single-brooded, July-early September. Can be disturbed from foliage of small pines and sometimes birches, preferring more isolated trees. Comes to light. **Hab** Glens, mountain valleys. **Fp** Probably mosses.

Catoptria pinella page 255 1313 (6301)

Common. Local in north-west England. **FL** 10-13mm. Forewing golden brown, with a broad white median longitudinal streak, edged reddish brown and bisected by an oblique reddish-brown line at about one-half, resulting  in two broad white patches. **Similar species** *C. permutatellus*. **FS** Single-brooded, late June-September. Comes to light. **Hab** Woodland rides and clearings, heathland, waste ground, marshes. **Fp** Tufted Hair-grass, cottongrass, other grasses, probably also sedges; in a silken tube among the host plant.

Catoptria margaritella page 255 1314 (6304)

Local. With the exception of parts of Devon and Cornwall, very scarce over southern and eastern England, where it is encountered only as a probable immigrant or wanderer. **FL** 10-12mm. Forewing light brown to reddish brown, with a broad white longitudinal  streak broadest at about four-fifths and tapering towards the apex, ending just short of the termen. **FS** Single-brooded, late June-early September. Easily disturbed by day and comes to light. Sometimes numerous where found. **Hab** Boggy heaths, mosses, moors, usually in upland situations. **Fp** Has been found among the moss *Campylopus flexuosus*; in a slight spinning.

Catoptria furcatellus page 255
1315 (6305)

Very local. Only two records from north Wales and not recently found in northern England. **FL** 9-11mm. Forewing brown, with a narrow dull white longitudinal streak tapering towards, but not reaching, the termen, and with the cilia white. **FS** Single-brooded, July-August. Makes short flights by day in calm conditions. **Hab** Mountains between 400-900m, but has been found up to c.1,100m, occurring at lower elevations further north. **Fp** Probably clubmoss and mosses.

Catoptria falsella page 255
1316 (6314)

Common. Local in Scotland. Possibly expanding its range. Recently recorded new to Ireland. **FL** 9-12mm. Forewing pale straw yellow, heavily speckled dark brown between the veins, with a white longitudinal streak, broadly and obliquely divided by a cross-line at about two-thirds; at about four-fifths is another cross-line, this strongly elbowed near the apex and double-angled near the dorsum. **FS** Single-brooded, June-September. Comes to light. **Hab** Found in a range of habitats, including gardens, woodland, coastal localities; inhabits thick moss-covered roofs of outbuildings. **Fp** Mosses, including *Syntrichia ruralis* and *Barbula* species; in a silk tube deep in the moss.

Thisanotia chrysonuchella page 255
1321 (6350)

Very local. Probably declining. **FL** 11-13mm. Forewing whitish, finely speckled black, with brown shading in a broad band along the costa and two narrower longitudinal bands towards the dorsum, with an oblique weakly elbowed brown cross-line at just beyond one-half merging with the bands of brown shading, and another thinner brown cross-line at about five-sixths which curves towards the costa, the whole wing having a weakly chequered appearance; the cilia are metallic. **FS** Single-brooded, late April-June. Readily disturbed by day, particularly in the afternoon, flies from dusk and comes to light. **Hab** Chalk downland, sandhills, cliffs, short-turfed grassland. **Fp** Sheep's-fescue, probably other grasses; on stem bases.

Pediasia fascelinella page 255
1322 (6352)

Rare. Coastal Suffolk and Norfolk. Occasionally found elsewhere, some representing probable scarce immigrants or wanderers from the East Anglian coast. **FL** 12-15mm. Forewing pale sandy brown, the veins paler, with ill-defined dark longitudinal streaks from the middle of the wing towards the dorsum and a series of brownish spots separated by paler veins at about one-half and two-thirds, these curved towards the costa; the cilia are chequered white and brown. **FS** Single-brooded, June-mid September. Comes to light. **Hab** Relatively stable parts of sand dunes. **Fp** Grasses, such as Sand Couch, Lyme-grass, possibly also hair-grasses; on the rootstock.

Pediasia contaminella page 255
1323 (6364)

Very local. Predominantly southern and eastern, with a population on Scilly. Possibly expanding its range. **FL** 10-14mm. **Rests in a declining posture.** Forewing dull sandy brown, greyish brown or reddish brown, variably speckled darker, usually with a small dark point at about one-half; the cross-lines are faint, often obscure, and angled towards the costa, with one just beyond one-half and another at about five-sixths, both oblique. A melanic form occurs. **Similar species** *Agriphila inquinatella*. **FS** Single-brooded, late June-early October. Comes to light. **Hab** Dry grasslands, including heathland, parkland, sand dunes. **Fp** Grasses, such as Sheep's-fescue; in a tube among grass tufts.

Pediasia aridella page 255   1324 (6367)

Very local. Coastal, rarely inland. Probably a vagrant on Shetland. **FL** 10-13mm. **Rests in a declining posture.** Forewing sandy brown, the veins paler, slightly glossy with a blackish longitudinal streak from the base to about one-half, where it meets an oblique blackish cross-line, with a sometimes obscure cross-line at about five-sixths, this elbowed towards the costa, and a chevron-shaped mark near the mid-point. The intensity of the blackish markings varies. **The hindwing has a dark spot or short line towards the termen.** **FS** Single-brooded, June-August. Comes to light. **Hab** Dry margins of saltmarshes, sand dune margins, dry grassland, also from a valley mire. **Fp** Common Saltmarsh-grass, Borrer's Saltmarsh-grass, possibly other grasses; on stem bases.

Platytes alpinella page 255   1325 (6377)

Rare. Predominantly coastal, south-east England to Yorkshire; very locally in south Devon and southern Ireland. Occasionally elsewhere as a wanderer or an immigrant. **FL** 9-11mm. Forewing brownish, the apex produced, with a fine white longitudinal streak most clearly defined in the basal two-thirds and with two oblique zigzag-shaped fine brown cross-lines, at one-half and at four-fifths, the outer cross-line outwardly edged whitish; the cilia are partly metallic. **FS** Single-brooded, late June-mid September. Flies from dusk, visits flowers of ragworts and comes to light. **Hab** Coastal sandhills, sandy shingle, sandy grassland. **Fp** Probably the moss *Syntrichia ruralis* var. *ruraliformis* and other mosses.

Platytes cerussella page 255   1326 (6376)

Local. Predominantly coastal but also inland, such as in Breckland in East Anglia. **FL** 5-8mm. Male ranges from sandy brown to dark greyish brown, with oblique reddish-brown cross-lines at one-half and four-fifths, these sometimes obscure, and with the cilia metallic. The female has a narrower forewing, whitish to pale yellowish brown, variably speckled brown, particularly towards the apex, with the cross-lines edged brownish, sometimes partly obscure. **FS** Single-brooded, late May-July, sometimes to early September. Readily disturbed by day from grasses and comes to light. Can be frequent where found. **Hab** Dry grasslands, including vegetated shingle, sand dunes, sandy heath, chalk and limestone grassland, rocky places near the sea. **Fp** Stiff grasses, such as fescues, possibly also Sand Sedge or other sedges; in a slight silk tube deep within a tussock.

Schoenobius gigantella page 256   1328 (6390)

Very local. Predominantly south-eastern, perhaps more frequent in coastal counties. **FL** 12-23mm. Male forewing rather broad with the apex blunt, light brownish to brown, sometimes tinted reddish or thinly speckled blackish, two small dark dots usually present at about two-thirds, and other dots variably present. Female forewing brown, relatively narrow compared with the male, the termen oblique, ranging from almost unmarked, sometimes with a blackish dot at about two-thirds along with other scattered dots, to heavy black shading forming a longitudinal streak from the base to the apex. **FS** Single-brooded, May-July, occasionally to mid September. Comes to light. **Hab** Large reedbeds, grazing levels, gravel pits. **Fp** Common Reed, Reed Sweet-grass; in young shoots.

Donacaula forficella page 256   1329 (6393)

Local. **FL** 11-17mm. Female generally larger than male. Male forewing pale straw yellow with a black dot at about two-thirds, a brownish longitudinal streak below the costa and a dark oblique apical streak, the streaks varying in intensity. Female forewing light yellowish brown, longer, strongly pointed, with the termen slightly concave, the markings similar to those of the male, again varying in intensity. **Similar species** *Chilo phragmitella* female. **FS** Single-brooded, late May-July, occasionally late August-September. Readily disturbed by day, flies from dusk and comes to light. **Hab** Marshes, fens, gravel pits, reedy ditches. **Fp** Common Reed, Reed Sweet-grass, bur-reeds, sedges; rolls together young shoots.

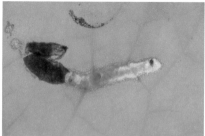

▲ Leaf mine of Brown China-mark *Elophila nymphaeata*.

▲ Larval case of Brown China-mark *Elophila nymphaeata*.

▲ Exposed larva of Brown China-mark *Elophila nymphaeata*.

▲ Larval case of Brown China-mark *Elophila nymphaeata* on pondweed.

Donacaula mucronella page 256   1330 (6394)

Local. Predominantly southern. **FL** 11-17mm. Male forewing dull greyish brown to brownish, female forewing more yellowish brown, with the apex pointed and an oblique termen. Both sexes have a broad pale coastal streak, edged below by a dark brown streak and extending to the wing apex. **FS** Single-brooded, May-July, occasionally to early September. Comes to light. **Hab** Marshes, fens, reedy ditches. **Fp** Common Reed, Reed Sweet-grass, Greater Pond-sedge and other sedges; in the stems just above the roots.

Brown China-mark *Elophila nymphaeata* page 256   1345 (6416)

Common. More local in Scotland. **FL** 12-16mm. Female larger and paler than the male. Forewing white, variably shaded light to dark brown, with three large white blotches and a broad white subterminal line interrupted by darker lines. **FS** Single-brooded, late May-early October. Readily disturbed by day from waterside vegetation and comes to light. **Hab** Edges of ponds, lakes, gravel pits, bogs, marshes, along slow-flowing rivers and canals. Occasionally elsewhere. **Fp** Probably on a range of water plants, including pondweeds, Frogbit, water-plantains, bur-reeds; aquatic, the young larva mines a leaf, later living in a floating case of two leaf fragments.

Water Veneer *Acentria ephemerella* page 256   1331 (6421)

Common. Very local in Scotland. **FL** 6-8mm. Forewing light grey. The female typically has rudimentary wings, although a fully-winged form occurs, this larger than the male. Could be overlooked as a caddis-fly. **FS** Possibly double-brooded in at least part of its range, late May-mid October. Flies at night over the water surface and comes to light. Can be abundant, occasionally swarming on warm nights. **Hab** Ponds, lakes, marshes. Disperses widely. **Fp** Canadian Waterweed, pondweeds, stoneworts, filamentous algae and possibly other water plants; aquatic, and has been found at a depth of up to 2m.

► Larva and larval case of Small China-mark *Cataclysta lemnata*.

Small China-mark *Cataclysta lemnata* page 256 1354 (6423)

Common. In Scotland, perhaps now found only at a site in Glasgow. **FL** 8-11mm. Male forewing shining white, with a minute dark spot at two-thirds, and the cilia chequered. Male hindwing white, with a conspicuous black band with bluish silvery dots bordering the termen. Female forewing suffused yellowish brown, with a darker, sometimes obscure spot at two-thirds. Female hindwing similar to male, with more brownish shading. **FS** Single-brooded, May-October. Readily disturbed by day from waterside vegetation and comes to light. **Hab** Ponds, ditches, grazing levels, open parts of reedbeds. Occasional elsewhere. **Fp** Duckweed, including Greater Duckweed, and possibly on a range of other aquatic plants; feeds below the water surface within a case made from fragments of the host plant.

Ringed China-mark *Parapoynx stratiotata* page 256 1348 (6425)

Local. Rare in western Scotland. **FL** 10-14mm. Male forewing whitish, shaded yellowish brown, with broken or diffused darker cross-bands at about one-third and at about two-thirds, the outer one curving around a dark-ringed spot beyond one-half. Male hindwing is white with a broken darker line. The female is usually larger and longer-winged, the

forewing heavily shaded orange-brown, brown or dark brown, with the cross-bands obscure but the darker-ringed white spot usually still obvious, and the hindwing similar to the male, with more brownish shading. **FS** Single-brooded, May-mid October. The male can be readily disturbed by day from waterside vegetation. Comes to light. **Hab** Margins of ponds and lakes, marshes, and along slow-flowing rivers, canals, drainage ditches. Occasionally elsewhere. **Fp** Pondweeds, Canadian Waterweed, Hornwort and other water plants; aquatic, spins leaves together and lives in an open web.

Beautiful China-mark *Nymphula nitidulata* page 256 1350 (6431)

Local. More local north from central Scotland. **FL** 10-12mm. Intensity of the markings variable, sometimes faint. Forewing shining white with brown markings, including three large, broadly rounded, white blotches in the middle part of the wing. **FS** Single-brooded, May-early September. Easily disturbed by day from waterside vegetation and comes to light. **Hab** Margins of rivers, streams, lakes, also fens and marshes. Occasionally elsewhere. **Fp** Bur-reed, Yellow Water-lily, possibly other water plants; feeds below the water surface within spun leaves or in the stem.

Cynaeda dentalis page 256 1359 (6446)

Rare. Coastal. Perhaps resident only from Devon to Kent and Suffolk. **FL** 10-14mm. Forewing creamy white, shaded light orangey brown, with a very strongly and irregularly toothed dark brown cross-band at about one-half; the cilia are strongly chequered. **FS** Single-brooded, with an occasional partial second brood, late May-August, September-October. Can be disturbed from the foodplant by day and comes to light. **Hab** Vegetated shingle, chalk and limestone cliffs, occasionally sand dunes. Historically, has been recorded inland on chalk downland. **Fp** Viper's-bugloss; internally on the stem and at the leaf bases. The hard pupal cocoon is covered with dried leaf fragments and attached to the surface of a dead leaf.

Garden Pebble *Evergestis forficalis* page 256 1356 (6497)

Common. More local in Scotland. **FL** 13-15mm. Forewing pale yellowish brown or straw-coloured, with several oblique, weakly curved or undulating darker cross-lines, that from the apex to the dorsum being strongest, and with two transversely placed dark brown spots just beyond one-half. **FS** Double-brooded, with an occasional third brood, late April-mid October. Comes to light. **Hab** Gardens, allotments, waste ground. **Fp** Various Brassicaceae, especially cultivated varieties of cabbage, Horse-radish, radishes, swedes; in a slight silk web among the leaves.

Evergestis limbata page 256 1356a (6500)

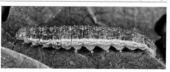

Very local. First found on mainland Britain in 1994, on the Isle of Wight. **FL** 10-11mm. Forewing yellow, with cross-lines at one-third and two-thirds, dark greyish brown beyond the outer cross-line; at about one-half is a kidney- or 8-shaped mark. **FS** Single-brooded, probably with a partial second brood in some seasons, June-August, occasionally into September. Easily disturbed by day and comes to light. **Hab** Open woodland, hedgerows, gardens. **Fp** Garlic Mustard, Hedge Mustard; on the underside of the leaves.

Evergestis extimalis page 256 1357 (6499)

Very local. Resident in the Channel Islands and probably some southern coastal counties and parts of eastern England, elsewhere an immigrant. **FL** 12-16mm. Forewing pale yellow, the cross-lines obscurely indicated by a series of small rusty dots, with a short oblique rusty brown apical streak. **FS** Single-brooded, possibly with a partial second brood in some years, late April-September, rarely into October. Comes to light. **Hab** Coastal, chalky habitats, waste ground. **Fp** Various Brassicaceae, especially Perennial Wall-rocket, Charlock, White Mustard; in a web on shoot tips, flowers and seed-pods.

Evergestis pallidata page 256 1358 (6501)

Local. Very local and scarce in northern England and Scotland. **FL** 12-14mm. Forewing pale yellow, glossy, lightly speckled brown, with brown cross-lines and, just before one-half, an almost circular outline adjoining a cross-line. **FS** Single-brooded, late May-September. Can be disturbed by day and comes to light. **Hab** Damp open woodland, marshy places, scrub. **Fp** Brassicaceae, chiefly Winter-cress; larvae are gregarious.

Pyrausta aurata page 257 1361 (6604)

Common. Very rare in Scotland.
FL 7-9mm. Forewing dark purple, usually with a single deep orange spot beyond the middle of the wing, sometimes with further small spots at the base and forming an interrupted cross-band from the dorsum at one-half to the costa at three-quarters. Hindwing with a short golden cross-band that rarely reaches the margins. **Similar species *P. purpuralis* and *P. ostrinalis* both have the** forewing with yellowish-orange or whitish markings, not deep orange, and the hindwing with pale yellow or whitish markings, not golden. *P. ostrinalis* is rather smaller and duller than *P. purpuralis*, has a narrower forewing with the costa usually slightly concave, and the cross-band on the underside of the hindwing reaches the costa. The underside marking in *P. ostrinalis* is also more clearly defined than in *P. purpuralis*. **FS** Double-brooded, March-early September. Flies in sunshine, rests on leaves and flowers of the foodplants, and comes to light.
Hab Grassland, especially chalk and limestone, gardens, wetlands. **Fp** Mints, Calamint, Marjoram, other garden Lamiaceae; larger larvae feed from a silken web.

Pyrausta purpuralis page 257 1362 (6605)

Common. More local in northern England; primarily a western species in Scotland.
FL 7-11mm. Forewing bright purple or reddish violet, with an oblique irregular yellowish-orange cross-band from the dorsum at one-half to the costa at three-quarters, sometimes broken into three spots, with further small spots below the costa. The underside of the hindwing with pale yellow or whitish spots, shading and a cross-band that does not reach the costa. **Similar species** *P. ostrinalis*, *P. aurata*. **FS** Double-brooded, end March-early September. Flies in sunshine, sometimes with *P. ostrinalis*, and comes to light.
Hab Grassland, especially chalk and limestone, coastal grassland, occasionally gardens. **Fp** Thymes, other Lamiaceae; between spun leaves.

Pyrausta ostrinalis page 257 1363 (6606)

Local. Primarily an eastern species in Scotland. **FL** 7-10mm. Forewing with the costa slightly concave, dull reddish violet, an oblique pale yellowish cross-band from the dorsum at one-half to the costa at three-quarters, with further small spots below the costa. The underside of the hindwing with whitish shading and a cross-band that reaches the costa. **Similar species** *P. aurata*, *P. purpuralis*. **FS** Double-brooded, mid April-mid June, end July-end August. Flies in sunshine, sometimes with *P. purpuralis*, and comes to light.
Hab Short grassland, especially chalk and limestone, coastal grassland. **Fp** Has been found on Thyme.

Pyrausta sanguinalis page 257 1364 (6599)

Rare. Formerly on the west coast of mainland Britain, but now restricted to parts of Ireland, Northern Ireland and the Isle of Man. **FL** 7-9mm. Forewing yellow, with a crimson cross-band from the middle of the dorsum, widening near the costa, merging with crimson along the costa and termen. **FS** Double-brooded, June, August. Flies in sunshine and comes to light. **Hab** Coastal sand dunes, limestone pavement. **Fp** Thymes; in a silken tube among flowers.

Pyrausta despicata page 257   1365 (6601)

Common. Local in north-west England. **FL** 7-9mm. Forewing mottled greyish brown, brown or sandy brown, with two dots near the middle of the wing and indistinct paler markings between the dots, and with slightly paler cross-bands before the termen, and along the termen. Darker forms have contrasting buff or yellowish markings, and are found especially in Shetland and Scilly. The female is usually smaller and darker, with the hindwing blackish. **Dark examples are similar to *Loxostege sticticalis* (not illustrated), an uncommon immigrant species which is much larger and without the cross-band before the termen. FS** Double-brooded, mid April-June, July-end September. Flies in sunshine and comes to light. **Hab** Grassland, heathland, sandhills, coastal shingle. **Fp** Plantains; gregarious, in galleries at the base of the leaves.

Pyrausta nigrata page 257   1366 (6613)

Local. **FL** 7-8mm. Forewing black, with a distinctly angled creamy white cross-band before the termen, thickened in places, and usually with other white marks in the wing, including a spot in the middle of the wing. The creamy white cross-band continues in the hindwing. **Similar species *P. cingulata* has a broader forewing with a white cross-band, nearly straight or gently curved, and lacks a white spot; also *Spoladea recurvalis* (not illustrated), a rare immigrant species, which is much larger, dark brown, with incomplete white cross-bands, and white marks in the cilia. Hab** Chalk grassland, less often limestone grassland. Flies in sunshine and comes to light. **FS** In two extended broods, mid April-October. **Fp** Thymes, Marjoram, other Lamiaceae; in a slight silken web under the leaves.

Pyrausta cingulata page 257   1367 (6595)

Local. Very local in Scotland. **FL** 7-8mm. Forewing brownish black with a narrow white cross-band which is almost straight or gently curved, and occasionally with some white scales on the dorsum towards the base. The white cross-line continues in the hindwing. **Similar species** *P. nigrata*. **FS** Double-brooded, end May-June, July-early September. Flies in sunshine and comes to light. **Hab** Coastal and limestone grasslands, sand dunes. **Fp** Wild Thyme, possibly Meadow Clary; in a slight silken web under the leaves.

Nascia cilialis page 257   1387 (6621)

Very local. Most widely found in parts of East Anglia. **FL** 11-13mm. Forewing slightly pointed, pale orangey yellow, the costa, termen and veins brown to reddish brown, with the cross-lines weak. **Similar species *Sclerocona acutellus* (not illustrated) is usually slightly larger and more uniform yellowish brown, with a white edge to the costa**; it is an adventive or rare immigrant, but may now be established in a reedbed on the Isle of Wight. **FS** Single-brooded, May-September, but usually occurring June-July. Can be disturbed from its resting site by day and comes to light. **Hab** Fens, water meadows. **Fp** Great Fen-sedge, Greater Pond-sedge, other small, narrow-leaved softer sedges; feeds openly on the leaves.

Sitochroa palealis page 257   1370 (6623)

Very local. Primarily resident in southern counties from Devon to Kent, and possibly parts of the east coast to Lincolnshire. Also an occasional immigrant that can lead to temporary colonisation. **FL** 13-16mm. Forewing pale sulphur yellow to whitish, the veins variably darker, with a central darker patch, although this can be obscure. **FS** Single-brooded, late May-mid September. Can be disturbed from vegetation by day and comes to light. **Hab** Chalk downland, Breckland, chalk and limestone quarries, grassy cliff tops, vegetated shingle. **Fp** Wild Carrot, Fennel, Moon Carrot; in a web in the seedhead.

Sitochroa verticalis page 257 — 1371 (6624)

Local. Probably resident only in Breckland in East Anglia, southern East Midlands, the London area and along the Thames Estuary. **FL** 12-15mm. Forewing light yellow with orangey yellow markings, the cross-line at about two-thirds gently curved and serrate, with another slightly broader cross-line adjacent towards the termen; just beyond one-half is an almost kidney-shaped mark, this preceded by a more dot-like mark. **The undersides of the wings have strongly contrasting dark markings. FS** Single-brooded, sometimes with a partial second brood, May-September. Easily disturbed by day and comes to light. **Hab** Fields, rough pasture, waste ground, gardens and coastal localities. **Fp** Perennial Wall-rocket, probably on a variety of other herbaceous plants; in a silk spinning among the flowers and developing seeds.

Small Magpie *Anania hortulata* page 258 — 1376 (6658)

Common. More local in central and northern Scotland. **FL** 13-16mm. Head and thorax yellow, black-spotted, the abdomen black and narrowly banded yellow. Forewing white, the costa, much of the basal area and other wing markings dark brownish grey. **FS** Single-brooded, May-September; also recorded in February, March, October and November. Easily disturbed by day and comes to light. **Hab** A range of habitats, including gardens, waste ground, hedgerows. **Fp** Common Nettle, occasionally on Lamiaceae, such as White Horehound, Black Horehound, woundworts, mints; in a rolled or spun leaf.

Anania lancealis page 258 — 1377 (6629)

Common. Rarely recorded in northern England. **FL** 13-16mm. Forewing relatively elongate, pale yellowish white, heavily clouded light greyish brown, with a strongly oblique greyish-brown curved and serrate cross-line at about two-thirds and a narrow crescent-shaped mark at about one-half, this preceded by a small dot. **FS** Single-brooded, June-August; has been recorded in mid May. Easily disturbed by day and comes to light. **Hab** Open wet woodland, ride edges, scrub, open marshy ground, sometimes in drier situations. **Fp** Hemp-agrimony, Hedge Woundwort, Wood Sage, ragworts; spins the margins of a lower leaf downwards, feeding from the underside near the leaf tip.

Anania coronata page 258 — 1378 (6631)

Common. Rarely recorded into southern Scotland. **FL** 11-13mm. Forewing dark greyish brown, finely speckled whitish, with a rectangular white spot at nearly one-half, a smaller and almost triangular white spot obliquely below and a large oval white blotch at about two-thirds; curving around the larger blotch to the costa is a series of small white blotches, these becoming larger nearer the costa. **Similar species *A. stachydalis* has a blunter forewing, with the white blotches towards the wing edge similarly sized.** **FS** Single-brooded, late May-September. Easily disturbed by day and comes to light. **Hab** Gardens, hedgerows, waste ground, woodland, scrubby situations. **Fp** Elder, viburnums, Lilac, privets; in a web on the underside of a leaf.

Anania perlucidalis page 258 — 1380 (6633)

Local. Expanding its range. **FL** 10-12mm. Forewing rounded, pearly, translucent whitish, with slightly darker, rather obscure cross-lines and a prominent dark almost crescent-shaped mark at one-half. **FS** Single-brooded, June-August. Occasionally disturbed by day and comes to light. **Hab** Wetland habitats, including fens, marshes, ditch banks, reedbeds, damp woodland. **Fp** Creeping Thistle, Marsh Thistle, probably other thistles; spins the margins of a lower leaf downwards, feeding from the underside near the leaf tip.

Anania stachydalis page 258   1384 (6632)

Rare. Predominantly southern. Probably overlooked. **FL** 11-12mm. Forewing dark greyish brown, finely speckled whitish, with indistinct darker cross-lines and a small whitish patch near the costa at about one-half, the cross-line at about two-thirds curving strongly around a large whitish blotch, which is finely bisected by a dark vein; beyond this cross-line is a row of whitish dots. **Similar species** *A. coronata*. **FS** Single-brooded, June-August. Can occasionally be disturbed by day and comes to light. **Hab** Hedgerows, woodland edges, damp woodland, shady tracks, overgrown ditches. **Fp** Woundworts, including Hedge Woundwort; spins the margins of a lower leaf downwards, feeding from the underside near the leaf tip.

Anania funebris page 258   1381 (6656)

Very local. Possibly declining. Still found in some south-eastern woodlands, parts of western England, Wales, western Scotland and southern and western Ireland. Also reported from Lincolnshire. **FL** 10-11mm. Thorax black, flanked yellow. Forewing black, with two large white blotches and occasionally an additional tiny white mark between the basal spot and the costa. **FS** Single-brooded, possibly with a partial second brood in the Burren, Ireland, May-September, but mainly June-July. Flies by day in a spinning motion that can be difficult to follow. **Hab** Woodland glades and margins, also rough hillsides and cliffs, particularly on limestone. **Fp** Goldenrod; on the flowers and leaves.

Anania verbascalis page 258   1382 (6655)

Very local. **FL** 11-13mm. Forewing deep brownish yellow, variably speckled brown, with brown cross-lines, the cross-line at about two-thirds strongly curved around a yellow quadrate mark, this mark more prominent on darker, more suffused examples. **FS** Single-brooded, predominantly mid June-July, but has also been noted to early September. Usually seen singly or in small numbers. Comes to light. **Hab** Open situations, including heathland, Breckland, coppiced woodland and woodland rides, open conifer plantations, vegetated shingle. **Fp** Wood Sage, possibly also Great Mullein; spins the margins of a lower leaf downwards, or two leaves together, feeding from the underside.

Anania terrealis page 258   1379 (6638)

Very local. Primarily a western species. Its occurrence in Ireland requires confirmation. **FL** 12-13mm. Forewing mauvish grey, speckled chocolate brown, darker towards the termen with the markings obscure; a darker cross-line at about two-thirds curves around a faint crescent-shaped mark. **FS** Single-brooded, June-July. Can be disturbed by day, flies naturally from dusk and comes to light. **Hab** Rocky coasts and hillsides. **Fp** Goldenrod; in a slight web on the underside of the lower leaves.

Anania crocealis page 258   1385 (6652)

Common. More local in Wales, north-west England, and very local in parts of western Scotland. **FL** 11-12mm. Forewing shades of yellow, paler examples weakly darker along the costa; the two brownish cross-lines are usually distinct, the outer at about two-thirds curves outwards around a small linear mark, this sometimes preceded by a small dot. **FS** Single-brooded, possibly with a second brood in parts of the south, late May-September. Easily disturbed from the foodplant by day and comes to light. **Hab** Marshy habitats, damp open areas, ditches, damp woodland and coastal localities, also dry, chalky banks and grassland. **Fp** Common Fleabane, Ploughman's-spikenard; large larvae feed in the heart of a shoot.

Anania fuscalis page 258

Common, although absent in some districts, being more local in northern parts of its range. **FL** 10-13mm. Forewing grey with a yellowish-grey gloss and two darker cross-lines, the outer at about two-thirds serrate and curving around a weak linear dark mark, the paler edging to the cross-line resulting in a small pale costal spot. **FS** Single-brooded, mid May-mid August. Easily disturbed from the foodplant by day and comes to light. **Hab** Meadows, marshes, open woodland, chalk and limestone grassland, northern moorlands, coastal localities. **Fp** Yellow-rattle, Common Cow-wheat; on flowers and seed capsules.

European Corn-borer *Ostrinia nubilalis* page 258

Local. Probably resident only in parts of southern England and East Anglia, elsewhere an immigrant. **FL** 13-18mm. Male forewing brown or greyish brown, with a prominent pale yellowish cross-band at about two-thirds; just before and just beyond one-half are two brownish dots, usually with a yellowish patch between them. Female forewing largely yellowish, with toothed cross-lines at about one-third and two-thirds, and two dark dots or marks just before and just beyond one-half. Comes to light. **FS** Single-brooded, possibly with a partial second brood, June-mid October. Has been recorded in December on the Channel Islands. **Hab** A wide range of habitats, including waste ground, road and railway cuttings, field margins, gardens. **Fp** Mugwort, also reported from Hop, Maize and Mallow; in the stem near ground level.

Bordered Pearl *Paratalanta pandalis* page 258

Very local. Possibly declining, but recently recorded new to Scotland. **FL** 12-14mm. Forewing slightly glossy, pale whitish yellow, pale greyish brown along the costa, veins and bordering the termen, with somewhat obscure cross-lines. **FS** Single-brooded, late May-July. Easily disturbed by day and comes to light. **Hab** Woodland glades on light soils, and open, unimproved grassland. **Fp** Wood Sage, Goldenrod, Wild Marjoram; in a portable case constructed of leaves.

Paratalanta hyalinalis page 258

Very local. Possibly declining. Southern counties, but formerly recorded north to Leicestershire. **FL** 14-16mm. Forewing slightly glossy, light yellow, shaded greyish yellow or orangey yellow along the costa, with brown cross-lines, the outer curving strongly around an oblong-shaped mark at about one-half, with another smaller mark preceding this at about one-third, but this can be obscure. **FS** Single-brooded, June-early September. Readily disturbed by day and comes to light. **Hab** Sheltered places on chalk downs and woodland clearings. **Fp** Common Knapweed, possibly also Great Mullein; in a silk gallery at the base of the plant.

Udea lutealis page 258

Common. Local within parts of its range. **FL** 11-12mm. Forewing pale yellowish white, darker along the costa, the markings yellowish brown and faint; the cross-line at about two-thirds strongly curves in the middle of the wing and around a sometimes obscure kidney-shaped mark at about one-half. **Similar species *U. decrepitalis*, particularly when worn, although *U. lutealis* has a lighter ground colour and lacks the darker banding towards the termen. Where both species occur together, *U. decrepitalis* starts flying much earlier in the season than *U. lutealis*.**

FS Single-brooded, June-September. Easily disturbed by day from low-growing vegetation, flies from dusk, visiting flowers such as thistles, and comes to light. **Hab** Woodland rides, chalk downland, marshes, rough fields and field margins, hedgerows, waste ground, overgrown gardens, allotments. **Fp** A wide range of herbaceous plants, including Wild Strawberry, Mugwort, Bramble, knapweeds, plantains; in a web on the underside of the lower leaves.

Udea fulvalis page 258 1389 (6533)

Very local. Currently resident only in parts of south Hampshire, east Dorset and the Isle of Wight. Also a scarce immigrant. **FL** 10-14mm. Forewing orangey brown, slightly darker beyond the cross-line at two-thirds, this line serrate and curved around a darker kidney-shaped mark, with a strong inward loop just above the dorsum. **FS** Single-brooded, July-mid September. Easily disturbed by day, flies naturally from dusk and comes to light. Can also be found feeding in numbers at Buddleia flowers. **Hab** Open ground, waste ground, gardens. **Fp** Various species of Lamiaceae, including Black Horehound, Cat-mint, Meadow Clary, White Dead-nettle; larva hard to find in the wild. Has been reared from Parsley growing in a greenhouse.

Udea prunalis page 258 1390 (6541)

Common. Local in Scotland. **FL** 11-13mm. Forewing greyish brown, darker on the costa and termen, the indistinct cross-line at two-thirds darker than the ground colour and slightly indented below a dark 8-shaped mark. **FS** Single-brooded, late May-August. Easily disturbed from scrub by day, and comes to light. **Hab** Thickets, hedgerows, woodland, gardens. **Fp** Feeds on a wide range of plants, including dead-nettles, Black Horehound, woundworts, Common Knapweed, Dog's Mercury, Alexanders, Common Nettle, Elder, Blackthorn; in spun leaves.

Udea decrepitalis page 259 1391 (6556)

Very local. A single record in Wales. **FL** 11-13mm. Forewing pale brown, slightly darker on the costa and towards the termen, with the cross-line at about two-thirds indistinct, serrate and sinuate, and an obscure brownish 8-shaped mark at about two-thirds. **Similar species** U. lutealis. **FS** Single-brooded, late May-August. Flies readily in the early evening towards dusk, also occasionally found sunning itself on Bracken. **Hab** Damp mountain ravines, hillsides, loch-side slopes. **Fp** Narrow Buckler-fern, Lemon-scented Fern, probably other ferns; in a slight web on the underside of a frond.

Udea olivalis page 259 1392 (6557)

Common. **FL** 12-13mm. Forewing brownish, the cross-lines indistinct, with a prominent, almost square-shaped white spot at about one-half and a scattering of other small white spots over the wing, particularly in the apical area. **FS** Single-brooded, mid May-August, sometimes into September. Easily disturbed from bushes by day and comes to light. **Hab** Woodland, hedgerows, bushy places, waste ground, gardens. **Fp** Feeds on a wide range of herbaceous plants, including Dog's Mercury, Yellow Archangel, Red Campion, Common Comfrey; in a spun or turned-down leaf.

Udea uliginosalis page 259 1393 (6550)

Very local. Recently recorded new to Ireland from Co. Mayo. **FL** 11-14mm. Forewing light greyish brown to pale reddish brown, virtually unmarked except for an often obscure small paler patch at about three-quarters. **FS** Single-brooded, June-July. Easily disturbed by day, when it usually flies uphill before settling again. Comes to light. **Hab** Grassy places or boggy ground on mountains, most frequently in grassy hollows beside streams. **Fp** Unknown, possibly on various herbs.

Rusty-dot Pearl *Udea ferrugalis* page 259 1395 (6531)

Common. Immigrant. Perhaps most frequent in the south. **FL** 9-11mm. Forewing orangey brown to chestnut, with a fine, sometimes slightly obscure, darker cross-line at about two-thirds which curves around and is strongly indented below a usually prominent squarish or kidney-shaped grey mark at about one-half; the apex has a short grey streak. **FS** Any month of the year, most often in late summer and autumn. Readily disturbed by day, comes to light and visits nectar sources. Numbers fluctuate annually, but can be numerous on occasion. **Hab** Can occur in any habitat, but usually most abundant in coastal localities. **Fp** Feeds on a wide variety of plants, including Hemp Agrimony, woundworts, burdocks, mints; in a spinning among leaves.

Mother of Pearl *Pleuroptya ruralis* page 259 1405 (6667)

Common. **FL** 15-17mm. Forewing pale yellowish brown, mottled greyish and with a pearly sheen, with darker cross-lines at one-third and two-thirds, the latter curving strongly around a small, elongate, sometimes weakly crescent-shaped mark, this preceded by a smaller round spot. **FS** Single-brooded, with a possible partial second brood, mid June-October. Readily disturbed by day from nettle

patches, flies from dusk and comes to light. **Hab** Gardens, waste ground, woodland, hedgerows, downland. **Fp** Mainly Common Nettle, also Wych Elm and English Elm; in a rolled leaf.

Mecyna flavalis page 259 1396 (6672)

Rare. Perhaps resident only in Wiltshire, the Isle of Wight, Hampshire, Sussex, Oxfordshire and Berkshire; a rare immigrant elsewhere. **FL** 11-14mm. Forewing yellow, usually darker along the costa, the cross-lines and markings greyish brown, with the cross-line at about two-thirds wavy, indented below the middle, and curving around an oblong or curved pale-centred mark, this preceded by a round pale-centred mark; the cilia are greyish brown, usually contrasting with the rest of the wing. **FS** Single-brooded, late July-August. Easily disturbed by day and comes to light. Can be abundant where found. **Hab** Short turf downland. **Fp** Unknown.

Mecyna asinalis page 259 1397 (6677)

Very local. Mostly coastal. **FL** 11-15mm. Forewing bluish grey to dove grey, rather elongate and pointed at the apex, with a large dark greyish-brown patch, sometimes obscure, which can extend to the costa, and a small dark blotch at one-third and one-half. **FS** At least two overlapping broods, May-mid November. Flies at night, visits flowers and comes to light. **Hab** Coastal scrub and cliffs, hedgerows inland, occasionally elsewhere. **Fp** Wild Madder; makes large whitish windows in the leaves by feeding from below, leaving the upper epidermis intact.

Agrotera nemoralis page 259
1410 (6680)

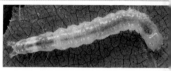

Rare. Perhaps restricted to a single site in Kent, but has occurred elsewhere in that county, with recent records from Cornwall, Dorset and Sussex, these probably relating to primary immigration. **FL** 9-11mm. Thorax yellowish, spotted orange. Forewing rather broad, with the basal area yellow, finely reticulated orange, the remainder of the wing light brown to orange-brown, darker near the basal patch, and with the cilia white, irregularly chequered blackish. **FS** Single-brooded, May-June; has been recorded in August. Hides among trees, from which it can be disturbed, flies at night and comes to light. **Hab** Hornbeam woods. **Fp** Hornbeam; between two leaves spun together.

Palpita vitrealis page 259
1408 (6690)

Immigrant. Typically found in southern coastal counties, rarely elsewhere. **FL** 13-15mm. Wings translucent white, the costal edge of the forewing yellowish brown. Flies at night, comes to light and is occasionally found at flowers, such as those of Buddleia and Ivy. **FS** June-early December, but usually late summer and autumn. Numbers fluctuate annually. **Hab** Potentially anywhere, but primarily coastal localities. **Fp** Egg-laying has been observed on Garden Privet in Britain, but the moth is not known to have bred in the wild in this country.

Dolicharthria punctalis page 259
1399 (6700)

Very local. Coastal. **FL** 10-12mm. A long-legged species. Forewing brown, sometimes rather paler along the middle of the dorsum, with darker brown indistinct cross-lines converging towards the dorsum and a prominent crescent- or V-shaped white mark at one-half. **FS** Single-brooded, June-early October. Can be disturbed by day, flies from dusk and comes to light. **Hab** Coastal, frequenting vegetated shingle, chalk and limestone downland and cliffs, sandhills. **Fp** Dead and decaying leaves of plants, such as Common Ragwort, bird's-foot-trefoils, knapweeds, plantains; in a slight silk spinning under a leaf on the ground.

Rush Veneer *Nomophila noctuella* page 259
1398 (6719)

Common. Immigrant. Usually most frequent in southern counties. **FL** 12-15mm. Forewing long and narrow, greyish brown with darker markings, to brownish or reddish brown with slightly darker and more obscure markings. In contrasting examples there is an 8-shaped mark just before one-half, with a kidney-shaped mark at about two-thirds. **Similar species** *N. nearctica* (not illustrated), a usually slightly larger American species, has been recorded only once in Britain. **FS** Has occurred in every month, with larger numbers usually found in the late summer and autumn. Easily disturbed by day and comes to light. Numbers fluctuate from year to year, but can be abundant. **Hab** Can occur anywhere, but typically most frequent in coastal localities. **Fp** Probably polyphagous and has been found on Selfheal, White Clover and Greater Plantain; in a silk-lined tunnel. Larvae are occasionally seen fully fed in moth traps, having crawled in overnight.

Dissection techniques

Although many of the micro-moths are distinctive and readily identifiable, particularly those in good condition or when reared from the larval foodstuff, many can be difficult to determine. This can be because there are superficially similar species or because individual examples are too worn, or in some cases even slightly worn, with characteristic features missing or obscured. Reference to the genitalia through dissection is a useful technique which can aid identification of these similar species or worn examples. This is made easier if the identity of the family is known, as many families or groups within families often have characteristic genitalia structures. When examining the genitalia it is still worth bearing in mind other clues to identity, such as habitat, potential foodplant and time of year, as these can help narrow down the possible candidates.

By way of an introduction to this technique, this guide includes genitalia photographs of a few frequently encountered couplets or groups of similar species, e.g. *Acleris ferrugana* and *A. notana* (Tortricidae) and the *Oegoconia* species (Autostichidae). In some cases, only one sex is illustrated in this guide, as the other sex can be very difficult to determine on genitalia characters. However, there are many sources of illustrations or images of micro-moth genitalia available elsewhere, for example some volumes of *The Moths and Butterflies of Great Britain and Ireland* series, the various volumes of *Microlepidoptera of Europe*, the two volumes of the *Tortricidae of Europe*, and the available volumes of the *Pyraloidea of Europe*. On occasion, illustrations/images are provided in papers or articles published in various entomological journals, for example the *Entomologist's Gazette* and *The Entomologist's Record and Journal of Variation*; this is most usually the case when species are added to the British fauna. In addition, there is the website of the Lepidoptera Dissection Group, www.dissectiongroup.co.uk, which covers a wide range of species.

Dissection techniques often take a bit of patience to acquire and frequently several attempts will be needed initially to gain competence. Therefore it is important to practise on specimens that can be expendable. Before undertaking your first dissection, various items of equipment, certain chemicals and a suitable space will be required. Townsend *et al.* (2010) provides a comprehensive guide to the tools and techniques required for genitalia dissection, and although that book focuses on macro-moths, the techniques are broadly similar. The Lepidoptera Dissection Group website also provides extensive tips on dissection and the equipment required.

A stable bench or table is the first essential, together with good lighting, such as that provided by an anglepoise lamp. Dissections of micro-moths require a binocular microscope, and one with zoom magnification is typically easier to use. To this can be added specialist lighting, i.e. fibre-optic lighting, but that provided by an anglepoise lamp is usually sufficient. Fine forceps, a selection of needles, brushes and probes, excavated glass blocks and glass covers, pipettes, glass tubes, glass slides and cover slips, and sticky labels are needed, along with the following chemicals: potassium hydroxide (KOH), euparal (the mountant), euparal essence, iso-propyl alcohol, a staining agent (such as chlorazol black or mercurochrome) and purified or distilled water. Needles and probes can be made at home, for example using old matchsticks and a variety of entomological pins, depending on how firm a needle or probe is required, and brushes can be supplemented or replaced by the pin-feather of a Snipe or Woodcock, if these can be obtained. Potassium hydroxide is caustic, and contact with skin and eyes should be avoided. The safety advice for this, and other chemicals used, is usually supplied with purchase and must be followed.

i Softening the abdomen

Usually, although not always, dissections are undertaken on dried and mounted (pinned) specimens. In these cases, the abdomen should be carefully removed from the specimen, taking care not to dislodge the hindwings. The abdomen is transferred into a glass tube containing a small amount of dilute potassium hydroxide; enough to cover the abdomen is all that is required. This can be left in place for several hours until the abdomen is soft, but timing is difficult and this method will require regular checking. Alternatively the tube and liquid can be warmed gently, for example using a water bath. In this method, the tube containing the potassium hydroxide and abdomen is placed in a glass beaker containing pre-heated water (for example from a kettle). Depending on the size of the abdomen and the heat of the water bath, it can be only a matter of minutes before the abdomen is soft, although it sometimes takes up to one hour in a cooler water bath.

ii Cleaning and staining
The softened abdomen is transferred to a watch glass containing purified (or distilled) water. The abdomen is then cleaned, removing external scales and internal fat and soft tissue, but taking care not to damage the genitalia structures. This is done under the microscope by gently brushing the abdomen. It may be necessary to transfer to a second watch glass with clean water to complete this process. If the abdomen is proving difficult to clean, then return to the potassium hydroxide for a few more minutes and repeat the process. After cleaning the abdomen, it can be transferred to the stain. This can also be done after the genitalia are removed (see below), although the separated structures can be small and easily lost in the stain.

iii Removing the genitalia
After the cleaning (and staining) process, the genitalia can be removed from the abdomen. This can require practice, but is sometimes surprisingly straightforward. The male structures may be separated by gently stroking the abdomen towards the genitalia, although the use of a pin at the base of the genitalia can aid this process. If the aedeagus (the structure functioning as a penis) needs to be removed from the rest of the structure, this should be undertaken at this stage. The female genitalia require a bit more care and need to be removed together with the ventral external opening (ostium). This will require removing the structure together, with the relevant abdominal segment, with the aid of a pin and, perhaps, fine forceps.

iv Slide preparation
It is useful to examine illustrations/images of the genitalia of any suspected species prior to this process as this will inform how the genitalia should be placed on the slide. The genitalia, together with the skin of the abdomen, are transferred to the alcohol. Usually this is done in stages, for example moving from 30% to 60%, about 30 seconds in each, and finally to 100% alcohol. Any difficult to remove scales usually become easier to brush clean during this process, although take care in the stronger alcohols as the abdomen and genitalia stiffen ('fix') through this process. The abdomen should be flattened dorso-ventrally through these stages. During the latter stages the genitalia should also be fixed in the position required, in the case of the males with the valves (clasping structures) spread out and fixed in the 100% alcohol.

The prepared abdomen and genitalia can be transferred to euparal essence prior to transferring to the euparal on a clean slide. The genitalia should be laid out: in the male with the aedeagus (if separated) close to the remaining structure, and in the female with the ostial opening facing upwards and avoiding unnecessary twisting of the ductus bursae (the tube from the ostium to the corpus bursae, which is a bag-like structure). The abdominal skin should be placed alongside. The cover slip should have one edge dipped into euparal essence and then applied over the preparation by placing at an angle and gently lowering to avoid any air bubbles forming. Excess euparal should be wiped away. The finished slide must be kept horizontal and covered to dry. This will take several months. The slide should be labelled with full data of the specimen (i.e. date, site, sex, name of the captor etc.), with a suitable cross-referencing label to the slide on the specimen.

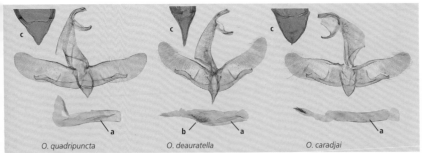

Genitalia preparations of *Oegoconia quadripuncta*, *O. deauratella* and *O. caradjai* (males only) **Autostichidae (p.123).** The preparations have been stained with mercurochrome. Inset above, saccus (**c**) enlarged. The genitalia of the females of *O. quadripuncta*, *O. deauratella* and *O. caradjai* are very similar; illustrations and a key describing the differences are given in Emmet & Langmaid (2002a). Differences may be seen under high magnification but there is variation within and between the females of these species and reliable separation on genitalia cannot always be achieved.

Feature	Oegoconia quadripuncta	Oegoconia deauratella	Oegoconia caradjai
a Aedeagus (all with projecting spine-tipped vesica)	With long basal bar	With long basal bar	With very small basal bar
b Aedeagus: patch of spines at $^2/_3$	Absent	Present	Absent
c Saccus shape	An equilateral triangle	A narrow, sharply pointed triangle	Shield shaped, with very small nipple at base

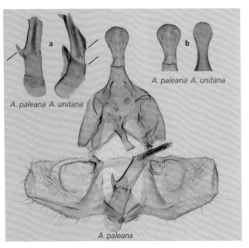

Genitalia preparation of *Aphelia paleana* (male only) Tortricidae (p.291). The main distinguishing features for determining the identification of *A. paleana* and *A. unitana* are seen in the aedeagus and uncus (**a** and **b**). The genitalia of the females of *A. paleana* and *A. unitana* are very similar and cannot be separated reliably by dissection.

Feature	Aphelia paleana	Aphelia unitana
a Aedeagus	Large thorn side projection, often with a smaller thorn at base	Several rows of small triangular projections
b Uncus	Broader, more rounded	Slightly narrower, less bulbous

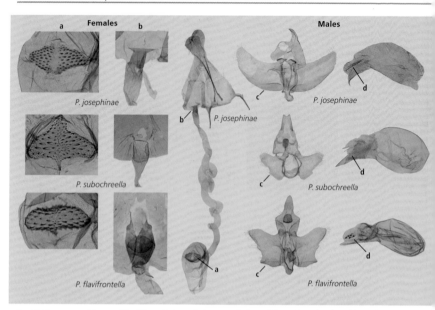

Genitalia preparations of the '*Pseudatemelia*' group: *P. josephinae*, *P. subochreella* and *P. flavifrontella* Lypusidae (p.132). The female preparations have been stained with chlorazol black and the males with mercurochrome. Insets on either side, aedeagus (**d**), signum (**a**) and ostium (**b**) enlarged. Clear-cut differences between the males can be seen in the shape of the valves (**c**) and in the small spines towards the tip of the aedeagus (**d**). In the females, the shapes of the signum (**a**) and the ostium (**b**) are diagnostic.

Feature	*Pseudatemelia josephinae*	*Pseudatemelia subochreella*	*Pseudatemelia flavifrontella*
Female			
a Signum	Central strip without spines	Shape triangular, central area with spines	Shape elliptical, central area with spines
b Ostium	Entrance shallow funnel with parallel sides below	Entrance oval, membrane creased on either side	Entrance with small triangular 'ears', front edge cleft
Male			
c Valves	Curved lower edge	Triangular with a small projection on lower edge	Triangular with a large projection on lower edge
d Aedeagus; spines at tip	Single	A few, fine	Several, distinct, more heavily tipped

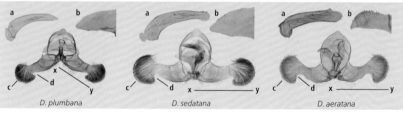

Genitalia preparations of the *Dichrorampha 'plumbana'* group (males only): *D. plumbana*, *D. sedatana* and *D. aeratana* Tortricidae (p.333). The preparations have been stained with mercurochrome. Inset above, *aedeagus* and *annellar* arm (**a** and **b**) enlarged. **See table opposite.** The genitalia of the females of *D. plumbana*, *D. sedatana* and *D. aeratana* are very similar; see Bradley, Tremewan & Smith (1979).

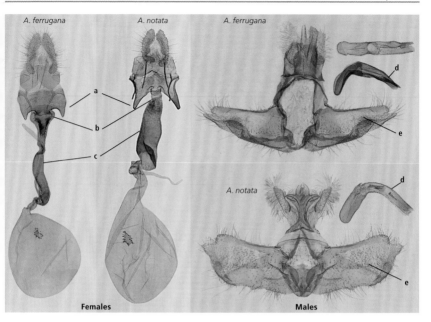

Genitalia preparations of *Acleris ferrugana* and *A. notana* Tortricidae (p.274). The female preparations (left) have been stained with chlorazol black and the males (right) with mercurochrome.

Feature	*Acleris ferrugana*	*Acleris notana*
Female		
a Ostial plate	Shorter lobes	Elongated lobes
b Colliculum	Bulbous	Narrow, parallel sided
c Ductus bursae	Slender	Broad
Male		
d Aedeagus	Side projection and 2 long cornuti	3 cornuti (1 pair and a single)
e Valves	Narrower and tapering	Broad, sides parallel

Males of the *Dichrorampha 'plumbana'* group

Feature	*Dichrorampha plumbana*	*Dichrorampha sedatana*	*Dichrorampha aeratana*
a Aedeagus	Tapered, without thorns at the tip	With two very small thorns at the tip *	With a single thorn at the tip
b Anellar arms	Triangular	Triangular	With a heavy covering of small spines, the upper edge more rounded
c Cucullus	Slightly smaller and less rounded than in *sedatana*	Larger and more rounded than in *plumbana*	Slightly larger and more rounded than in *plumbana*
d Indentation between the sacculus and cucullus	Narrow	Narrow	Broad
x-y Length of valve	Less than 1.20mm, average 1.10mm	Greater than 1.30mm, average 1.45mm	Greater than 1.20mm, average 1.35mm

*This is best noted during dissection as, depending on the orientation, they may be difficult to see when set in preparation.

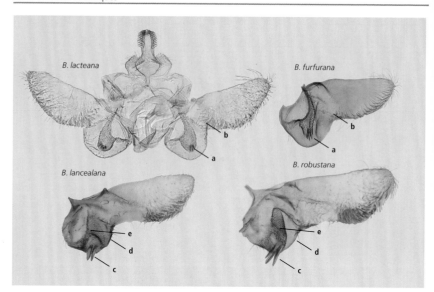

Genitalia preparation showing the males of the 'Bactra' group Tortricidae (pp.293-4). *B. lacteana* in full and right-hand valves for *B. furfurana*, *B. lancealana* and *B. robustana*. The *B. robustana* valve has been lightly stained with mercurochrome. **Using the shape of the valves, depending on whether or not the spines extend beyond the edge of the sacculus, the *Bactra* group comprises two similar pairs with *B. lacteana* and *B. furfurana* being one set and *B. lancealana* and *B. robustana* the other.** Female genitalia are not given; these are figured in Bradley, Tremewan & Smith (1979) and Razowski (2003). The main 'critical' features are seen in the round structure, the sacculus, at the base of the valves, and care needs to be taken when preparing the dissection to ensure that this does not get compressed.

Feature	*Bactra lacteana*	*Bactra furfurana*
a Spines in the sacculus	2 - 4	5 - 9
b Area at junction of sacculus and blade of valve	More fine spines than *furfurana*	Relatively few spines
	Bactra lancealana	*Bactra robustana*
c Set of blade-like spines projecting from the depression of the sacculus	2 - 7	3 - 4 slightly larger and stouter than *lancealana*
d Edge of sacculus	Straighter than *robustana*	More rounded than *lancealana*
e Spine above depression	A small spine, sometimes difficult to see	A more robust spine

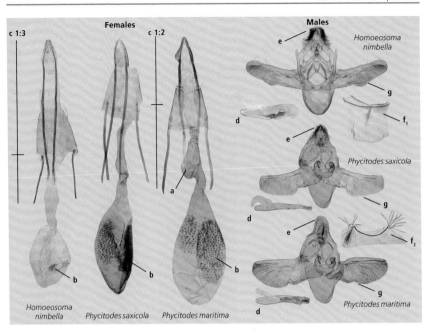

Genitalia preparations of the 'Phycitodes' group: Homoeosoma nimbella, Phycitodes saxicola and Phycitodes maritima Pyralidae (pp.361-2). The female preparations (left) have been stained with chlorazol black and the males (right) with mercurochrome.

Feature	Homoeosoma nimbella	Phycitodes saxicola	Phycitodes maritima
Female			
a Ductus bursae	Not bulbed	Not bulbed	Bulbed (a)
b Signum	Relatively small, oval	Two dentate patches of approximately similar size	Two unequal sized dentate patches
c Proportion anterior : posterior apophyses	1 : 3	1 : 2	1 : 2
Male			
d Aedeagus	Broad, cornutus a corrugated band	Thin, tapering to tip, vesica with minute teeth	Slightly curved, vesica spined
e Uncus	Very 'bristly'	'Bristly'	Rounded and smooth
f 8th abdominal segment	Rim chitinised without scales (f₁)	With long scales (f₂)	With long scales (f₂)
g Valves	Rear edge of valves with central lobe	Rear edge of valves with central lobe	Rear edge of valves with double lobe

A Checklist of the micro-moths of Britain and Ireland

The taxonomic list of British and Irish micro-moths below follows the hierarchy adopted by *Fauna Europaea* (FE). Within genus, however, the order of the species largely follows Bradley (2000), with subsequent changes included where appropriate. There have been many recent changes in nomenclature; where there have been changes to genus and species names, the name in Bradley is given in square brackets below the current useage. Within Pterophoridae there are differences in both the taxonomic order and nomenclature used by Hart (2011) and that by FE; for consistency we follow FE, except within the genus *Stenoptilia*, where Hart recognises more species in Britain than listed in FE.

Also, there have been some small changes to spellings of species names, authors and dates, which are incorporated without listing the entry in Bradley in square brackets. There are several species on the British list which are not in FE; the positions of these within the Checklist have been guided by other authorities or interpolated.

Species **emboldened** are either covered in the guide by species accounts, or are included within Similar species sections, where key identification features are given.

English names have been included where listed in Bradley (2000) and for the plume moths, Pterophoridae, following Hart (2011).

MICROPTERIGIDAE
Micropterix Hübner
 tunbergella (Fabricius, 1787)
 mansuetella Zeller, 1844
 aureatella (Scopoli, 1763)
 aruncella (Scopoli, 1763)
 calthella (Linnaeus, 1761)

ERIOCRANIIDAE
Dyseriocrania Spuler
 [= *Eriocrania* Zeller]
 subpurpurella (Haworth, 1828)
Heringocrania Kuznetzov
 [= *Eriocrania* Zeller]
 unimaculella (Zetterstedt, 1839)
Paracrania Zagulajev
 [= *Eriocrania* Zeller]
 chrysolepidella (Zeller, 1851)
Eriocrania Zeller
 sparrmannella (Bosc, 1791)
 salopiella (Stainton, 1854)
 cicatricella (Zetterstedt, 1839)
 sangii (Wood, 1891)
 semipurpurella (Stephens, 1835)

NEPTICULIDAE
Enteucha Meyrick
 acetosae (Stainton, 1854)
Stigmella Schrank
 aurella (Fabricius, 1775)
 splendidissimella (Herrich-Schäffer, 1855)
 auromarginella (Richardson, 1890)
 pretiosa (Heinemann, 1862)
 aeneofasciella (Herrich-Schäffer, 1855)

 dryadella (Hofmann, 1868)
 filipendulae (Wocke, 1871)
  [= *ulmariae* (Wocke, 1879)]
 poterii (Stainton, 1857)
 lemniscella (Zeller, 1839)
 continuella (Stainton, 1856)
 speciosa (Frey, 1858)
 sorbi (Stainton, 1861)
 plagicolella (Stainton, 1854)
 salicis (Stainton, 1854)
 obliquella (Heinemann, 1862)
 zelleriella (Snellen, 1875)
 myrtillella (Stainton, 1857)
 trimaculella (Haworth, 1828)
 assimilella (Zeller, 1848)
 floslactella (Haworth, 1828)
 carpinella (Heinemann, 1862)
 tityrella (Stainton, 1854)
 incognitella (Herrich-Schäffer, 1855)
 perpygmaeella (Doubleday, 1859)
 ulmivora (Fologne, 1860)
 hemargyrella (Kollar, 1832)
 paradoxa (Frey, 1858)
 atricapitella (Haworth, 1828)
 ruficapitella (Haworth, 1828)
 suberivora (Stainton, 1869)
 roborella (Johansson, 1971)

 svenssoni (Johansson, 1971)
 samiatella (Zeller, 1839)
 basiguttella (Heinemann, 1862)
 tiliae (Frey, 1856)
 minusculella (Herrich-Schäffer, 1855)
 anomalella (Goeze, 1783)
 centifoliella (Zeller, 1848)
 spinosissimae (Waters, 1928)
 viscerella (Stainton, 1853)
 malella (Stainton, 1854) Apple Pigmy
 catharticella (Stainton, 1853)
 hybnerella (Hübner, 1796)
 mespilicola (Frey, 1856)
 oxyacanthella (Stainton, 1854)
 pyri (Glitz, 1865)
 aceris (Frey, 1857)
 nylandriella (Tengström, [1848])
 magdalenae (Klimesch, 1950)
 † *desperatella* (Frey, 1856)
 † *torminalis* (Wood, 1890)
 regiella (Herrich-Schäffer, 1855)
 crataegella (Klimesch, 1936)
 prunetorum (Stainton, 1855)
 betulicola (Stainton, 1856)
 microtheriella (Stainton, 1854)
 luteella (Stainton, 1857)

sakhalinella Puplesis, 1984
glutinosae (Stainton, 1858)
alnetella (Stainton, 1856)
lapponica (Wocke, 1862)
confusella (Wood & Walsingham, 1894)
Trifurcula Zeller
headleyella (Stainton, 1854)
subnitidella (Duponchel, [1843])
immundella (Zeller, 1839)
squamatella Stainton, 1849
beirnei Puplesis, 1984
cryptella (Stainton, 1856)
eurema (Tutt, 1899)
Bohemannia Stainton
quadrimaculella (Boheman, 1853)
auriciliella (de Joannis, 1908)
pulverosella (Stainton, 1849)
Ectoedemia Busck
decentella (Herrich-Schäffer, 1855)
sericopeza (Zeller, 1839)
louisella (Sircom, 1849)
argyropeza (Zeller, 1839)
turbidella (Zeller, 1848)
hannoverella (Glitz, 1872)
intimella (Zeller, 1848)
agrimoniae (Frey, 1858)
spinosella (de Joannis, 1908)
angulifasciella (Stainton, 1849)
atricollis (Stainton, 1857)
arcuatella (Herrich-Schäffer, 1855)
rubivora (Wocke, 1860)
erythrogenella (de Joannis, 1908)
occultella (Linnaeus, 1767)
minimella (Zetterstedt, 1839)
heckfordi van Nieukerken, Lǎstůvka & Lǎstůvka, 2009
quinquella (Bedell, 1848)
heringella (Mariani, 1939)
albifasciella (Heinemann, 1871)
subbimaculella (Haworth, 1828)
heringi (Toll, 1934)
atrifrontella (Stainton, 1851)
amani Svensson, 1966
longicaudella Klimesch, 1953
septembrella (Stainton, 1849)
weaveri (Stainton, 1855)

OPOSTEGIDAE
Opostega Zeller
salaciella (Treitschke, 1833)
† *spatulella* Herrich-Schäffer, 1855
Pseudopostega Kozlov
auritella (Hübner, [1813])
crepusculella (Zeller, 1839)

HELIOZELIDAE
Antispila Hübner
metallella ([Denis & Schiffermüller], 1775)
treitschkiella (Fischer von Röslerstamm, 1843)
Heliozela Herrich-Schäffer
sericiella (Haworth, 1828)
resplendella (Stainton, 1851)
hammoniella Sorhagen, 1885

ADELIDAE
Nemophora Hoffmannsegg
fasciella (Fabricius, 1775)
minimella ([Denis & Schiffermüller], 1775)
cupriacella (Hübner, [1819])
metallica (Poda, 1761)
degeerella (Linnaeus, 1758)
Adela Latreille

cuprella ([Denis & Schiffermüller], 1775)
Cl *violella* ([Denis & Schiffermüller], 1775)
reaumurella (Linnaeus, 1758)
croesella (Scopoli, 1763)
Cauchas Zeller
[= *Adela* Latreille]
rufimitrella (Scopoli, 1763)
fibulella ([Denis & Schiffermüller], 1775)
Nematopogon Zeller
swammerdamella (Linnaeus, 1758)
schwarziellus Zeller, 1839
pilella ([Denis & Schiffermüller], 1775)
metaxella (Hübner, [1813])
magna (Zeller, 1878)

INCURVARIIDAE
Incurvaria Haworth
pectinea Haworth, 1828
masculella ([Denis & Schiffermüller], 1775)
oehlmanniella (Hübner, 1796)
praelatella ([Denis & Schiffermüller], 1775)
Phylloporia Heinemann
bistrigella (Haworth, 1828)

PRODOXIDAE
Lampronia Stephens
capitella (Clerck, 1759) **Currant Shoot Borer**
flavimitrella (Hübner, [1817])
luzella (Hübner, [1817])
corticella (Linnaeus, 1758) **Raspberry Moth**
morosa Zeller, 1852
fuscatella (Tengström, [1848])
pubicornis (Haworth, 1828)

TISCHERIIDAE
Tischeria Zeller
ekebladella (Bjerkander, 1795)
dodonaea Stainton, 1858
Coptotriche Walsingham
[= *Emmetia* Leraut]
marginea (Haworth, 1828)
heinemanni (Wocke, 1871)
† *gaunacella* (Duponchel, [1843])
angusticollella (Duponchel, [1843])

PSYCHIDAE
Diplodoma Zeller
laichartingella Goeze, 1783
[= *herminata* (Geoffroy, 1785)]
Narycia Stephens
duplicella (Goeze, 1783)
[= *monilifera* (Geoffroy, 1785)]
Dahlica Enderlein
triquetrella (Hübner, [1813])
inconspicuella (Stainton, 1843)
Lesser Lichen Case-bearer
lichenella (Linnaeus, 1761) **Lichen Case-bearer**
Taleporia Hübner
tubulosa (Retzius, 1783)
Bankesia Tutt
conspurcatella (Zeller, 1850)
[= *douglasii* (Stainton, 1854)]
Luffia Tutt
lapidella (Goeze, 1783)
ferchaultella (Stephens, 1850)
Bacotia Tutt

claustrella (Bruand, 1845)
[= *sepium* (Speyer, 1846)]
Proutia Tutt
betulina (Zeller, 1839)
Psyche Schrank
casta (Pallas, 1767)
crassiorella Bruand, 1851
Epichnopterix Hübner
plumella ([Denis & Schiffermüller], 1775)
Whittleia Tutt
[= *Epichnopterix* Hübner]
retiella (Newman, 1847)
Acanthopsyche Heylaerts
atra (Linnaeus, 1767)
Canephora Hübner
* **hirsuta** (Poda, 1761)
Thyridopteryx Stephens
A *ephemeraeformis* (Haworth, 1803)
Pachythelia Westwood
villosella (Ochsenheimer, 1810)
Sterrhopterix Hübner
fusca (Haworth, 1809)

TINEIDAE
Myrmecozela Zeller
ochraceella (Tengström, [1848])
Tenaga Clemens
† *nigripunctella* (Haworth, 1828)
[= *pomiliella* Clemens, 1862]
Eudarcia Clemens
richardsoni (Walsingham, 1900)
Infurcitinea Spuler
argentimaculella (Stainton, 1849)
captans (Gozmány, 1960)
albicomella (Stainton, 1851)
Ischnoscia Meyrick
borreonella (Millière, 1874)
Stenoptinea Dietz
†**l(?)** *cyaneimarmorella* (Millière, 1854)
Dryadaula Meyrick
A *pactolia* Meyrick, 1902
Morophaga Herrich-Schäffer
choragella ([Denis & Schiffermüller], 1775)
Triaxomera Zagulajev
parasitella (Hübner, 1796)
fulvimitrella (Sodoffsky, 1830)
Archinemapogon Zagulajev
yildizae Koçak, 1981
Nemaxera Zagulajev
betulinella (Paykull, 1785)
Nemapogon Schrank
granella (Linnaeus, 1758) **Corn Moth**
cloacella (Haworth, 1828) **Cork Moth**
inconditella (Lucas, 1956)
wolffiella Karsholt & Nielsen, 1976
variatella (Clemens, 1859)
ruricolella (Stainton, 1849)
clematella (Fabricius, 1781)
picarella (Clerck, 1759)
A *falstriella* (Bang-Haas, 1881)
Triaxomasia Zagulajev
caprimulgella (Stainton, 1851)
Cephimallota Bruand
[= *Cephitinea* Zagulajev]
† *crassiflavella* Bruand, 1851
* *colongella* Zagulajev, 1964
Trichophaga Ragonot
tapetzella (Linnaeus, 1758)
Tapestry Moth

A *mormopis* Meyrick, 1935
Tineola Herrich-Schäffer
  bisselliella (Hummel, 1823)
  Common Clothes Moth
Tinea Linnaeus
  columbariella Wocke, 1877
  pellionella Linnaeus, 1758 **Case-bearing Clothes Moth**
  A *lanella* Pierce & Metcalfe, 1934
  A *translucens* Meyrick, 1917
  dubiella Stainton, 1859
  flavescentella Haworth, 1828
  pallescentella Stainton, 1851
  Large Pale Clothes Moth
  semifulvella Haworth, 1828
  trinotella Thunberg, 1794
  A *fictrix* Meyrick, 1914
  A *murariella* Staudinger, 1859
Ceratophaga Petersen
  A *orientalis* (Stainton, 1878)
  A *haidarabadi* Zagulajev, 1966
Niditinea Petersen
  fuscella (Linnaeus, 1758) **Brown-dotted Clothes Moth**
  striolella (Matsumura, 1931)
Monopis Hübner
  laevigella ([Denis & Schiffermüller], 1775) **Skin Moth**
  weaverella (Scott, 1858)
  obviella ([Denis & Schiffermüller], 1775)
  crocicapitella (Clemens, 1859)
  imella (Hübner, [1813])
  monachella (Hübner, 1796)
  fenestratella (Heyden, 1863)
Setomorpha Zeller
  A *rutella* Zeller, 1852 Tropical Tobacco Moth
Lindera Blanchard
  A *tessellatella* Blanchard, 1852
Ateliotum Zeller
  A *insularis* (Rebel, 1896)
Haplotinea Diakonoff & Hinton
  A *ditella* (Pierce & Metcalfe, 1938)
  insectella (Fabricius, 1794)
Opogona Zeller
  A *sacchari* (Bojer, 1856)
  omoscopa (Meyrick, 1893) Detritus Moth
  A *antistacta* Meyrick, 1937
Oinophila Stephens
  v-flava (Haworth, 1828) Yellow V Moth
Euplocamus Latreille
  * *anthracinalis* (Scopoli, 1763)
Psychoides Bruand
  verhuella Bruand, 1853
  filicivora (Meyrick, 1937)

ROESLERSTAMMIIDAE
Roeslerstammia Zeller
  * ***pronubella*** ([Denis & Schiffermüller], 1775)
  erxlebella (Fabricius, 1787)

BUCCULATRICIDAE
Bucculatrix Zeller
  cristatella (Zeller, 1839)
  nigricomella (Zeller, 1839)
  maritima Stainton, 1851
  humiliella Herrich-Schäffer, 1855
  * *artemisiella* Herrich-Schäffer, 1855
  A *chrysanthemella* Rebel, 1896
  frangutella (Goeze, 1783)
  albedinella (Zeller, 1839)
  cidarella (Zeller, 1839)

thoracella (Thunberg, 1794)
ulmella Zeller, 1848
ulmifoliae Hering, 1931
bechsteinella (Bechstein & Scharfenberg, 1805)
demaryella (Duponchel, 1840)

GRACILLARIIDAE
Parectopa Clemens
  ononidis (Zeller, 1839)
Micrurapteryx Spuler
  * *kollariella* (Zeller, 1839)
Caloptilia Hübner
  cuculipennella (Hübner, 1796)
  populetorum (Zeller, 1839)
  elongella (Linnaeus, 1761)
  betulicola (Hering, 1928)
  rufipennella (Hübner, 1796)
  azaleella (Brants, 1913) **Azalea Leaf Miner**
  alchimiella (Scopoli, 1763)
  robustella Jäckh, 1972
  stigmatella (Fabricius, 1781)
  falconipennella (Hübner, [1813])
  semifascia (Haworth, 1828)
  hemidactylella ([Denis & Schiffermüller], 1775)
Gracillaria Haworth
  [= *Caloptilia* Hübner]
  syringella (Fabricius, 1794)
Aspilapteryx Spuler
  tringipennella (Zeller, 1839)
Euspilapteryx Stephens
  [= *Eucalybites* Kumata]
  auroguttella Stephens, 1835
Calybites Hübner
  phasianipennella (Hübner, [1813])
Povolnya Kuznetzov
  [= *Caloptilia* Hübner]
  leucapennella (Stephens, 1835)
Acrocercops Wallengren
  brongniardella (Fabricius, 1798)
Dialectica Walsingham
  * *imperialella* (Zeller, 1847)
  I *scalariella* (Zeller, 1850)
Leucospilapteryx Spuler
  omissella (Stainton, 1848)
Callisto Stephens
  denticulella (Thunberg, 1794)
  coffeella (Zetterstedt, 1839)
Parornix Spuler
  [= *Deltaornix* Kuznetzov]
  loganella (Stainton, 1848)
  betulae (Stainton, 1854)
  fagivora (Frey, 1861)
  carpinella (Frey, 1863)
  anglicella (Stainton, 1850)
  devoniella (Stainton, 1850)
  scoticella (Stainton, 1850)
  alpicola leucostola Pelham-Clinton, 1964
  finitimella (Zeller, 1850)
  torquillella (Zeller, 1850)
Phyllonorycter Hübner
  harrisella (Linnaeus, 1761)
  roboris (Zeller, 1839)
  heegeriella (Zeller, 1846)
  tenerella (de Joannis, 1915)
  kuhlweiniella (Zeller, 1839)
  quercifoliella (Zeller, 1839)
  messaniella (Zeller, 1846)
  platani (Staudinger, 1870)
  muellerella (Zeller, 1839)
  oxyacanthae (Frey, 1856)
  sorbi (Frey, 1855)

mespilella (Hübner, [1805])
blancardella (Fabricius, 1781)
hostis Triberti, 2007
  [= *cydoniella* ([Denis & Schiffermüller], 1775)]
junoniella (Zeller, 1846)
spinicolella (Zeller, 1846)
cerasicolella (Herrich-Schäffer, 1855)
lantanella (Schrank, 1802)
corylifoliella (Hübner, 1796)
leucographella (Zeller, 1850)
  Firethorn Leaf Miner
salictella (Zeller, 1846)
  [= *viminiella* (Sircom, 1848)]
viminetorum (Stainton, 1854)
salicicolella (Sircom, 1848)
dubitella (Herrich-Schäffer, 1855)
hilarella (Zetterstedt, 1839)
cavella (Zeller, 1846)
ulicicolella (Stainton 1851)
scoparella (Zeller, 1846)
staintoniella (Nicelli, 1853)
maestingella (Müller, 1764)
coryli (Nicelli, 1851) Nut Leaf Blister Moth
esperella (Goeze, 1783)
  [= *quinnata* (Geoffroy, 1785)]
strigulatella (Lienig & Zeller, 1846)
rajella (Linnaeus, 1758)
distentella (Zeller, 1846)
anderidae (Fletcher, 1885)
quinqueguttella (Stainton, 1851)
nigrescentella (Logan, 1851)
insignitella (Zeller, 1846)
lautella (Zeller, 1846)
schreberella (Fabricius, 1781)
ulmifoliella (Hübner, [1817])
emberizaepenella (Bouché, 1834)
scabiosella (Douglas, 1853)
tristrigella (Haworth, 1828)
stettinensis (Nicelli, 1852)
froelichiella (Zeller, 1839)
nicellii (Stainton, 1851)
klemannella (Fabricius, 1781)
trifasciella (Haworth, 1828)
acerifoliella (Zeller, 1839)
joannisi (Le Marchand, 1936)
  [= *platanoidella* (de Joannis, 1920)]
geniculella (Ragonot, 1874)
comparella (Duponchel, [1843])
sagitella (Bjerkander, 1790)
Cameraria Chapman
  ohridella Deschka & Dimic, 1986
  Horse-chestnut Leaf Miner
Phyllocnistis Zeller
  saligna (Zeller, 1839)
  ramulicola Langmaid & Corley, 2007
  unipunctella (Stephens, 1834)
  xenia Hering, 1936

YPONOMEUTIDAE
Scythropia Hübner
  crataegella (Linnaeus, 1767)
  Hawthorn Moth
Yponomeuta Latreille
  evonymella (Linnaeus, 1758) **Bird-cherry Ermine**
  padella (Linnaeus, 1758) **Orchard Ermine**
  malinellus Zeller, 1838 **Apple Ermine**
  cagnagella (Hübner, [1813])
  Spindle Ermine
  rorrella (Hübner, 1796) **Willow Ermine**

irorella (Hübner, 1796)
plumbella ([Denis & Schiffermüller], 1775)
sedella Treitschke, 1832
Euhyponomeuta Toll
 † *stannella* (Thunberg, 1788)
Zelleria Stainton
 hepariella Stainton, 1849
 l *oleastrella* (Millière, 1864)
Kessleria Nowicki
 † *fasciapennella* (Stainton, 1849)
 saxifragae (Stainton, 1868)
Pseudoswammerdamia Friese
 combinella (Hübner, 1786)
Swammerdamia Hübner
 caesiella (Hübner, 1796)
 passerella (Zetterstedt, 1839)
 pyrella (Villers, 1789)
 compunctella Herrich-Schäffer, 1855
Paraswammerdamia Friese
 albicapitella (Scharfenberg, 1805)
 nebulella (Goeze, 1783)
 [= *lutarea* (Haworth, 1828)]
Cedestis Zeller
 gysseleniella Zeller, 1839
 subfasciella (Stephens, 1834)
Ocnerostoma Zeller
 piniariella Zeller, 1847
 friesei Svensson, 1966

ARGYRESTHIIDAE
Argyresthia Hübner
 laevigatella Herrich-Schäffer, 1855
 * *illuminatella* Zeller, 1839
 glabratella (Zeller, 1847)
 praecocella Zeller, 1839
 arceuthina Zeller, 1839
 abdominalis Zeller, 1839
 dilectella Zeller, 1847
 aurulentella Stainton, 1849
 ivella (Haworth, 1828)
 trifasciata Staudinger, 1871
 cupressella Walsingham, 1890
 Cypress Tip Moth
 brockeella (Hübner, [1813])
 goedartella (Linnaeus, 1758)
 pygmaeella ([Denis & Schiffermüller], 1775)
 sorbiella (Treitschke, 1833)
 curvella (Linnaeus, 1761)
 retinella Zeller, 1839
 glaucinella Zeller, 1839
 spinosella Stainton, 1849
 conjugella Zeller, 1839 **Apple Fruit Moth**
 semifusca (Haworth, 1828)
 pruniella (Clerck, 1759) **Cherry Fruit Moth**
 bonnetella (Linnaeus, 1758)
 albistria (Haworth, 1828)
 semitestacella (Curtis, 1833)

PLUTELLIDAE
Plutella Schrank
 l *xylostella* (Linnaeus, 1758)
 Diamond-back Moth
 porrectella (Linnaeus, 1758)
 haasi Staudinger, 1883
Rhigognostis Zeller
 senilella (Zetterstedt, 1839)
 annulatella (Curtis, 1832)
 incarnatella (Steudel, 1873)
Eidophasia Stephens
 messingiella (Fischer von Röslerstamm, 1840)

GLYPHIPTERIGIDAE
Orthotelia Stephens
 sparganella (Thunberg, 1788)
Digitivalva Gaedike
 perlepidella (Stainton, 1849)
 pulicariae (Klimesch, 1956)
Acrolepiopsis Gaedike
 assectella (Zeller, 1839) **Leek Moth**
 betulella (Curtis, 1838)
 marcidella (Curtis, 1850)
Acrolepia Curtis
 autumnitella Curtis, 1838
Glyphipterix Hübner
 simpliciella (Stephens, 1834)
 Cocksfoot Moth
 schoenicolella Boyd, 1858
 equitella (Scopoli, 1763)
 forsterella (Fabricius, 1781)
 haworthana (Stephens, 1834)
 fuscoviridella (Haworth, 1828)
 thrasonella (Scopoli, 1763)

YPSOLOPHIDAE
Ypsolopha Latreille
 mucronella (Scopoli, 1763)
 nemorella (Linnaeus, 1758)
 dentella (Fabricius, 1775)
 Honeysuckle Moth
 † *asperella* (Linnaeus, 1761)
 scabrella (Linnaeus, 1761)
 horridella (Treitschke, 1835)
 lucella (Fabricius, 1775)
 alpella ([Denis & Schiffermüller], 1775)
 sylvella (Linnaeus, 1767)
 parenthesella (Linnaeus, 1761)
 ustella (Clerck, 1759)
 sequella (Clerck, 1759)
 vittella (Linnaeus, 1758)
Ochsenheimeria Hübner
 taurella ([Denis & Schiffermüller], 1775)
 urella Fischer von Röslerstamm, 1842
 vacculella Fischer von Röslerstamm, 1842 **Cereal Stem Moth**

PRAYDIDAE
Atemelia Herrich-Schäffer
 torquatella (Lienig & Zeller, 1846)
Prays Hübner
 fraxinella (Bjerkander, 1784) **Ash Bud Moth**
 A *citri* (Millière, 1873)
 peregrina Agassiz, 2007
 A *oleae* (Bernard, 1788)

HELIODINIDAE
Heliodines Stainton
 † *roesella* (Linnaeus, 1758)

BEDELLIIDAE
Bedellia Stainton
 somnulentella (Zeller, 1847)

LYONETIIDAE
Leucoptera Hübner
 laburnella (Stainton, 1851)
 Laburnum Leaf Miner
 spartifoliella (Hübner, [1813])
 lathyrifoliella (Stainton, 1866)
 orobi (Stainton, 1869)
 lotella (Stainton, 1859)
 malifoliella (Costa, 1836) **Pear Leaf Blister Moth**
 † *sinuella* Reutti, 1853
Lyonetia Hübner

 l *prunifoliella* (Hübner, 1796)
 clerkella (Linnaeus, 1758) **Apple Leaf Miner**

DOUGLASIIDAE
Tinagma Zeller
 ocnerostomella (Stainton,1850)
 balteolella (Fischer von Röslerstamm, [1841])

AUTOSTICHIDAE
Oegoconia Stainton
 quadripuncta (Haworth, 1828)
 deauratella (Herrich-Schäffer, 1854)
 caradjai Popescu-Gorg & Căpuşe, 1965
Symmoca Hübner
 A *signatella* Herrich-Schäffer, 1854

BLASTOBASIDAE
Blastobasis Zeller
 [= *Auximobasis* Walsingham]
 adustella Walsingham, 1894
 [= *lignea* Walsingham, 1894]
 lacticolella Rebel, 1940
 [= *decolorella* (Wollaston, 1858)]
 phycidella (Zeller, 1839)
 * *normalis* (Meyrick, 1918)
 rebeli Karsholt & Sinev, 2004
 vittata (Wollaston, 1858)
Hypatopa Walsingham
 binotella (Thunberg, 1794)

OECOPHORIDAE
Bisigna Toll
 procerella ([Denis & Schiffermüller], 1775)
Schiffermuelleria Hübner
 [= *Schiffermuellerina* Leraut]
 grandis (Desvignes, 1842)
Denisia Hübner
 subaquilea (Stainton, 1849)
 similella (Hübner, 1796)
 albimaculea (Haworth, 1828)
 † *augustella* (Hübner, 1796)
Metalampra Toll
 italica Baldizzone, 1977
Endrosis Hübner
 sarcitrella (Linnaeus, 1758) **White-shouldered House-moth**
Hofmannophila Spuler
 pseudospretella (Stainton, 1849)
 Brown House-moth
Borkhausenia Hübner
 fuscescens (Haworth, 1828)
 † *minutella* (Linnaeus, 1758)
Crassa Bruand
 [= *Batia* Stephens]
 tinctella (Hübner, 1796)
 unitella (Hübner, 1796)
Batia Stephens
 lunaris (Haworth, 1828)
 internella Jäckh, 1972
 lambdella (Donovan, 1793)
Epicallima Dyar
 † *formosella* ([Denis & Schiffermüller], 1775)
Esperia Hübner
 sulphurella (Fabricius, 1775)
Dasycera Stephens
 [= *Esperia* Hübner]
 oliviella (Fabricius, 1794)
Oecophora Latreille
 bractella (Linnaeus, 1758)
Alabonia Hübner
 geoffrella (Linnaeus, 1767)

Harpella Shrank
  A *forficella* (Scopoli, 1973)
Tachystola Meyrick
  acroxantha (Meyrick, 1885)
Pleurota Hübner
  bicostella (Clerck, 1759)
  CI *aristella* (Linnaeus, 1767)
Aplota Stephens
  palpella (Haworth, 1828)
Barea Walker
  asbolaea (Meyrick, 1884)

LYPUSIDAE
Pseudatemelia Rebel
  josephinae (Toll, 1956)
  flavifrontella ([Denis &
    Schiffermüller], 1775)
  subochreella (Doubleday, 1859)
Amphisbatis Zeller
  incongruella (Stainton, 1849)

CHIMABACHIDAE
Diurnea Haworth
  fagella ([Denis & Schiffermüller],
    1775)
  lipsiella ([Denis & Schiffermüller],
    1775)
Dasystoma Curtis
  salicella (Hübner, 1796)

PELEOPODIDAE
Carcina Hübner
  quercana (Fabricius, 1775)

ELACHISTIDAE
Perittia Stainton
  [= *Mendesia* de Joannis]
  obscurepunctella (Stainton, 1848)
  * *farinella* (Thunberg, 1794)
Stephensia Stainton
  brunnichella (Linnaeus, 1767)
Elachista Treitschke
  [= *Biselachista* Traugott-Olsen &
  Nielsen]
  [= *Cosmiotes* Clemens]
  regificella Sircom, 1849
  † *geminatella* (Herrich-Schäffer,
    1855)
  tengstromi Kaila, Bengtsson, Šulcs &
    Junnilainen, 2001
  gleichenella (Fabricius, 1781)
  biatomella (Stainton, 1848)
  poae Stainton, 1855
  atricomella Stainton, 1849
  kilmunella Stainton, 1849
  eskoi Kyrki & Karvonen, 1985
  alpinella Stainton, 1854
  luticomella Zeller, 1839
  albifrontella (Hübner, [1817])
  nobilella Zeller, 1839
  apicipunctella Stainton, 1849
  subnigrella Douglas, 1853
  orstadii Palm, 1943
  pomerana Frey, 1870
  humilis Zeller, 1850
  canapennella (Hübner, [1813])
  rufocinerea (Haworth, 1828)
  maculicerusella (Bruand, 1859)
  argentella (Clerck, 1759)
  triatomea (Haworth, 1828)
  collitella (Duponchel, [1843])
  subocellea (Stephens, 1834)
  triseriatella Stainton, 1854
  cahorsensis Traugott-Olsen, 1992
  bedellella (Sircom, 1848)
  littoricola Le Marchand, 1938

obliquella Stainton, 1854
  [= *megerlella* (Hübner, [1810])]
  cingillella (Herrich-Schäffer, 1855)
  unifasciella (Haworth, 1828)
  gangabella Zeller, 1850
  subalbidella Schläger, 1847
  adscitella Stainton, 1851
  bisulcella (Duponchel, [1843])
  trapeziella Stainton, 1849
  cinereopunctella (Haworth, 1828)
  serricornis Stainton, 1854
  scirpi Stainton, 1887
  eleochariella Stainton, 1851
  utonella Frey, 1856
  albidella Nylander, 1848
  freyerella (Hübner, [1825])
  consortella Stainton, 1851
  stabilella Stainton, 1858
Semioscopis Hübner
  avellanella (Hübner, [1793])
  steinkellneriana ([Denis &
    Schiffermüller], 1775)
Luquetia Leraut
  lobella ([Denis & Schiffermüller],
    1775)
Levipalpus Hannemann
  hepatariella (Lienig & Zeller, 1846)
Exaeretia Stainton
  ciniflonella (Lienig & Zeller, 1846)
  allisella Stainton, 1849
Agonopterix Hübner
  heracliana (Linnaeus, 1758)
  ciliella (Stainton, 1849)
  cnicella (Treitschke, 1832)
  purpurea (Haworth, 1811)
  subpropinquella (Stainton, 1849)
  putridella ([Denis & Schiffermüller],
    1775)
  nanatella (Stainton, 1849)
  alstromeriana (Clerck, 1759)
  propinquella (Treitschke, 1835)
  arenella ([Denis & Schiffermüller],
    1775)
  kuznetzovi Lvovsky, 1983
  kaekeritziana (Linnaeus, 1767)
  bipunctosa (Curtis, 1850)
  pallorella (Zeller, 1839)
  ocellana (Fabricius, 1775)
  assimilella (Treitschke, 1832)
  atomella ([Denis & Schiffermüller],
    1775)
  scopariella (Heinemann, 1870)
  umbellana (Fabricius, 1794)
  nervosa (Haworth, 1811)
  carduella (Hübner, [1817])
  liturosa (Haworth, 1811)
  conterminella (Zeller, 1839)
  curvipunctosa (Haworth, 1811)
  astrantiae (Heinemann, 1870)
  angelicella (Hübner, [1813])
  yeatiana (Fabricius, 1781)
  capreolella (Zeller, 1839)
  rotundella (Douglas, 1846)
Depressaria Haworth
  † *discipunctella* Herrich-Schäffer,
    1854
  daucella ([Denis & Schiffermüller],
    1775)
  ultimella Stainton, 1849
  radiella (Goeze, 1783) **Parsnip
  Moth**
    [= *heracli* Retzius, 1783]
    [= *pastinacella* (Duponchel, 1838)]
  pimpinellae Zeller, 1839
  badiella (Hübner, 1796)

  pulcherrimella Stainton, 1849
  douglasella Stainton, 1849
  sordidatella Tengström, [1848]
  * *emeritella* Stainton, 1849
  albipunctella ([Denis &
    Schiffermüller], 1775)
  olerella Zeller, 1854
  chaerophylli Zeller, 1839
  † *depressana* (Fabricius, 1775)
  silesiaca Heinemann, 1870
Telechrysis Toll
  tripuncta (Haworth, 1828)
Hypercallia Stephens
  citrinalis (Scopoli, 1763)
Ethmia Hübner
  terminella Fletcher, 1938
  dodecea (Haworth, 1828)
  quadrillella (Goeze, 1783)
  bipunctella (Fabricius, 1775)
  * *pusiella* (Linnaeus, 1758)
  pyrausta (Pallas, 1771)
Blastodacna Wocke
  hellerella (Duponchel, 1838)
  atra (Haworth, 1828) **Apple Pith
  Moth**
Spuleria Hofmann
  flavicaput (Haworth, 1828)
Dystebenna Spuler
  stephensi (Stainton, 1849)
Chrysoclista Stainton
  lathamella Fletcher, 1936
  linneella (Clerck, 1759)

STATHMOPODIDAE
Stathmopoda Herrich-Schäffer
  pedella (Linnaeus, 1761)
  A *diplaspis* Meyrick, 1887
  A *auriferella* (Walker, 1864)

BATRACHEDRIDAE
Batrachedra Herrich-Schäffer
  praeangusta (Haworth, 1828)
  pinicolella (Zeller, 1839)
  I(?) *parvulipuncta* Chrétien, 1915

COLEOPHORIDAE
Augasma Herrich-Schäffer
  † *aeratella* (Zeller, 1839)
Metriotes Herrich-Schäffer
  lutarea (Haworth, 1828)
Goniodoma Zeller
  limoniella (Stainton, 1884)
Coleophora Hübner
  † *albella* (Thunberg, 1788)
  lutipennella (Zeller, 1838)
  gryphipennella (Hübner, 1796)
  flavipennella (Duponchel, [1843])
  serratella (Linnaeus, 1761)
  coracipennella (Hübner, 1796)
  prunifoliae Doets, 1944
  spinella (Schrank, 1802) **Apple &
  Plum Case-bearer**
  milvipennis Zeller, 1839
  adjectella Herrich-Schäffer, 1861
  badiipennella (Duponchel, [1843])
  alnifoliae Barasch, 1934
  limosipennella (Duponchel, [1843])
  hydrolapathella Hering, 1921
  siccifolia Stainton, 1856
  trigeminella Fuchs, 1881
  fuscocuprella Herrich-Schäffer,
    1855
  lusciniaepennella (Treitschke, 1833)
  idaeella Hofmann, 1869
  vitisella Gregson, 1856
  glitzella Hofmann, 1869

arctostaphyli Meder, 1934
violacea (Ström, 1783)
juncicolella Stainton, 1851
orbitella Zeller, 1849
binderella (Kollar, 1832)
potentillae Elisha, 1885
ahenella Heinemann, [1876]
albitarsella Zeller, 1849
trifolii (Curtis, 1832)
alcyonipennella (Kollar, 1832)
frischella (Linnaeus, 1758) **Clover
  Case-bearer**
mayrella (Hübner, [1813])
deauratella Lienig & Zeller, 1846
amethystinella Ragonot, 1885
  [= *fuscicornis* Zeller, 1847]
conyzae Zeller, 1868
calycotomella Stainton, 1869
lineolea (Haworth, 1828)
hemerobiella (Scopoli, 1763)
lithargyrinella Zeller, 1849
laricella (Hübner, [1817]) **Larch
  Case-bearer**
chalcogrammella Zeller, 1839
tricolor Walsingham, 1899
lixella Zeller, 1849
ochrea (Haworth, 1828)
albidella ([Denis & Schiffermüller],
  1775)
anatipennella (Hübner, 1796)
  Pistol Case-bearer
currucipennella Zeller, 1839
ibipennella Zeller, 1849
betulella Heinemann, [1876]
kuehnella (Goeze, 1783)
  [= *palliatella* (Zincken, 1813)]
vibicella (Hübner, [1813])
conspicuella Zeller, 1849
† *vibicigerella* Zeller, 1839
pyrrhulipennella Zeller, 1839
serpylletorum Hering, 1889
vulnerariae Zeller, 1839
albicosta (Haworth, 1828)
saturatella Stainton, 1850
genistae Stainton, 1857
discordella Zeller, 1849
niveicostella Zeller, 1839
pennella ([Denis & Schiffermüller],
  1775)
solitariella Zeller, 1849
silenella Herrich-Schäffer, 1855
galbulipennella Zeller, 1838
striatipennella Nylander, 1848
inulae Wocke, [1876]
follicularis (Vallot, 1802)
trochilella (Duponchel, [1843])
linosyridella Fuchs, 1880
gardesanella Toll, 1953
ramosella Zeller, 1849
peribenanderi Toll, 1943
paripennella Zeller, 1839
therinella Tengström, [1848]
asteris Mühlig, 1864
argentula (Stephens, 1834)
virgaureae Stainton, 1857
  [= *obscenella* Herrich-Schäffer,
  1855]
saxicolella (Duponchel, [1843])
sternipennella (Zetterstedt, 1839)
adspersella Benander, 1939
versurella Zeller, 1849
squamosella Stainton, 1856
pappiferella Hofmann, 1869
vestianella (Linnaeus, 1758)
atriplicis Meyrick, [1928]

deviella Zeller, 1847
aestuariella Bradley, 1984
salinella Stainton, 1859
albicans Zeller, 1849
  [= *artemisiella* Scott, 1861]
artemisicolella Bruand, [1855]
otidipennella (Hübner, [1817])
† *antennariella* Herrich-Schäffer,
  1861
sylvaticella Wood, 1892
lassella Staudinger, 1859
taeniipennella Herrich-Schäffer,
  1855
glaucicolella Wood, 1892
tamesis Waters, 1929
alticolella Zeller, 1849
maritimella Newman, 1873
adjunctella Hodgkinson, 1882
caespititiella Zeller, 1839
salicorniae Heinemann & Wocke,
  [1876]
clypeiferella Hofmann, 1871
wockeella Zeller, 1849

MOMPHIDAE
Mompha Hübner
  miscella ([Denis & Schiffermüller],
  1775)
  langiella (Hübner, 1796)
  terminella (Humphreys &
  Westwood, 1845)
  locupletella ([Denis &
  Schiffermüller], 1775)
  raschkiella (Zeller, 1839)
  conturbatella (Hübner, [1819])
  ochraceella (Curtis, 1839)
  lacteella (Stephens, 1834)
  propinquella (Stainton, 1851)
  divisella Herrich-Schäffer, 1854
  bradleyi Riedl, 1965
  jurassicella (Frey, 1881)
  sturnipennella (Treitschke, 1833)
  subbistrigella (Haworth, 1828)
  epilobiella ([Denis &
  Schiffermüller], 1775)

SCYTHRIDIDAE
Scythris Hübner
  grandipennis (Haworth, 1828)
† *fuscoaenea* (Haworth, 1828)
  fallacella (Schläger, 1847)
  crassiuscula (Herrich-Schäffer,
  1855)
  picaepennis (Haworth, 1828)
  siccella (Zeller, 1839)
  empetrella Karsholt & Nielsen, 1976
  limbella (Fabricius, 1775)
  cicadella (Zeller, 1839)
  potentillella (Zeller, 1847)
  inspersella (Hübner, [1817])
  A *sinensis* (Felder & Rogenhofer,
  1875)

COSMOPTERIGIDAE
Pancalia Stephens
  leuwenhoekella (Linnaeus, 1761)
  schwarzella (Fabricius, 1798)
Euclemensia Grote
† *woodiella* (Curtis, 1830)
Limnaecia Stainton
  phragmitella Stainton, 1851
Cosmopterix Hübner
  zieglerella (Hübner, [1810])
† *schmidiella* Frey, 1856
  orichalcea Stainton, 1861
  scribaiella Zeller, 1850

pulchrimella Chambers, 1875
  lienigiella Zeller, 1846
Pyroderces Herrich-Schäffer
  Cl *argyrogrammos* (Zeller, 1847)
Anatrachyntis Meyrick
  A *badia* (Hodges, 1962)
  A *simplex* (Walsingham, 1891)
Sorhagenia Spuler
  rhamniella (Zeller, 1839)
  lophyrella (Douglas, 1846)
  janiszewskae Riedl, 1962

GELECHIIDAE
Aristotelia Hübner
† *subdecurtella* (Stainton, 1859)
  ericinella (Zeller, 1839)
  brizella (Treitschke, 1833)
Chrysoesthia Hübner
  drurella (Fabricius, 1775)
  sexguttella (Thunberg, 1794)
Xystophora Wocke
  pulveratella (Herrich-Schäffer, 1854)
Isophrictis Meyrick
  striatella ([Denis & Schiffermüller],
  1775)
Metzneria Zeller
  littorella (Douglas, 1850)
  lappella (Linnaeus, 1758)
  aestivella (Zeller, 1839)
  metzneriella (Stainton, 1851)
  neuropterella (Zeller, 1839)
  aprilella (Herrich-Schäffer, 1854)
Apodia Heinemann
  bifractella (Duponchel, [1843])
Ptocheuusa Heinemann
  paupella (Zeller, 1847)
Psamathocrita Meyrick
† *osseella* (Stainton, [1860])
  argentella Pierce & Metcalfe, 1942
Argolamprotes Benander
  micella ([Denis & Schiffermüller],
  1775)
Monochroa Heinemann
  cytisella (Curtis, 1837)
  tenebrella (Hübner, [1817])
  lucidella (Stephens, 1834)
  palustrellus (Douglas, 1850)
  tetragonella (Stainton, 1885)
  conspersella (Herrich-Schäffer,
  1854)
  hornigi (Staudinger, 1883)
  niphognatha (Gozmány, 1953)
  suffusella (Douglas, 1850)
  lutulentella (Zeller, 1839)
  elongella (Heinemann, 1870)
  arundinetella (Boyd, 1857)
  moyses Uffen, 1991
  divisella (Douglas, 1850)
Eulamprotes Bradley
  atrella ([Denis & Schiffermüller],
  1775)
  immaculatella (Douglas, 1850)
  unicolorella (Duponchel, [1843])
  wilkella (Linnaeus, 1758)
Bryotropha Heinemann
  basaltinella (Zeller, 1839)
  dryadella (Zeller, 1850)
  umbrosella (Zeller, 1839)
  affinis (Haworth, 1828)
  similis (Stainton, 1854)
  senectella (Zeller, 1839)
  boreella (Douglas, 1851)
  galbanella (Zeller, 1839)
  * *figulella* (Staudinger, 1859)
  desertella (Douglas, 1850)

terrella ([Denis & Schiffermüller],
1775)
politella (Stainton, 1851)
domestica (Haworth, 1828)
Recurvaria Haworth
nanella ([Denis & Schiffermüller],
1775)
leucatella (Clerck, 1759)
Coleotechnites Chambers
piceaella (Kearfott, 1903)
Exotelia Wallengren
dodecella (Linnaeus, 1758)
Stenolechia Meyrick
gemmella (Linnaeus, 1758)
Parachronistis Meyrick
albiceps (Zeller, 1839)
Teleiodes Sattler
vulgella ([Denis & Schiffermüller],
1775)
wagae (Nowicki, 1860)
luculella (Hübner, [1813])
flavimaculella (Herrich-Schäffer,
1854)
sequax (Haworth, 1828)
Teleiopsis Sattler
diffinis (Haworth, 1828)
Carpatolechia Câpuse
decorella (Haworth, 1812)
notatella (Hübner, [1813])
proximella (Hübner, 1796)
alburnella (Zeller, 1839)
fugitivella (Zeller, 1839)
Pseudotelphusa Janse
scalella (Scopoli, 1763)
paripunctella (Thunberg, 1794)
Xenolechia Meyrick
aethiops (Humphreys & Westwood,
1845)
Altenia Sattler
scriptella (Hübner, 1796)
Gelechia Hübner
rhombella ([Denis &
Schiffermüller], 1775)
scotinella Herrich-Schäffer, 1854
senticetella (Staudinger, 1859)
sabinellus (Zeller, 1839)
sororculella (Hübner, [1817])
muscosella Zeller, 1839
cuneatella Douglas, 1852
hippophaella (Schrank, 1802)
nigra (Haworth, 1828)
turpella ([Denis & Schiffermüller],
1775)
Psoricoptera Stainton
gibbosella (Zeller, 1839)
Mirificarma Gozmány
mulinella (Zeller, 1839)
lentiginosella (Zeller, 1839)
Sophronia Hübner
semicostella (Hübner, [1813])
† *humerella* ([Denis &
Schiffermüller], 1775)
Chionodes Hübner
fumatella (Douglas, 1850)
distinctella (Zeller, 1839)
Aroga Busck
velocella (Duponchel, 1838)
Neofriseria Sattler
peliella (Treitschke, 1835)
singula (Staudinger, 1876)
Prolita Leraut
sexpunctella (Fabricius, 1794)
solutella (Zeller, 1839)
Athrips Billberg
tetrapunctella (Thunberg, 1794)

rancidella (Herrich-Schäffer, 1854)
Cotoneaster Webworm
mouffetella (Linnaeus, 1758)
Gnorimoschema Busck
† *streliciella* (Herrich-Schäffer, 1854)
Tuta Kieffer & Jørgensen
A *absoluta* (Meyrick, 1917)
Scrobipalpa Janse
acuminatella (Sircom, 1850)
pauperella (Heinemann, 1870)
murinella (Duponchel, [1843])
suaedella (Richardson, 1893)
salinella (Zeller, 1847)
instabilella (Douglas, 1846)
nitentella (Fuchs, 1902)
obsoletella (Fischer von
Röslerstamm, [1841])
atriplicella (Fischer von Röslerstamm,
[1841])
ocellatella (Boyd, 1858) Beet Moth
samadensis (Pfaffenzeller, 1870)
[= *samadensis plantaginella*
(Stainton, 1883)]
artemisiella (Treitschke, 1833)
Thyme Moth
† *stangei* (Hering, 1889)
clintoni Povolný, 1968
costella (Humphreys & Westwood,
1845)
Scrobipalpula Povolný
† *diffluella* (Frey, 1870)
tussilaginis (Stainton, 1867)
Phthorimaea Meyrick
A *operculella* (Zeller, 1873) Potato
Tuber Moth
Caryocolum Gregor & Povolný
vicinella (Douglas, 1851)
alsinella (Zeller, 1868)
viscariella (Stainton, 1855)
marmorea (Haworth, 1828)
fraternella (Douglas, 1851)
proxima (Haworth, 1828)
blandella (Douglas, 1852)
blandelloides Karsholt, 1981
junctella (Douglas, 1851)
tricolorella (Haworth, 1812)
blandulella (Tutt, 1887)
kroesmanniella (Herrich-Schäffer,
1854)
† *huebneri* (Haworth, 1828)
Syncopacma Meyrick
sangiella (Stainton, 1863)
larseniella Gozmány, 1957
cinctella (Clerck, 1759)
taeniolella (Zeller, 1839)
albifrontella (Heinemann, 1870)
† *vinella* (Bankes, 1898)
albipalpella (Herrich-Schäffer,
1854)
suecicella (Wolff, 1958)
l *polychromella* (Rebel, 1902)
Aproaerema Durrant
anthyllidella (Hübner, [1813])
Anacampsis Curtis
temerella (Lienig & Zeller, 1846)
populella (Clerck, 1759)
blattariella (Hübner, 1796)
Mesophleps Hübner
† *silacella* (Hübner, 1796)
Anarsia Zeller
spartiella (Schrank, 1802)
A *lineatella* Zeller, 1839 Peach Twig
Borer
Hypatima Hübner
rhomboidella (Linnaeus, 1758)

Nothris Hübner
† *verbascella* ([Denis &
Schiffermüller], 1775)
congressariella (Bruand, 1858)
Neofaculta Gozmány
ericetella (Geyer, [1832])
Dichomeris Hübner
marginella (Fabricius, 1781)
Juniper Webber
juniperella (Linnaeus, 1761)
ustalella (Fabricius, 1794)
† *derasella* ([Denis & Schiffermüller],
1775)
alacella (Zeller, 1839)
Brachmia Hübner
blandella (Fabricius, 1798)
inornatella (Douglas, 1850)
Helcystogramma Zeller
rufescens (Haworth, 1828)
lutatella (Herrich-Schäffer, 1854)
Acompsia Hübner
cinerella (Clerck, 1759)
schmidtiellus (Heyden, 1848)
Pexicopia Common
malvella (Hübner, [1805])
Hollyhock Seed Moth
Platyedra Meyrick
subcinerea (Haworth, 1828)
Sitotroga Heinemann
A *cerealella* (Olivier, 1789)
Angoumois Grain Moth
Thiotricha Meyrick
subocellea (Stephens, 1834)

ALUCITIDAE
Alucita Linnaeus
hexadactyla Linnaeus, 1758
Tweny-plume Moth

PTEROPHORIDAE
Agdistis Hübner
meridionalis (Zeller, 1847) **Cliff
Plume**
bennetii (Curtis, 1833) **Saltmarsh
Plume**
Cl *tamaricis* (Zeller, 1847) Tamarisk
Plume
Platyptilia Hübner
Ir *tesseradactyla* (Linnaeus, 1761)
Irish Plume
calodactyla ([Denis &
Schiffermüller], 1775) **Goldenrod
Plume**
gonodactyla ([Denis &
Schiffermüller], 1775) **Triangle
Plume**
isodactylus (Zeller, 1852) **Hoary
Plume**
Gillmeria Tutt
[= Platyptilia Hübner]
ochrodactyla ([Denis &
Schiffermüller], 1775) **Tansy
Plume**
pallidactyla (Haworth, 1811)
Yarrow Plume
Amblyptilia Hübner
acanthadactyla (Hübner, [1813])
Beautiful Plume
punctidactyla (Haworth, 1811)
Brindled Plume
Stenoptilia Hübner
† *pneumonanthes* (Büttner, 1880)
Gentian Plume
millieridactyla (Bruand, 1861)
Saxifrage Plume

zophodactylus (Duponchel, 1840)
Dowdy Plume
bipunctidactyla (Scopoli, 1763)
Twin-spot Plume
islandicus (Staudinger, 1857)
Mountain Plume
annadactyla Sutter, 1988 Small
Scabious Plume
inopinata Bigot & Picard, 2002
Scarce Plume
[= *aridus* (Zeller, 1847)]
scabiodactylus (Gregson, 1869)
Gregson's Plume
pterodactyla (Linnaeus, 1761)
Brown Plume
Cnaemidophorus Wallengren
rhododactyla ([Denis &
Schiffermüller], 1775) **Rose Plume**
Marasmarcha Meyrick
lunaedactyla (Haworth, 1811)
Crescent Plume
Oxyptilus Zeller
† *pilosellae* (Zeller, 1841) Downland
Plume
parvidactyla (Haworth, 1811)
Small Plume
Crombrugghia Tutt
[= *Oxyptilus* Zeller]
distans (Zeller, 1847) **Breckland
Plume**
l *laetus* (Zeller, 1847) **Scarce Light
Plume**
Capperia Tutt
britanniodactylus (Gregson, 1869)
Wood Sage Plume
Buckleria Tutt
paludum (Zeller, 1839) **Sundew
Plume**
Pterophorus Schäffer
pentadactyla (Linnaeus, 1758)
White Plume Moth
Porrittia Tutt
[= *Pterophorus* Schäffer]
galactodactyla ([Denis &
Schiffermüller], 1775) **Spotted
White Plume**
Merrifieldia Tutt
leucodactyla ([Denis &
Schiffermüller], 1775) **Thyme
Plume**
tridactyla (Linnaeus, 1758)
Western Thyme Plume
[= *tridactyla phillipsi* Huggins,
1955]
baliodactylus (Zeller, 1841) **Dingy
White Plume**
Wheeleria Tutt
[= *Pterophorus* Schäffer]
spilodactylus (Curtis, 1827)
Horehound Plume
Pselnophorus Wallengren
heterodactyla (Müller, 1764)
Short-winged Plume
Oidaematophorus Wallengren
lithodactyla (Treitschke, 1833)
Dusky Plume
Hellinsia Tutt
[= *Ovendenia* Tutt]
[= *Euleioptilus* Bigot & Picard]
lienigianus (Zeller, 1852) **Mugwort
Plume**
osteodactylus (Zeller, 1841) **Small
Goldenrod Plume**

chrysocomae (Ragonot, 1875)
Scarce Goldenrod Plume
tephradactyla (Hübner, [1813])
Plain Plume
carphodactyla (Hübner, [1813])
Citron Plume
Adaina Tutt
microdactyla (Hübner, [1813])
Hemp-agrimony Plume
Emmelina Tutt
monodactyla (Linnaeus, 1758)
Common Plume
argoteles (Meyrick, 1922)
Reedbed Plume

SCHRECKENSTEINIIDAE
Schreckensteinia Hübner
festaliella (Hübner, [1819])

EPERMENIIDAE
Phaulernis Meyrick
dentella (Zeller, 1839)
fulviguttella (Zeller, 1839)
Epermenia Hübner
farreni (Walsingham, 1894)
profugella (Stainton, 1856)
falciformis (Haworth, 1828)
insecurella (Stainton, 1854)
chaerophyllella (Goeze, 1783)
aequidentellus (Hofmann, 1867)

CHOREUTIDAE
Anthophila Haworth
fabriciana (Linnaeus, 1767) **Nettle-
tap**
Prochoreutis Diakonoff & Heppner
sehestediana (Fabricius, 1776)
myllerana (Fabricius, 1794)
Tebenna Billberg
l *micalis* (Mann, 1857)
Choreutis Hübner
pariana (Clerck, 1759) **Apple Leaf
Skeletonizer**
diana (Hübner, [1822])

TORTRICIDAE
Phtheochroa Stephens
inopiana (Haworth, 1811)
schreibersiana (Frölich, 1828)
sodaliana (Haworth, 1811)
rugosana (Hübner, [1799])
Hysterophora Obraztsov
maculosana (Haworth, 1811)
Cochylimorpha Razowski
alternana (Stephens, 1834)
straminea (Haworth, 1811)
Phalonidia Le Marchand
manniana (Fischer von
Röslerstamm, 1839)
affinitana (Douglas, 1846)
gilvicomana (Zeller, 1847)
curvistrigana (Stainton, 1859)
Gynnidomorpha Turner
minimana (Caradja, 1916)
permixtana ([Denis &
Schiffermüller], 1775)
vectisana (Humphreys &
Westwood, 1845)
alismana (Ragonot, 1883)
luridana (Gregson, 1870)
Agapeta Hübner
hamana (Linnaeus, 1758)
zoegana (Linnaeus, 1767)
Eugnosta Hübner
* *lathoniana* (Hübner, [1800])
Eupoecilia Stephen

angustana (Hübner, [1799])
angustana thuleana Vaughan,
1880
ambiguella (Hübner, 1796) **Vine
Moth**
Commophila Hübner
aeneana (Hübner, [1800])
Aethes Billberg
tesserana ([Denis & Schiffermüller],
1775)
rutilana (Hübner, [1817])
hartmanniana (Clerck, 1759)
piercei Obraztsov, 1952
† *margarotana* (Duponchel, 1836)
williana (Brahm, 1791)
cnicana (Westwood, 1854)
rubigana (Treitschke, 1830)
smeathmanniana (Fabricius, 1781)
margaritana (Haworth, 1811)
dilucidana (Stephens, 1852)
francillana (Fabricius, 1794)
fennicana (Hering, 1924)
beatricella (Walsingham, 1898)
l *bilbaensis* (Rössler, 1877)
Cochylidia Obraztsov
implicitana (Wocke, 1856)
heydeniana (Herrich-Schäffer,
1851)
subroseana (Haworth, 1811)
rupicola (Curtis, 1834)
Cochylis Treitschke
roseana (Haworth, 1811)
flaviciliana (Westwood, 1854)
dubitana (Hübner, [1799])
molliculana Zeller, 1847
hybridella (Hübner, [1813])
atricapitana (Stephens, 1852)
pallidana Zeller, 1847
nana (Haworth, 1811)
Falseuncaria Obraztsov & Swatschek
ruficiliana (Haworth, 1811)
degreyana (McLachlan, 1869)
Spatalistis Meyrick
bifasciana (Hübner, [1787])
Tortrix Linnaeus
viridana Linnaeus, 1758 **Green
Oak Tortrix**
Aleimma Hübner
loeflingiana (Linnaeus, 1758)
Acleris Hübner
bergmanniana (Linnaeus, 1758)
forsskaleana (Linnaeus, 1758)
holmiana (Linnaeus, 1758)
laterana (Fabricius, 1794)
comariana (Lienig & Zeller, 1846)
Strawberry Tortrix
caledoniana (Stephens, 1852)
sparsana ([Denis & Schiffermüller],
1775)
rhombana ([Denis & Schiffermüller],
1775) **Rhomboid Tortrix**
aspersana (Hübner, [1817])
ferrugana ([Denis & Schiffermüller],
1775)
notana (Donovan, 1806)
shepherdana (Stephens, 1852)
schalleriana (Linnaeus, 1761)
variegana ([Denis & Schiffermüller],
1775) **Garden Rose Tortrix**
permutana (Duponchel, 1836)
kochiella (Goeze, 1783)
[= *boscana* (Fabricius, 1794)]
logiana (Clerck, 1759)
umbrana (Hübner, [1799])

hastiana (Linnaeus, 1758)
cristana ([Denis & Schiffermüller], 1775)
hyemana (Haworth, 1811)
lipsiana ([Denis & Schiffermüller], 1775)
rufana ([Denis & Schiffermüller], 1775)
lorquiniana (Duponchel, 1835)
abietana (Hübner, [1822])
maccana (Treitschke, 1835)
literana (Linnaeus, 1758)
emargana (Fabricius, 1775)
effractana (Hübner, [1799])
Neosphaleroptera Réal
  nubilana (Hübner, [1799])
Exapate Hübner
  congelatella (Clerck, 1759)
Tortricodes Guenée
  alternella ([Denis & Schiffermüller], 1775)
Eana Billberg
  argentana (Clerck, 1759)
  osseana (Scopoli, 1763)
  incanana (Stephens, 1852)
  penziana (Thunberg, 1791)
    [= *penziana bellana* (Curtis, 1826)]
  penziana colquhounana (Barrett, 1884)
Cnephasia Curtis
  longana (Haworth, 1811)
  A *gueneana* (Duponchel, 1836)
  communana (Herrich-Schäffer, 1851)
  conspersana Douglas, 1846
  stephensiana (Doubleday, 1849) **Grey Tortrix**
  asseclana ([Denis &Schiffermüller], 1775) **Flax Tortrix**
  pasiuana (Hübner, [1799])
  pumicana (Zeller, 1847)
  genitalana Pierce & Metcalfe, 1922
  incertana (Treitschke, 1835) **Light Grey Tortrix**
Sparganothis Hübner
  pilleriana ([Denis & Schiffermüller], 1775)
Platynota Clemens
  A *rostrana* (Walker, 1863)
Eulia Hübner
  ministrana (Linnaeus, 1758)
Pseudargyrotoza Obraztsov
  conwagana (Fabricius, 1775)
Ditula Stephens
  angustiorana (Haworth, 1811) **Red-barred Tortrix**
Epagoge Hübner
  grotiana (Fabricius, 1781)
Paramesia Stephens
  gnomana (Clerck, 1759)
Periclepsis Bradley
  cinctana ([Denis & Schiffermüller], 1775)
Philedone Hübner
  gerningana ([Denis & Schiffermüller], 1775)
Capua Stephens
  vulgana (Frölich, 1828)
Philedonides Obraztsov
  lunana (Thunberg, 1784)
Homona Walker
  A *coffearia* (Nietner, 1861) Camellia Tortrix
    [= *menciana* (Walker, 1863)]
Archips Hübner

oporana (Linnaeus, 1758)
podana (Scopoli, 1763) **Large Fruit-tree Tortrix**
† *betulana* (Hübner, 1787)
crataegana (Hübner, [1799]) **Brown Oak Tortrix**
xylosteana (Linnaeus, 1758) **Variegated Golden Tortrix**
rosana (Linnaeus, 1758) **Rose Tortrix**
A *argyrospila* (Walker, 1863)
A *semiferanus* (Walker, 1863)
Choristoneura Lederer
  diversana (Hübner, [1817])
  hebenstreitella (Müller, 1764)
  † *lafauryana* (Ragonot, 1875)
Argyrotaenia Stephens
  ljungiana (Thunberg, 1797)
Ptycholomoides Obraztsov
  aeriferana (Herrich-Schäffer, 1851)
Ptycholoma Stephens
  lecheana (Linnaeus, 1758)
Pandemis Hübner
  corylana (Fabricius, 1794) **Chequered Fruit-tree Tortrix**
  cerasana (Hübner, 1786) **Barred Fruit-tree Tortrix**
  cinnamomeana (Treitschke, 1830)
  heparana ([Denis & Schiffermüller], 1775) **Dark Fruit-tree Tortrix**
  dumetana (Treitschke, 1835)
Syndemis Hübner
  musculana (Hübner, [1799])
  musculana musculinana (Kennel, 1899)
Lozotaenia Stephens
  forsterana (Fabricius, 1781)
  * *subocellana* (Stephens, 1834)
Cacoecimorpha Obraztsov
  pronubana (Hübner, [1799]) **Carnation Tortrix**
Aphelia Hübner
  viburnana ([Denis & Schiffermüller], 1775) **Bilberry Tortrix**
  paleana (Hübner, [1793]) **Timothy Tortrix**
  unitana (Hübner, [1799])
Dichelia Guenée
  A *histrionana* (Frölich, 1828)
Clepsis Guenée
  senecionana (Hübner, [1819])
  rurinana (Linnaeus, 1758)
  spectrana (Treitschke, 1830) **Cyclamen Tortrix**
  A? *coriacana* (Rebel, 1894)
  consimilana (Hübner, [1817])
  * *trileucana* (Doubleday, 1847)
  * *melaleucanus* (Walker, 1863)
Epiphyas Turner
  postvittana (Walker, 1863) **Light Brown Apple Moth**
Lozotaeniodes Obraztsov
  formosana (Frölich, 1830)
Adoxophyes Meyrick
  orana (Fischer von Röslerstamm, 1834) **Summer Fruit Tortrix**
  A *privatana* (Walker, 1863)
Epichoristodes Diakonoff
  A *acerbella* (Walker, 1864) African Carnation Tortrix
Olindia Guenée
  schumacherana (Fabricius, 1787)
Isotrias Meyrick
  rectifasciana (Haworth, 1811)

Bactra Stephens
  furfurana (Haworth, 1811)
  lancealana (Hübner, [1799])
  lacteana Caradja, 1916
  robustana (Christoph, 1872)
  Ir *venosana* (Zeller, 1847)
Endothenia Stephens
  gentianaeana (Hübner, [1799])
  oblongana (Haworth, 1811)
  marginana (Haworth, 1811)
  pullana (Haworth, 1811)
  ustulana (Haworth, 1811)
  nigricostana (Haworth, 1811)
  ericetana (Humphreys & Westwood, 1845)
  quadrimaculana (Haworth, 1811)
Eudemis Hübner
  profundana ([Denis & Schiffermüller], 1775)
  porphyrana (Hübner, [1799])
Pseudosciaphila Obraztsov
  branderiana (Linnaeus, 1758)
Apotomis Hübner
  semifasciana (Haworth, 1811)
  infida (Heinrich, 1926)
  lineana ([Denis & Schiffermüller], 1775)
  turbidana Hübner, [1825]
  betuletana (Haworth, 1811)
  capreana (Hübner, [1817])
  sororculana (Zetterstedt, 1839)
  sauciana (Frölich, 1828)
  sauciana grevillana (Curtis, 1835)
Orthotaenia Stephens
  undulana ([Denis & Schiffermüller], 1775)
Hedya Hübner
  pruniana (Hübner, [1799]) **Plum Tortrix**
  nubiferana (Haworth, 1811) **Marbled Orchard Tortrix**
  ochroleucana (Frölich, 1828)
  salicella (Linnaeus, 1758)
Metendothenia Diakonoff
  atropunctana (Zetterstedt, 1839)
Celypha Hübner
  striana ([Denis & Schiffermüller], 1775)
  rosaceana Schläger, 1847
  rufana (Scopoli, 1763)
  woodiana (Barrett, 1882)
  cespitana (Hübner, [1817])
  rurestrana (Duponchel, [1843])
  lacunana ([Denis & Schiffermüller], 1775)
  † *doubledayana* (Barrett, 1872)
  rivulana (Scopoli, 1763)
  aurofasciana (Haworth, 1811)
Phiaris Hübner
  [= *Olethreutes* Hübner]
  metallicana (Hübner, [1799])
  schulziana (Fabricius, 1776)
  palustrana (Lienig & Zeller, 1846)
  micana ([Denis & Schiffermüller], 1775)
    [= *olivana* (Treitschke, 1830)]
  obsoletana (Zetterstedt, 1839)
Pristerognatha Obraztsov
  † *penthinana* (Guenée, 1845)
Cymolomia Lederer
  I *hartigiana* (Saxesen, 1840) Plumbeous Spruce Tortrix
Argyroploce Hübner
  [= *Olethreutes* Hübner]
  arbutella (Linnaeus, 1758)

Stictea Guenée
[= *Olethreutes* Hübner]
  mygindiana ([Denis &
    Schiffermüller], 1775)
Olethreutes Hübner
  arcuella (Clerck, 1759)
Piniphila Falkovitsh
  bifasciana (Haworth, 1811)
Lobesia Guenée
  occidentis Falkovitsh, 1970
  reliquana (Hübner, [1825])
  A *botrana* ([Denis & Schiffermüller],
    1775) European Vine Moth
  abscisana (Doubleday, 1849)
  littoralis (Humphreys & Westwood,
    1845)
Eucosmomorpha Obraztsov
  albersana (Hübner, [1813])
Enarmonia Hübner
  formosana (Scopoli, 1763) **Cherry**
    Bark Moth
Ancylis Hübner
  achatana ([Denis & Schiffermüller],
    1775)
  comptana (Frölich, 1828)
  unguicella (Linnaeus, 1758)
  uncella ([Denis & Schiffermüller],
    1775)
  geminana (Donovan, 1806)
  diminutana (Haworth, 1811)
  subarcuana (Douglas, 1847)
  mitterbacheriana ([Denis &
    Schiffermüller], 1775)
  upupana (Treitschke, 1835)
  obtusana (Haworth, 1811)
  laetana (Fabricius, 1775)
  tineana (Hübner, [1799])
  unculana (Haworth, 1811)
  badiana ([Denis & Schiffermüller],
    1775)
  paludana Barrett, 1871
  myrtillana (Treitschke, 1830)
  apicella ([Denis & Schiffermüller],
    1775)
Eriopsela Guenée
  quadrana (Hübner, [1813])
Thiodia Hübner
  * *torridana* (Lederer, 1859)
  citrana (Hübner, [1799])
Rhopobota Lederer
  ustomaculana (Curtis, 1831)
  naevana (Hübner, [1817]) **Holly**
    Tortrix
  stagnana ([Denis & Schiffermüller],
    1775)
  myrtillana (Humphreys &
    Westwood, 1845)
Spilonota Stephens
  ocellana ([Denis & Schiffermüller],
    1775) **Bud Moth**
  laricana (Heinemann, 1863)
Acroclita Lederer
  subsequana (Herrich-Schäffer,
    1851)
Gibberifera Obraztsov
  † *simplana* (Fischer von
    Röslerstamm, 1836)
Epinotia Hübner
  pygmaeana (Hübner, [1799])
  subsequana (Haworth, 1811)
  subocellana (Donovan, 1806)
  bilunana (Haworth, 1811)
  ramella (Linnaeus, 1758)
  demarniana (Fischer von
    Röslerstamm, 1840)

  immundana (Fischer von
    Röslerstamm, 1839)
  tetraquetrana (Haworth, 1811)
  nisella (Clerck, 1759)
  tenerana ([Denis & Schiffermüller],
    1775) **Nut Bud Moth**
  nigricana (Herrich-Schäffer, 1851)
  nemorivaga (Tengström, [1848])
  tedella (Clerck, 1759)
  fraternana (Haworth, 1811)
  signatana (Douglas, 1845)
  granitana (Herrich-Schäffer, 1851)
  nanana (Treitschke, 1835)
  rubiginosana (Herrich-Schäffer,
    1851)
  cruciana (Linnaeus, 1761) **Willow**
    Tortrix
  mercuriana (Frölich, 1828)
  crenana (Hübner, [1817])
  abbreviana (Fabricius, 1794)
  trigonella (Linnaeus, 1758)
  maculana (Fabricius, 1775)
  sordidana (Hübner, [1824])
  caprana (Fabricius, 1798)
  brunnichana (Linnaeus, 1767)
  solandriana (Linnaeus, 1758)
Zeiraphera Treitschke
  ratzeburgiana (Saxesen, 1840)
    Spruce Bud Moth
  rufimitrana (Herrich-Schäffer,
    1851)
  isertana (Fabricius, 1794)
  griseana (Hübner, [1799]) **Larch**
    Tortrix
Crocidosema Zeller
  plebejana Zeller, 1847
Phaneta Stephens
  pauperana (Duponchel, [1843])
Pelochrista Lederer
  caecimaculana (Hübner, [1799])
Eucosma Hübner
  aspidiscana (Hübner, [1817])
  rubescana (Constant, 1895)
    [= *catoptrana* (Rebel 1903)]
  conterminana (Guenée, 1845)
  tripoliana (Barrett, 1880)
  aemulana (Schläger, 1849)
  lacteana (Treitschke, 1835)
  I *metzneriana* (Treitschke, 1830)
  campoliliana ([Denis &
    Schiffermüller], 1775)
  pupillana (Clerck, 1759)
  hohenwartiana ([Denis &
    Schiffermüller], 1775)
  parvulana (Wilkinson, 1859)
  fulvana (Stephens, 1834)
  cana (Haworth, 1811)
  obumbratana (Lienig & Zeller,
    1846)
Gypsonoma Meyrick
  aceriana (Duponchel, [1843])
  sociana (Haworth, 1811)
  dealbana (Frölich, 1828)
  oppressana (Treitschke, 1835)
  minutana (Hübner, [1799])
  † *nitidulana* (Lienig & Zeller, 1846)
Epiblema Hübner
  grandaevana (Lienig & Zeller,
    1846)
  turbidana (Treitschke, 1835)
  foenella (Linnaeus, 1758)
  scutulana ([Denis & Schiffermüller],
    1775)
  cirsiana (Zeller, 1843)
  cnicicolana (Zeller, 1847)

  sticticana (Fabricius, 1794)
  costipunctana (Haworth, 1811)
Notocelia Hübner
  [= *Epiblema* Hübner]
  cynosbatella (Linnaeus, 1758)
  uddmanniana (Linnaeus, 1758)
    Bramble Shoot Moth
  trimaculana (Haworth, 1811)
  rosaecolana (Doubleday, 1850)
  roborana ([Denis & Schiffermüller],
    1775)
  incarnatana (Hübner, [1800])
  tetragonana (Stephens, 1834)
Pseudococcyx Swatschek
  posticana (Zetterstedt, 1839)
  turionella (Linnaeus, 1758) **Pine**
    Bud Moth
  A *tessulatana* (Staudinger, 1871)
Retinia Guenée
  resinella (Linnaeus, 1758) **Pine**
    Resin-gall Moth
Clavigesta Obraztsov
  sylvestrana (Curtis, 1850)
  purdeyi (Durrant, 1911) **Pine Leaf-**
    mining Moth
Gravitarmata margarotana
  (Heinemann, 1863)
Rhyacionia Hübner
  buoliana ([Denis & Schiffermüller],
    1775) **Pine Shoot Moth**
  pinicolana (Doubleday, 1849)
  pinivorana (Lienig & Zeller, 1846)
    Spotted Shoot Moth
  logaea Durrant, 1911 **Elgin Shoot**
    Moth
Dichrorampha Guenée
  petiverella (Linnaeus, 1758)
  alpinana (Treitschke, 1830)
  flavidorsana Knaggs, 1867
  plumbagana (Treitschke, 1830)
  senectana Guenée, 1845
  sequana (Hübner, [1799])
  acuminatana (Lienig & Zeller, 1846)
  consortana Stephens, 1852
  simpliciana (Haworth, 1811)
  sylvicolana Heinemann, 1863
  montanana (Duponchel, [1843])
  vancouverana McDunnough, 1935
    [= *gueneeana* Obraztsov, 1953]
  plumbana (Scopoli, 1763)
  sedatana Busck, 1906
  aeratana (Pierce & Metcalfe, 1915)
Cydia Hübner
  strobilella (Linnaeus, 1758) **Spruce**
    Seed Moth
  ulicetana (Haworth, 1811)
    [= *succedana* ([Denis &
    Schiffermüller], 1775)]
  medicaginis (Kuznetzov, 1962)
    Alfalfa Moth
  microgrammana (Guenée, 1845)
  servillana (Duponchel, 1836)
  nigricana (Fabricius, 1794) **Pea**
    Moth
  millenniana (Adamczewski, 1967)
  fagiglandana (Zeller, 1841)
  splendana (Hübner, [1799])
  pomonella (Linnaeus, 1758)
    Codling Moth
  I *amplana* (Hübner, [1799])
  A *deshaisiana* (Lucas, 1858) Jumping
    Bean Moth
  I *inquinatana* (Hübner, [1799])
  † *leguminana* (Lienig & Zeller, 1846)
  cognatana (Barrett, 1874)

pactolana (Zeller, 1840)
illutana (Herrich-Schäffer, 1851)
indivisa (Danilevsky, 1963)
cosmophorana (Treitschke, 1835)
coniferana (Saxesen, 1840)
conicolana (Heylaerts, 1874)
A *injectiva* (Heinrich, 1926)
corollana (Hübner, [1823])
Lathronympha Meyrick
strigana (Fabricius, 1775)
Thaumatotibia Zacher
A *leucotreta* (Meyrick, 1913) False
Codling Moth
Selania Stephens
leplastriana (Curtis, 1831)
Grapholita Treitschke
caecana Schläger, 1847
compositella (Fabricius, 1775)
internana (Guenée, 1845)
pallifrontana Lienig & Zeller, 1846
gemmiferana Treitschke, 1835
janthinana (Duponchel, 1835)
tenebrosana Duponchel, [1843]
funebrana Treitschke, 1835 **Plum
Fruit Moth**
A *molesta* (Busck, 1916) **Oriental
Fruit Moth**
lobarzewskii (Nowicki, 1860)
lathyrana (Hübner, [1813])
jungiella (Clerck, 1759)
lunulana ([Denis & Schiffermüller],
1775)
orobana Treitschke, 1830
Pammene Hübner
splendidulana (Guenée, 1845)
luedersiana (Sorhagen, 1885)
obscurana (Stephens, 1834)
agnotana Rebel, 1914
giganteana (Peyerimhoff, 1863)
argyrana (Hübner, [1799])
ignorata Kuznetzov, 1968
albuginana (Guenée, 1845)
suspectana (Lienig & Zeller, 1846)
spiniana (Duponchel, [1843])
populana (Fabricius, 1787)
aurita Razowski, 1991
regiana (Zeller, 1849)
trauniana ([Denis & Schiffermüller],
1775)
fasciana (Linnaeus, 1761)
herrichiana (Heinemann, 1854)
germmana (Hübner, [1799])
ochsenheimeriana (Lienig & Zeller,
1846)
rhediella (Clerck, 1759) **Fruitlet
Mining Tortrix**
gallicana (Guenée, 1845)
aurana (Fabricius, 1775)
Strophedra Herrich-Schäffer
weirana (Douglas, 1850)
nitidana (Fabricius, 1794)

PYRALIDAE
Aphomia Hübner
[= *Arenipses* Hampson]
sociella (Linnaeus, 1758) **Bee Moth**
zelleri (de Joannis, 1932)
A *sabella* (Hampson, 1901)
Paralipsa Butler
A *gularis* (Zeller, 1877) Stored Nut
Moth
Corcyra Ragonot
A *cephalonica* (Stainton, 1866)
Rice Moth
Achroia Hübner

grisella (Fabricius, 1794) **Lesser
Wax Moth**
Galleria Fabricius
mellonella (Linnaeus, 1758) **Wax
Moth**
Synaphe Hübner
punctalis (Fabricius, 1775)
Pyralis Linnaeus
lienigialis (Zeller, 1843)
farinalis (Linnaeus, 1758) **Meal
Moth**
A *manihotalis* Guenée, 1854
A *pictalis* (Curtis, 1834) Painted
Meal Moth
Aglossa Latreille
caprealis (Hübner, [1809]) **Small
Tabby**
pinguinalis (Linnaeus, 1758) **Large
Tabby**
A *dimidiatus* (Haworth, 1809) Tea
Tabby
A *ocellalis* Lederer, 1863
Hypsopygia Hübner
[= *Orthopygia* Ragonot]
costalis (Fabricius, 1775) **Gold
Triangle**
glaucinalis (Linnaeus, 1758)
Endotricha Zeller
flammealis ([Denis &
Schiffermüller], 1775)
A *consobrinalis* Zeller, 1852
Cryptoblabes Zeller
bistriga (Haworth, 1811)
A *gnidiella* (Millière, 1867)
Trachonitis Zeller
* *cristella* (Hübner, 1796)
Salebriopsis Hannemann
albicilla (Herrich-Schäffer, 1849)
Elegia Ragonot
I *fallax* (Staudinger, 1881)
similella (Zincken, 1818)
Ortholepis Ragonot
betulae (Goeze, 1778)
Matilella Leraut
[= *Pyla* Grote]
fusca (Haworth, 1811)
Moitrelia Leraut
[= *Pempelia* Hübner]
obductella (Zeller, 1839)
Pempeliella Caradja
ornatella ([Denis & Schiffermüller],
1775)
Delplanqueia Leraut
[= *Pempeliella* Caradja]
dilutella ([Denis & Schiffermüller],
1775)
Catastia Hübner
* *marginea* ([Denis &
Schiffermüller], 1775)
Sciota Hulst
hostilis (Stephens, 1834)
adelphella (Fischer von
Röslerstamm, 1836)
I *rhenella* (Zincken, 1818)
Selagia Hübner
A *argyrella* ([Denis & Schiffermüller],
1775)
Pima Hulst
boisduvaliella (Guenée, 1845)
Etiella Zeller
AI *zinckenella* (Treitschke, 1832)
Oncocera Stephens
semirubella (Scopoli, 1763)
Pempelia Hübner

palumbella ([Denis &
Schiffermüller], 1775)
genistella (Duponchel, 1836)
Dioryctria Zeller
abietella ([Denis & Schiffermüller],
1775)
schuetzeella Fuchs, 1899
sylvestrella (Ratzeburg, 1840)
simplicella Heinemann, 1863
Phycita Curtis
roborella ([Denis & Schiffermüller],
1775)
Hypochalcia Hübner
ahenella ([Denis & Schiffermüller],
1775)
Epischnia Hübner
asteris Staudinger, 1870
[= *bankesiella* Richardson, 1888]
Nephopterix Hübner
angustella (Hübner, 1796)
Acrobasis Zeller
[= *Conobathra* Meyrick]
[= *Trachycera* Ragonot]
I *tumidana* ([Denis &
Schiffermüller], 1775)
repandana (Fabricius, 1798)
consociella (Hübner, [1813])
suavella (Zincken, 1818)
advenella (Zincken, 1818)
marmorea (Haworth, 1811)
Apomyelois Heinrich
bistriatella (Hulst, 1887)
[= *bistriatella subcognata* (Ragonot,
1887)]
A *ceratoniae* (Zeller, 1839) Locust
Bean Moth
Eurhodope Hübner
† *cirrigerella* (Zincken, 1818)
Myelois Hübner
circumvoluta (Fourcroy, 1785)
Thistle Ermine
Gymnancyla Zeller
canella ([Denis & Schiffermüller],
1775)
Ancylodes Ragonot
I *pallens* Ragonot, 1887
Zophodia Hübner
AI *grossulariella* (Hübner, [1809])
Eccopisa Zeller
I *effractella* Zeller, 1848
Assara Walker
terebrella (Zincken, 1818)
Euzophera Zeller
cinerosella (Zeller, 1839)
pinguis (Haworth, 1811)
A *osseatella* (Treitschke, 1832)
AI *bigella* (Zeller, 1848)
Nyctegretis Zeller
lineana (Scopoli, 1786)
Ancylosis Zeller
I *cinnamomella* (Duponchel, 1836)
oblitella (Zeller, 1848)
Homoeosoma Curtis
nebulella ([Denis & Schiffermüller],
1775)
sinuella (Fabricius, 1794)
nimbella (Duponchel, 1837)
Phycitodes Hampson
binaevella (Hübner, [1813])
saxicola (Vaughan, 1870)
maritima (Tengström, [1848])
Vitula Ragonot
I *edmandsii* (Packard, 1864)
biviella (Zeller, 1848)

Plodia Guenée
  A *interpunctella* (Hübner, [1813])
    Indian Meal Moth
Ephestia Guenée
  elutella (Hübner, 1796) **Cacao Moth**
  unicolorella woodiella Richards & Thompson, 1932
    [= *parasitella unicolorella* Staudinger, 1881]
  A *kuehniella* Zeller, 1879
    Mediterranean Flour Moth
Cadra Walker
  [= *Ephestia* Guenée]
  A *cautella* (Walker, 1863) Dried Currant Moth
  A *figulilella* (Gregson, 1871) Raisin Moth
  A *calidella* (Guenée, 1845) Dried Fruit Moth
Rhodophaea Guenée
  [= *Pempelia* Hübner]
  formosa (Haworth, 1811)
Mussidia Ragonot
  AI *nigrivenella* Ragonot, 1888
Anerastia Hübner
  lotella (Hübner, [1813])

CRAMBIDAE

Scoparia Haworth
  subfusca Haworth, 1811
  pyralella ([Denis & Schiffermüller], 1775)
  ambigualis (Treitschke, 1829)
  basistrigalis Knaggs, 1866
  ancipitella (La Harpe, 1855)
Eudonia Billberg
  [=*Dipleurina* Chapman]
  lacustrata (Panzer, 1804)
  pallida (Curtis, 1827)
  alpina (Curtis, 1850)
  murana (Curtis, 1827)
  truncicolella (Stainton, 1849)
  lineola (Curtis, 1827)
  angustea (Curtis, 1827)
  delunella (Stainton, 1849)
  mercurella (Linnaeus, 1758)
Euchromius Guenée
  I *ocellea* (Haworth, 1811)
  A *cambridgei* (Zeller, 1867)
Chilo Zincken
  phragmitella (Hübner, [1805])
Friedlanderia Agnew
  [= *Haimbachia* Dyer]
  I *cicatricella* (Hübner, [1824])
Calamotropha Zeller
  paludella (Hübner, [1824])
Chrysoteuchia Hübner
  culmella (Linnaeus, 1758)
Crambus Fabricius
  pascuella (Linnaeus, 1758)
  * *leucoschalis* Hampson 1898
  silvella (Hübner, [1813])
  uliginosellus Zeller, 1850
  ericella (Hübner, [1813])
  hamella (Thunberg, 1788)
  pratella (Linnaeus, 1758)
  lathoniellus (Zincken, 1817)
  perlella (Scopoli, 1763)
Agriphila Hübner
  selasella (Hübner, [1813])
  straminella ([Denis & Schiffermüller], 1775)
  tristella ([Denis & Schiffermüller], 1775)

  inquinatella ([Denis & Schiffermüller], 1775)
  latistria (Haworth, 1811)
  I *poliellus* (Treitschke, 1832)
  geniculea (Haworth, 1811)
Catoptria Hübner
  permutatellus (Herrich-Schäffer, 1848)
  I *ostheIderi* (de Lattin, 1950)
  * *speculalis* Hübner, [1825]
  pinella (Linnaeus, 1758)
  margaritella ([Denis & Schiffermüller], 1775)
  furcatellus (Zetterstedt, 1839)
  falsella ([Denis & Schiffermüller], 1775)
  I *verellus* (Zincken, 1817)
  I *lythargyrella* (Hübner, 1796)
Chrysocrambus Bleszynski
  I *linetella* (Fabricius, 1781)
  I *craterella* (Scopoli, 1763)
Thisanotia Hübner
  chrysonuchella (Scopoli, 1763)
Pediasia Hübner
  fascelinella (Hübner, [1813])
  contaminella (Hübner, 1796)
  aridella (Thunberg, 1788)
Platytes Guenée
  alpinella (Hübner, [1813])
  cerussella ([Denis & Schiffermüller], 1775)
Ancylolomia Hübner
  I *tentaculella* (Hübner, 1796)
Schoenobius Duponchel
  gigantella ([Denis & Schiffermüller], 1775)
Donacaula Meyrick
  forficella (Thunberg, 1794)
  mucronella ([Denis & Schiffermüller], 1775)
Agassiziella Yoshi
  A *angulipennis* (Hampson, 1891)
Elophila Hübner
  [= *Synclita* Lederer]
  nymphaeata (Linnaeus, 1758) **Brown China-mark**
  A *diffluals* (Snellen, 1880)
  A *melagynalis* (Agassiz, 1978)
  A *manilensis* (Hampson, 1917)
  A *obliteralis* (Walker, 1859)
Acentria Stephens
  ephemerella ([Denis & Schiffermüller], 1775) **Water Veneer**
Cataclysta Hübner
  lemnata (Linnaeus, 1758) **Small China-mark**
Parapoynx Hübner
  [=*Nymphula* Schrank]
  [=*Oligostigma* Guenée]
  stratiotata (Linnaeus, 1758) **Ringed China-mark**
  A *fluctuosalis* (Zeller, 1852)
  A *obscuralis* (Grote, 1881)
  A *diminutalis* Snellen, 1880
  A *crisonalis* (Walker, 1859)
  A *polydectalis* (Walker, 1859)
  A *bilinealis* (Snellen, 1876)
Nymphula Schrank
  nitidulata (Hufnagel, 1767) **Beautiful China-mark**
    [= *stagnata* (Donovan, 1806)]
Musotima Meyrick
  A *nitidalis* (Walker, 1866)

Metaxmeste Hübner
  * *phrygialis* (Hübner, 1796)
Cynaeda Hübner
  dentalis ([Denis & Schiffermüller], 1775)
Evergestis Hübner
  forficalis (Linnaeus, 1758) **Garden Pebble**
  limbata (Linnaeus, 1767)
  extimalis (Scopoli, 1763)
  pallidata (Hufnagel, 1767)
Eustixia Hübner
  A *pupula* Hübner, [1823]
Hellula Guenée
  I *undalis* (Fabricius, 1781) Old World Webworm
Paracorsia Marion
  I *repandalis* ([Denis & Schiffermüller], 1775)
Loxostege Hübner
  †I *sticticalis* (Linnaeus, 1761)
Pyrausta Schrank
  aurata (Scopoli, 1763)
  purpuralis (Linnaeus, 1758)
  ostrinalis (Hübner, 1796)
  sanguinalis (Linnaeus, 1767)
  despicata (Scopoli, 1763)
  nigrata (Scopoli, 1763)
  cingulata (Linnaeus, 1758)
Uresiphita Hübner
  I *gilvata* (Fabricius, 1794)
    [= *polygonalis* ([Denis & Schiffermüller], 1775)]
  I *reversalis* (Guenée, 1854)
Nascia Curtis
  cilialis (Hübner, 1796)
Sitochroa Hübner
  palealis ([Denis & Schiffermüller], 1775)
  verticalis (Linnaeus, 1758)
Anania Hübner
  [= *Eurrhypara* Hübner]
  [= *Perinephela* Hübner]
  [= *Phlyctaenia* Hübner]
  [= *Algedonia* Lederer]
  [= *Ebulea* Doubleday]
  [= *Opsibotys* Warren]
  hortulata (Linnaeus, 1758) **Small Magpie**
  lancealis ([Denis & Schiffermüller], 1775)
  coronata (Hufnagel, 1767)
  perlucidalis (Hübner, [1809])
  stachydalis (Zincken, 1821)
  funebris (Ström, 1768)
  verbascalis ([Denis & Schiffermüller], 1775)
  terrealis (Treitschke, 1829)
  crocealis (Hübner, 1796)
  fuscalis ([Denis & Schiffermüller], 1775)
Sclerocona Meyrick
  acutellus (Eversmann, 1842)
Psammotis Hübner
  I *pulveralis* (Hübner, 1796)
Ostrinia Hübner
  nubilalis (Hübner, 1796) **European Corn-borer**
Paratalanta Meyrick
  pandalis (Hübner, [1825]) **Bordered Pearl**
  hyalinalis (Hübner, 1796)
Udea Guenée
  lutealis (Hübner, [1809])
  fulvalis (Hübner, [1809])

Checklist of the micro-moths of Britain and Ireland

prunalis ([Denis & Schiffermüller], 1775)
decrepitalis (Herrich-Schäffer, 1848)
olivalis ([Denis & Schiffermüller], 1775)
uliginosalis (Stephens, 1834)
* *alpinalis* ([Denis & Schiffermüller], 1775)
I **ferrugalis** (Hübner, 1796) **Rusty-dot Pearl**
Leucinodes Guenée
  A *vagans* Tutt, 1890
  A *orbonalis* Guenée, 1854 Eggplant Borer
Pleuroptya Meyrick
  [= *Herpetogramma* Lederer]
  ruralis (Scopoli, 1763) **Mother of Pearl**
  A *aegrotalis* (Zeller, 1852)
Mecyna Doubleday
  flavalis ([Denis & Schiffermüller], 1775)
  asinalis (Hübner, [1819])

Agrotera Schrank
  nemoralis (Scopoli, 1763)
Diasemia Hübner
  I† *reticularis* (Linnaeus, 1761)
  A *accalis* (Walker, 1859)
Diasemiopsis Munroe
  I *ramburialis* (Duponchel, 1834)
Duponchelia Zeller
  AI *fovealis* Zeller, 1847
Spoladea Guenée
  AI *recurvalis* (Fabricius, 1775)
Palpita Hübner
  I *vitrealis* (Rossi, 1794)
Hodebertia Leraut
  I *testalis* (Fabricius, 1794)
Sceliodes Guenée
  AI *laisalis* (Walker, 1859)
Dolicharthria Stephens
  punctalis ([Denis & Schiffermüller], 1775)
Antigastra Lederer
  I *catalaunalis* (Duponchel, 1833)
Nomophila Hübner

I **noctuella** ([Denis & Schiffermüller], 1775) **Rush Veneer**
I *nearctica* Munroe, 1973
Diplopseustis Meyrick
  AI *perieresalis* (Walker, 1859)
Diaphania Hübner
  A *indica* (Saunders, 1851) Melonworm
Cydalima Lederer
  [=*Diaphania* Hübner]
  A *perspectalis* (Walker, 1859) Boxworm Moth or Box Tree Moth
Conogethes Meyrick
  A *punctiferalis* (Guenée, 1854) Yellow Peach Moth
Maruca Walker
  AI *vitrata* (Fabricius, 1787) Mung Moth
Herpetogramma Lederer
  * *centrostrigalis* (Stephens, 1834)
  I *licarsisalis* (Walker, 1859) Grass Webworm

Index

Textual references are shown in normal type. Illustrations and photographs are in italics. Bold type is used to indicate the main text and plate illustrations.

Index of common names

Photographic credits

Abbreviations used: l = left, r = right, t = top, m = middle, b = bottom, tr = top right, tl = top left, br = bottom right, bl = bottom left, uml = upper middle left, umr = upper middle right, lml = lower middle left, lmr = lower middle right, um = upper middle, lm = lower middle, bm = bottom middle, ubl = upper bottom left

Jan Bailey: 15; **Andy Banthorpe**: 83; **Adrian Bicker**: 56; **Peter Bolson** 115(b); **Rudolf Bryner**: 57(br), 262(l,r), 286(b), 290(b), 309(t,b), 314, 319, 327, 338(t), 347 ; **Patrick Clement**: 46(l), 53(br), 73(br), 76(l), 89(r), 91(r), 131(tl), 156(bl,br), 289, 298, 299(b), 301(b), 341(l), 358(r); **Hugh Clifford**: 377(b); **Peter Costen**: 21; **Rob Edmunds**: 52(tl,tr), 155(tl,uml), 156(lml), 173(tr), 200(inset); **Dave Green**: 21(t); **Bob Heckford**: 132, 134(bl,br), 323(l); **Ian Kimber**: 193(l), 280(b), 282, 385; **Gabi Krumm**: 350; **Patrick Le Mao**: 378(b); **Andy Mackay**: 54; **Kevin McCabe**: 61(tl); **Heidrun Melzer**: 134(tl,tr), 147(l), 352, 359(b), 386(t); **Tymo Muus** : 167; **Mark Parsons**: 22(b), 23(l), 46(r), 47(r), 51(tl,tr), 57(bl), 65(ml), 73(tr,tl,bl), 78, 81(tl), 87(l), 90(lmr,bl,br), 91(l), 93, 95(tl,tr,b,inset), 101, 154(lmr), 155(tr,lml,bl,br), 156(tl,umr), 157(bl), 164(t), 169(b), 174(tl), 188, 194(l,r), 196, 292(b), 294(l), 301(t), 348, 355, 356(l,r); **Franz Prelicz**: 198; **Mark Skevington**: 92; **Hannah Sterling**: 19; **Phil Sterling**: 22(tl), 23(r), 49(tr,uml,br), 51(bl,br), 53(uml), 55(l), 57(tr), 61(r), 63(tl,b), 65(tl), 71(tl,tr,m,bm,br), 73(mr), 76(r), 81(tr,bl), 85(r), 86(l),

87(r), 90(tl,umr), 95(ml,mr), 97,106, 108, 111(l), 121(b), 122, 123, 127, 128(l), 131(bl), 138, 142(r), 148, 154(tl,tr,umr,bl), 155(lmr), 156(uml), 157(tl,tr,uml,umr,lml), 163(r), 165(l,r), 169(tl), 170(t), 173(lml,lmr,ubl), 174(umr,br), 192, 193(r), 200(main), 201, 203(r), 269, 273, 276(t), 285(main, inset), 249(r), 315(l,r), 335(l), 336(b), 342(l,r), 351(t,b), 359(tr), 368(t,um,lm,b), 370(t,b), 373(t,b), 378(m), 386(m,b); **Ben Smart**: 22(tr), 49(tl,lml,mr,bl), 52(uml,umr,lml, lmr,bl,br), 53(tl,tr,umr,lml,lmr), 55(r), 57(tl), 61(bl), 63(tr), 65(b), 67(tl,tr,br), 72, 73(ml), 79, 80(l,r), 81(br), 85(l), 86(r), 88(l,r), 89(l), 90(tr,uml,lml), 96, 98(l,r), 99, 103(tl,tr,b), 105(l,r), 107, 111(t,br), 112, 113, 114, 115(tl), 116, 117, 118, 119(main, inset), 121(t), 124, 125, 128(r), 131(r), 133, 135, 136, 137(l,r), 140, 141, 142(l), 143, 145(l,r), 147(r), 149, 151, 154(br,uml), 155(umr), 156(tr,lmr), 157(lmr,br), 159, 160, 161, 163(l), 164(b), 169(tr), 170(b), 172, 173(tl,uml,umr,br, bl),174(tr,uml,lml,lmr), 186, 189, 190, 195, 197(t,b), 203(l), 268, 271, 272, 275(r), 276(b), 279, 280(t), 284, 286(t), 288(t,b), 290(t,m), 291(t), 292(t), 295, 297, 299(t), 305, 306, 311, 312, 317, 318(l,r), 321, 322, 323(r), 325, 326(t,b), 333, 335(b), 336(t), 338(b), 341(r), 345, 346, 349, 357(l,r), 359(tl), 361, 362, 369, 372, 376(tl,tr,bl,br), 377(t), 378(t), 379, 381, 382, 383; **Oliver Wadsworth**: 53(bl), 55(t), 65(tr,mr), 67(bl), 71(bl), 154(lmr), 174(bl), 275(l), 340(t,b), 358(l); **Will Woodrow**: 291(b)